THE CONSTRUCTION LAW LIBRARY FROM WILEY LAW PUBLICATIONS

ALTERNATIVE CLAUSES TO STANDARD CONSTRUCTION CONTRACTS
   James E. Stephenson, Editor

ARBITRATION OF CONSTRUCTION DISPUTES
   Michael T. Callahan, Barry B. Bramble, and Paul M. Lurie

ARCHITECT AND ENGINEER LIABILITY: CLAIMS AGAINST DESIGN PROFESSIONALS
   Robert F. Cushman and Thomas G. Bottum, Editors

CONDOMINIUM AND HOMEOWNER ASSOCIATION LITIGATION
   Wayne S. Hyatt and Philip S. Downer, Editors

CONSTRUCTION ACCIDENT PLEADING AND PRACTICE
   Turner W. Branch, Editor

CONSTRUCTION BIDDING LAW
   Robert F. Cushman and William J. Doyle, Editors

CONSTRUCTION CLAIMS AND LIABILITY
   Michael S. Simon

CONSTRUCTION DEFAULTS: RIGHTS, DUTIES, AND LIABILITIES
   Robert F. Cushman and Charles A. Meeker, Editors

CONSTRUCTION DELAY CLAIMS
   Barry B. Bramble and Michael T. Callahan

CONSTRUCTION ENGINEERING EVIDENCE
   Loren W. Peters

CONSTRUCTION FAILURES
   Robert F. Cushman, Irvin E. Richter, and Lester E. Rivelis, Editors

CONSTRUCTION INDUSTRY CONTRACTS: LEGAL CITATOR AND CASE DIGEST
   Wiley Law Publications Editorial Staff

CONSTRUCTION INDUSTRY FORMS (TWO VOLUMES)
   Robert F. Cushman and George L. Blick, Editors

CONSTRUCTION LITIGATION FORMBOOK
David M. Buoncristiani, John D. Carter, and Robert F. Cushman

CONSTRUCTION LITIGATION: REPRESENTING THE CONTRACTOR
Robert F. Cushman, John D. Carter, and Alan Silverman, Editors

CONSTRUCTION LITIGATION: REPRESENTING THE OWNER (SECOND EDITION)
Robert F. Cushman, Kenneth M. Cushman, and Stephen B. Cook, Editors

CONSTRUCTION LITIGATION: STRATEGIES AND TECHNIQUES
Barry B. Bramble and Albert E. Phillips, Editors

DIRECTORY OF CONSTRUCTION INDUSTRY CONSULTANTS
Robert F. Cushman and George L. Blick, Editors

HANDLING FIDELITY AND SURETY CLAIMS
Robert F. Cushman and Charles H. Stamm, Editors

HAZARDOUS WASTE DISPOSAL AND UNDERGROUND CONSTRUCTION LAW
Robert F. Cushman and Bruce W. Ficken, Editors

PROVING AND PRICING CONSTRUCTION CLAIMS
Robert F. Cushman and David A. Carpenter, Editors

SWEET ON CONSTRUCTION INDUSTRY CONTRACTS
Justin Sweet

PROVING AND PRICING CONSTRUCTION CLAIMS

ROBERT F. CUSHMAN, Esquire
DAVID A. CARPENTER
Editors

Wiley Law Publications
JOHN WILEY & SONS
New York • Chichester • Brisbane • Toronto • Singapore

Copyright © 1990 by John Wiley & Sons, Inc.

All rights reserved. Published simultaneously in Canada.

Reproduction or translation of any part of this work beyond that permitted by Section 107 or 108 of the 1976 United States Copyright Act without the permission of the copyright owner is unlawful. Requests for permission or further information should be addressed to the Permissions Department, John Wiley & Sons, Inc.

This publication is designed to provide accurate and authoritative information in regard to the subject matter covered. It is sold with the understanding that the publisher is not engaged in rendering legal, accounting, or other professional service. If legal advice or other expert assistance is required, the services of a competent professional person should be sought. *From a Declaration of Principles jointly adopted by a Committee of the American Bar Association and a Committee of Publishers.*

Library of Congress Cataloging in Publication Data

Proving and pricing construction claims / Robert F. Cushman, David A. Carpenter, editors.
     p.   cm. — (Construction law library)
  ISBN 0-471-50913-2. — ISBN 0-471-52724-6 (custom bdg.)
   1. Construction contracts—United States. 2. Breach of contract—United States. 3. Construction industry—Law and legislation—United States. 4. Damages—United States. I. Cushman, Robert Frank, 1931- . II. Carpenter, David A. (David Allen), 1951- .
III. Series.
KF902.P76   1990
343.73'07869—dc20
[347.3037869]   90-12004
                                                              CIP

Printed in the United States of America

10 9 8 7 6 5 4 3 2 1

PREFACE

Although there are those who believe that most claims on construction projects are avoidable or easily remedied by good planning, documentation, and management—and perhaps they are—nevertheless, and unfortunately, most projects are developed without good planning, good documentation, and good management. This reality, coupled with the complexity of today's sophisticated construction projects augmented by the pressure on both owners and contractors to get the most bang for the least buck, leads to one alarming conclusion: construction claims on most construction projects are an inevitable reality.

This is a focused handbook for lawyers, contractors, consultants, and others who prepare, present, and defend construction claims against owners and construction claims against contractors and design professionals. It covers within one volume each and every type of construction claim and clarifies issues of entitlement as well as factual and legal causation in each claim situation.

This book is intended to serve as a reference work for those who must identify their claim and preserve their right to pursue it. It is a pathway for those who must determine the contractual remedy granting clause in their public or private contract that gives the basis for pursuing the claim or, alternatively, are obliged to develop a breach of contract, tort, or constructive trust theory of recovery. It is a guide to those who first must produce the facts to establish that their additional cost was incurred as a result of the action of the other party. It is a road map to the methods of both calculating and proving the amount of the damages sustained with reasonable certainty. It is a practical development of what evidence should be presented in the litigation forum and a guideline of how best to present that evidence.

The material in this book is presented in a manner that is both instructive and readable to those whose knowledge ranges from slight to extensive. It is a strategic guide to the practicing lawyer who sues or defends both owners and contractors as well as a legal treatise that includes recent developments in case law.

The coauthors have been carefully selected from among a most experienced group of construction litigators and claims specialists with years of experience in prosecuting and defending construction claims. The breadth and depth of their vast experience becomes most apparent as they develop the key issues and problems involved in proving and pricing claims. They concisely, yet thoroughly, identify and explain solutions to commonly faced issues and problems.

We believe that through the efforts of these coauthors, we have a handbook that will be invaluable to the construction and legal communities. We are indebted to the coauthors for their thorough work, and extend to them our thanks and appreciation for their willingness to share their stores of knowledge and experience.

We are confident that the judgment of our profession will be that the distilled expertise of these coauthors has created a most useful, complete, and definitive work which, for many, there has long been a great need.

Philadelphia, Pennsylvania ROBERT F. CUSHMAN
February 1990 DAVID A. CARPENTER

ABOUT THE EDITORS

Robert F. Cushman is a partner in the national law firm of Pepper, Hamilton & Scheetz and a recognized specialist and lecturer on all phases of construction and real estate law. He serves as legal counsel to numerous trade associations and construction, development, and bonding companies. Mr. Cushman is editor and coauthor of several books about construction, including the following John Wiley Publications: *Construction Litigation: Representing the Contractor; Architect and Engineer Liability: Claims Against Design Professionals; Construction Defaults: Rights, Duties, and Liabilities; Construction Failures;* and *Construction Bidding Law.* A member of the Pennsylvania bar, he is admitted to practice before the United States Supreme Court and the Court of Appeals for the Federal Circuit. Mr. Cushman has served as executive vice-president and general counsel of the Construction Industry Foundation, as counsel to the American Construction Owners Association, and as regional chairman of the Public Contract Law Section of the American Bar Association. He is permanent chairman of the Andrews Conference Group's Construction Litigation and Hazardous Waste superconferences. Mr. Cushman is a charter member of the American College of Construction Attorneys.

David A. Carpenter is national director of Litigation Services and regional partner-in-charge of the Business Investigation Services (BIS) Group of Coopers & Lybrand. The BIS Group is responsible for performing a wide range of analytical and planning services frequently in connection with bankruptcies/troubled companies, litigation, and insurance claims. As regional partner-in-charge, Mr. Carpenter is responsible for overall supervision of this practice area in the firm's Atlantic Region. As national director of Litigation Services, he provides executive direction for the firm's services in litigation matters.

Mr. Carpenter is a graduate of Bucknell University with a B.S. in Business Administration, of Temple University with an M.B.A., and Rutgers University with a J.D. He is also a member of Beta Gamma Sigma (the National Business Honorary), the Banking Administration Institute, and the Association of Insolvency Accountants. He is a frequent writer and speaker on financial and business topics.

SUMMARY CONTENTS

Short Reference List		xxiii
PART I	**PRECLAIM AND TAX CONSIDERATIONS**	
Chapter 1	Winning Strategies for Proving and Pricing Claims	3
Chapter 2	Tax Considerations in Litigating and Settling Claims	41
PART II	**INTERNATIONAL CLAIMS**	
Chapter 3	Resolution of International Claims in the United States	57
Chapter 4	Resolution of International Claims Outside the United States	71
PART III	**CONTRACTOR CLAIMS AGAINST OWNERS**	
Chapter 5	Delay Claims	99
Chapter 6	Disruption Claims	141
Chapter 7	Acceleration Claims	167
Chapter 8	Differing Site Conditions Claims	199
Chapter 9	Changes in Scope Claims	243
Chapter 10	Termination Claims from Owner's and Contractor's Perspective	277
Chapter 11	Payment Delay Claims	323
PART IV	**OWNER CLAIMS AGAINST CONTRACTORS AND DESIGN PROFESSIONALS**	
Chapter 12	Cost Claims	351
Chapter 13	Causes of Action and Possible Recoveries	371
Chapter 14	Construction Delay Claims	425
Chapter 15	Claims for Lost Profit	451
Table of Cases		475
Index		509

DETAILED CONTENTS

Short Reference List

PART I   PRECLAIM AND TAX CONSIDERATIONS

Chapter 1   Winning Strategies for Proving and Pricing Claims

Overton A. Currie, Esquire
Neal J. Sweeney, Esquire
Smith, Currie & Hancock
Atlanta, Georgia

Knute P. Kurtz, CPA
Coopers & Lybrand
Atlanta, Georgia

§ 1.1	Life in the Construction Industry: Avoiding Risks and Preserving Awards
§ 1.2	Involving Reputable and Reliable Participants in the Project
§ 1.3	Defining Rights, Responsibilities, and Risks: The Parties and Their Contracts
§ 1.4	—Contract Framework
§ 1.5	—Standard Contract Forms and Key Contract Provisions
§ 1.6	—Interpreting and Applying Express and Implied Terms
§ 1.7	—Contract Modification
§ 1.8	The Arbitration Alternative
§ 1.9	—Time and Costs of Arbitration
§ 1.10	—Selection of Arbitrators
§ 1.11	—Informality and Limited Appeals in Arbitration
§ 1.12	—Enforceability of Agreements to Arbitrate
§ 1.13	—Special Problems Involving Multiple Parties
§ 1.14	Planning Ahead—Avoiding and Preparing for Claims Through Proper Management and Documentation
§ 1.15	—Prudent and Responsible Estimating
§ 1.16	—Establishing Standard Operating Procedures
§ 1.17	—Establishing Lines of Communication
§ 1.18	—Project Documentation
§ 1.19	—Cost-Accounting Records
§ 1.20	—Monitoring the Work Through Scheduling
§ 1.21	Effective Claim Development

DETAILED CONTENTS

	§ 1.22	—Early Claim Recognition and Preparation
	§ 1.23	—Early Involvement of Experts and Attorneys
	§ 1.24	—Use of Demonstrative Evidence
	§ 1.25	—Components of a Well-Prepared Claim Document
	§ 1.26	Importance of Calculating and Proving Damages
	§ 1.27	—Basic Damage Principles
	§ 1.28	—Methods of Pricing Claims
	§ 1.29	—Project Cost Reviews and Audits
	§ 1.30	Pursuing Negotiation and Settlement
	§ 1.31	Conclusion

Chapter 2 Tax Considerations in Litigating and Settling Claims

John J. Hopkins, CPA
Anthony C. Weiss
Coopers & Lybrand
Philadelphia, Pennsylvania

§ 2.1	Characterizing Damage Recoveries to the Plaintiff for Tax Purposes
§ 2.2	—Substantiation Requirement
§ 2.3	—Timing Issues
§ 2.4	Tax Considerations to the Defendant—Damage Payments
§ 2.5	—Tax Treatment of Loss Amount
§ 2.6	—Timing Issues
§ 2.7	Deductibility of Legal Fees

Part II INTERNATIONAL CLAIMS

Chapter 3 Resolution of International Claims in the United States

Alan Silverman
Coopers & Lybrand
San Francisco, California

Fredric H. Weisberg, Esquire
Cummings & Lockwood
Stamford, Connecticut

§ 3.1	Introduction
§ 3.2	Currency Exchange Issues Affecting Damage Awards
§ 3.3	Calculating Damages for Obligations Stated in United States Dollars
§ 3.4	Calculating Damages for Foreign Denominated Obligations
§ 3.5	The Emerging Rule
§ 3.6	Effect of Arbitration Clause on Currency Fluctuations
§ 3.7	Contract Language to Assign Risks of Currency Fluctuations

DETAILED CONTENTS

§ 3.8	Forum Selection and Choice of Law Considerations Regarding Arbitration Clauses
§ 3.9	Arbitration Instructions
§ 3.10	Inflation
§ 3.11	Interest Rates
§ 3.12	Conclusion

Chapter 4 — **Resolution of International Claims Outside the United States**

Richard A. Shadbolt, Esquire
McKenna & Co.
London, England

Christopher J. Lemar, FCA
Coopers & Lybrand
London, England

A LAWYER'S VIEWPOINT

§ 4.1	Introduction
§ 4.2	Choice of Law Considerations
§ 4.3	—Validity and Structure of Contract
§ 4.4	—Performance
§ 4.5	—Construction
§ 4.6	—Forum
§ 4.7	—Enforcement
§ 4.8	Types of Claims
§ 4.9	Approach to Entitlement
§ 4.10	—Price
§ 4.11	—Loss/Expense or Additional Costs
§ 4.12	—Quantum Meruit
§ 4.13	—Compensatory Damages
§ 4.14	—Punitive or Exemplary Damages
§ 4.15	—Fines
§ 4.16	—Contribution
§ 4.17	—Liquidated Damages and Penalties
§ 4.18	—Other Approaches

MATTERS FOR THE EXPERT ACCOUNTANT

§ 4.19	The Accountant and the Legal Environment
§ 4.20	The Style of the Evidence
§ 4.21	The Language Barrier
§ 4.22	Financial Evidence
§ 4.23	Verification of Claims
§ 4.24	The Global Approach to Damages

§ 4.25	Foreign Exchange Rate
§ 4.26	Purely Economic Loss
§ 4.27	Proving Future Losses
§ 4.28	Unjust Enrichment
§ 4.29	Interest

PART III CONTRACTOR CLAIMS AGAINST OWNERS

Chapter 5 Delay Claims

Robert F. Cushman, Esquire
James D. Hollyday, Esquire
Pepper, Hamilton & Scheetz
Philadelphia, Pennsylvania

Frederic R. Miller, CPA
Vincent J. Kiernan
Coopers & Lybrand
Washington, D.C.

§ 5.1	Introduction

FROM THE ATTORNEY'S PERSPECTIVE

§ 5.2	Delays and Cause of Delay
§ 5.3	Excusable and Nonexcusable Delay
§ 5.4	Compensable and Noncompensable Delay
§ 5.5	Concurrent Delay
§ 5.6	Burden of Proof and Apportionment in Delay Cases
§ 5.7	Apportionment in Liquidated Damages Cases
§ 5.8	Computing the Delay
§ 5.9	Contractor's Right to Finish Early
§ 5.10	Pricing the Claim
§ 5.11	Increased Material and Labor Costs (Escalation)
§ 5.12	Costs of Idle Construction Equipment
§ 5.13	Home Office Overhead
§ 5.14	Prejudgment Interest and Financing Costs
§ 5.15	Loss of Profits

FROM THE ACCOUNTANT'S PERSPECTIVE

§ 5.16	Cost Escalation
§ 5.17	General Conditions Costs
§ 5.18	Equipment Costs
§ 5.19	Unabsorbed Home Office Overhead
§ 5.20	Interest/Cost of Capital
§ 5.21	Summary
§ 5.22	Bibliography

DETAILED CONTENTS

Chapter 6   **Disruption Claims**

Donald G. Gavin, Esquire
Daniel E. Toomey, Esquire
Robert J. Smith, Esquire
Wickwire Gavin
Vienna, Virginia

Frederic R. Miller, CPA
Coopers & Lybrand
Washington, D.C.

§ 6.1	Introduction
§ 6.2	Delay versus Disruption
§ 6.3	Types of Disruption
§ 6.4	—Noncompensable Disruptions
§ 6.5	—Compensable Disruptions
§ 6.6	Effects of Disruption
§ 6.7	Nature of Disruption Damages
§ 6.8	Considerations in Proving Cause and Effect of Disruptions
§ 6.9	Considerations in Pricing Disruption Claims
§ 6.10	—Should Cost Estimates
§ 6.11	—Industry Standards and Handbooks
§ 6.12	—Time and Motion Studies
§ 6.13	—Expert Opinion
§ 6.14	—Total Cost and Modified Total Cost Methods
§ 6.15	—Jury Verdict Method
§ 6.16	Defenses
§ 6.17	Summary

Chapter 7   **Acceleration Claims**

Michael F. Albers, Esquire
Robert L. Meyers III, Esquire
Jones, Day, Reavis & Pogue
Dallas, Texas

Mark J. Hosfield, CPA
Coopers & Lybrand
Chicago, Illinois

INTRODUCTION

§ 7.1	Acceleration in General
§ 7.2	Acceleration Claims
§ 7.3	Acceleration in Fact
§ 7.4	Constructive Acceleration

PROVING ACCELERATION

§ 7.5	Identifying Acceleration in the Project

§ 7.6	Proof of Acceleration	
§ 7.7	Elements of an Acceleration Claim	
§ 7.8	—Acceleration Orders	
§ 7.9	—Acceleration Costs	
§ 7.10	—Time Extensions	
§ 7.11	Acceleration Provisions	
§ 7.12	Documentation	

PROVING DAMAGES

§ 7.13	Acceleration Damages
§ 7.14	Legally Recognized Measures of Damages
§ 7.15	Using Available Information
§ 7.16	Analysis of Actual Costs
§ 7.17	Reasonableness of Damage Figures
§ 7.18	Profit and Overhead Calculations
§ 7.19	Mitigation and Cost Savings
§ 7.20	Presentation of Evidence

Chapter 8   Differing Site Conditions Claims

Julian F. Hoffar, Esquire
Donna S. McCaffrey, Esquire
John B. Tieder, Jr., Esquire
Watt, Tieder, Killian & Hoffar
Vienna, Virginia

Mark A. Taylor, CPA
Coopers & Lybrand
Salt Lake City, Utah

§ 8.1	Introduction

CONTRACTUAL DIFFERING SITE CONDITIONS

§ 8.2	Contract Provisions
§ 8.3	Typical Clauses
§ 8.4	Type 1 Conditions
§ 8.5	—Contract Data versus Informational Data
§ 8.6	—Nature of Contract Indications
§ 8.7	—Lack of Indications
§ 8.8	—Reliance
§ 8.9	—Material Difference
§ 8.10	—Examples of Type 1 Differing Site Conditions
§ 8.11	Type 2 Conditions
§ 8.12	—Examples of Type 2 Differing Site Conditions
§ 8.13	Reverse Differing Site Conditions
§ 8.14	Notice Requirements
§ 8.15	Effect of Site Investigation and Other Disclaimer Clauses

DIFFERING SITE CONDITIONS WITHOUT CLAUSE
- § 8.16 Remedies in Absence of Contract Clause
- § 8.17 Breach of Warranty
- § 8.18 Misrepresentation and Fraud
- § 8.19 Effect of Disclaimer Clauses

FACTUAL PROOF OF DIFFERING CONDITIONS CLAIM
- § 8.20 Proving a Claim from the Prebid Stage
- § 8.21 Proving a Claim from the Bid Preparation Stage
- § 8.22 Proving a Claim from the Encountered Conditions Phase
- § 8.23 Use of Expert Testimony

PROOF OF DAMAGES
- § 8.24 Impact of a Differing Site Condition
- § 8.25 Impacted Productivity
- § 8.26 —Computation of Damages
- § 8.27 —The Productivity Benchmark
- § 8.28 —The Causal Connection
- § 8.29 —No Unimpacted Period
- § 8.30 —Selecting the Productivity Benchmark
- § 8.31 Other Approaches to Proving Damages
- § 8.32 Other Costs to Consider

Chapter 9 **Changes in Scope Claims**

Louis R. Pepe, Esquire
Pepe & Hazard
Hartford, Connecticut

Glenn Haese
Coopers & Lybrand
Hartford, Connecticut

- § 9.1 Introduction
- § 9.2 Formal Change Orders
- § 9.3 Constructive Change Orders
- § 9.4 Cardinal Changes
- § 9.5 Notice and Writing Requirements
- § 9.6 Pricing a Change Claim
- § 9.7 Claim Preparation
- § 9.8 Elements of Proof of a Claim
- § 9.9 —Extra and Additional Work
- § 9.10 —Authorization
- § 9.11 —Reason for the Extra Work
- § 9.12 —Value of the Extra Work
- § 9.13 Methods of Calculating Costs

§ 9.14	Types of Recoverable Costs
§ 9.15	—Labor Costs
§ 9.16	—Materials Costs
§ 9.17	—Equipment Costs
§ 9.18	—Bonds, Insurance, and Operational Costs
§ 9.19	Changes and the Performance Bond
§ 9.20	Changes and the Fast-Track Contract

Chapter 10　Termination Claims from Owner's and Contractor's Perspective

Jesse B. Grove III, Esquire
Thelen, Marrin, Johnson & Bridges
San Francisco, California

Craig M. Jacobsen, CPA
Coopers & Lybrand
Salt Lake City, Utah

§ 10.1	Introduction

TERMINATIONS FOR CONVENIENCE

§ 10.2	Overview
§ 10.3	The Federal Compensation Scheme
§ 10.4	Danger of Unit Price Settlements
§ 10.5	Termination Clauses in Private Contracts
§ 10.6	Conclusion

OVERVIEW OF TERMINATIONS FOR DEFAULT

§ 10.7	The Nonbreaching Party's Options
§ 10.8	Considerations to Weigh Before Acting
§ 10.9	The Difference Between Termination and Rescission
§ 10.10	When Termination and Rescission Are Favorable
§ 10.11	Legal Prerequisites to Termination or Rescission
§ 10.12	When Termination or Rescission Has Been Allowed or Denied
§ 10.13	Procedures for Rescission
§ 10.14	Deciding Whether to Carry On, Terminate, or Rescind

TERMINATIONS FOR DEFAULT FROM OWNER'S VIEWPOINT

§ 10.15	Basic Damage Remedy
§ 10.16	—Before Work Begins
§ 10.17	—During the Project
§ 10.18	—At Project's Conclusion
§ 10.19	Extended Duration and Overhead Costs Incurred by the Owner
§ 10.20	Delay in Use
§ 10.21	Lost Profits

DETAILED CONTENTS

§ 10.22　Liquidated Damages
§ 10.23　Special Liquidated Damages Issues
§ 10.24　—Both Liquidated and Actual Damages Can Be Recovered
§ 10.25　—Damages Continue to Accrue after Termination
§ 10.26　Rescission and Restitution
§ 10.27　Owner's Recovery When It Defaults

TERMINATIONS FOR DEFAULT FROM CONTRACTOR'S VIEWPOINT

§ 10.28　Basic Damage Remedy
§ 10.29　—Before Work Begins
§ 10.30　—During the Project
§ 10.31　—Importance of Maintaining Good Records
§ 10.32　—Loss Contracts
§ 10.33　—Upon Full Completion of the Contract
§ 10.34　Other Damages Available
§ 10.35　Offsets Against the Contractor
§ 10.36　Rescission and Restitution
§ 10.37　—When Rescission Is the Better Option
§ 10.38　—When Rescission Is Not Favorable to the Contractor
§ 10.39　—How Restitution Is Measured for Contractors
§ 10.40　Contractor's Recovery When It Is in Default
§ 10.41　Recoveries by Parties Not in Privity with the Owner

Chapter 11　Payment Delay Claims

　　　W. Robert Buxton, Esquire
　　　Robert L. Ivey, Esquire
　　　Louis D. Victorino, Esquire
　　　Pillsbury, Madison & Sutro
　　　San Francisco, California

　　　James T. Schmid, CPA
　　　Hemming Morse & Co.
　　　San Francisco, California

§ 11.1　Introduction

CONTRACT TERMS

§ 11.2　Contract Payments Provisions
§ 11.3　Important Payments Clause Issues
§ 11.4　Common Payment Terms
§ 11.5　—AIA Payments Clauses
§ 11.6　—AGC Subcontract Payments Clauses
§ 11.7　—FAR Payments Provisions
§ 11.8　—Suggested Payments Clause

INTEREST DAMAGES

§ 11.9 Interest as a Cost
§ 11.10 —Theoretical Basis for Interest Cost
§ 11.11 —Interest as a Collectible Damage
§ 11.12 Proving Interest Cost Damages
§ 11.13 —Traceability (Causation)
§ 11.14 Calculating Interest for Commercial Contracts
§ 11.15 Calculating Interest for Federal Government Contracts
§ 11.16 Determining the Proper Interest Rate

OTHER DAMAGES

§ 11.17 Foreseeability of Other Payment Delay Damages
§ 11.18 Increased Bonding Cost
§ 11.19 Loss of Other Work
§ 11.20 Collection Fees and Credit Checks
§ 11.21 Costs of Defending Subcontractors' Suits
§ 11.22 Delayed Payment Checklists

PART IV   OWNER CLAIMS AGAINST CONTRACTORS AND DESIGN PROFESSIONALS

Chapter 12   Cost Claims

Paul W. Pocalyko, CPA
Coopers & Lybrand
Philadelphia, Pennsylvania

James L. Shea, Esquire
Venable, Baetjer & Howard
Baltimore, Maryland

§ 12.1 Introduction
§ 12.2 Methods of Investigating Contractor-Claimed Costs
§ 12.3 Contractor's Books and Records
§ 12.4 Contractor's Relationship with Subcontractors and Material Suppliers
§ 12.5 Problems with the Collection of Records
§ 12.6 How Are Contractor Costs Overstated?
§ 12.7 Was the Work Properly Billed as Cost-Plus Rather Than Under a Fixed Fee Term?
§ 12.8 Did the Contractor Perform Efficiently?
§ 12.9 Are the Claimed Material Costs Reasonable?
§ 12.10 Are the Alleged Labor Costs Excessive?
§ 12.11 Are the Office and Administrative Costs Excessive?
§ 12.12 How to Judge a Reasonable Profit Margin
§ 12.13 Were There Premature Payments?

DETAILED CONTENTS

§ 12.14 Owner's Involvement in Cost-Plus Contracting
§ 12.15 Nature and Extent of the Surety's Obligations

Chapter 13 Causes of Action and Possible Recoveries

Steven J. Comen, Esquire
Hinckley, Allen, Snyder & Comen
Boston, Massachusetts

Martin E. Stauffer, CPA
Coopers & Lybrand
Hartford, Connecticut

§ 13.1 Introduction and Hypothetical Construction Failure

CLAIMS AGAINST THE ARCHITECT
§ 13.2 Breach of Warranties
§ 13.3 —Architect's Defenses
§ 13.4 UCC
§ 13.5 Negligence
§ 13.6 —Architect's Defenses
§ 13.7 Strict Liability
§ 13.8 Application of Liability Rules to Hypothetical

CLAIMS AGAINST THE OWNER
§ 13.9 Breach of Warranties
§ 13.10 —Owner's Defenses
§ 13.11 UCC
§ 13.12 Negligence
§ 13.13 Strict Liability
§ 13.14 —Owner's Defenses
§ 13.15 Application of Liability Rules to Hypothetical

CLAIMS AGAINST THE GENERAL CONTRACTOR
§ 13.16 Breach of Warranties
§ 13.17 —Contractor's Defenses
§ 13.18 UCC
§ 13.19 Negligence
§ 13.20 Strict Liability
§ 13.21 Application of Liability Rules to Hypothetical

CLAIMS AGAINST A MANUFACTURER/SUPPLIER
§ 13.22 Breach of Warranties and UCC
§ 13.23 —Manufacturer/Supplier Defenses
§ 13.24 Negligence
§ 13.25 Strict Liability
§ 13.26 Application of Liability Rules to Hypothetical

CLAIMS AGAINST THE SURETY, INSURER, AND LENDER
- § 13.27 Types of Bonds
- § 13.28 All Risk Policies
- § 13.29 Indemnification Clauses
- § 13.30 Subcontractor's Payment or Performance Bond Construed to Extend Coverage for Defects and Failures
- § 13.31 Claims Against the Lender

DAMAGES AVAILABLE TO OWNER AS PLAINTIFF
- § 13.32 Damages Generally
- § 13.33 Recovery of the Cost of Corrective Work
- § 13.34 Liquidated Damages Clause
- § 13.35 Recovery of the Cost to Complete
- § 13.36 Recovery of Interest Rate Differential
- § 13.37 Recovery of Land Value
- § 13.38 Recovery for Loss of Permanent Financing Commitment
- § 13.39 Recovery of Lost Revenues
- § 13.40 Recovery for Loss of Use
- § 13.41 Recovery of Additional Supervision Costs
- § 13.42 Lost Profits
- § 13.43 Recovery of Additional Professional Services
- § 13.44 Recovery of Attorneys' Fees
- § 13.45 Recovery of Depreciation
- § 13.46 Recovery for Loss of Professional Reputation

DAMAGES AVAILABLE TO CONTRACTOR AS PLAINTIFF
- § 13.47 Compliance with Contractual Notice Provisions
- § 13.48 Determining Actual Damages
- § 13.49 Recovery of Damages for Defective Plans and Specifications
- § 13.50 —Economic Loss Doctrine
- § 13.51 Recovery Despite No Damages for Delay Clause
- § 13.52 Recovery for Defective Performance Specifications
- § 13.53 Recovery of Consequential Costs and Damages

OTHER CONSIDERATIONS
- § 13.54 Statutes of Limitations and Repose
- § 13.55 Claims Preparation and Settlement Considerations

Chapter 14 Construction Delay Claims

John N. Chapin, Jr.
Coopers & Lybrand
Chicago, Illinois

DETAILED CONTENTS

Stephen D. Hurst, Esquire
Hinshaw, Culbertson, Moelmann, Hoban & Fuller
Chicago, Illinois

§ 14.1   Introduction

DEVELOPING THE CLAIM
§ 14.2   Preparing the Claim
§ 14.3   Identifying Potential Defendants
§ 14.4   Obtaining Construction Documentation
§ 14.5   Identifying the Causes of Delay
§ 14.6   Focusing the Claim on the Responsible Party

DEVELOPING LEGAL THEORIES OF LIABILITY
§ 14.7   Compensable and Noncompensable Delays
§ 14.8   Anticipating Potential Defenses
§ 14.9   Mitigating Damages

DEVELOPING ELEMENTS OF DAMAGES
§ 14.10   Contracts with Liquidated Damages Provisions
§ 14.11   Contracts without Liquidated Damages Provisions
§ 14.12   —Direct Damages
§ 14.13   —Consequential Damages

DEVELOPING LITIGATION STRATEGIES
§ 14.14   Selecting a Forum
§ 14.15   Presenting the Claim
§ 14.16   Conclusion
§ 14.17   Bibliography

Chapter 15   Claims for Lost Profit

Herman M. Braude, Esquire
Braude & Margulies
Washington, D.C.

George D. Sergio, CPA
WWF Paper Corp.
Bala Cynwyd, Pennsylvania

§ 15.1   Introduction
§ 15.2   Foreseeability
§ 15.3   Proximate Cause
§ 15.4   Reasonable Certainty
§ 15.5   Calculation of Profits
§ 15.6   —Impairment of Corporate Value
§ 15.7   Unestablished Business versus Established Business
§ 15.8   Proof Techniques

§ 15.9 Hypothetical Lost Profits Case
§ 15.10 Conclusion

Table of Cases

Index

SHORT REFERENCE LIST

Short Reference	*Full Title*
ADR	Alternate Dispute Resolution Technique(s)
AGC	Associated General Contractors
AIA	American Institute of Architects
AICPA Guide	Guide for Prospective Financial Statements, published by the American Institute of Certified Public Accountants
Comment, *Remedies-Lost Profits*	Comment, *Remedies-Lost Profits as Contract Damages for an Unestablished Business: The New Business Role Becomes Outdated,* 56 N.C.L. Review 695 (1978)
CPM	Critical Path Method
Dunn	Dunn, Recovery of Damages for Lost Profits (1978)
EJCDC	Engineers Joint Contract Documents Committee
FAR	Federal Acquisition Regulations
FTCA	Federal Tort Claims Act
McCormick, Damages	C. McCormick, Damages (1935)
NSPE	National Society for Professional Engineers
Restatement	Restatement (Third) of the Foreign Relations Law of the United States § 823 (1987)
UCC	Uniform Commercial Code

PART I
PRECLAIM AND TAX CONSIDERATIONS

CHAPTER 1

WINNING STRATEGIES FOR PROVING AND PRICING CLAIMS

Overton A. Currie, Esquire
Neal J. Sweeney, Esquire
Knute P. Kurtz, CPA

Overton A. Currie, senior partner of the Atlanta law firm of Smith, Currie & Hancock, founded and heads the firm's 50-lawyer construction law department. Mr. Currie has served the American Bar Association as national chairman of the Public Contract Section, the Committee on Subcontracts, and the Construction Litigation Committee of the Litigation Section. He earned a Master of Law degree from Yale University as valedictorian and remained on the Yale Law School faculty as a research associate until returning to private practice in 1959. His experience includes both private and public projects across the nation and overseas. He currently serves as a national director of The American Arbitration Association.

Neal J. Sweeney is a partner in the Atlanta law firm of Smith, Currie & Hancock. He practices exclusively in the area of construction law and litigation, with further concentration in federal, state, and local contracting. A graduate of Rutgers University, Mr. Sweeney received a law degree from George Washington University. He has handled claims, litigation, and arbitration involving many federal agencies and numerous other public and private owners and has lectured and written extensively on construction law topics.

Knute P. Kurtz is a partner in the Business Investigation Services Group in the Atlanta Office of Coopers & Lybrand. He specializes in litigation and reorganization services to the real estate and construction industries and in environmental issues. He also has an extensive background in providing audit services to real estate and construction industry companies in both the Atlanta and New Orleans Offices of Coopers & Lybrand. Mr. Kurtz is a member of the American Institute of Certified Public Accountants and the Georgia and Louisiana Society of Certified Public Accountants. He graduated from Nicholls State University in 1977 with a Bachelors Degree in Accounting.

§ 1.1 Life in the Construction Industry: Avoiding Risks and Preserving Awards
§ 1.2 Involving Reputable and Reliable Participants in the Project
§ 1.3 Defining Rights, Responsibilities, and Risks: The Parties and Their Contracts
§ 1.4 —Contract Framework
§ 1.5 —Standard Contract Forms and Key Contract Provisions
§ 1.6 —Interpreting and Applying Express and Implied Terms
§ 1.7 —Contract Modification
§ 1.8 The Arbitration Alternative
§ 1.9 —Time and Costs of Arbitration
§ 1.10 —Selection of Arbitrators
§ 1.11 —Informality and Limited Appeals in Arbitration
§ 1.12 —Enforceability of Agreements to Arbitrate
§ 1.13 —Special Problems Involving Multiple Parties
§ 1.14 Planning Ahead—Avoiding and Preparing for Claims Through Proper Management and Documentation
§ 1.15 —Prudent and Responsible Estimating
§ 1.16 —Establishing Standard Operating Procedures
§ 1.17 —Establishing Lines of Communication
§ 1.18 —Project Documentation
§ 1.19 —Cost-Accounting Records
§ 1.20 —Monitoring the Work Through Scheduling
§ 1.21 Effective Claim Development
§ 1.22 —Early Claim Recognition and Preparation
§ 1.23 —Early Involvement of Experts and Attorneys
§ 1.24 —Use of Demonstrative Evidence
§ 1.25 —Components of a Well-Prepared Claim Document

§ 1.26 Importance of Calculating and Proving Damages
§ 1.27 —Basic Damage Principles
§ 1.28 —Methods of Pricing Claims
§ 1.29 —Project Cost Reviews and Audits
§ 1.30 Pursuing Negotiation and Settlement
§ 1.31 Conclusion

§ 1.1 Life in the Construction Industry: Avoiding Risks and Preserving Awards

Construction is a large, volatile industry. It requires tremendous capital outlays but generally offers low rates of return, particularly in relation to the amount of risk imposed. The construction industry is affected by the same business cycles and economic influences that affect other industries. But construction imposes an additional element of risk and volatility that generally does not exist in other major industries. That element is the manner in which disputes and claims are woven through the fiber of the construction process.

Construction is a dispute-prone industry, and claims are a fact of life. Even successful projects have claims. That does not mean there is something inherently wrong with the industry or its participants. Claims are a natural outgrowth of a complex and highly competitive process during which the unexpected often happens. Careful organization and coordination of numerous parties is required, and it may then be outside parties who control many of the circumstances and events that generate claims. The potential for claims cannot be ignored. It is naive and bad business to do so. The responsible owner, contractor, subcontractor, or designer, as part of doing business, recognizes the need to anticipate claims and to develop effective and affirmative strategies for dealing with them.

The best way to handle claims is to anticipate them and avoid them as much as possible. Despite the uniqueness of each project and its participants, there are certain recurring problems which generate disputes. The past is prologue and some of those recurring problems can be avoided, their impact mitigated, or, at a minimum, some preparation made for the dispute if it should occur. Of course, too much focus on eliminating all risks and anticipating claims and disputes can also create problems and a paralysis that can impair one's ability to do business effectively. A certain element of risk must be recognized and accepted. Risk can only be mitigated, not eliminated.

The chapters of this book, written by recognized authorities in the field, provide detailed insight and instruction on pricing and proving specific types of claims that are frequently encountered. Each of the authors agrees that the successful prosecution of claims requires that foresight and planning be applied long before the facts and circumstances giving rise to the claim actually occur. This chapter focuses on those measures that can be taken at the outset of a project to avoid or effectively prepare for successful prosecution of claims when they cannot be avoided. Of course, there are no "sure things," and construction claims are no exception. Nevertheless, common sense, planning, skill, and experience can help identify those strategies for approaching construction claims that will increase the likelihood of success.

§ 1.2 Involving Reputable and Reliable Participants in the Project

Construction is a cooperative enterprise involving numerous entities and disciplines, from design professionals, the owner or developer, and lender through the prime contractor, subcontractors, and suppliers, each with an integral function. A failure by any participant to properly perform can mean disaster for the entire project and the rest of the participants.

The initial choice of project participants can dictate the destiny of the project and is one of the first steps in avoiding claims. Many headaches and possible losses can be avoided simply by investigating the past performance record of the other parties, rather than looking solely at the lowest price or the opportunity to obtain some work. By dealing with reputable companies and individuals with a proven ability to perform, by running credit checks, and by inquiring about the experience of others with that particular company, bad risks and big mistakes can be avoided.

Naturally, the owner is in the best position to control the selection of project participants because under most contracting schemes the owner selects the designer and the contractor. The owner can also have a significant impact on subcontractor selection. The prime contractor, who stands in a similar position to the owner, has the greater control over the selection of subcontractors, as does the architect or lead designer over the selection of its subconsultants. On the other hand, the prime contractor cannot select the owner or developer of a project. The contractor, however, does decide with whom it wants to do business, as do subcontractors. When the choice of a specific party would create a risk of claims or disputes to the extent that any reasonable return is jeopardized, that party should not be used on the project, regardless of price. Sometimes the risks of a project or of doing business with a particular owner are simply too great and prudence dictates forgoing certain opportunities. Higher volume and backlog figures are meaningless if they engender unnecessary risks and do not translate to profit. Despite the time crunch and euphoria often associated with the beginning of a project, everyone involved should consider certain key factors when selecting project participants: the financial condition of the parties, their qualification for bonds, evidence of their technical skills,

§ 1.2 PROJECT PARTICIPANTS

and their reputation in the industry. Even a cursory investigation of potential project participants may yield clues to future problems.

Money, if not "the root of all evil," is the source of many disputes and claims that arise on a construction project. The financial resources of the owner are probably of paramount concern and an underfinanced owner virtually dooms any project. Although the possibility of lien rights might provide some comfort, if the owner goes under, the likelihood of full and complete payment to the contractor is pretty dim. Considering the preeminent influence of the owner on the success or failure of a project, contractors and architects are wise to subject the owner's background to an informal prequalification process similar to that used for other project participants. This precaution can confirm whether the owner has the capacity to meet its commitments. In addition to other independent sources that may be available to obtain information about the owner's finances, subparagraph 2.2.1 of the standard General Conditions published by the American Institute of Architects, AIA Document A201 (1987 edition), allows the contractor to demand reasonable evidence of the owner's ability to finance the work. Subparagraph 2.1.2 also requires the owner to provide written information regarding title to the property as well as any changes in title, if requested by the contractor.

The financial condition of contractors and subcontractors is also extremely important. A subcontractor who has insufficient working capital may bring a myriad of problems, such as slow deliveries of materials, as suppliers grow concerned about the subcontractor's ability to pay. This can have a ripple effect on other work. Similarly, a contractor needing cash flow may front-end-load his bid and his pay requests. The early overpayment caused by front-end-loading may result in the contractor's default on the latter part of the project as contract funds run out. Unfortunately, there are few, if any, effective remedies against an unbonded, insolvent contractor. The typical action to recover completion costs is often pointless when the default resulted directly from the contractor's financial problems.

An obvious source of financial protection for owners is to require payment and performance bonds. Bonding serves two purposes. First, the contractor's competence and financial well-being are endorsed by the surety's underwriting department, which is also trying to avoid bad risks. If a contractor is incapable of obtaining bonding, it means sureties have a grave concern about its ability to complete a project. That warning is probably best heeded. Second, and more directly, the bonds represent a financial guarantee. A performance bond usually means that, if the contractor defaults and fails to complete, the surety will complete performance or pay damages up to the limit of the penal sum of the bond. In contrast, a labor and material payment bond helps assure the owner that labor and materials will be paid for and creates alternatives to the filing of liens on the project.

Even if provided, payment and performance bonds are not a cure-all. It is also necessary to carefully consider the financial stability of the surety itself. In recent years, too many sureties have themselves gone bankrupt. Moreover, even solvent, well-financed sureties are far from an automatic source of relief. Claims under the bond can themselves be the subject of lengthy disputes and litigation.

Of course, there are concerns about technical qualifications that go beyond money. For example, licensing requirements provide some protection from incompetent and inexperienced contractors, particularly in the skilled trades such as electrical and mechanical. Licensing should be deemed a bare minimum requirement, however, and not an endorsement of qualifications for any type of work authorized by a particular license. Inquiries into the contractor's experience on particular types and sizes of projects should also be pursued. There is a big difference between installing plumbing in a lowrise apartment building and installing the mechanical systems for a major health care facility. The owner's technical capabilities and qualifications to handle a particular type or size of project are also relevant. However, the owner's shortcomings may be offset by the association of capable consultants.

More subjective reports about other parties should also be considered, but perhaps given lesser weight. For example, engaging a subcontractor with a reputation for shoddy or defective work may result in the prime's having to remedy unsatisfactory work at its own expense. A particularly litigious owner may refuse to negotiate a settlement in the event of a dispute in this type of situation, forcing the contractor into more expensive arbitration or court battles.

Even if an owner, designer, or contractor appear to have the qualifications and established track record to pass muster, it is still important to consider which personnel they will assign to the particular project. Companies can become too successful, causing them to be stretched too thin, with all their capable and experienced personnel assigned and consumed by other projects. The choice of a company is certainly important, but the individuals representing that company who will execute responsibilities and perform work in the field are no less important.

§ 1.3 Defining Rights, Responsibilities, and Risks: The Parties and Their Contracts

A written contract generally provides the foundation for each of the numerous relationships and binds the disparate project participants into a cohesive force to get the job built. Keeping those participants together requires in part anticipating issues and events that might create disputes and detract from the goal of prompt and cost-effective completion. This is done by a combination of allocating risks among the parties, so it is clear who will have to bear the burden if the risk becomes reality, and providing mechanisms for resolving disputes when the risk allocation is not clear or there is disagreement. A well-drafted contract is another important element in effectively managing a project and avoiding or efficiently dealing with claims.

Clarity, common sense, and precision should be employed in the drafting of contract language. Such efforts will limit later uncertainty and misunderstanding among the parties and the need to refer to some third-party decisionmaker (in court or arbitration) to determine how the contract will be interpreted. Unreasonable and overly burdensome terms should be avoided because they can unnecessarily drive up the cost of the work through inflated contingencies and may be difficult to enforce. On the other hand, such strict terms cannot be ignored in an unrealistically optimistic view that they will not be enforced or that circumstances requiring their implementation will not arise on the project. The parties must grapple with the tough issues raised by their conflicting interests in the contract preparation stage or face the prospect of much more serious disagreements and disputes during the performance of the contract.

§ 1.4 —Contract Framework

Establishing the contract framework for the project is a threshold decision that must be made by the owner. The selection depends on a variety of factors, including the owner's needs and its expertise and capabilities. Construction projects have traditionally been designed, bid, built, and paid for within a framework of strictly defined roles, relationships, and procedures. This has proven satisfactory for many construction projects, but perceived weaknesses in the traditional method have led to consideration and use of new, alternative methods, such as various forms of construction management, multiprime contracting, and design/build.[1] The new methods have provided many advantages, but the manner in which they diverge from clearly defined practices and roles requires careful attention in the contract drafting phase to make certain that the advantages in their use are not lost through unanticipated problems and disputes.

§ 1.5 —Standard Contract Forms and Key Contract Provisions

There are a number of available standard contract forms which establish the various relationships on a construction project. The documents

---

[1] *See generally Construction Management and Design-Build/Fast Tract Construction from the Perspective of a General Contractor,* 46 Law & Contemp. Probs. (1983).

published by the American Institute of Architects (AIA) are by far the most widely used and most generally accepted. The Associated General Contractors of America (AGC) and the Engineers Joint Contract Documents Committee (EJCDC) also publish contract documents. Even with 1987 revisions, the provisions of the AIA documents are generally well understood by developers, architects, contractors, lenders, and others involved in the construction process. These common forms permit all parties to focus on critical variables when negotiating construction transactions and obviate the need to start from scratch with each new construction transaction.

The AIA documents are fairly well integrated, with the terms of the various contracts coordinated with and complementing each other.[2] This consistency enhances the reliability of the AIA documents. The AIA documents have the advantage of familiarity and acceptance in the industry but do not necessarily meet the needs of each and every project, and some modification may be required for each specific situation. Moreover, it must be recognized that the AIA documents were drafted by an association that strongly promotes the architect's interests, often at the expense of the owner and contractor.

Whether reliance is placed on a standard form, a custom-drafted contract, or some combination of the two, certain contract provisions are of critical importance in anticipating, avoiding, and resolving claims. They are:

Payment
Time for Completion and Time Extensions
Damages for Delay
Changes in The Work
Termination for Default and for Convenience
Changed Conditions
Dispute Resolution
Insurance.

Careful attention should also be paid to the use of liquidated damages or no damages for delay clauses as well as exculpatory and indemnity provisions, which can weigh heavily in the resolution of claims. It is also worthwhile to consider whether the parties intend for Article 2 of the Uniform Commercial Code, which governs the sale of goods, to apply to their construction contract. An extensive discussion of these provisions is beyond the scope of this chapter.[3] However, arbitration, as a

---

[2] *See generally* J. Sweet, Sweet on Construction Industry Contracts (John Wiley & Sons 1987).

[3] *See generally* O. Currie, J. Stephenson, P. Beck, & R. Hafer, *Construction Contracts,* in Negotiating Real Estate Transactions (John Wiley & Sons 1988).

means of resolving construction claims, is discussed separately in §§ 1.8 through 1.13.

§ 1.6 —Interpreting and Applying Express and Implied Terms

Parties to a construction contract must recognize that the print within the four corners of the contract is not the limit of a contract's reach or application. The contract carries with it the baggage of industry trade and custom, the past dealing of the parties, and a set of principles for applying and interpreting the contract and further defining the relationships between the parties. Those considerations can have every bit as much consequence as the printed words on the contract document. Accordingly, whenever possible, a contract should be drafted and then performed with an understanding of the principles of contract interpretation.

One fundamental aspect of contract interpretation is that every contract contains implied obligations in addition to the obligations expressly enumerated in the contract. Perhaps the most important implied contract obligation in the construction context is the duty of cooperation. This duty to cooperate manifests itself in the form of an obligation on the part of the owner not to hinder or delay the contractor's performance.[4] It also encompasses the obligation to coordinate the activities of parallel prime contractors.[5]

Another very important implied obligation, known as the *Spearin doctrine,* is that the party furnishing the plans and specifications (that is, the owner) impliedly warrants their adequacy and sufficiency.[6] Under the Spearin doctrine the contractor may recover for delays, extra work, disruption, and constructive changes when there are errors in the plans and specifications.[7]

§ 1.7 —Contract Modification

Contracts typically provide that their provisions cannot be modified without the written agreement of both parties. Despite the presence of such a

---

[4] *See, e.g.,* Coatesville Contractors v. Borough of Ridley, 509 Pa. 553, 506 A.2d 862 (1986).

[5] *See, e.g.,* Baldwin-Lima-Hamilton Corp. v. United States, 434 F.2d 1371 (Ct. Cl. 1970).

[6] *See, e.g.,* Ordinance Research, Inc. v. United States, 609 F.2d 462 (Ct. Cl. 1979); United States v. Spearin, 248 U.S. 132, 136 (1918).

[7] *See generally* USA Petroleum Corp. v. United States, 821 F.2d 622 (Fed. Cir. 1987); La Crosse Garment Mfg. Co. v. United States, 432 F.2d 1377 (Ct. Cl. 1970); Chaney & James Constr. Co. v. United States, 421 F.2d 728 (Ct. Cl. 1970); R.M. Hollingshead v. United States, 111 F. Supp. 285 (Ct. Cl. 1953).

stipulation, parties can also modify the terms of their written agreement orally or by their conduct.[8] The issue of modification frequently arises in situations in which formal written notice is required but, despite the contractor's failure to properly provide it, the owner repeatedly acts as if the notice had been given. Similarly, when the contract states that written work or change orders are required before a contractor is entitled to additional compensation, an established pattern of payment without the contractor's having first obtained a written order may serve to constitute a modification or waiver of this contract requirement.[9] Waiver of written notice and change order requirements, however, should never be presumed, and all efforts should be undertaken to comply with them.

§ 1.8 The Arbitration Alternative

The issue of arbitration is generally contemplated at the conclusion of the claims process. Arbitration is sought when the parties are unable to resolve the claim between themselves. The availability of arbitration as a means of resolving construction claims and disputes, however, must be planned and provided for at the outset of the project, in the contract itself. Otherwise, that alternative will not exist and if the parties cannot agree they will have to look to the courts. Generally, no particular form or special words are necessary to establish an agreement to arbitrate a dispute between the parties. However, there must be a clear indication from the contract language that the parties intended the disputed issue to be subject to arbitration.

To avoid any dispute about the scope of the agreement to arbitrate, it is best to expressly state that all disputes arising from the contract will be arbitrated. For example, the most widely used arbitration clause, contained in AIA Document A201 (1987), provides:

> 4.5.1 **Controversies and Claims Subject to Arbitration.** Any controversy or claim arising out of or related to the contract, or the breach thereof, shall be settled by arbitration in accordance with the Construction Industry Arbitration Rules of the American Arbitration Association.

Arbitration is not new to the construction industry. Contract clauses providing for arbitration of disputes have been commonplace in the construction industry for many years. Although arbitration is generally perceived as a way to avoid the delays and problems associated with

---

[8] Certified Corp. v. Hawaii Teamsters & Allied Workers, Local 996, 597 F.2d 1269 (10th Cir. 1979).

[9] W.E. Garrison Grading Co. v. Piracci Constr. Co., 27 N.C. App. 725, 221 S.E.2d 512 (1975).

litigation, time and many tests have demonstrated the drawbacks as well as advantages of relying on arbitration as a means of resolving construction claims.

§ 1.9 —Time and Costs of Arbitration

Avoiding delays normally associated with the courts and crowded dockets and trial calendars is one of the most often-cited benefits of arbitration. On balance, and particularly for smaller claims, arbitration does provide a faster resolution. Given the right set of circumstances, however, arbitration can also be excruciatingly time-consuming and expensive. A dispute over the existence, scope, or validity of an arbitration clause agreement itself can engender a protracted court proceeding and appeal before there is a determination of whether and to what extent the parties should proceed with arbitration. In addition, arbitration of larger claims involving multiple parties can match the delays and complexity of the most arcane court proceeding.

Although ultimately dependent on the parties involved and the scope and complexity of the issues, the cost of arbitration may be less than that of a comparable court proceeding. Generally, a shortened period for resolution will keep costs down, but there are certain costs that cannot be avoided. These costs include those usually incurred in trial preparation, such as the examination and analysis of documents, legal research, the use of experts, and the development of demonstrative evidence.

The costs that can be avoided in arbitration usually involve certain trade-offs. For example, if it is not provided for by agreement or statute, discovery is often not available in arbitration.[10] The costs of discovery, which can be substantial, can thus be avoided. The lack of discovery, however, means less preparation for and knowledge of the opposition's case. It can also possibly lead to a less focused hearing, which may then require more time and increased costs. In addition, there are costs that are unique to arbitration, such as arbitrators' fees, which can be substantial, as well as administrative fees and the cost of meeting rooms.

§ 1.10 —Selection of Arbitrators

The qualifications and fairness of each arbitrator are essential to the viability of arbitration as a means of resolving disputes. From a partisan vantage point, the selection of arbitrators can be a major factor in the

---

[10] *See* Tupman, *Discovery and Evidence in U.S. Arbitration: The Prevailing Views,* Arb. J. (Mar. 1989).

success or failure of a claim. Many factors must be considered in selecting the arbitrator or arbitration panel. Some of the most important considerations must be addressed long before any claim arises or construction begins, in the drafting of the arbitration clause. Some basic but fairly strategic considerations are the number of arbitrators and the manner in which they are selected.

The fact that arbitrators generally have more expertise in the construction industry than a judge or jury is generally cited as one of the major advantages of arbitration. Of course, this may not be an advantage if the arbitrator's experience and background are contrary to the claimant's. An engineer may view all contractor claims with a jaundiced eye, for example. Despite her expertise, that engineer is not desirable to the contractor-claimant. Conversely, a contractor-arbitrator may be reluctant to enforce a hefty liquidated damages provision as part of an owner's claim.

In terms of numbers, the basic choice is between a single arbitrator and a three-member panel. A single arbitrator costs less and will probably simplify scheduling hearings. On the other hand, a three-member panel would probably be more balanced. The three-member panel generally decides by majority vote, that is, two of the three panel members can render a decision over the objection or dissent of the third member.

Parties to an arbitration are free to agree to their own procedures for arbitrator selection. The procedure is often set forth in the arbitration clause of the contract. Provided both sides agree, the procedures can also be established or changed at the time the claim is submitted to arbitration. The method for selecting arbitrators set forth in the American Arbitration Association (AAA) Construction Industry Arbitration Rules is most widely employed. The AAA procedures provide for selection of one or three neutral and unbiased arbitrators either by mutual agreement between the parties or by administrative appointment if no agreement can be reached. An alternative to the AAA procedures that is sometimes used to select a three-member panel allows each party to select an arbitrator sympathetic to their side. The third arbitrator is then appointed by the two partisan arbitrators and is expected to be neutral. As a practical matter, this process makes the neutral arbitrator the "swing vote" that decides the arbitration.

§ 1.11 —Informality and Limited Appeals in Arbitration

The emphasis on the technical expertise of arbitrators usually involves a substantial de-emphasis of legal principles, technicalities, and procedures, including the right of appeal. This aspect of arbitration is often cited as a positive, but the contrary can be argued as well.

In arbitration proceedings, strict rules of evidence do not apply and arbitrators are generally liberal in their acceptance of evidence. This permits an easier and faster presentation of records, correspondence, documents, photographs, and live testimony. In fact, the AAA Rules encourage arbitrators to accept any and all evidence that may shed light on the dispute. In the more relaxed environment of arbitration, substantive defenses such as statute of limitations, no damage for delay clauses, and notice requirements may be given little weight.

In addition to the arbitrators' being granted considerable latitude in their conduct of hearings and rendering of awards, the right of appeal by means of a challenge of an award is extremely limited in scope. This limited scope of judicial review of arbitration awards is yet another trade-off. Although it certainly curtails a party's rights as compared to the scope of an appeal of a jury verdict, the limited scope tends to reduce the number and length of appeals from arbitration awards. This is, of course, in contrast to the court system with its lengthy and expensive appellate procedures.

§ 1.12 —Enforceability of Agreements to Arbitrate

In addition to the arbitration clause in the contract, a party's right to arbitration depends upon its ability to go to court and enforce the agreement to arbitrate. At common law, the courts were jealous of their jurisdiction and protective of a person's right of access to the courts. Today, however, the overburdened court system has made alternative methods of dispute resolution a necessity, and most states have attempted to broaden the right to arbitration either by statute or by court decision.

In addition to state laws, construction arbitration agreements are often enforceable under the Federal Arbitration Act,[11] which is the expression of a strong federal policy favoring arbitration.[12] The Federal Arbitration Act only applies if the arbitration clause is in a contract "evidencing a transaction involving commerce," meaning interstate commerce. Transactions of the type generally involved in a large construction project often satisfy this interstate commerce requirement and come within the scope of the Federal Arbitration Act.[13] If the interstate commerce requirement is met, the Federal Arbitration Act must be enforced, even in state court, and it

---

[11] 9 U.S.C. §§ 1–15.

[12] J.S. & H. Constr. Co. v. Richmond County Hosp. Auth., 473 F.2d 212 (4th Cir. 1973).

[13] *See, e.g.,* Pennsylvania Eng'g Corp. v. Islip Resource Recovery Agency, 710 F. Supp. 456 (E.D.N.Y. 1986); Electronic & Missile Facilities v. United States, 306 F.2d 554 (5th Cir. 1962), *rev'd on other grounds,* 374 U.S. 167 (1963); Metro Indus. Painting Corp. v. Terminal Constr. Co., 287 F.2d 382 (2d Cir. 1961).

preempts and supercedes all contrary and inconsistent state law.[14] However, state law incorporated into the contract by a choice of law provision may still affect the manner in which the arbitration proceeds, even if the Federal Arbitration Act were applicable.[15]

§ 1.13 —Special Problems Involving Multiple Parties

A recurring issue in the administration of construction arbitrations is the consolidation of a number of separate arbitrations and multiple parties on the same project into one proceeding. The potential for problems and the desirability of consolidation need to be considered at the time the arbitration clause is being drafted and the contract signed, long before any claim develops.

Although the consolidation of court proceedings involving numerous parties is common, few construction contracts presently provide for such consolidated proceedings in arbitration. Even without contractual authorization, however, some courts have required consolidation as an expeditious means to resolve construction disputes.[16] However, the traditional and majority rule appears to be that, without express contractual consent to multiparty arbitrations, courts will not require consolidation.[17] In *Stop and Shop Co. v. Gilbane Building Co.*,[18] the court insisted that "[i]f multiparty arbitration is to become a standard procedure, arbitration clauses and the rules and procedures of the AAA and other concerned organizations should be redrawn to provide for it."

The arbitration clause contained in AIA Document A201 (1987), Article 4.5.5, imposes strict limits on consolidation and prohibits consolidation of the owner's claim against the architect in any arbitration between the owner and the contractor. The AIA arbitration clause does, however, allow for consolidated proceedings involving parallel contractors. The

---

[14] Perry v. Thomas, 482 U.S. 483 (1987); Southland Corp. v. Keating, 465 U.S. 1 (1984); Moses H. Cone Memorial Hosp. v. Mercury Constr. Corp., 460 U.S. 1 (1983).

[15] Volt Information Sciences v. Stanford Univ., 109 S. Ct. 1248 (1989).

[16] *See* Episcopal Hous. Corp. v. Federal Ins. Co., 273 S.C. 181, 255 S.E.2d 451 (1979); Exber, Inc. v. Sletten Constr. Co., 92 Nev. 721, 558 P.2d 517 (1976); Grover-Diamond Assocs. v. American Arbitration Ass'n, 297 Minn. 324, 211 N.W.2d 787 (1973); James Stuart Polsek & Assocs. v. Bergen County Iron Works, 142 N.J. Super. 516, 362 A.2d 63 (1976).

[17] Consolidated Pac. Eng'g v. Greater Anchorage Area Borough, 563 P.2d 251 (Alaska 1979); Cumberland Perry Vocational Technical School Auth. v. Bogar & Bink, 261 Pa. Super. 350, 396 A.2d 433 (1978).

[18] 364 Mass. 325, 304 N.E.2d 429, 432 (1973).

prohibition against joining the architect can put the owner in the unenviable position of having to defend in arbitration the contractor's claim for the architect's defective design, without being able to compel the architect to join the same arbitration as a party. This places the burden of defending the architect's design or conduct on the owner. The prohibition against consolidation also causes the owner to run the risk of inconsistent results as well as extra expenses from duplicative arbitration proceedings. For example, an owner could lose to a contractor's claim of defective design in one arbitration, only to fail to convince a separate arbitration panel that the design was defective and that the architect should therefore reimburse the owner for its loss to the contractor. In order for the owner to avoid this situation, it must be addressed in advance in the arbitration clauses of its contracts with the contractor and the architect.

§ 1.14 Planning Ahead—Avoiding and Preparing for Claims Through Proper Management and Documentation

The prudent and realistic contractor designs and utilizes systems and procedures to manage, monitor, and document the work and progress on the project. These systems serve two important functions. First, they ensure an adequate flow of information to facilitate proper project control and coordination, including adjustments needed to respond to unexpected circumstances. Second, they aid in the compilation of an accurate and complete record of job conditions and problems and their impact on the project.

The contractor certainly bears the bulk of responsibility during construction as it installs the work and generally controls the means and methods employed to do so. The architect and the owner should not, however, abdicate their responsibility and oversight by totally removing themselves from the construction process in an effort to insulate themselves from liability. They cannot avoid all liability; moreover, some interaction and monitoring of the construction are always required of the owner and architect and are in their interests. If the owner or the architect become too removed from the construction, they can neither anticipate nor promptly address problems requiring their assistance. The degree of activity and monitoring will depend on the type and terms of the contract involved, but should not be so active or intrusive as to constitute interference and disruption of the contractor's work.

Although it may be somewhat unpleasant to begin a project with an eye to possible future claims, a failure to adopt such prudent management procedures almost ensures that disputes will develop.

§ 1.15 —Prudent and Responsible Estimating

Efforts to effectively manage work on the project and avoid claims should begin for the contractor before it even mobilizes or reaches the site. Many risks and claims arise not in the field but in the estimating department. Prudent estimating and bidding can avoid a host of performance problems and claims. A project that starts out in the hole because of bad estimating generally cannot climb out. Instead, the hole gets bigger and deeper, expanding the problem and drawing more parties into it.

Failure at any level to accurately perceive and then price the scope of the work or the risks associated with it results in unnecessary losses and difficulties that tend to ripple throughout the project. Estimates and bids should be supported by worksheets and backup documentation of sufficient detail under the circumstances. Such backup and the entire estimating process should be subject to standard forms and procedures and management review to ensure their accuracy.

Overly optimistic estimates based on vague or incomplete designs should be avoided or at least clearly identified and qualified as such. Performance specifications that often entail much more responsibility and cost than is initially apparent is often another soft spot in the estimating effort. The zeal involved in selling the project or submitting an early guaranteed maximum price to satisfy the owner must be balanced with caution against establishing an unrealistic budget or inflated expectations that, regardless of any contractual significance, are bound to cause disappointment, distrust, and disagreement if they are not met.

This scrutiny must be applied to bids received from subcontractors as well as those generated in-house. It is not always the case that the contractor may have recourse against the subcontractor. More importantly, the contractor has its own obligations to the owner and other subcontractors and cannot evade that responsibility or liability simply by pointing to the subcontractor who is unable to perform because of an estimating error.

Owners should likewise beware of a contractor bid that sounds too good to be true—it probably is! Success in initially enforcing a mistaken or reckless bid can reap bitter returns later in the project if the contractor's cash and capital run short.

§ 1.16 —Establishing Standard Operating Procedures

Construction projects that are run by the seat of the pants are accidents waiting to happen. Every project should have formalized, standard operating procedures with which all project personnel are familiar. The procedures should identify the specific authority and areas of responsibilities for each project staff position. Ideally, these should be standardized within a company and consistent from project to project. Standard job descriptions can then be used to define the roles of the individuals on a particular project. The standard procedures should cover responsibility for processing change orders and extra work, purchasing and receiving, project documentation, and costs and accounting.

As the project team is being assembled and mobilized and the standard procedures are adjusted, defined, and implemented for the particular project, it is a good idea to reexamine the efforts on the project in terms of estimating, scheduling, procurement, cost accounting, and the like before construction begins. This reassessment can serve as an additional safeguard for the early identification and correction of problems that might otherwise have a serious impact on the project at some later date if left undetected.

§ 1.17 —Establishing Lines of Communication

The ability of the parties on the project to establish and maintain constructive lines of communications is essential to the success of any project. Prosecution of the work must be recognized as a cooperative effort that demands a team approach rather than adversarial conflict. The owner, the architect, and the prime contractor must establish some method to both discover what is going on at the jobsite (for example, what problems the subcontractors are having and the problems they anticipate) and relay suggestions, recommendations, and requirements as to how these problems can be avoided or solved. Satisfactory communications can be achieved only if the parties have personnel who can develop pleasant and confident working relationships with one another. The subcontractor's workers should have sufficient confidence in the prime contractor's on-site personnel so that they will not hesitate to report difficulties and seek the prime contractor's recommendations as to how those difficulties can be avoided or resolved. Efforts to develop this confidence should begin at the preconstruction conference and should continue throughout the contract period.

An important procedural aid to establishing and maintaining the required lines of communications is regular job meetings. Weekly, biweekly, or at least monthly meetings should be regularly scheduled and held. The participants and the frequency of the meetings will depend on their purpose and the status or level of activity on the job. Field coordination meetings should involve the project superintendent, subcontractor superintendents, and key foremen. Brief but regular meetings like this can aid the process of coordinating and scheduling the work on a firsthand basis. They can also help identify problem areas and information needed for progress before a situation becomes critical.

Regular meetings between the architect's staff and the contractor are helpful for keeping up on the status of submittals, shop drawings, and areas requiring clarifications. Meetings between the contractor, architect, and owner should also take place, but probably on a less frequent basis. These meetings can be used to apprise the owner of important developments and to work out contractual issues. Further, the parties can discuss problems that are not being worked out on a more operational level and require the owner's intervention. The contractor should be wary of allowing the owner to get too far removed from the construction effort.

§ 1.18 —Project Documentation

Paperwork on a construction project can be overwhelming, but it is essential. The contractor typically generates and maintains the bulk of the documentation on a construction project, but all the participants have an interest in it. Contracts often require that the contractor maintain certain documentation, with copies and/or access available to the owner and others.

Project documentation creates an accessible history of the project that serves two roles: (1) planning and managing the project and (2) aiding in resolving claims and disputes. It must be organized and maintained in such a manner that it is a help and not a hindrance to effective project management and prosecution or defense of claims. Routine and uniformity are two essentials to an effective system of project documentation. The procedures should not only be standardized for the project but for the company as a whole. Only with that level of emphasis and indoctrination can all the fruits and benefits of a system be reaped.

The system and procedures must be written down. The length and level of detail of the written description will vary with the size and complexity of the project. Some description may be obviated by the use of computerized tracking systems, such as for cost accounting and tracking submittals. Regardless of how extensive the procedures are, it is imperative that they be clear and specific. If they are vague and general, allowing for personal interpretation and selective application, there will be no system at all. Instead, a hodgepodge of personal record-keeping and filing systems will result.

Just putting procedures on paper is not enough. The procedures must be reviewed with all levels of personnel who will be responsible for implementation so they are understood, used, and enforced. The critical importance of project documentation must be emphasized and that emphasis maintained throughout the project and from project to project.

§ 1.18 PROJECT DOCUMENTATION

Certain basic information should be maintained and organized in separate files:

The contract, including all its components, and all change orders or amendments, including a bid or original set of project plans and specifications

All documents, worksheets, and forms associated with the original bid estimate and subsequent revisions

Subcontractor or vendor files, including bids, subcontracts, or purchase orders, together with changes, and correspondence

Project schedules, including the original schedule and all updates

Insurance requirements and information for all parties.

The standard procedures relating to documentation should also address the creation, maintenance, and organization of specific types of documentation:

Correspondence

Procedures for date-stamping, copying, routing, filing, and indexing incoming and outgoing correspondence should be the responsibility of secretarial or clerical support staff to perform in accordance with standard procedures. A copy of all correspondence should go in a master correspondence file. The party responsible for responding to or acting on incoming correspondence should be identified.

As a matter of routine, the project management should be drilled on the importance of complying with technical notice requirements in the contract. Discussion with other parties should likewise be confirmed in writing to the involved parties with copies to the file. Such confirmation will help immediately resolve any misunderstanding that might exist and also preserve the substance of the discussion if there is a dispute at some later date.

Meeting Notes

Regular job coordination meetings between the various parties to the project, on a cumulative basis, probably cover more issues and contribute more to the exchange of information necessary to complete the work than all the correspondence on the project. What occurs at such meetings is therefore of great importance. Someone should be designated to maintain the minutes or notes for each meeting, preferably the same person at each meeting. That person should record the subjects covered, the nature of the discussion, the future actions to be taken and by whom. The name, title, and affiliation of each participant should be listed. The notes should be concise but informative. The items discussed should be indexed or designated in a manner so that they can be located for future reference. The notes should then be distributed to all participants and those affected on a regular basis.

A word processor or personal computer can be valuable for updating regular meeting notes, as certain items will likely remain open to discussion through several meetings. At the opening of each regular meeting, the notes from the previous meeting can be reviewed to confirm their accuracy and the mutual understanding of the participants. By identifying those items that remain outstanding, the previous week's minutes can also serve as an agenda for the current meeting.

Jobsite Logs or Daily Reports

Jobsite logs or daily reports are generally maintained by the project superintendent and can provide the best record of what happens in the field. They help keep management and office personnel informed of progress and problems. In the event of a claim, they are often among the most helpful documents in recreating the progress on the job and as-built schedules.

The daily log or report must be a part of the superintendent's daily routine. If it is too burdensome, it either will be ignored or will detract from the superintendent's primary function of getting the job built. Key information should be elicited briefly and concisely, requiring as little narrative as possible. The information covered should include:

Manpower, preferably broken down by subcontractors
Equipment used and idle
Major work activities
Any delays or problems
Areas of work not available
Safety and accidents
Oral instructions and informal meetings
A brief weather summary
Jobsite visitors.

The burden on the superintendent can be eased and the information maintained in a more organized manner by using a standard form. The process can be further expedited by simply allowing the superintendent to dictate entries and having the report typed by office staff.

All key project personnel such as foremen, project engineers, and project managers should also be encouraged to maintain personal daily logs

§ 1.18   PROJECT DOCUMENTATION   23

and procedures established to facilitate this effort. The information they record should be similar to the job log or daily report but not as extensive or detailed.

These types of routine, contemporaneous descriptions of work progress, site conditions, labor and equipment usage, and the contractor's ability (or inability) to perform its work can provide valuable information necessary to accurately reconstruct the events of the project in preparation of a claim. In maintaining these reports or logs, project personnel must be consistent in recording the events and activities on the job, particularly those relating to claims or potential claims. The failure to record an event, once the responsibility of a daily report or log is undertaken, carries with it the implication that the event did not occur or was insignificant and also threatens the credibility of the entire log.

Standard Forms and Status Logs

There is a constant flow of information between the project participants by means of a variety of media. Drawings are revised, shop drawings are submitted, reviewed, and returned, field orders and change orders are issued, questions are asked, and clarifications are provided. Cumulatively and individually, these bits and pieces of information are essential for building the job and also for reconstructing the progress of events on paper in the event of a claim. The standard procedures must include the means for providing, eliciting, recording, and tracking this mass of data so that it can be used during the course of the job and efficiently retrieved in an after-the-fact claim setting.

Routine transmittal forms should be customized to address specific, routine types of communications in order to expedite the process, but also to ensure required information is provided. For example, separate specialized forms can be prepared for transmittal of shop drawings and submittals, requests for clarifications, drawing revisions, and, of course, field orders and change orders. When possible, the forms should provide space for responses, including certain standard responses that simply can be checked off or filled in. At a minimum, the forms should identify the individual sender, the date issued, and specific and self-descriptive references to the affected or enclosed drawings, submittal, or specification. If a response is requested by a certain date, that date should be identified on the form.

Ideally, each discrete type of communication or specialized form should be numbered or somehow identified in a chronologically sequential manner based on the date it is initiated. Shop drawings and submittals, however, are best identified by specification section, with a suffix added to indicate resubmittals. This provides a basis for easy reference and orientation. Copies of the completed forms should be maintained in binders in reverse chronological/numerical order. Although various project staff members may require working copies, a complete master file should be maintained as a complete reference source and historical document.

In order to maintain the status of and track these numerous and varied communications, which can number many thousands, logs should be maintained. These logs need only address key information such as number assigned, date, and a self-descriptive reference. Proposed change orders and change order logs should also identify any increase or decrease in contract amount as well as time extensions. Such logs can be kept on personal computers using inexpensive, commercially available software or even on a word processor to expedite updating. Logs should be maintained for internal record-keeping and also for distribution to other parties on the project. The logs serve as a reminder of outstanding items and can highlight action required to keep the work progressing.

The contractor should use standard forms and procedures for communications with subcontractors as well as the owner and architect. Ideally, subcontractors should be encouraged to standardize their communications so there will be a more integrated approach for the entire project.

Photographs and Video Tapes

Photographs and video tapes are helpful, easy, and inexpensive means to monitor, depict, and preserve conditions of the work as those conditions change and the work progresses. They are particularly helpful in claims situations. One approach, incorporated in many contracts, is to accumulate a periodic pictorial diary of the job through a series of weekly or monthly photographs showing significant milestones in the construction. This encourages personnel to take photographs of site conditions on a routine basis, perhaps concentrating on problem areas and those areas associated with crucial construction procedures and scheduling. Photographs are also the best evidence of defective work or problem conditions that are cured or covered up and cannot be viewed later.

A professional photographer may be needed at times or required by contract but generally jobsite personnel with some instruction are more than capable of handling the photography chores. Cameras that produce quality photographs and negatives should be used. However, it is also a good idea to use as a backup a self-developing camera that allows the party responsible for taking photographs to check the content and clarity of the photos while she is still at the site and before conditions are altered.

Pictures should always be identified regarding time, date, location, conditions depicted, personnel present, and the photographer. This should be done at the time the photograph is taken if a self-developing camera is used. Otherwise, a log should be kept as the photos are taken and the log should be immediately checked when the photos are developed and the appropriate entries made on the back of the prints. Unless this information is correlated to specific photographs, the utility of the entire effort can be substantially undermined. Negatives should also be retained in an organized, retrievable manner.

Videocassette recorders have become relatively inexpensive and easy to operate. In some situations, a videotape can be considerably more informative than a still photograph, such as when attempting to depict an activity or the overall status of the project. Static conditions, however, are best photographed. The availability of a contemporaneous narrative as part of the video can give the viewer a much better idea of what is being depicted and why. A monthly videotape is an excellent way of preserving and presenting evidence. Again, properly trained jobsite personnel can operate the video recorder and later testify in conjunction with the showing of the tape.

§ 1.19 —Cost-Accounting Records

The use of effective cost-accounting methods and the maintenance of appropriate cost records can minimize many of the proof problems inherently associated with construction claims. Unfortunately, even though a claimant may be able to prove that an event has occurred entitling it to additional compensation, it will be able to recover only that amount of damages that it can prove with reasonable certainty. Proving the actual dollars lost is crucial to the claim.

Cost-accounting systems utilized by contractors vary drastically in their level of sophistication. More often than not, the accounting function suffers from a lack of priority by senior management until a dispute arises and the claim development process begins. The procedures described in **§ 1.18** to effectively capture and document events or occurrences to prove liability are only half the battle. Without effective accounting systems, the development of a clear, concise, and winning claim is haphazard at best. In addition, accounting rules for the construction industry are sometimes subjective and, in many cases, different conclusions can be reached on a single set of circumstances by accounting professionals and business people. With this in mind, it is important that accounting policies and procedures be documented and their application be appropriately monitored for compliance by management. This approach tends to improve the consistency in the manner in which items pertaining to all contracts are accounted for and, therefore, improves the credibility of the way in which matters are treated in developing a claim. Further, and perhaps of primary importance, a good cost-accounting system is invaluable in providing timely information to management for decisionmaking purposes and for monitoring financial performance. Examples of areas that warrant clearly documented policies and procedures are:

1. **Treatment of direct costs.** Direct costs consist of direct materials, direct labor, subcontract costs, and other miscellaneous direct costs such as bonding and equipment rentals, that are directly related to and can be specifically attributed to an individual contract.
2. **Treatment of the cost of major items or equipment.** The recovery of the cost of major items or equipment, whether owned or leased by the contractor, is an important element of cost-accounting procedures inherent in the construction industry. The American Institute of Certified Public Accountants' industry audit and accounting guide, *Construction Contractors,* provides some authoritative support as to methods of accomplishing such allocations, although substantial diversity exists within the industry.
3. **Treatment of contract overhead.** Contract overhead consists of the cost of items necessary for the proper and timely completion of the job but which cannot necessarily be attributed directly to the contract, such as supervision and project accounting. These costs must be allocated to the job in some fashion. There are many acceptable methodologies employed in the industry to accomplish the rational allocation of overhead costs. Allocation methodologies are by nature very subjective, theoretical, and often complex, and they are beyond the scope of this chapter. The key issues are to ensure that all relevant costs are captured, that the allocation method bears a sense of reality from a business standpoint, and that the method employed is consistently followed.

A carefully planned and well-documented accounting system, together with the appropriate internal controls and specific attention to the above areas, can make the claim process go smoothly. A detailed discussion of these matters is beyond the scope of this book. The advice of knowledgeable accounting and claims professionals should be sought in connection with implementing any system.

§ 1.20 —Monitoring the Work Through Scheduling

The prime contractor should continuously monitor the work of all subcontractors to determine that each is meeting its deadlines so that the work of other trades can proceed as originally scheduled. The owner must perform the same task when multiple prime contractors are

§ 1.20 MONITORING THROUGH SCHEDULING

involved. Even if the contractor has primary scheduling responsibility, which is usually the case, the owner should nonetheless continuously monitor the progress of the work and the scheduling effort. Most prime contracts require the preparation of a bar chart or progress schedule, which provides the easiest means of monitoring the work. The Critical Path Method (CPM) schedule required by the prime contract on many large projects can be even more valuable as a scheduling tool if properly developed, updated, and utilized.

The input of subcontractors and all project participants in the development and updating of any project schedule is critical to its usefulness. As a practical matter, a schedule that is developed without the input of the parties actually performing the work may result in an unworkable product, and the schedule as an instrument of coordination will be wasted. By getting the parties to participate in the preparation of the schedule, it becomes a much more meaningful and productive project management device. In addition, through its involvement, each party has in effect admitted what was reasonable and expected of it. If a party later fails to perform or follow the schedule, its ability to dispute the relevance of the project schedule and what was required of that party can be substantially reduced.

A project schedule can be a double-edged sword for the prime contractor, particularly if it is a CPM that shows the interrelationship of all activities and trades. A properly developed schedule can be used to demonstrate how a subcontractor is behind schedule and how its delayed performance is impacting the entire project.[19] Conversely, a subcontractor may also use a project schedule against the prime contractor to show how the subcontractor reasonably expected and planned to proceed with the work and how that plan was disrupted by the prime contractor, another subcontractor, or the owner, for which the affected subcontractor is entitled to additional compensation.[20] If the schedule is not properly maintained, updated, and enforced in order to reflect the actual progress of the work or the parties' contractual obligations, it may be dismissed by a court or arbitrators as merely representing "theoretical aspirations rather than practical contract requirements."[21] The heavy use of scheduling information and analysis in resolving claims underscores the importance of preparing and maintaining through updates a realistic schedule that secures subcontractor involvement and agreement.

---

[19] Illinois Structural Steel Corp. v. Pathman Constr. Co., 23 Ill. App. 3d 1, 318 N.E.2d 232 (1974).

[20] *See* United States *ex rel.* R.W. Vaughn Co. v. F.D. Rich Co., 439 F.2d 895 (8th Cir. 1971).

[21] *Id.*

§ 1.21 Effective Claim Development

Construction claims, by their nature, are often tedious and complex. They generally involve facts and circumstances that stretch over months and years, rather than one catastrophic instant in time. The many facts and circumstances involved in the claim often ripple throughout the project in ways not so readily perceived, touching other parties and compounding the complexity of the situation. All these factors make the early formation and pursuit of a strategy an important step toward winning a construction claim.

The first step in developing a claim strategy is to decide upon the objective or goal the strategy will be geared to achieve. Simply identifying the goal as "winning" is not enough. What constitutes winning a construction dispute is a variable. Is a claim being pursued simply to recoup a pecuniary loss? Or, is the claim being pursued to satisfy some principle or to set an important example or precedent so that the financial recovery may be of no importance? A practical and pragmatic approach generally prevails over thoughts of pursuing construction claims and litigation on principle. Most construction claims are best reduced to economic considerations of cost, risk, and potential recovery. Otherwise, a long and costly process of dispute and litigation can result.

Consequently, the amount of the claim or maximum likely recovery will dictate the level of effort and expense prudently invested in the claim effort. There is no exact formula for preparing for a successful claim, but there do exist useful procedures and fundamental techniques that provide a favorable basis for successful communication, documentation, negotiation, and, if necessary, arbitration or trial.

§ 1.22 —Early Claim Recognition and Preparation

Before there can be any preparation or prosecution of a claim, the claim must be recognized. Early recognition is necessary to ensure that notice requirements are met and that evidence needed to support the claim is preserved. Familiarity with the contract requirements is needed in order to recognize claims and to avoid unknowingly providing or accepting a nonconforming quantity, quality, or method of performance. Consequently, all jobsite personnel should be familiar with the contract terms, including the plans and specifications, the general conditions, and special provisions so they can evaluate the performance actually demanded as compared to the performance specified in the contract. In-house educational programs to enhance this ability and better equip job personnel to identify and handle possible claim situations should also be considered.

§ 1.22 EARLY CLAIM RECOGNITION

Of course, a claim should not be asserted for every minor incident or disagreement. Conversely, trying too hard to get along and never filing a claim is not good business either. Part of an effective program of identifying claims requires targeting those incidents that are sufficiently meritorious and substantial to justify the cost of preparing and prosecuting them. Filing claims with little merit or significance will merely waste resources and squander credibility.

Once a determination is made that a claim merits prosecution, comprehensive preparation and organization should be promptly undertaken. The facts, evidence, and documents bearing on the claim should be assembled, organized, and reviewed when they are fresh and before they are lost or forgotten. This preparation should be undertaken with an eye toward resolving the claim in the formal setting of an arbitration or litigation while still seeking early resolution through informal and less onerous means. If early resolution is not achieved, complete preparation at an early stage provides important insight for developing a claim strategy and a foundation that can be relied upon, subject to revision, as prosecution of the claim continues.

The first step in claim preparation should be an exhaustive investigation of the claimant's own records and sources of information about the project and the claim. Project records are generally voluminous. Although the review of the records must be sufficient in scope to cover the documents relevant to the claim and anticipated defenses, it must also be sufficiently focused and specific enough to avoid inundating the claim preparation process with unnecessary and irrelevant documents. There are certain categories of documents that almost always merit some consideration, such as the contract and all change orders, pay applications, daily logs or reports, bonds and insurance policies, correspondence files, and internal memoranda. Some further organization of the documents may be required beyond that used during construction in order to make individual documents relating to the claim more readily accessible.

Although documentation is certainly critical in any construction claim, it is not everything. At trial, although the documents will be admitted into evidence, the witnesses and their words, perceptions, and recollections will gain the most attention. Those individual resources should not be overlooked in the claim preparation process, but they often are. The more remote claim preparation is from the people actively involved in the field, the more likely there will be unpleasant surprises and inconsistencies as the claim is subjected to greater scrutiny in discovery or at trial. Consequently, the field personnel involved should be interviewed to confirm that management's secondhand understanding about the facts and circumstances of the claim is accurate and complete. To the extent possible, project personnel should be utilized to staff and assist in the claim preparation effort. At a minimum, field personnel should be given the opportunity to review the claim at various stages of preparation, and certainly before it is submitted, to confirm they can vouch for its accuracy.

§ 1.23 —Early Involvement of Experts and Attorneys

Construction claims often require the assistance of experts to help solve problems and to assemble and analyze the facts. Part of a program of prompt and cost-effective claim preparation requires considering involving attorneys experienced with construction claims and other technical experts at an early stage. Of course, the use of outside assistance will depend on the size and complexity of the claim, but in most claims such early involvement will facilitate prompt resolution or better preparation for trial and will be worth the investment. Scrimping on experienced and qualified legal and technical support for a claim can prove very costly in the long run.

An attorney experienced in construction claims and litigation can be deemed an expert whose advice and guidance at the early stages of a claim are often desirable. The construction attorney who will ultimately be charged with presenting the claim to a judge, jury, or arbitrators should be consulted to ensure that the claim's supporting documentation and evidence are being assembled and preserved in a manner consistent with favorable resolution of the claim in a formal proceeding. An experienced construction attorney can often suggest competent technical consultants in specialized areas such as accounting and scheduling, and thereby help avoid the expense and frustration of relying on an individual who does not have the proper qualifications to testify in the case. Involvement of an attorney does not presuppose resort to litigation or arbitration. On the contrary, it is simply another element of comprehensive preparation that can contribute to the early resolution of claims.

If it is possible to involve an expert during the actual construction phase, when a claim is merely a probability, that option should be considered. At such an early stage, the expert may be able to suggest ways of mitigating damages or reducing the impact of an injurious condition. The expert may also be able to recommend methods of preserving evidence and creating demonstrative evidence for use during negotiation, arbitration, or trial. Further, testimony based on firsthand observation of the construction will generally be more credible and persuasive than testimony based solely on secondhand input. There is the risk, however, that an expert's involvement in the project may jeopardize the credibility of any future testimony as the expert goes from neutral observer to active and adversarial participant. This risk can only be weighed on a case-by-case basis.

Construction is a complex process, involving a broad spectrum of scientific and technical disciplines. Resolution of construction claims likewise requires a variety of experts to serve as consultants in the claim preparation or to testify as expert witnesses. It is important to note and recognize such specialization and to ensure that the proper expert is consulted for the particular problem at hand. If a claim merits prosecution, it also merits the best and most qualified technical expert in that subject who is reasonably available.

The immediate concern with technical qualifications must be balanced with the need to have a witness who can provide persuasive testimony in the formal forum selected for resolution of claims. A respected technical expert may be understood by a technically oriented arbitration panel, but be incomprehensible to a lay jury. This type of counterbalancing must be considered and addressed early in the claim preparation process to avoid difficult surprises and dramatic adjustments in the later stages of the claim, immediately before or during trial.

Delay is frequently a factor in construction claims. Hence, scheduling analysis and scheduling experts are often involved in resolution of claims. Beyond their scheduling expertise, scheduling consultants must have a detailed, working knowledge of the construction process so that their analysis reflects the practical problems and difficulties experienced on the construction site and not merely a computer-generated abstraction. Likewise, it is essential that the scheduling expert involved in a claim be provided access to contemporaneous project documentation that accurately reflects the manner in which the work proceeded. Costly and complex "as-built" scheduling analyses presented in support of claims can be severely undermined if the dates used in the analysis conflict with those contained in project documentation, such as daily reports or monthly schedule updates. The scheduling expert's task and effectiveness can be substantially enhanced and such problems avoided through the maintenance of project documentation as described in § **1.18**.

Certified public accountants who are familiar with the construction industry and its financial and accounting practices can also contribute significantly to quantifying and proving the financial consequence of the technical problems that generated the claim. Their involvement is discussed separately in § **1.29**.

§ 1.24 —Use of Demonstrative Evidence

Demonstrative evidence has the special advantage of presenting in pictorial form abstract, complicated, and extensive facts. It can clarify or explain oral testimony or documentary narrative in concrete terms. In addition, demonstrative evidence adds interest and avoids the tedium of a relentless one-dimensional recitation of facts. The simplicity and clarity that demonstrative evidence can provide are particularly effective in large, highly technical, and complex construction claims. However, the utility of demonstrative evidence should not be overlooked in smaller, more straightforward disputes.

Demonstrative evidence can range from photographs and videos to charts summarizing facts or making comparisons, such as a chart comparing as-planned manpower to as-built manpower in order to graphically depict an overrun. Charts and graphs are often used in connection with scheduling presentations, again usually comparing the as-planned schedule to the as-built schedule, with a focus on the factors that created problems or delays. By displaying this information in an attractive, visual way in combination with other written and oral presentations, the claim can be advanced in a more compelling and persuasive manner. The goal of the demonstrative presentation is lost if it is not clear and understandable and firmly supported by the facts.

Discussions of the importance and usefulness of demonstrative evidence as a means of persuasion generally are found in trial advocacy materials. But there is no need to hold such a powerful and effective tool in reserve until trial. Demonstrative evidence should be developed and used to simplify the claim and persuade the other side as soon as possible.

§ 1.25 —Components of a Well-Prepared Claim Document

Simplicity to promote prompt understanding, while making the claim interesting and appear to be well supported, is the key to effective claim preparation and presentation. One means by which to synthesize the claim and this approach is to use a *claim document,* a written synopsis of the claim that can be presented to the opposition at the early stages of the dispute. Like essentially every other aspect of claim preparation, the claim document serves two, alternate purposes. First, its immediate and primary goal is to bring about a prompt and satisfactory resolution of the claim. Failing that, the second purpose of the claim document is to provide a blueprint or plan for further prosecution of the claim.

The claim document provides an opportunity for the claimant to explain its grievance in a complete and comprehensive fashion. The process of preparing the claim document is an important step in developing a claim strategy because it requires the claimant to refine and synthesize the claim from beginning to end. The claim document should be viewed as telling a story. It should have a clear and definite theme that can be readily communicated, understood, and remembered. The theme should be the strongest argument supporting the claimant's theory of recovery.

§ 1.25 COMPONENTS OF A CLAIM DOCUMENT

There will certainly be a considerable quantity of facts gathered in support of the claim, but trying to present and argue each and every one of these facts will simply overwhelm and confuse the reader, and the claim document as a tool of persuasion will be a failure. When multiple and unrelated claims are presented in one document, the document must be structured to emphasize the strongest claim.

The primary communicative component of the claim document is the factual narrative. Although this narrative will certainly focus on the claimant's point of view, it should not be expressed in overly argumentative or combative terms. Instead, to the extent possible, the facts presented should be permitted to speak for themselves. The writing style should be clear and precise, but should not read like technical specifications. It is, after all, a story and not simply a recital of a string of facts. The narrative should be comprehensive and logically organized, so it can be used as a resource throughout negotiations and further prosecution of the claim. If the complexity of the matter is such that the narrative is exceedingly long, an executive summary should be prepared.

The factual narrative is often followed by a written discussion of the applicable legal principles that support and illustrate the theories on which the claim is based. Assistance from an attorney experienced in construction claims is generally required in order to fashion the legal arguments and otherwise to ensure that the factual narrative is presented in a manner consistent with applicable legal principles. The need for or extent of a legal discussion is generally geared to the expertise or experience of the ultimate decisionmaker for the opposition. For example, in federal construction contracts, certain theories of entitlement are so firmly established and recognized on all levels that no or only a limited legal discussion is required. In other situations, the legal discussion may be a crucial element in causing the other side to recognize its liability and exposure. A one-time owner may have no idea of what a differing site condition is or why the contractor should be paid for it. On the other hand, claims against local governmental entities which contract regularly may be ruled upon by elected officials who also require an education about construction law before they can be expected to recognize the need to settle a claim.

As discussed in greater detail in §§ **1.26** through **1.29** and throughout this book, pricing the claim and supporting such calculations is every bit as important as establishing liability for the claim. The claim document must recognize the importance of damages and include a specific dollar figure and fairly detailed cost analysis and breakdown. Supporting information and sources should be identified and appended if not too voluminous.

Finally, the claim document should be used to showcase and highlight the most persuasive and impactive documentary and demonstrative evidence. The most potent documents should be quoted or even reproduced in their entirety in the body of the narrative. Those documents that do not merit incorporation into the text, but which are referenced and support the claim, can be included in an indexed appendix that is cross-referenced with and organized like the factual narrative. In this manner, the narrative can be reviewed without having to sift through every bit of paper, but that backup is readily available should further review be desired.

In addition to documents, charts, graphs, drawings, and photographs, other demonstrative and visual evidence should be incorporated into the claim document to the extent practical. Similarly, consideration should be given to including relevant reports by experts as attachments to the claim document to be used as exhibits, with appropriate references to and quotes from the reports in the narrative.

In certain situations, the nature of the claim or the character or capacity of the opposition may counsel against submitting an extensive claim document. The opposition may not have the financial resources or genuine interest in resolving the claim by negotiation, which is a primary goal of the claim document. Instead, opponents may seek a one-way flow of information, that is, they are willing to receive a detailed presentation of the claim, but unwilling to explain or document any response, defense, or counterclaim until trial. The claimant must evaluate whether pursuing the "race for disclosure" by providing a claim document will ultimately eliminate the roadblocks to negotiation and settlement, or simply better equip the opposition to defend the claim, without a commensurate benefit to the claimant. Generally, but not always, a sound, well-documented and well-prepared claim should be able to withstand and be improved through feedback from the opposition's scrutiny. Moreover, even if the judgment is made that an extensive claim document should not be submitted, that conclusion does not necessarily mean that a claim document should not be prepared for internal use to better prepare for whatever proceeding may follow. Of course, if formal claim submission is mandated by the contract, such a requirement should be fulfilled, although the extent of the submission may vary.

§ 1.26 Importance of Calculating and Proving Damages

The issue in construction disputes that generally receives the most attention and focus is liability. Does a differing site condition exist? Who caused the delay and is it compensable? However, the issue of damages or costs flowing from the events that give rise to liability is no less important. Too often, the issue of calculating and proving damages is given a backseat, with little precision or scrutiny applied until the eve of trial. That approach can result in an entirely misguided claim effort, missed opportunities for settlement, and loss at trial or in arbitration. An early analysis of damages can help determine whether a claim really exists and the best means of preparing and positioning the claim for the affirmative recovery sought.

The problem of calculating and proving damages can be substantially reduced by initiating proper cost accounting at the time the claim is identified. Accounting measures can be established to segregate costs generated by the events giving rise to the claim and carefully maintain separate records. If such a procedure is followed, proof of damages can be reduced to little more than the presentation of evidence of separate accounts. Unfortunately, this ideal situation seldom exists; either the problem is not recognized in time to set up separate accounting procedures, the maintenance of separate accounts is simply not possible because of an inability to isolate costs, or no attempt is made to establish the requisite procedures. These circumstances necessitate the development of some formula that is sufficiently reliable to permit the court or arbitrators to allow its use as proof of damage. If settlement is being sought, the claimant must likewise convince the other side of the validity and reliability of its damage calculations. The following chapters present specific approaches and alternatives for pricing and proving claims of the nature frequently encountered. However, certain principles and possible approaches are common to all construction claims.

§ 1.27 —Basic Damage Principles

The law of damages is compensatory in nature. In contracts, the goal is to reimburse the claimant for all "losses caused and gains prevented" by the other party's breach.[22] Tort claims, which in construction generally involve allegations of negligence and misrepresentation, may offer a broader scope of damages, but many courts have limited tort damages to cases involving either personal injury or property damage, denying recovery for the purely economic loss that is typically the subject of construction claims.[23]

Damages in construction claims, like other contract claims, are of two basic types: direct and consequential. Although punitive damages may also be available in the extreme case, they are by far the exception. *Direct damages* are those which result from the direct, natural, and immediate impact of the breach and are recoverable when proven.[24] In the contractor's case, such damages may include the cost of items such as idle labor

---

[22] J. Calamari & J. Perillo, The Law of Contracts § 327 (West 1970).
[23] *See, e.g.,* State v. Mitchell Constr. Co., 108 Idaho 335, 699 P.2d 1349 (1985).
[24] Spang Indus. v. Aetna Casualty & Sur. Co., 512 F.2d 365 (2d Cir. 1975).

and machinery, material and labor escalations, and extended jobsite and home office overhead associated with delay. The owner's direct damages are generally those costs incurred in completing or correcting the contractor's work and the cost of delay, which is either its actual costs in terms of lost rent, loss of use, or liquidated damages.

Consequential damages do not flow directly from the alleged breach but are an indirect source of losses. The most frequently sought types of consequential damages in connection with delay claims are lost profits (stemming from reduced bonding capacity), interest on tied-up capital, and damage to business reputation. These are more difficult to prove because the causal link between such damages and the act constituting the breach is often tenuous and uncertain.

§ 1.28 —Methods of Pricing Claims

There are two basic methods for pricing construction claims. The simplest method is the *total cost method.* The other, more complicated but more widely accepted, method is the discrete or *segregated cost method.*

Total Cost Method

A total cost claim is simply what the name implies. It essentially seeks to convert a standard fixed-price construction contract into a cost-reimbursement arrangement. The contractor's total out-of-pocket costs of performance are tallied and marked up for overhead and profit. Payments already made to that contractor are deducted from that amount, and the total cost price is achieved. Of course, this approach can be refined or adjusted to meet particular needs and circumstances, but the basic components and approach remain: costs associated with the basis for the claim are not segregated. The total cost method is well suited for impact-disruption claims when the segregation of costs is virtually impossible.

The total cost approach, although preferred by claimants because of the ease of computation, is generally discouraged by the courts because it requires the contractor to be virtually fault-free and is fraught with uncertainties. Numerous court decisions have therefore established fairly rigorous requirements for the presentation of total cost claims. In order for the total cost method to yield an accurate and reliable figure, the contractor's original price must have been accurate and correct, overruns in performance cannot be the result of any performance problems caused by the contractor or its subcontractors, and those costs actually incurred must be reasonable. Even when these conditions are met, the contractor must demonstrate that the use of a segregated cost approach was not feasible.

Of course, if another approach is not feasible because of the contractor's own poor record-keeping, the validity of the total cost approach is further questioned.

Owners are likely to view contractor total cost claims with even greater suspicion and distrust than the courts, so much so that the credibility of the entire claim and the claimant can be undermined. The difficulties of establishing the prerequisites for use of a total cost calculation in court combined with the skepticism it can generate counsel against use of the total cost method whenever possible.

Segregated Cost Method

The segregated cost method of pricing claims is certainly more difficult than the total cost method, but it is a far more accurate, reliable, and persuasive way of presenting damages. Under this approach, the additional costs associated with the events or occurrences that gave rise to the claim are segregated from those incurred in the normal course of performance of the contract. For example, on an extra work claim, the pricing would reflect an allocation (actual or estimated) for the additional labor, materials, and equipment used in performing the extra work. If the project was delayed, costs of field overhead and home office overhead would also be calculated.

§ 1.29 —Project Cost Reviews and Audits

Even if the segregated cost method is to be employed to price and prove the claim, a good starting point is to summarize the project from a financial standpoint by making a total cost calculation. This will establish the contractor's overall profit or loss position on the entire project. Although the overall profit or loss may not ensure or bar recovery on a discrete claim issue, it is nonetheless important. As a practical matter, a claimant's out-of-pocket loss can be the most compelling evidence to support recovery, whether through settlement or at trial. Conversely, if a substantial profit was earned on the project but an overrun was incurred in one area as a result of some action by the other party, it may not make sense to pursue the claim.

Once the overall profitability of the project has been evaluated, it is a good idea to audit the contract. The size of the claim dictates whether an internal audit by accounting personnel is adequate or if an outside certified public accountant (CPA) is required. Generally, it is best to get outside, specialized accounting expertise involved early. Allowing an outside CPA to perform the audit provides a number of significant benefits. The scrutiny of an audit will identify any problems in the contractor's accounting that can be corrected or at least taken into account without the embarrassment of a surprise discovery by the other side. An outside audit will further refine and organize the contractor's cost information so that it is more readily usable in developing a segregated cost approach to the claim. The CPA doing the audit should be the one to testify, if necessary, in support of damages, so the audit will give that CPA greater credibility and also allow the CPA to provide greater assistance in the development of the claim. Depending on the size and complexity of the claim and the condition of the project records, such audit exercises are relatively inexpensive and generally extremely cost-effective.

If the initial audit confirms the accuracy and reliability of the claimant's accounting records, the claim should be priced and presented in a format consistent with those records and audit results. This will limit the opportunity for inconsistency and confusion. The endorsement of an outside CPA will also lend considerable credibility to the claim. In addition, this level of preparation and organization allows the claimant to invite an audit of the job and claim costs by the opposition. Construction contracts with the federal government and most public contracts provide for audits of claims, as do some private contracts. An audit can be used offensively by the claimant if it will confirm that the costs were actually incurred. That confirmation will give the claimant greater credibility and hopefully limit the scope of the dispute.

§ 1.30 Pursuing Negotiation and Settlement

Although claims and disputes are sometimes results of the construction process, they need not and should not dominate the process. When claims and disputes cannot be avoided, efforts should be redoubled to resolve them as quickly as possible. The complexity, time, and cost of arbitration and litigation naturally favor negotiation and settlement. An approach favoring prompt resolution should be part of a claims policy and project personnel and management should be indoctrinated and trained accordingly. Although contract provisions regarding notice of claims and other technical requirements should be complied with, other lines of communication on the project should not be overlooked as a means of bringing a claim to quick settlement. It is far easier and less expensive to resolve problems in the field, where they arose, than in the courtroom. Even if early settlement is not achieved, the negotiations force the claimant to examine seriously the merits of its claim and also reveal the strengths and weaknesses of the claim at an early stage.

Comprehensive and careful preparation greatly enhances the likelihood of early resolution and settlement. A party attempting to settle a claim should know its own case intimately and should have anticipated as many of the opposing party's points as possible without the benefit of discovery. Use of a well-prepared claim document, as discussed in § 1.25, is helpful both as a starting point and as a reference during settlement discussions. People with firsthand, detailed knowledge of the underlying facts are also an essential part of any negotiating effort. There is simply no substitute for the person who lived with the project's problems on a daily basis.

§ 1.31 Conclusion

Winning strategies in proving and pricing construction claims begin with a recognition that claims are best avoided. Identifying the recurring causes of claims permits the necessary planning and preparation to steer clear of major risks or to handle claims responsibly when they cannot be avoided. The same policies and procedures that aid in limiting claims also contribute to comprehensive and effective preparation of claims. Because the potential for claims cannot be ignored, skillful and determined management is needed both before and during construction to handle the threats and challenges that claims present.

CHAPTER 2

TAX CONSIDERATIONS IN LITIGATING AND SETTLING CLAIMS

John J. Hopkins, CPA
Anthony C. Weiss

John J. Hopkins is a partner in the Philadelphia office of Coopers & Lybrand and specializes in real estate and related tax matters. John has made presentations to many national, local, and university-sponsored professional groups and is a frequent contributor to national and local publications. He has a Bachelor of Science degree in Business Administration from Drexel University, a Masters of Science degree in Taxation from Villanova University's School of Law, and is a Certified Public Accountant in Pennsylvania.

Anthony C. Weiss is a real estate tax manager in the Philadelphia office of Coopers & Lybrand. He has a Bachelor o Science degree in Business Administration from Pennsylvania State University and a J.D. and Masters in Taxatior from Villanova. Mr. Weiss is a member of the Bars of Nev Jersey and Pennsylvania.

§ 2.1   Characterizing Damage Recoveries to the Plaintiff for Tax Purposes

§ 2.2   —Substantiation Requirement

§ 2.3 —Timing Issues
§ 2.4 Tax Considerations to the Defendant—Damage Payments
§ 2.5 —Tax Treatment of Loss Amount
§ 2.6 —Timing Issues
§ 2.7 Deductibility of Legal Fees

§ 2.1 Characterizing Damage Recoveries to the Plaintiff for Tax Purposes

In determining the tax treatment of damages received for breach of a construction contract, it is immaterial whether the taxpayer recovers damages amicably or by resolving the dispute through litigation or settlement.[1] It is the nature of the underlying claim, not the manner of collection, that affects the tax treatment.[2]

It is a well-established principle that amounts received for reimbursement of lost profits are includable in the taxpayer's gross income.[3] The rationale is that, because the profits would have been taxable income, and the damage proceeds are considered a substitute for the profits, they therefore are taxable in a like manner.[4] If the profits would have been considered ordinary income, the recovery representing lost profits is also taxable as ordinary income.

If the recovery received represents damages for injury to capital, such recovery will be treated as a return of capital.[5] An example of damages treated as a return of capital is found in a revenue ruling issued by the Internal Revenue Service.[6] In this revenue ruling, corporation M agreed to construct a plant for corporation P for $250 million. As a result of a change in the law with regard to environmental safeguards, a dispute arose between M and P over M's obligation to provide stricter safeguards at no additional cost.

The parties eventually agreed that M was responsible for delivering a plant that met the stricter environmental standards. M paid P $40 million for the estimated cost to satisfy the stricter environmental standards rather than completing the plant's construction. The revenue ruling concluded that the payment of $40 million from M to P represented the

---

[1] Spangler v. Commissioner, CA-9, 63-2 U.S. Tax Cas. (CCH) ¶ 9777, 323 F.2d 913 (1963), *cert. denied,* 364 U.S. 825.

[2] *Id.*

[3] Raytheon Prod. Corp. v. Commission, CA-1, 44-2 U.S. Tax Cas. (CCH) ¶ 9424, 144 F.2d 110, *cert. denied,* 323 U.S. 779.

[4] *Id.*

[5] *Id.*

[6] Rev. Rul. 81-277, 1981-2 C.B. 14.

present damages P had incurred because of the breach of contract by M. P received no economic gain as a result of the $40 million payment and was merely made whole under the contract. P was restored to the position that it would have been if M had fulfilled the terms of the contract, and thus the $40 million was a return of capital.

If the damages received are a return of capital, the basis of the asset must be considered in identifying whether the recovery is taxable. There will be no income recognition to the taxpayer unless damages received exceed the basis of the underlying asset. The Internal Revenue Code (Code) provides generally that the *basis* of property shall be the cost of such property.[7] Federal Income Tax Regulations further provide that *cost* is the amount paid for such property in cash or other property.[8] Also, pursuant to § 1016(a) of the Code, a taxpayer must reduce his basis in an asset by the amount of damages treated as a return of capital.

If damages received do exceed the taxpayer's basis in the underlying asset, the taxpayer will recognize income equal to the excess of the amount received over the basis of the asset. Whether the income is capital gain or ordinary income depends on the nature of the underlying asset.[9] If it is a capital asset, the Code provides that the taxpayer will have capital gain as if there had been a sale or exchange of the asset and it had been held for a period greater than six months.[10]

The issue that usually arises is whether there has been a sale or exchange. There appears to be no clear authority on whether injury to capital is to be treated as a sale or exchange. The Internal Revenue Service took the position in a Revenue Ruling that the receipt of damages in excess of the basis of a capital asset is ordinary income.[11] Contrary to this ruling however, are Tax Court cases which have concluded that, if proceeds represent damages for injury to capital assets, they are taxable as capital gain to the extent of any excess over basis.[12] In neither the ruling nor the court cases is there a supporting discussion on the sale or exchange issue.

Although the taxpayer may prove injury to a capital asset, he will only avoid current taxation to the extent he has a basis in the injured asset. It should be noted that a taxpayer frequently receives payment for injury to the goodwill of his business. A taxpayer's basis in goodwill is generally only acquired through the purchase of an existing business, and thus in many instances the basis is zero. Consequently, a taxpayer cannot defer

---

[7] I.R.C. § 1012.
[8] Treas. Reg. § 1.1012-1(a).
[9] I.R.C. § 1001.
[10] I.R.C. §§ 1201, 1221, & 1222.
[11] Rev. Rul. 68-378, 1968-2 C.B. 301.
[12] Bresler v. Commissioner, 65 T.C. 182 (1975).

recognition of income through cost basis reduction. The full amount of the recovery is a capital gain.

Because there is a significant difference in tax treatment between lost profit and return of capital, it is imperative that the taxpayer take the proper steps to ensure he receives the tax treatment that he desires. Failure to show support for the desired outcome may result in adverse tax consequences to the taxpayer.

§ 2.2 —Substantiation Requirement

Section 2.1 focused on the characterization of payments and the related tax consequences. However, a taxpayer will not achieve the desired results unless the claim is properly substantiated.

The focal point for determining the tax characterization of proceeds in a settlement or an adjudication of a claim is the complaint.[13] The complaint must contain facts to support the tax characterization that is alleged. If the complaint is not factually supported, then the tax characterization contended by the Internal Revenue Service will be upheld by the courts.

For example, a taxpayer received a recovery in settlement of a lawsuit.[14] In the complaint in the original suit for damages, it was alleged that there was injury to goodwill and loss of profits. This contention would have resulted in favorable tax consequences to the taxpayer. However, the court rejected the taxpayer's assertions and held that the entire amount received by the taxpayer in settlement of its suit was for lost profits and nothing was allocable to goodwill. The court reasoned that "a mere allegation in the complaint for injury to goodwill is not sufficient in itself to establish that the settlement represented at least in part a recovery for that damage in the face of a substantial showing that the recovery was, on the contrary, entirely for lost profits."

In another case, however, the Internal Revenue Service's contentions were not upheld.[15] The issue involved was also whether a recovery in settlement of litigation was for harm to goodwill or loss of profits. The IRS had contended that, because the taxpayer failed to meet his burden of proof, the recovery should be fully taxable as lost profits. This time, however, the government was not successful. The court held that when IRS determinations are arbitrary and excessive, taxpayers are required neither

---

[13] Rev. Rul. 85-98, 1985-21, C.B. 51.

[14] Phoenix Coal Co. v. Commissioner, CA-2, 56-1 U.S. Tax Cas. (CCH) ¶ 9366, 231 F.2d 420.

[15] Durkee R. J. v. Commissioner, CA-6 (*rev'g and rem'g* TC); 47-1 U.S. Tax Cas. (CCH) ¶ 9279, 162 F.2d 184, *aff'g* 14 T.C.M. 96 (1955).

to establish the correct amount that might lawfully be charged against them nor to pay a tax liability not owed.[16]

With regard to substantiation, the distinction between a settlement agreement and a court adjudication is important. For example, there was a case involving an out-of-court settlement in which the taxpayer claimed that a portion of the recovery pursuant to his settlement was allocable to loss of goodwill.[17] The court, however, upheld the assertion of the Internal Revenue Service that there was no proof of actual loss to goodwill in this tax case nor in the negotiations leading to the settlement. Moreover, the IRS was not bound by the terms of the settlement agreement.

In contrast, however, a case may be resolved through a court adjudication. The court judgment may allocate the recovery between lost profits and harm to goodwill. If so, the Tax Court seems more inclined to follow the same allocation.[18]

A way of avoiding a problem with the Internal Revenue Service is to address the characterization of potential recoveries when the dispute arises. Moreover, if the parties do settle, they should expressly agree to an allocation at the time they are negotiating the settlement.[19]

§ 2.3 —Timing Issues

Once the amount and character of income has been determined, the next issue is the tax year in which such income is taxable. The Code provides for permissible methods of income inclusion.

One acceptable method is the *cash receipts and disbursements method.* Under this method all items that constitute gross income—cash, negotiable instruments, or property—are to be included in the taxable year in which the income is actually or constructively received. Income is constructively received in the taxable year during which it is credited to a taxpayer's account, set apart for him, or otherwise made available to draw upon it.[20]

If, for example, a contractor with a December 31 year end receives a check for damages on December 30, the contractor could not defer recognition of income simply by avoiding cashing the check or by giving it back to the owner and asking him to hold it until the next year.

---

[16] *Id.*

[17] Basle v. Commissioner, CA-3, 58-2 U.S. Tax Cas. (CCH) ¶ 9748; 256 F.2d 581, *aff'g* 16 T.C.M. 745 (1957).

[18] Specialty Eng'g Co. v. Commissioner, 12 T.C. 1173 (1949).

[19] Major v. Commissioner, 76 T.C. 239 (1981).

[20] Treas. Reg. § 1.446-1 (c)(1).

Generally, when a plaintiff receives damages, they are included in the year of actual receipt. However, if a plaintiff has a question on whether there has been constructive receipt, then the Code, Treasury Regulations, and case law should be consulted.

Another acceptable method for determining the year of inclusion is the *accrual method* of accounting. Under this theory, income is included for the taxable year when all events have occurred that fix the right to receive the income and the amount can be determined with reasonable accuracy.[21] For example, a contractor sued to collect damages, and the amount was determined in settlement of litigation in the same year. Under the accrual method, the damages would be includable in that year even though the proceeds have not actually been received, assuming that all the events have occurred that fix the right to receive the income.

A third accounting method specifically applicable to contractors is the *long-term contract method.* Prior to 1986, income from a long-term contract was includable in gross income under the percentage of completion method or the completed contract method.[22] However, for a contract entered into after February 28, 1986, taxpayers are required to use a modified percentage of completion method.[23] Two exceptions to this requirement are (1) a home construction contract and (2) any other construction contract, completed within a two-year period, which has average annual gross receipts not exceeding $10 million for three taxable years preceding the taxable year in which the contract was entered.[24]

If a contractor is required to use this new method, then a portion of the litigation proceeds must be accounted for under the percentage of completion method and a portion under the completed contract method.[25] The proportion of income accountable under the percentage of completion method varies depending upon when the contract was entered into.

Generally, under the *percentage of completion method,* a portion of the profit earned on a contract is reported in each of the years in which work on the contract is performed.[26] The portion of the profit to be reported is determined by comparing the costs incurred in a particular period to all costs expected to be incurred in performing the contract.[27] This percentage is then applied to the estimated contract profit to determine the profit for the applicable year. Under the *completed contract method* of accounting, the revenues and costs associated with a particular contract are

---

[21] Treas. Reg. § 1.446-1 (c)(1)(ii).
[22] Treas. Reg. § 1.451-3 (a)(1) (proposed amendments).
[23] I.R.C. § 460.
[24] I.R.C. § 460 (e)(1).
[25] I.R.C. § 460 (a).
[26] Treas. Reg. § 1.451-3 (c)(1), (2), & (3).
[27] *Id.*

§ 2.3 TIMING ISSUES FOR PLAINTIFF   47

accumulated and reported for tax purposes only at the completion of the entire contract.[28] The ability to use the completed contract method has been substantially curtailed with the passage of the Tax Reform Act of 1986.

In addition, it should be noted that there are certain rules that govern when a contractor tenders the subject matter of a long-term contract to the party with whom it is contracting and a dispute has arisen because the contractor is requesting an increase in the contract price.[29] The regulations that apply only to contractor claims are as follows:

1. The entire amount of the gross contract price shall be included in gross income in the taxable year the contract is completed (without regard to the dispute). All costs that are properly allocable to the contract and that have been incurred prior to the end of the taxable year in which the contract is completed (without regard to the dispute) shall be deducted in the same year.

2. Any item of income that was not included in gross income pursuant to **1.** above shall be included in gross income in the taxable year in which the dispute is resolved. Any expenses relating to the contract and incurred in the year subsequent to the completed contract shall be deducted from gross income in the taxable year in which the expense is incurred.

3. The term "gross contract price" means the original stated price of the contract with any modifications to which the parties have agreed as of the end of the taxable year. For example, the term includes any amount that the taxpayer is claiming by virtue of changes in the specifications of the contract which the other parties to the contract have agreed is proper, but it does not include any amount that the contractor is claiming that is disputed by the other parties to the contract. However, no amount is excluded from the gross contract price solely because a party refuses to pay that amount when due. Thus, for example, if the parties to a contract agree that the gross contract price is $100,000, but a party refuses to pay $60,000 of the amount when due, that refusal does not prevent the gross contract price from being $100,000.[30]

An example of the application of these rules is as follows. S, a calendar year taxpayer utilizing the completed contract method of accounting, constructs a building for B pursuant to a long-term contract. Under the terms of the contract, S is entitled to receive $100,000 upon completion of the

---

[28] Treas. Reg. § 1.451-3 (d)(1) & (2).
[29] Treas. Reg. § 1.451-3(d)(3).
[30] *Id.*

building. S finishes construction of the building in 1989 at a cost of $105,000. B examines the building in 1989 and agrees that it meets his specifications. However, as of the end of 1989, S and B are unable to agree as to the merits of S's claim for an additional $10,000 for certain items S alleges are changes in contract specifications and B alleges are within the scope of the contract's original specifications. Under these circumstances, S must include in income in 1989 the gross contract price of $100,000 and must deduct from gross income in that year the $105,000 of costs. In 1990 the dispute is resolved by a payment to S of $2,000 on his claim. S must then include this $2,000 in gross income in 1990.

§ 2.4 Tax Considerations to the Defendant—Damage Payments

The income tax ramifications to defendants in construction claims revolve around the characterization of payments that a defendant makes for legal services and costs associated with defending the claim and actual damage costs either awarded or agreed to in a settlement of the claim. The subject of legal expenses and costs is addressed separately in § 2.7. With respect to payments made on the actual merits of a construction claim, whether made in settlement of that claim or adjudicated, the income tax issues may be separated into two categories: (1) the tax treatment of the loss amount and (2) the timing of that treatment—that is, when, if at all, the loss is recognized for tax purposes.

§ 2.5 —Tax Treatment of Loss Amount

A defendant of a construction claim must address the tax consequences of damage payments, property surrendered, or amounts expended in response to a plaintiff's demand. The tax result varies depending upon the nature of the claim and falls into one of the following possibilities:

1. A deductible item
2. A nondeductible item
3. A capital item, not currently deductible, but added to basis.

Federal income tax laws provide a deduction for all ordinary and necessary expenses paid or incurred in carrying on a trade or business.[31] The requirement that the expenditure be characterized as ordinary may cause concern regarding the deductibility of legal damages. The Supreme Court, however, has stated that "ordinary" in this context does not mean that the

---

[31] I.R.C. § 162.

payments must be habitual or normal in the sense that the same taxpayer has to make them often.

> A lawsuit affecting the safety of a business may happen once in a lifetime. Nonetheless, the expense is an ordinary one because we know from experience that payments for such a purpose, whether the amount is large or small, are the common and accepted means of defense against attack. The situation is unique in the life of the individual affected, but not in the life of the group, the community, of which he is a part.[32]

Accordingly, as a general rule, business expenditures, including damage payments, in any amount qualify as ordinary and necessary expenses and are tax deductible, unless the payment relates to a capital item or is made nondeductible pursuant to another income tax law.

The general rules permitting a tax deduction for damage payments apply to both compensatory and punitive damages. For example, the IRS has ruled that amounts paid by a corporation as punitive damages that arose as a result of a civil lawsuit against the company for breach of contract and fraud in connection with the ordinary conduct of its business activities are deductible as ordinary and necessary business expenses.[33]

A damage payment by a defendant may in certain circumstances have to be capitalized. The income tax laws provide that expenditures that add to the value or substantially prolong the life of property or adapt it to a new or different use are not deductible. These expenditures are capital in nature and may be added to the cost basis of the affected property. Thus, for example, expenditures incurred in defending or perfecting title to property, in recovering property, or in developing or improving property are considered to be a part of the cost of the property and are not deductible expenses.[34]

If the capital expenditure may be depreciated or amortized, the loss of a current deduction as a result of capitalization of the payment will be offset over time as the cost is recovered through depreciation or amortization.

When a taxpayer is constructing property for another and accounting for the construction project in accordance with the long-term contract method, any payments that the contractor has to make with respect to that contract, including damage payments, should be capitalized into the cost of the contract.[35] Such costs will be recovered and essentially expensed when the contract is completed or on a partial basis as the contract is performed, depending upon the type of contract and when it was entered into.[36]

---

[32] Thomas H. Welch v. Helvering, 3 U.S. Tax Cas. (CCH) ¶ 290 U.S. 111 (1933).
[33] Rev. Rul. 80-211, 1980-2 C.B. 57.
[34] Stand. Fed. Tax Rep. (CCH) ¶ 1348.468 (1989).
[35] I.R.C. § 460(c).
[36] I.R.C. § 460.

For contractors using the long-term contract method, detailed regulations have been issued regarding the treatment of disputes that occur on or after the time a contractor tenders the subject matter of the long-term contract to the customer.[37] When, after tendering the subject matter of the contract, a customer seeks to reduce the original contract price or seeks to have additional work performed, the following rules are applicable:

1. If the amount in dispute is sufficiently large and affects so much of the contract price that it is not possible to ascertain whether a profit or a loss will ultimately be realized on the contract, no income or deductions shall be included in, or deducted from, gross income until the taxable year in which the dispute is resolved by the contractor agreeing to perform additional work.[38] The dispute is not considered resolved until the taxable year in which the additional work is completed.[39]
2. If the contractor is assured of a profit on the contract regardless of the outcome of the dispute, the gross contract price, reduced by the amount in dispute, must be included in the contractor's gross income in the taxable year in which the work is tendered, and all costs allocable to the contract must be deducted in that year.[40]
3. If the contractor is assured of a loss on the contract regardless of the outcome of the dispute, the gross contract price, reduced by the amount in dispute, must be included in the contractor's gross income in the taxable year in which the work is tendered, and all costs allocable to the contract, reduced by the amount in dispute, must be deducted in that year. All other costs that are properly allocable to the contract shall be deducted in the taxable year in which they are incurred.[41]

The foregoing rules may be illustrated by the following examples. A, a calendar year taxpayer using the completed contract method, constructs a building for B pursuant to a long-term contract. According to the contract, the gross contract price is $5 million. A finishes construction of the building in 1989 at a cost of $4,900,000. B examines the building, is dissatisfied with the job, and demands that A either make certain alterations or reduce the contract price. The amount reasonably in dispute is $200,000. Because the dispute affects so much of the contract price that A is unable to determine whether a profit or loss will ultimately be realized

---

[37] Treas. Reg. § 1.451-3(d).
[38] Treas. Reg. § 1.451-3(d)(ii).
[39] Treas. Reg. § 1.451-3(d)(vi).
[40] Treas. Reg. § 1.451-3(d)(iii) & (iv).
[41] Treas. Reg. § 1.451-3(d)(v).

on the contract, A would not include any portion of the contract price in gross income and would not deduct any costs that are properly allocable to the contract until the taxable year in which the dispute is resolved.

Alternatively, assume the amount in dispute was just $50,000 and the dispute was resolved unfavorably to A. A must recognize $4,950,000 ($5 million contract price minus the $50,000 disputed amount) of the contract price as gross income in 1989, and $4,900,000 of costs must be deducted from A's gross income in 1989. If the dispute were ultimately resolved in 1990 by A's performing additional work at a cost of $30,000 and A and B agreeing that the contract price is not to be reduced, then in 1990 A must include an additional $50,000 in gross income and must deduct an additional $30,000 from gross income.

Finally, it is possible that a payment in the nature of a damages expenditure could be both nondeductible and not capitalized. The income tax laws deny a deduction in certain circumstances in which Congress believes that the allowance of a deduction would frustrate public policy. Therefore, deductions are not permitted for items such as payments of illegal bribes and kickbacks,[42] fines and penalties,[43] and two-thirds of antitrust damages.[44] Since the Supreme Court's decision in *Commissioner v. Tellier*,[45] wherein a deduction was allowed to a securities firm executive for his expenses incurred in defending himself even though he was convicted of criminal violations of the Securities Act of 1933, denial of a deduction on public policy grounds has been reserved for the specific situations outlined above. For example, payments made by taxpayers pursuant to nonconformance penalty assessments made by the Environmental Protection Agency for failure to satisfy certain emission standards are deductible business expenses.[46] Likewise, a payment of a penalty assessed for an Occupational Safety and Health Agency violation presumably would also be deductible.

§ 2.6 —Timing Issues

Once a taxpayer has determined that an expenditure in the nature of damages either is deductible or can be capitalized, the final question is when to deduct or capitalize the expenditure. Initially, the answer to this question is based upon the taxpayer's overall method of accounting. Most taxpayers use either the cash or accrual method of accounting. When the

---

[42] I.R.C. § 162(c).
[43] I.R.C. § 162(f).
[44] I.R.C. § 162(g).
[45] 66-1 U.S. Tax Cas. (CCH) ¶ 9319, 383 U.S. 687.
[46] Rev. Rul. 88-46, 1988-24 C.B. 5.

cash method is used, a deduction or capitalization of the expenditure is not permitted until the expenditure is actually made in cash or its equivalent.[47] Under the accrual method of accounting, an item is deductible when the item meets three tests: (1) all of the events which determine the fact of the liability must have occurred; (2) the amount of the liability must be determinable with reasonable accuracy; and (3) economic performance with respect to the item must have occurred.[48]

Based upon these tests, as a general rule no deduction is allowable when there is a contest regarding a liability. A special provision of the federal income tax laws, however, allows for the deduction of a contested liability if the following requirements are met:

1. The taxpayer contest is an asserted liability
2. The taxpayer transfers money or other property to provide for the satisfaction of the asserted liability
3. The contest with respect to the asserted liability exists after the time of transfer
4. But for the fact that the liability is contested, a deduction would be allowed for the taxable year of the transfer (or for an earlier taxable year), and economic performance with respect to the item has occurred.[49]

The required transfer may be made by conveying money to the person asserting the liability, an escrow agent, or a trustee.[50]

Taxpayers need to be careful in timing their tax deductions for damage payments. Generally Accepted Accounting Principles (GAAP) require a liability to be estimated and reserved for in the period in which it is realized. Tax principles, as discussed herein generally do not permit the deduction of a reserve amount until it is more readily ascertainable in both its nature and amount. Accordingly, a tax deduction at the same time as a financial statement deduction for these types of claims is in most cases unusual.

§ 2.7 Deductibility of Legal Fees

In almost every construction claim case attorneys are called upon to represent the parties to the lawsuit. Whether the case is settled prior to court

---

[47] Treas. Reg. § 1.461-1(a)(1).
[48] Treas. Reg. § 1.461-1(a)(2); I.R.C. § 461(h).
[49] I.R.C. § 461(f).
[50] Treas. Reg. § 1.461-2(c).

§ 2.7 DEDUCTIBILITY OF LEGAL FEES

adjudication or after litigation, legal fees will be expended by both the plaintiff and defendant.

Depending on the taxpayer's overall method of accounting, legal fees will be deductible in the year they are incurred or paid, or over a period of years. The Code provides authority for deducting the fees in the current year if they constitute ordinary and necessary expenses of carrying on a trade or business.[51] Whether the fee constitutes an ordinary and necessary expense is a question of fact. For example, if a contractor sues for payment pursuant to a contract to construct a building and incurs legal fees to collect the money, then the fees would be deductible in the year they are paid or incurred.[52]

The Code also provides authority for deducting legal fees in the year they are paid or incurred if the fee is incurred in the production or collection of income or in the management, conservation, or maintenance of property held for the production of income.[53]

If legal fees are not fully deductible in one taxable year, they could possibly be deductible, proportionately, over a period of years. This means that the fees are capitalized or added to the cost basis of an asset and therefore increase its basis. Basis recovery is achieved via depreciation or amortization.

The issue of current deduction in contrast to capitalization was addressed in a Tax Court case in which a dispute arose with a construction company during the construction period.[54] The contractor had sued the defendant for not paying the full contract price. The parties eventually agreed to a settlement. However, the customer had incurred legal fees to defend himself. The issue brought before the court was whether the customer could deduct all his legal fees in the one year or whether he had to capitalize them. The court ruled that the crucial test in resolving this issue is to look at the origin of the claim. If the claim being litigated is related to the acquisition of a capital asset itself, then the legal fees incurred must be capitalized.

The court concluded that the origin of the claim litigated in this case resulted from the acquisition of a capital asset because it arose out of work performed by contractors in the construction of the asset. The court's rationale was that, because the litigation involved claims that were in part determinative of the cost of the asset, the legal fees incurred in resisting these claims must also be considered to be part of the asset.

---

[51] I.R.C. § 162.
[52] Wineland J.D. v. Commissioner, 10 T.C.M. 919 (1951).
[53] I.R.C. § 212.
[54] Smerling Enters., Inc. v. Commissioner, 29 T.C.M. 1412 (1970).

PART II
INTERNATIONAL CLAIMS

CHAPTER 3

RESOLUTION OF INTERNATIONAL CLAIMS IN THE UNITED STATES

Alan Silverman
Fredric H. Weisberg, Esquire

Alan Silverman is a partner in the San Francisco office of Coopers & Lybrand. He is a member of the California Society of Certified Public Accountants' Committee on Litigation Consulting Services, in addition to being a certified management consultant. Mr. Silverman has served as a consultant and expert witness in a wide variety of litigation since 1973. He has specialized in high technology litigation and international disputes. He was formerly director of litigation analysis for IBM.

Fredric H. Weisberg received his B.A. degree from the University of Vermont and his J.D. degree from the Cornell Law School. Mr. Weisberg is the chairman of the litigation department at Cummings & Lockwood, one of New England's oldest and largest law firms, with offices in Greenwich, Stamford, and Hartford, Connecticut, as well as in Naples and Palm Beach, Florida. Mr. Weisberg specializes in trials and appeals involving corporate and commercial matters. He is an active member of various American, Connecticut, and Stamford regional bar association committees and has served as president of the latter organization.

§ 3.1 Introduction
§ 3.2 Currency Exchange Issues Affecting Damage Awards

§ 3.3   Calculating Damages for Obligations Stated in United States Dollars
§ 3.4   Calculating Damages for Foreign Denominated Obligations
§ 3.5   The Emerging Rule
§ 3.6   Effect of Arbitration Clause on Currency Fluctuations
§ 3.7   Contract Language to Assign Risks of Currency Fluctuations
§ 3.8   Forum Selection and Choice of Law Considerations Regarding Arbitration Clauses
§ 3.9   Arbitration Instructions
§ 3.10 Inflation
§ 3.11 Interest Rates
§ 3.12 Conclusion

§ 3.1 Introduction

Contract clauses relating to dispute resolution are seldom deal breakers during the negotiation of complex commercial dealings between nationals of different countries. Yet, as the world becomes an ever-smaller marketplace, it becomes increasingly important for contracting parties to consider how they will resolve an irreconcilable breakdown in their relationship.

In recent years, parties to international contracts have begun to look more and more to alternate dispute resolution techniques (ADR) to settle their differences rather than to traditional litigation. The reasons for this trend are in large part the perceived economies of time and money available with ADR as well as the discomfort many foreign nationals feel with the legal system in the United States and its liberal discovery process. As parties begin to rely more frequently on private vehicles to resolve their differences, negotiation of the dispute resolution terms of commercial contracts has become increasingly more important.

Issues that receive little attention during the contract negotiation stage, but which can prove to be the most difficult and expensive elements of an international claim being adjudicated by litigation or ADR in a United States forum, are those related to the basic economic differences between the countries in which the parties reside. In particular, the parties need to consider:

1. Exchange rates and how currencies have fluctuated during the life of the contract and any subsequent dispute
2. Inflation and how it might affect any delay
3. Interest rates and how they will affect any judgment.

It is common to consider inflation and interest in calculating claims in the United States. These factors take on a much greater significance in international disputes because of the widely differing rates prevalent in different countries. Because most lawyers practicing in the United States seldom deal with currency fluctuation, that issue will be addressed first.

§ 3.2 Currency Exchange Issues Affecting Damage Awards

Over the course of the past century, a substantial body of common-law precedent has evolved to address currency exchange issues. Most of these precedents date back to periods of volatility in currency exchange markets such as the years between World War I and World War II.[1] Problems arising from currency fluctuations increased after 1973 when the Bretton-Woods Agreement, which had stabilized national currency values, ended. Since that time, the floating of currencies has led to significant fluctuation in the exchange rates of the world's major currencies. The situation in the United States during the 1980s has been exacerbated by the volume of international trade and the increased susceptibility of the United States dollar to depreciation. As a result, companies entering into international commercial agreements must take into consideration the risks of currency fluctuation as a factor influencing not only the value of the goods or services transferred but also the amount of damages awarded if the contract is not fulfilled. Therefore, the parties should attempt to minimize this latter risk by the careful drafting of contracts. See § 3.7.

§ 3.3 Calculating Damages for Obligations Stated in United States Dollars

Despite legislative changes and varying judicial treatment of exchange rate issues, in the United States it has been generally agreed that, when an obligation is denominated in United States dollars rather than in a foreign currency, and damages are to be derived from the application of domestic law, damages must be computed only in United States dollars. The rule has been stated and restated throughout the decades that:

> [N]o account should be taken of the change in value of the currency of the forum (in terms of the foreign currency) unless there is an *express agreement* to that effect.... Thus on an obligation to pay $10,000 without more, the damages in an American court would always be $10,000 (with

---

[1] *See* Note, *Foreign Moneys In Domestic Courts,* 35 Colum. L. Rev. 360 (1935) (noting "violent and varied" monetary fluctuation following worldwide conflicts); Note, *The Rate of Exchange in the Law of Damages,* 22 Colum. L. Rev. 217 (1922) (noting increased interest in foreign exchange issues as a result of "unprecedented fluctuations" of exchange rates).

interest, of course), irrespective of when or where the money should have been paid.[2]

In discussing the complex issues which arise in connection with an obligation payable in a foreign currency, the same commentator stated flatly, "our problem does not arise with an obligation to pay currency of the forum abroad."[3] In other words, and as a general rule: "Changes in the value of domestic money are judicially ignored."[4]

The Supreme Court has held unequivocally that an obligation for a payment of a sum expressed in any currency remains unaffected by depreciation of that currency relative to some other currency.[5] The Court held that "[a]n obligation in terms of the currency of a country takes the risk of currency fluctuations and whether creditor or debtor profits by the change the law takes no account of it."[6] The "quite obvious" reason for the consistent treatment accorded an obligation payable in dollars is that

> Courts under our nationalistic system of economy and law, by their inherent limitations, . . . do not award a judgment for damages in currency other than that of the country in which they have jurisdiction. Aside from that they can apply the ordinary rules of damages.[7]

In dealing with United States dollars obligations, American courts have generally calculated damages at the time of the breach, thus disregarding any appreciation or depreciation of the United States dollar against foreign currencies. If the injured party's national currency appreciates or declines against the dollar between the time of breach and the time of payment, no adjustment is made. As a result, the injured party may receive either more, or less, in damages for the breach than if there had been full performance.

Illustration. A is an American company that contracts with B, a Japanese contractor, to have B build a factory in New York. The contract is denominated in United States dollars. A breaches the contract by failing to make the final contract payment. B files for arbitration, wins, and ultimately receives payment. In New York, under the traditional rule, the injured party, B, will receive payment in United States dollars for the

---

[2] Note, *The Rate of Exchange in the Law of Damages,* 22 Colum. L. Rev. 217, 243 (1922).

[3] *Id.*

[4] Note, *Foreign Moneys In Domestic Courts,* 35 Colum. L. Rev. 360 (1935).

[5] Deutsche Bank Filiale Nurnberg v. Humphrey, 272 U.S. 517 (1926).

[6] *Id.* at 519; *see also* P.T. Perusahaan Pelayaran Samudera Trikora Lloyd v. Salzachtal, 373 F. Supp. 267 (E.D.N.Y. 1974) (when an obligation is payable in United States currency, it is of no consequence that the dollar's decline would decrease the value of the plaintiff's recovery).

[7] Note, *The Rate of Exchange in the Law of Damages,* 22 Colum. L. Rev. 217, 243 (1922).

§ 3.4 OBLIGATIONS IN OTHER CURRENCIES

damages calculated at the time of breach. If B has suffered damages in yen, those damages will be converted to United States dollars at the exchange rate prevailing at the time of breach. If the yen appreciates 20 percent against the United States dollar between breach and payment, no adjustment will be made. In effect, because the judgment is in United States dollars calculated at the time of breach, the injured party may believe that it has received 20 percent less in damages when the dollars are converted into yen. B would argue then that the payment in dollars would not leave B whole because B must pay its expenses in yen, not in dollars.

This apparent problem is ameliorated to some extent in those cases in which the prevailing party is entitled to an award of statutory interest, which "is compensation for the use of money."[8] It has been argued that the legal presumption underlying an award of prejudgment interest should preclude an additional recovery for a loss incurred because the prevailing party did not receive payments from the defaulting party when expected. This presumption is explained as follows:

> Upon the assumption that the value of the use of the principal sum to the creditor during the default of the debtor would equal interest at the statutory rate, the law properly measures his loss of such use at the statutory rate.[9]

Thus, perhaps claims for losses as a result of exchange rate fluctuations are, in essence, claims for the loss of the use of monies owed to the prevailing party who would have converted the monies into its own currency. If, in fact, this claim is for the loss of opportunity to invest the proceeds of the award in the currency of the prevailing party, then it would not be unreasonable to argue that an award of prejudgment interest is designed to compensate for this element of damage.

Of course, the corollary is also true. That is, if damages are awarded in United States dollars calculated as of the date of breach and the currency of the prevailing party has depreciated since that time, one might argue that the successful litigant has reaped a windfall as a result of the exchange rate fluctuation.

§ 3.4 Calculating Damages for Foreign Denominated Obligations

When obligations are incurred in a foreign currency, American courts have traditionally found it necessary to convert the debt into United

---

[8] Bulk Oil (U.S.A.) v. Sun Oil Trading Co., 697 F.2d 481, 485 (2d Cir. 1983).
[9] Rachlin & Co. v. Tra-Mor, Inc., 33 A.D.2d 370, 308 N.Y.S.2d 153, 157 (1970).

States dollars. There has been, however, a split of authority over the issue of the time at which the rate of exchange should be fixed under such circumstances.

Some courts have held that the exchange rate prevailing at the time of breach should govern. Others hold that conversion should be made as of the date of judgment. In a given case, the choice of the date on which the conversion should be made may well have a significant impact upon the measure of damages.

Illustration. A, a United States contractor, agrees to build a factory in Holland for B. B, a Dutch corporation, agrees to pay A for the building in Dutch guilders. On February 1, B breaches the agreement by failing to pay the final sum of 100 guilders due at the completion of the project. According to contractual provisions, A brings a claim for arbitration in the United States on June 1. On November 1, the arbitrator finds that A is entitled to damages equal to 100 guilders. On December 1, B is prepared to pay A 100 guilders pursuant to the arbitral award. During this period, the Dutch guilder has depreciated substantially against the United States dollar as follows:

February 1	$1 = 10 guilder
June 1	$1 = 12 guilder
November 1	$1 = 13 guilder
December 1	$1 = 15 guilder

Depending on the date the currency conversion is made (breach date, claim date, judgment date, or payment date), any one of four conversion rates might apply in determining the award. In this illustration, if the award were premised upon the breach date value of the contract in guilders, the award of 100 guilders on December 1 will mean that, when the American contractor receives its award in United States dollars, it will receive one-third fewer dollars than it would have received if there had been no breach and if it had received its payment when due on February 1.

The United States Supreme Court has suggested that the choice of breach date or judgment date should depend upon the jurisdiction in which the obligation to pay arose. Accordingly, if the obligation arose in the United States, then the exchange rate applied should be that prevailing as of the date of the breach. If the obligation arose under foreign law, then the calculation should be made as of the time of the judgment.[10]

---

[10] Hicks v. Guinness, 269 U.S. 71, 80 (1925); Deutsche Bank Filiale Nurnberg v. Humphrey, 272 U.S. 517, 519 (1926).

The *Restatement (Second) of Conflict of Laws* adopts the view that, if the cause of action is governed by the law of a country other than the United States, then the applicable exchange rate is that as of the date of the judgment or award. The rationale for this position is that the obligation should be converted into the currency of the foreign state as of the time when the courts of that jurisdiction would have done so.[11]

§ 3.5 The Emerging Rule

Some courts and commentators believe that the traditional rules can leave the injured party less than whole as a result of currency fluctuations. Accordingly, a new rule has been emerging that challenges the traditional American breach-day rule.[12] This so-called equitable approach involves permitting courts in the United States to award damages in currencies other than United States dollars and to base the damages on calculations made at the time of judgment if necessary to adequately compensate the injured party. The premise of this approach is that damage awards are traditionally intended to put the injured party "in as good a position as if the other party had fully performed."[13]

This approach was recently given expression in the *Restatement (Third) of the Foreign Relations Law of the United States* (*Restatement*):

Judgments on Obligations in Foreign Currency:

Law of the United States

(1) Courts in the United States ordinarily give judgment on causes of action arising in another state, or denominated in a foreign currency, in United States dollars, but they are not precluded from giving judgment in the currency in which the obligation is denominated or the loss was incurred.

(2) If, in a case arising out of a foreign currency obligation, the court gives judgment in dollars, the conversion from foreign currency to dollars is to be made at such rate as to make the creditor whole and to avoid rewarding a debtor who has delayed in carrying out the obligation.[14]

---

[11] Restatement (Second) Conflict of Laws § 144 comment d (1971).

[12] The new, equitable approach has already been adopted by English courts in Miliangoes v. George Frank (Textiles) Ltd., [1975] 3 All E.R. 801 (H.L.). In its narrow holding, the House of Lords held that a plaintiff "was entitled to claim and obtain judgment for the amount of the debt expressed in the currency of a foreign country if the proper law of the contract was the law of that country and the money of account and payment [in the contract] was that of the same country." In this case, the Lords allowed payment of a judgment in a foreign currency or its sterling equivalent at the time of enforcing the judgment.

[13] *See, e.g.,* U.C.C. § 1-106 (1987).

[14] § 823 (1987).

In comment c, the *Restatement* emphasizes that the goal of a damages award is to place the injured party "in a position as close as possible to that in which he would have been if the obligation had been carried out." Accordingly, comment c states the *equitable rule* in the following terms:

> In general, if the foreign currency has depreciated since the injury or breach, judgment should be given at the rate of exchange applicable on the date of injury or breach; if the foreign currency has appreciated since the injury or breach, judgment should be given at the rate of exchange applicable on the date of judgment or the date of payment.

This approach provides three main innovations. First, the new rule would provide for payment in currencies other than United States dollars. This represents a major departure from the traditional view that courts in the United States may render judgment only in United States dollars.

Second, the calculation of damages would not be restricted to the currency denomination of the contract. Rather, a judge or arbitrator would consider damages that occur in the injured party's currency even if the contract is not in that denomination.

Illustration. A, an American construction company, subcontracts with B, a French company, to do part of a construction project in Holland. B is to be paid in United States dollars. A breaches the contract with B. In determining the damages award to B, the American arbitrator may consider not only the lost profit in United States dollars (to be converted at the breach date or judgment date exchange rate, depending on currency fluctuations), but also any reasonably foreseeable losses suffered by B in French francs and other currencies.

For example, if B has had to pay a work force that was hired specifically for the Dutch project and cannot be otherwise employed, this loss may occur in French francs or Dutch guilders, depending on the nationality of the employees. In addition, if B must pay a penalty for the cancellation of its order for German steel, this loss may occur in German marks.

Under the equitable approach, the arbitrator may make the award for each of these damages in the currency in which the loss was incurred, regardless of the fact that the contract was denominated in United States dollars.

Third, in calculating damages, the arbitrator will consider fluctuations between the currency denominated in the contract and the currency of the injured party, recognizing that, had payment been made, the receiving party would probably have converted the payment to its national currency. For example, when there is no breach, an American corporation is likely to seek conversion of foreign currency to United States dollars upon receipt of payment. When a breach has occurred, an injured American corporation should be awarded the equivalent foreign currency to assure the receipt of the same amount of United States dollars, thus compensating for the fluctuation of currency between the time of breach and the time of payment.

One of the best examples of the evolution of the law on this subject is the situation in New York, whose law historically provided:

> In all judgments or decrees rendered by any court for any debt, damages or costs, in all executions issued thereupon and in all accounts arising from proceedings in courts, the amount shall be computed, as near as may be, in dollars and cents, rejecting lesser fractions; and no judgment, or other proceeding, shall be considered erroneous for such omissions.[15]

The traditional New York common law rule required that judgments entered in a foreign country and rendered in a foreign currency must be converted into United States dollars at the exchange rate prevailing on the breach date. The breach date was deemed to be the date on which judgment was rendered abroad.

In *Competex, S.A. v. Labow*,[16] however, the court subjected this rule to serious criticism. The Second Circuit noted, "[if] we were free to choose a conversion rule, we would select either the judgment-day or the payment-day rule."[17]

Subsequently, in 1987, the New York legislature enacted Judiciary Law § 27(b), authorizing the rendition of judgments in a foreign currency and providing that any such judgment be converted into United States dollars "at the rate of exchange prevailing on the date of entry of the judgment or decree." Some believe that this legislative change evidenced an intent to abandon New York's long-held day-of-breach rule even in cases seeking a judgment in dollars in favor of an approach which many feel is more in line with the general principle that the party injured by a breach of contract should, to the extent possible and fair, be put in the position in which it would have been if no breach had occurred.[18]

---

[15] N.Y. Jud. Law § 27.

[16] 783 F.2d 333 (2d Cir. 1986).

[17] *Id.* at 339. For further criticism of the day-of-breach rule, *see* Note, *Exchange Rate Selection for Foreign Money Obligations: Time to Recognize the Equitable Approach*, 25 Colum. J. Transnat'l L. 169, 179 (1986).

[18] In connection with this amendment, *see also* 18 N.Y.U. J. Int'l L. & Pol. 791 (1986).

§ 3.6 Effect of Arbitration Clause on Currency Fluctuations

The present uncertainty in the American law relating to international contractual disputes necessitates that parties drafting contracts with foreign nationals or in a foreign currency be very careful and very specific in their drafting of contractual language. In general, in contracts that provide for arbitration, arbitration clauses and contractual language must be very clear and specific because of the difficulty of overturning an arbitrator's award. The need to be very specific in these matters is even greater when the law is in the process of transition. If there is no clear, explicit governing law, the arbitrator's discretion will be afforded great latitude. This can lead to sometimes surprising results. Consequently, it is essential that parties use explicit language to limit such discretion when it comes to the risks of currency fluctuation if a contract involves parties of different nations or a contract provides for payment for work in one country in a different nation's currency.

The equitable powers of arbitrators would seem to be enhanced when the law is not definite.[19] In the case of currency fluctuations, an arbitrator might be inclined to award damages for currency fluctuation to an injured party. In American forums, the UCC § 1-106 definition of awardable damages could justify such an award. As a result, if the parties do not wish the arbitrator to award foreign exchange damages, the contractual language and arbitration instructions must specifically reflect this.

§ 3.7 Contract Language to Assign Risks of Currency Fluctuations

In general, the contract language should very specifically state the method of calculating damages and should assign the risks of currency fluctuations. There are many ways to accomplish this, among which are these possibilities:

1. Damages from a breach of this contract by the owner will be calculated in [specify currency] based on the harm at the time of breach plus interest calculated [specify the rate or method of determining the rate]. Damages from a breach of this contract by the contractor will be calculated in [specify currency] based on the harm at the time of breach plus interest calculated [specify the rate or method of determining the rate].
2. This is a [specify currency] contract. No currency fluctuation of the [specify currency] will be considered in awarding damages to the

---

[19] American Arbitration Association Rule 43 gives arbitrators broad equity powers.

injured party. The lost opportunity of a party to convert the denominated currency into any other currency is beyond this agreement and not to be considered in the calculation of damages. [It should be noted that standard contractual language barring consequential damages for a breach may not prevent an arbitrator from awarding foreign exchange losses. A specific reference, such as this clause, may be necessary.]

3. A breach by either party to this agreement will not entitle the injured party to any protection from currency fluctuations between the time of breach and the payment of damages.
4. The breaching party shall bear the risk of any currency fluctuation. If the currency of the injured party has depreciated since the breach, judgment should be given at the rate of exchange applicable on the date of breach; if the currency of the injured party has appreciated since the breach, judgment should be given at the rate of exchange applicable on the day of [payment/judgment].[20]

If the contract involves a series of transactions, the calculation of damages may be complicated by the possibility of multiple breaches. Parties should indicate in the contract whether the injured party's damages from the breach will be converted for the purposes of calculating damages according to the foreign exchange rate at the time of the initial breach for all of the transactions or converted separately at the time of each breach.

Illustration. A, an American corporation, contracted with B, a French construction company, to have B build a corporate office park consisting of four buildings in New York. The contract stipulated that B would receive payment in dollars but that B's damages from a breach would be measured in francs, converted to dollars at the time of breach. Each building had a separate cost and construction schedule. After the completion of Building 1, A breached the agreement by firing B. B was to receive payment for Building 2 on January 1, for Building 3 on July 1, and for Building 4 on December 31. The value of the dollar appreciated against the franc during this period as follows:

January 1	$1 = 3 francs
July 1	$1 = 4 francs
December 31	$1 = 5 francs

For the purpose of calculating damages, it is significant whether the damages from A's breach will be calculated using the exchange rate at the time

---

[20] Based on Restatement (Third) of the Foreign Relations Law of the United States § 823 comment c (1987).

of the initial breach for all of the damages or whether separate exchange rate conversions will be made for each breach.

§ 3.8 Forum Selection and Choice of Law Considerations Regarding Arbitration Clauses

Although forum selection and choice of law are beyond the scope of this chapter, it may be helpful to consider the following issues related to foreign exchange losses in the context of these subjects.

1. Does the forum state allow judgments in foreign currency?
2. Does the forum state apply the traditional or the equitable rule to calculate damages?
3. If the forum is outside the United States, is the forum state a signatory nation of the United Nations Convention on the Recognition and Enforcement of Foreign Arbitral Awards? The Convention makes contract clauses to arbitrate future contract disputes enforceable in the signatory states. It becomes important to know whether the forum state is a party to the Convention when enforcement of the award will have to be sought in a state other than the forum state.[21] The United States, for example, has restricted its adherence to the Convention to accept only final arbitral awards of other states that have ratified the Convention.
4. What are the forum state's laws on assigning costs and attorneys' fees?

§ 3.9 Arbitration Instructions

The present uncertainty of the law in the area of foreign currency fluctuations means that parties should be very specific in setting limits and guidelines for the arbitrator in their instructions. If the parties would like foreign currency fluctuations to be beyond the arbitrator's powers, the choice of language should be very specific. For example, the contract should state:

> The arbitrator may only award damages in [specify currency]. If it is necessary to consider damages in another currency, the conversion rate to be used will be the rate at the time of [breach/judgment/payment].

In addition, it is possible to specifically limit the arbitrator's discretion by including a contractual provision that determines the method of

---

[21] A partial list of signatory nations can be found at 9 U.S.C.A. § 201.

§ 3.10 INFLATION

calculating damages in the event of a breach (see § 3.7). Such contractual language should be accompanied by a specific clause in the arbitrator's instructions, such as:

> The arbitrator shall have no power to award consequential damages. In addition, the arbitrator shall have no power to adjust an award based on currency fluctuations except as stipulated in the original agreement.

§ 3.10 Inflation

The concept of allowing for inflation in claims for damage in the United States is not unusual. For example, claims arising from permanent personal injury always have an estimate of future lost earnings or future medical expenses. In such a damages award, an estimate is being made of the rate of growth of future income or costs.

The effect of inflation on international construction disputes is quite different, particularly as it applies to delay claims. Consider the following circumstances.

Illustration. A United States contractor A was awarded a contract to construct a hydroelectric dam and plant in Argentina. The owner of the project was the Argentine government, and the contract required A to employ a large percentage of local labor and use a large percentage of local material. The contract was a lump-sum contract awarded in pesos. It called for construction to start on January 1, 1985, and to be completed on January 1, 1988. There were to be periodic payments in pesos, based on construction milestones.

Various political problems caused the project to be delayed. The facility was finally complete on January 1, 1989. Many of the milestones that generated interim payments were also one year late, although the full payments were made in pesos as the stages were complete, that is, one year late. During the one-year delay the wages of local workers rose by 100 percent, an inflation rate unheard of in the United States. Each hour of labor cost A twice as many pesos as would have been the case if the project had proceeded on schedule. Similarly the local material costs doubled in one year. If the contract had not addressed inflation during owner-caused delays, A could have received payment in pesos which would buy only half as much as originally planned.

In calculating the impact on A, it is necessary to consider where A was planning to find the money to pay the local workers. If the plan was to transfer dollars from the United States, then exchange rate issues must be considered. The extreme inflation in Argentina may have caused a decline in the peso against the dollar, allowing A to transfer less than 100 percent more dollars to meet the 100 percent greater payroll. Also, if any part of the interim payment could be repatriated to the United States and converted to dollars at a more favorable rate than was planned, this would tend to mitigate the negative effect of inflation on the delay.

§ 3.11 Interest Rates

The concept of allowing for the time value of money is also not new to the United States court or to the arbitration system. Many states have statutory rates for prejudgment interest, and it is common for claims of future economic damage to be discounted back to the date of the judgment. If statutory rates do not apply, it is common for awards to be based on an interest rate that will compensate the injured party for the loss of the use of the money but will assume a conservative use of the funds. Interest based on one-year treasury bills is common; interest based on the potential appreciation if the funds had been invested in junk bonds is not.

Illustration. Consider the case of a United States company A, which contracted to have an office building built in Japan by a Japanese contractor B, with payment due in yen. A fails to make the final payment which was due on January 1, 1987. The matter is taken to arbitration and an American arbitration tribunal awards B 100 million yen, to be paid on January 1, 1988. B has lost the use of 100 million yen for one year, and an award in an American tribunal might consider a 10 percent interest based on T-bills. But interest rates in Japan in the same period may have been running at half the United States rate, meaning that B would need only 105 million yen to be whole, rather than 110 million.

§ 3.12 Conclusion

When advising a client who wishes to enter into a contract with a foreign party, it is important to think about:

1. The venue where future disputes might be resolved
2. The currency in which payment is to be made
3. The currency in which any claims are to be resolved
4. The effect of currency fluctuations on the delay of any payment or award
5. The effect of inflation on any delay of performance or payment
6. The interest rate to be used to compensate for the time value of money.

CHAPTER 4

RESOLUTION OF INTERNATIONAL CLAIMS OUTSIDE THE UNITED STATES

Richard A. Shadbolt, Esquire
Christopher J. Lemar, FCA

Richard A. Shadbolt is an English lawyer and a senior partner in the law firm of McKenna & Co., which has offices in London, Brussels, Bahrain, Hong Kong, Singapore, and Tokyo. He is a graduate of King's College University of London and qualified as a Solicitor in 1968. He is an international associate of the American Bar Association and lectures occasionally in the United States on international construction projects. Mr. Shadbolt was responsible for drafting one of the British standard forms of building contract. He has been involved in many projects in Europe, Africa, and the Middle East. His clients include leading names in the international construction field, particularly owners, consultants, and contractors, and his experience covers drafting and negotiating project documentation, the essential contractual guidance during the construction phase, and the resolution of disputes.

Christopher J. Lemar graduated from Cambridge University in England with a degree in Economics. He is a Fellow of the Institute of Chartered Accountants in England and Wales, a member of The British Academy of Experts, and a partner in the firm of Coopers & Lybrand. His background in audits and financial investigations of many kinds led him into litigation claims assignments in 1981. Since then Mr. Lemar has specialized in this branch of accountancy, particularly in claims for compensation and damages and in construction disputes. He has given expert testimony in both the United Kingdom and the International Court of Justice in The Hague.

CLAIMS OUTSIDE THE UNITED STATES

A LAWYER'S VIEWPOINT

§ 4.1 Introduction
§ 4.2 Choice of Law Considerations
§ 4.3 —Validity and Structure of Contract
§ 4.4 —Performance
§ 4.5 —Construction
§ 4.6 —Forum
§ 4.7 —Enforcement
§ 4.8 Types of Claims
§ 4.9 Approach to Entitlement
§ 4.10 —Price
§ 4.11 —Loss/Expense or Additional Costs
§ 4.12 —Quantum Meruit
§ 4.13 —Compensatory Damages
§ 4.14 —Punitive or Exemplary Damages
§ 4.15 —Fines
§ 4.16 —Contribution
§ 4.17 —Liquidated Damages and Penalties
§ 4.18 —Other Approaches

MATTERS FOR THE EXPERT ACCOUNTANT

§ 4.19 The Accountant and the Legal Environment
§ 4.20 The Style of the Evidence
§ 4.21 The Language Barrier
§ 4.22 Financial Evidence
§ 4.23 Verification of Claims
§ 4.24 The Global Approach to Damages
§ 4.25 Foreign Exchange Rate
§ 4.26 Purely Economic Loss
§ 4.27 Proving Future Losses
§ 4.28 Unjust Enrichment
§ 4.29 Interest

A LAWYER'S VIEWPOINT

§ 4.1 Introduction

Any contribution to a textbook for use principally by practitioners in the United States on the subject of claims outside the United States can only provide the briefest glimpse of the rest of the world. The greatest mistake that a practitioner could make, confronted with an international claim for the first time, is to assume that all his knowledge and experience of construction claims in the United States will be valid and useful elsewhere. The second greatest mistake would be to assume that none of it will have any relevance.

In addition to using his usual skill and experience, in the international field a practitioner will first have to ask himself some basic questions, the answers to which at home he may often be able to take for granted, and to obtain and verify information and advice. In all cases, whatever his expertise in construction claims at home, foreign advice will be required. The background and experience of the practitioner in the United States will be brought to bear in evaluating that information and advice and translating it into a clear and consistent plan of action for the conduct of the claim.

This chapter therefore aims to draw attention to some areas that may need to be explored in the conduct of foreign claims. In many countries more detailed expositions are available which apply to their own particular jurisdictions, and in all countries legal and accountancy advice will be necessary.

§ 4.2 Choice of Law Considerations

In order to make basic decisions about the conduct of claims, a lawyer needs to ascertain which law or legal system governs such matters. In some cases investigation will reveal a single answer, but on some occasions the attorney may discover alternatives, giving rise to a deliberate choice. These choices may enhance (or may inadvertently prejudice) the chances of success. Some jurisdictions may have aspects of law that favor a particular situation or a particular argument; different systems of law provide different procedures which may have a bearing on the methods used for proving and pricing the claim.

Some of the main points at which such a question should be raised in the mind of the adviser are discussed in §§ 4.3 through 4.7.

§ 4.3 —Validity and Structure of Contract

Many construction claims arise out of a contract. Most, but not all, jurisdictions share the concept that a binding agreement made between competent parties is enforceable by operation of law, but in detail, the laws differ.[1] Laws differ widely regarding the necessary formal requirements for

---

[1] For example, in the U.S.S.R. and other centrally planned economies, the concept exists, but more as an instrument of the planning process than for the private benefit of the

a valid contract. Some countries require that to be enforceable all contracts must be in writing. Some require that all contracts should be in writing and signed by the contracting parties. Some require that the authority of the signatory should be established, either in the document itself or by separate documentary evidence. Some permit oral contracts. Some countries (such as England), although firm in the disciplines to be applied once a contract has been established, are relatively relaxed about the particular method by which the contract was made or evidenced. Almost all jurisdictions distinguish between mere agreement and enforceable contracts.

Also, many countries require a duty or tax to have been paid on the document before it becomes a valid contract or can be used as evidence of a contract. If the duty has not been correctly paid, the contract may be invalid or may not be allowed as evidence in legal proceedings.

These issues are generally regarded as procedural rather than substantive and as such are usually decided by the procedural law of the country in which the proceedings take place. Substantive questions are almost certainly to be decided by reference to the proper law of the contract, that is, the law that is found to govern the contract.

The substantive law considerations, such as the necessary elements for the formation of a contract, are often intended to ascertain the intentions of the parties, particularly the intention to create a legally enforceable relationship. Mutuality of contract is a question on which considerable differences exist; however, all countries identify certain elements that are essential to the creation of a contract and these must be ascertained in each case.

So, a fundamental question for an attorney to investigate if it is alleged that the claim arose from a contract is whether there is indeed a contract at all. This must be done by reference to the law of an appropriate jurisdiction. This in turn requires the adviser to ascertain, insofar as a certain answer can be obtained, the law by which the question is to be judged.

§ 4.4 —Performance

Although there may be a general tendency for legal theory to move in the direction of applying one system of law to all aspects of a contract, this may not be the case in a particular jurisdiction. Thus, in some places, given particular facts, one law may be used to judge a contract's validity and another, its performance. The most common theory regarding performance is that the law of the place of the performance of the contract

---

parties. In the case of an agreement between a Soviet party and a foreigner, the concept appears to apply as stated although the legal basis of the distinction between domestic Soviet contracts and those between a Soviet party and a foreigner remains unclear.

§ 4.5   CONSTRUCTION

decides whether the contract has been duly performed. A common test is to ascertain that country with which the contract (significantly, its performance) has the closest connection. Some countries, however, although adhering to that theory, may at the same time apply a different law to other aspects of the contract.[2]

It is, however, a sensible starting point for any practitioner to look to the law of the country of performance of the contract, but he must ask himself each time whether the particular question is one which falls within that general rule.

§ 4.5   —Construction

Serious confusion can arise on questions of construction or interpretation of contractual documents if the language of the documents is not the same as that in which legal proceedings are conducted. On most occasions, there may be no choice but to submit to a particular jurisdiction or a particular procedure, but a wise counsellor will try to ensure that difficult matters of interpretation or construction of documents are dealt with in a forum whose language is the same as that in which those documents were originally expressed. The adviser should be cautious of all translations when the issue is one of interpretation of obligation or liability under a contract, and should ensure particularly that noncontractual translations of documents are not used in the process of proving a claim. Only original documentation should be used, with suitably authenticated translations prepared for the particular purpose of the proceedings.

Of course, fine questions of interpretation and arguments founded on sophisticated reasoning will not be impressive to a forum whose natural language is not that of the contract or of the document out of which the argument arises.

Careful consideration, therefore, not only of language, but also of the law which is to be applied to questions of interpretation will have to be undertaken. As can be seen, to apply the well-tried common-law rules of interpretation available, say, in England, to a contract that was originally in French and signed by parties from Japan and Brazil becomes unreal. To miss the chance to choose, if it had been available, to conduct that dispute in the French courts would likely affect a client's chances of success, although one can never be certain whether that effect will be adverse. It will certainly make the proving of the claim more cumbersome.

---

[2] Even sophisticated jurisdictions may have great difficulty with contracts in which each party performs in a different country. This has resulted in difficulties with contracts in Arab countries, for example.

§ 4.6 —Forum

Perhaps the most important choice to be made is the forum for the resolution of disputes. In reality, most international construction contracts (contracts in which one of the parties is a foreigner to the country in which the project is located) provide for arbitration in the event of disputes. A careful choice, well-informed and thought-out, at the time the contract is made will do much to reduce difficulty and surprises in handling claims later. Clear provisions referring disputes to arbitration in a precise way and according to specified rules will mean that from the outset the parties are sure by what method claims will eventually be adjudged. This extends a considerable degree of certainty to methods of calculation, evidence, the principles to be applied to financial claims, and so on.

Conversely, the simple absence of any arbitration provision, or even worse, the inclusion of an incomplete or ambiguous provision, may lead to a potentially serious obstacle in the way of the claim later. Although in most sophisticated jurisdictions it will be possible to ascertain, with the help of lawyers, in which forum disputes are to be resolved, in many countries the uncertainty that proceedings may turn out to be abortive may persist. Indeed, in an extreme case, the obstacle presented may deter the client from the pursuit of legal proceedings at all.

In short, a careful choice at the time a contract is made and an express provision in the contract itself, even when there may seem to be an obvious answer, will do much to reduce difficulty at a later stage. In making the choice, the procedures of the forum must be a vital consideration because it is a choice also about methods, the degree of sophistication, the availability of local advice, the admissibility of expert evidence, and the analysis of the claim. The language in which the proceedings will be conducted must follow this choice, except when arbitrators using one language are chosen but the arbitration takes place in another country. As is obvious, a difference of forum may well provide a different financial conclusion to a claim depending on the rules of procedure to be applied.

A further serious consideration on the question of forum is the costs and expenses of conducting legal proceedings. The expense of legal proceedings sometimes acts as a deterrent and is at least an undeniable risk to a client. He may therefore feel more reassured when embarking on legal proceedings against a reluctant defendant if he were able to choose a country in which he knows no order for costs can be made against him. Although many countries give a successful claimant some contribution to his costs of conducting the proceedings, the extent of that contribution can vary widely.

Forum shopping is now a sophisticated global exercise, and it is nowhere more evident than in the international construction field.

§ 4.7 —Enforcement

In most cases it will seem obvious that a judgment or award rendered by a competent forum will be best enforced in the home country of the party against whom it was given. Nevertheless, there may be opportunities to choose alternative jurisdictions for enforcement. The international construction business, being truly international, presents opportunities to take steps to enforce judgments against assets that are located outside of the home country of the defeated party. The questions of the home country, the country in which disputes are to be resolved, and the country in which any judgment or award is best enforced in practice are closely linked. However, a prudent adviser should keep open all possible alternative venues for enforcement because the enforcement methods used in some countries are considerably easier than in others. This also involves a review of reciprocal enforcement arrangements between countries. It should be remembered also that enforcement may not necessarily be conducted in the same currency as is used in the judgment or award (see **Chapter 3**).

As can be seen, difficult questions can arise on many aspects of the contract. When alternatives exist, a choice one way or the other will decide the methods that are to be used in proving and pricing the claim. In making a choice, therefore, a prudent attorney should always regard the difficulties inherent in construction claims—that is, the complexity and causation—which can be seriously multiplied by a careless choice of inappropriate jurisdiction.

§ 4.8 Types of Claims

The types of claims (that is, their broad legal basis) internationally are familiar to experienced practitioners and arise out of one or more of the following.

Express Contract Terms

Most construction contracts contain provisions permitting the recovery of additional sums of money in specific circumstances. Similarly, most of those contracts contain some express stipulation as to the method of pricing or proving these claims, although it can only reasonably be expected that the provision will be in general terms. The principle of such a claim is that it arises because of the express terms of the contract, whether or not a corresponding claim could have been made under the general law.

Breach of Contract

Whether or not a claim can be brought within an express contract provision permitting extra payment, when a contract exists the failure of either party to perform it constitutes a breach of contract. In almost all foreign jurisdictions that accept the concept of enforceability of a contract between private parties, the concept of damages or compensation for breach of that contract also applies; otherwise, methods of enforcement of the contract are likely to be ineffective. The question the practitioner should therefore ask is whether the particular set of circumstances gives rise to an argument that the contract has been broken rather than duly performed. If the answer is yes, then a financial claim in the nature of damages is usually applicable.

Termination of Contract

A contract that is brought to an end before it is duly and finally performed is to be distinguished from a contract that continues to exist but is not duly performed. International construction contracts frequently contain provisions for automatic termination in some circumstances, or for the exercise of a right to terminate, perhaps on default, by one of the contracting parties. Many jurisdictions provide a remedy when the contract has become impossible to perform and is thus terminated, or when the conduct of one party is so serious as to give rise to an election by the other party either to continue or to terminate the contract. The financial outcome of such a situation will differ significantly from one jurisdiction to another and in many instances depends not only on the operation of contractual provisions but also on the general law in the appropriate jurisdiction. This question is directly linked to the question of the law applicable to the performance of the contract and to the interpretation of the obligations and liabilities arising under it. Many countries have relevant legislation, even in countries where a common law system operates.

Tort

Common law systems label as "torts" the broad category of causes of action which are available to claimants in the general law. Although the name by which such causes of action, and their exact foundation, may differ from country to country, there still exist the broad principles of liability for negligent or careless action, for fraudulent acts, or for such things as the wrongful retention of possession of physical items. Many claims in the international construction field are founded in this broad body of law, and comparisons are impossible in a book of this nature. However, a wise practitioner always asks whether the financial outcome of the claim would be affected by the choice of basing the claim on the contract or on the general body of tort law. Even in common-law countries, some statutory interference may have occurred, and the recoverable amount of a claim may have been affected by statute law providing some precise method of calculation. Lawyers (and philosophers) are familiar with the difference between a financial outcome based (as in contract) on the idea that damages are intended to restore a party to the financial position it would have been in had the contract not been broken and the idea (as in the common-law tort of negligence) that the injured party is entitled to be put into the position for the future that he would have been in had the tort not been committed.

Statutory Provisions

Given particular circumstances, statutory provisions for compensation or claim may be available to an injured claimant. For example, in countries that find themselves at war, special legislation is often enacted to deal with compensation for certain types of, usually physical, damage.[3] Such special statutes, although usually applying to compensation funds made available by the government, sometimes also provide for the limitation or special calculation of claims between private parties. Similar principles may apply to other areas such as, for example, the statutory provisions that are to be found in many countries regarding nuclear incidents and liability for them.[4] Also, countries have statutory provisions enabling claims to be made as a consequence of criminal acts even when no criminal conviction has been or can be obtained. Special methods of calculation are usually prescribed for that type of statutory compensation scheme. Government schemes that go beyond being compensatory, and move into the area of subsidy, are entirely outside the scope of this book.

Breach of Statutory Duty

In many countries, the law has established particular liabilities and obligations on parties involved in construction. Decennial liability is common in countries with codified laws.[5] The laws provide for breach and remedies

---

[3] For example, in Iraq in recent years.

[4] It is because of this type of legislation, which often provides for government compensation, that many insurance policies exclude or give special attention to nuclear risks.

[5] *Decennial liability,* that is, a liability for a period of 10 years, seems to have originated in France during Napoleonic times. It has been adopted in many countries, particularly those that have drawn on the "Code Napoleon" (still the basic civil code in France), for example, Egypt. It is likely to be extended further within the European community.

for it. In addition, however, a type of claim can sometimes be made, not because a statute itself provides any particular claimant personally with a remedy, but because the defendant has a general duty to comply with his statutory obligations.[6] In this instance, there may be a difference between the calculation of the financial outcome of a claim made under a statute when compared with the liability for a general breach of statutory duty. In practice, however, the latter type of claim is likely to produce a more restricted opportunity, but this area is worth investigation if it appears to the adviser that a particular statutory enactment may have a bearing. In the field of environmental law and public utilities law, which are so often closely associated with the construction process, this opportunity to claim, with its consequent special position regarding proof and pricing of the claim, is one which should not be overlooked.

Extracontractual Claims

This heading is used to identify a category of claim that may arise out of an exhausted relationship of contract between two parties. In practice, the construction process is invariably conducted as a result of agreement (and usually therefore contract) between the parties. However, the parties may be mutually mistaken as to the existence or nature of their contract, or they may depart mutually so far from their original agreement that the rules provided by that contract can no longer be said to apply. Thus, although there may no longer be, or may never have been, a contract on the basis of which the parties' respective financial positions can be adjudged, and no breach could therefore arise, and no other culpable conduct can be said to have happened, it would be wrong for the parties to be said to have no relationship at all to provide for financial recovery or adjustments. Common-law systems have responded to this type of situation by introducing such concepts as quantum meruit and quasi contract. It is beyond the competence of the author to illustrate the same point with examples from other systems, but they may well exist, and on occasions may be recognizable to the American practitioner as being in the broad area of unjust enrichment.

Similarities can also be found between the areas covered by such general concepts and those dealt with under some codified European systems that provide for financial adjustment of contractual positions when disequilibrium can be said to have entered into the economic relationship

---

[6] Recent cases in the European Community have been concerned with public procurement procedures. These procedures are imposed by community directive and adopted into national laws. Although they impose obligations on public authorities regarding procurement procedures, they do not generally provide remedies to aggrieved parties. A recent directive, however, does envisage such an action: the "Staerebelt" case, Commission v. Kingdom of Denmark OJ/89.

between the parties or when the strict application of the contract terms would produce an unreasonable result. Some European countries provide for this to be remedied by the intervention of the judicial process, and for reasonable adjustment to be made either financially or by the reduction of contractual obligations.[7] Other countries do not excuse performance merely because it has become more expensive to perform;[8] however, they may do so only when it has become impossible.[9]

Within the scope of this book, no more can be done than to recommend that the practitioner ask himself whether remedies of this type are available in each case. A practitioner familiar only with the concept of contract and its breach or tort may overlook this type of possibility.

International Law

It may be possible to bring a particular claim within the ambit of some provision of international law, such as a treaty obligation or agreed scheme between governments for the settlement of disputes or claims.[10] By nature such claims are likely to be permissible for limited circumstances or financial extent. A stringent approach is likely to be applied by a forum in the interests of impartiality and the maintenance of the acceptability of its findings, particularly to governmental parties whose continued acquiescence is required for the future of the arrangement.

§ 4.9 Approach to Entitlement

Before embarking on a large-scale exercise of obtaining expert advice and evidence on a quantum of claims, it is important to be clear on what conceptual basis recovery can be expected. Contractors all too often fall into the trap of producing elaborate calculations and estimations of costs

---

[7] For example, in West Germany, the supreme court applies a concept known as "Wegfall der Geschaefts Grundlage," that is, that the underlying assumption on which the contract was based has ceased to exist or has failed. The details of the concept are much debated but it can be asserted that due to the changed circumstances, the affected party cannot legitimately be expected to put up with such a risk in the course of business. For example, a change in value of 60 percent post-contract has been found to be sufficient to justify an intervention by the court to adjust the contract terms.

[8] For example, in England, the principle can be found as early as the case of *Paradine v. Jane* (1647) Aleyn 26.

[9] For example, in France. But *cf.* Belgian law, which on this point appears to distinguish between contracts between private parties and those involving public authorities. In the latter case, it is accepted that a court may intervene either in the public interest or because of specific statutory authority to do so in certain types of contracts.

[10] Such arrangements may occur on the cessation of hostilities between states or as part of the rationalization of the economy of a debtor nation.

when the true basis of recovery is something different. Indeed, settlement negotiations may fail simply because a contractor's basis of calculation is inconsistent with the entitlement to recovery. An attorney becoming involved in an international construction dispute should therefore consider the underlying philosophy and purpose of the financial remedy available. The more obvious bases are discussed in §§ 4.10 through 4.18.

§ 4.10 —Price

Many claims are simple in calculation and involve the recovery of an agreed price or an ascertained figure under a contract. This should apply to items in the nature of fees or charges that are calculable by some agreed formula or method. In such cases the question is whether the full calculable amount is recoverable or whether a deduction or discounting is necessary for some reason. The essence of such a claim is that the amount is ascertainable or calculable by a process of arithmetic, not judgment, and that therefore the basis of the calculated figure is no longer open to challenge because of the existence of a binding contract. A claim for quantum meruit (often akin to a claim for price) involves a slightly different approach (see § 4.12).

§ 4.11 —Loss/Expense or Additional Costs

Many contracts in the international field contain provisions that offer recovery of extra payment to contractors (and sometimes others) in given circumstances. Leaving aside any discussion of the circumstances in which such claims can be made, such provisions commonly provide that, if the circumstances should arise, a contractor becomes entitled to recover the loss and/or expense, or the extra cost which he incurred by reason of the circumstances. Although at first sight such a contractual provision may appear to coincide with an entitlement to damages for breach, at least insofar as the extent and nature of the recovery itself is concerned, this is not always the case. In some jurisdictions such a provision may have the effect of enabling a contractor to recover extra payments even though a corresponding claim made as a claim for damages would fail because of, for example, lack of clear causal link or lack of proof. Such provisions need to be considered on their own, and particular care must be taken in interpreting them in the light of the proper law of the contract and the forum for the resolution of disputes. A provision such as this may well be interpreted differently by an arbitrator than by a judge, and from one jurisdiction to another.

A legal argument arises as to whether such clauses are additional to, or in substitution for, a general right to recover in damages. On analysis, most such clauses are intended to cover specific circumstances when there is either no question of culpability or breach or such may be difficult to prove. The parties therefore may be said to be expressing a clear intention that the risk of such circumstances (as distinct from the blame) is to be on one or other party, frequently the building owner. In principle, recovery should not be duplicated, and care needs to be exercised in practice to avoid that.

Extreme care should be exercised in analyzing the wording of such clauses because inadvertently extensive claims may be permitted. Wording such as costs "which are howsoever attributable to or consequent on or the result of or in anyway whatsoever connected with" the particular circumstances may lead to extensive recovery. This wording appears in the FIDIC Conditions of Contract for Civil Engineering Construction.[11]

§ 4.12 —Quantum Meruit

Reference was made in § 4.8 to the extracontractual type of claim. In principle, a quantum meruit claim is comparable with a claim for an unpaid price. *Quantum meruit* is simply an unascertained price because the concept is that a price ought to be paid and the question becomes one of assessing a reasonable price. A claim for quantum meruit does not, as a matter of principle, aim to indemnify a seller but merely to provide him with a reasonable payment for what he has done. That payment may or may not cover his expense; therefore it is not restitutionary in essence.

Further, what is a reasonable price to one party (for example, the attributable costs plus an addition for overheads and profit) may be unreasonable to the other party, so that some balance must often be struck. References, with greater or lesser degrees of sophistication, are often made to comparisons with market prices, in which case the necessary evidence may become a matter of economics as much as a matter of actual proof.

§ 4.13 —Compensatory Damages

Apart from claims for an unpaid price or the operation of particular contractual provisions (for example, for variations), the most likely situations to confront a claimant arise from the breach or termination of a contract or from tortious interference. In such cases the general principle in many

---

[11] These are the internationally known Conditions of Contract prepared by Federation Internationale des Ingenieurs-Conseils (FIDIC). The wording quoted appears in the 1977 and 1987 editions. These Conditions of Contract are almost certainly the most widely used standard conditions in use in the building and civil engineering industries internationally.

systems of law is that the aim of a financial remedy is to compensate the victim, not necessarily to provide restitution. *Restitution* involves a more general element of making good the damage that has been done. *Compensation* may be more concerned with reimbursement against expense, merely to restore the injured party's financial position had the event not occurred. In rare cases, a retributive approach may be applied; this approach is nearer to punitive damages (see § 4.14).

Both compensatory and restitutionary approaches may be encountered, sometimes both in the same jurisdiction. In the common-law system, for example, both approaches sometimes are adopted as a result of the same cause of action and may be reflected in such concepts as the difference between general and special damages. It should be added, however, that the common law does not normally allow general damages (that is, restitutionary damages) for breach of contract.

§ 4.14 —Punitive or Exemplary Damages

Less common, and less likely to be encountered in the construction field, are claims for damages that are punitive or exemplary in nature. Indeed, a practitioner from the United States is unlikely to encounter such claims except as a defendant to claims made by governmental authorities, and then only in rare circumstances. In some countries (for example, in the European community) there is a tendency to encourage private litigants to enforce public obligations so that, for example, obligations imposed upon public authorities in the field of public procurement can be the subject of proceedings which have the effect of ensuring enforcement or compensation, and are conducted by private litigants. Alongside such remedies, it is common to find that penalties may be imposed either on the parties or on the authorities concerned, but it seems at least a possibility that, in the future, when a financial remedy is available to an affected litigant, some element of the financial award will be exemplary in nature.

§ 4.15 —Fines

Different from punitive or exemplary damages are fines or charges usually imposed by operation of specific provisions of the law. Such sanctions may not be criminal in nature but rather would be administratively imposed, and, as with punitive or exemplary damages, the practitioner is unlikely to encounter them except as a defendant to proceedings brought by governmental authorities. The aim of such fines may not always be to impose a criminal penalty for an offense, but rather to provide compensation for administrative matters. Although in most countries many such matters are dealt with by criminal prosecution, giving the opportunities for defense, in some cases the fines are imposed by administrative decision, which may or may not be open to challenge by legal or administrative proceedings.

§ 4.16 —Contribution

Specific enactments may provide precise calculations for particular circumstances. Thus, for example, in some countries, originally because of the absence of available insurance coverage, acts of terrorism causing damage to property may be the subject of claims against government funds. Decisions as to the amount of compensation may be administrative in nature and may or may not be subject to review by legal proceedings or other appeals. Perhaps the occasion most likely to involve an American attorney is in the field of employment law. Many countries provide statutory schemes for limited compensation to employees in circumstances such as improper dismissal or early retirement and other socially based schemes.

§ 4.17 —Liquidated Damages and Penalties

Common-law systems are averse to the concept of parties imposing penalties on each other in their contracts. Elaborate concepts such as the English law relating to liquidated damages have therefore been created to justify the enforcement of such contractual provisions by the courts. In practice, it is almost impossible to distinguish between liquidated damages and penalties. In those countries using a common-law system it is always worth checking whether a liquidated damages or penalty provision is enforceable. In many countries penalties are treated with less caution.

§ 4.18 —Other Approaches

The international business community is, on the whole, a reluctant litigant because of uncertainties, some of which are self-imposed. As a result, alternative approaches to dispute resolution are increasingly popular, carrying with them different methods of financial adjustment between the parties.[12] These sometimes import concepts available in particular jurisdictions (for

---

[12] Alternative dispute resolution (ADR) is a subject that needs no introduction to the United States attorney. Its methods vary and thus the financial basis applied may also vary, a practice that is true in the international context as well.

example, *amiable compositeur*[13] or a finding *ex aequo et bono*[14]). They may also be based on a stipulation that broad principles, perhaps described as natural justice, can and should be applied. In such cases strict proof may not be necessary and a "global approach" may be applied with less hesitation. Certainly less formal requirements for proving and pricing claims will be applied.

The global approach, which is broadly comparable to the "total cost approach" familiar to United States practitioners of construction law, is not impossible even in countries requiring proof of damages. This approach is not so much a different method of calculation, but rather the only approach that can be applied in an instance when damages are clear but calculating them carries evidentiary difficulties.

The attitude of the English courts, for example, to the quality of evidence supporting damage has been succinctly put:

> The fact that damages cannot be assessed with certainty does not relieve the wrongdoer of the necessity of paying damages.[15]

> Where the precise evidence is obtainable, the court naturally expects to have it; where it is not, the court must do the best it can.[16]

> As much certainty and particularity must be insisted on, both in pleading and proof of damages, as is reasonable, having regard to the circumstances and to the nature of the acts themselves by which the damage is done.[17]

Thus, the English courts will not fail to give judgment merely because strict proof of financial matters is not forthcoming. They will apply a broader approach to calculation or assessment of damage. This has resulted in the expression "global approach," in which the strict link between cause and effect is not insisted upon.

---

[13] The expression *amiable compositeur* derives from French law. When used in contracts internationally and when French law does not govern the relationship between the parties, it is not absolutely certain how a person appointed to act as amiable compositeur must conduct himself and therefore at what point his activities may be overridden by the courts. In general, however, it may be assumed that the expression intends that the appointee should act as an impartial intermediary doing his best to establish common ground and thus a resolution. This necessarily involves compromise of financial claims.

[14] The use of the expression *ex aequo et bono* in a contract may be taken as an intention merely that the person on whom the obligation is placed should conduct himself generally in an equitable manner, showing good faith. This may result in a broader, less stringent approach to financial matters.

[15] Chaplin v. Hicks, [1911] 2 K.B. 786 (C.A.).

[16] Biggin & Co. v. Permanite, Ltd., [1951] 2 K.B. 314 (C.A.).

[17] Ratcliffe v. Evans, [1892] 2 Q.B. 524 (C.A.).

MATTERS FOR THE EXPERT ACCOUNTANT

§ 4.19 The Accountant and the Legal Environment

For the accountant, many of the techniques to be applied in proving and pricing claims outside the United States are no different from those that underlie the work inside the United States. The tools of the trade are essentially the same. The differences arise in the emphasis given to these various tools. They must be shaped to suit the local environment in which they are to be applied. This environment may be influenced by law, government bureaucracy, accounting convention, and general business practice.

In describing the role of the expert accountant in claims outside the United States, it is not the purpose of this chapter to categorize the differing features of his work on a country-by-country basis. That would entail a book in itself. In these few pages we have sought to introduce the features of the accountants' work in which emphasis may be different from country to country. We hope that, in doing so, we will provide signposts to the areas that will require particular attention when venturing into unknown territory in claims resolution.

First and foremost the accountant must establish the legal framework in which the claim is founded. This may be contract, tort, equity, statutory law, or international law.

Because most construction claims are initially founded in contract, the accountant must make himself aware of the terms of the contract under which the claim allegedly arose. Ideally, he should read the contract for himself. On many occasions this is impractical, and in such circumstances a clear account of the relevant terms should be obtained.

The contract may well describe how damages are to be calculated. But even if the terms are not so specific, they may well describe the law within which the contract is to be applied, such as the law of England, for example. When the contract itself is not specific on the method for calculating damages, the legal domicile of the contract may be of help. It is essential that the expert accountant ensure he is properly briefed on the legal position before setting out on his task.

Outside the bounds of contract, claims are typically founded in tort or equitable principles. Again, the accountant must establish whether any local code or precedent exists to establish, or perhaps limit, the claim.

It is not for the accountant to be conversant with each system. What he must do at the outset of any assignment in a foreign jurisdiction is obtain guidance from lawyers with relevant local experience with the legal framework in which his evidence is to be prepared.

§ 4.20 The Style of the Evidence

Little could be worse in a jury trial than for the expert to present the jury with a 150-page closely written report with a dozen or so appendixes full of figures. Experts who are used to testifying before juries in the United States learn quickly that the only realistic way to proceed is to simplify, to use day-to-day analogy, and to present clear and colorful charts. Many lessons could be learned from this approach by experts who practice outside the United States and who generally do not have juries to contend with in commercial litigation. However, to adopt the simplistic approach to a judge or arbitrator may seem patronizing or even naive and could be as unpopular with the judge as a voluminous text is to a jury. The key to holding the court's attention is to know the type of audience being addressed and to develop a style to suit the assignment.

Litigation in the United Kingdom, like that in America, is conducted on the adversarial approach. The expert for the most part is expected to prepare a report in advance of the hearing or arbitration. This is often exchanged with the expert's report from the opposing camp. Each expert then submits himself to cross-examination on his opinions in open court. More often than not, new financial evidence will be introduced during the hearing that the accounting expert will have to deal with at short notice.

Conversely, in continental Europe, where the inquisitorial system applies, the practice of submitting written reports in accordance with established timetables is more rigorously followed. It is not uncommon in such courts for the judge or judges to appoint their own expert whom they may themselves examine in due course at the hearing. There are signs that this approach is beginning to find favor in commercial courts in the United Kingdom by some judges. In these circumstances, both parties are entitled to cross-examine the court's expert. Many parties decide to field their own expert in response to the court expert's findings. In these circumstances greater importance is attached to written submissions than has traditionally been the case in the adversarial system in the United Kingdom, where oral evidence is a key feature.

§ 4.21 The Language Barrier

The language in which a court conducts its business is of course a feature of international litigation. It is surprising how often this can be overlooked by an expert in preparing his evidence. The accountant-author was once involved in a case before an international tribunal in The Hague. Evidence was permitted to be given in either English, Swedish, or Farsi. It was more than usually important for the oral evidence to be given slowly, clearly, and with particular care to keeping specialist jargon to an absolute minimum. The use of colloquial comments must be avoided, because when they are translated they can take on an entirely different flavor.

The language issue can of course also be a barrier to the settlement of international disputes. Misunderstandings may arise over contract terms, the meanings of interparty letters, the interpretation of experts' opinions, and so on. The accounting expert should be alert to such possibilities both when drafting his own treatise and when reviewing the opponent's evidence.

§ 4.22 Financial Evidence

The expert accountant's work can often be divided into two main categories. The first follows the more traditional audit approach; it is the review and examination of detailed financial data and relates to the project's actual revenue and costs. It may involve skills of data analysis and presentation and often involves interpretation of that financial data. The second is the consideration of consequential losses, forecasts, projections, and so on. These tend to come within the category of economic loss, discussed in **§ 4.26**.

Assignments to calculate or evaluate a financial claim involve obtaining data from different sources, analyzing those data, and drawing conclusions from them. The nature and extent of data required depend on the individual case. No common list can be produced.

The quality and reliability of financial evidence can, however, vary substantially from country to country. The level of public accountability by businesses and the tax environment in individual countries may both play an important part in establishing the quality of financial evidence. When venturing into a new jurisdiction, the expert should take steps to establish the accounting environment into which he is moving. In some countries, it is not uncommon to find that a company maintains two sets of accounting records: one for the auditors and one for the tax authorities. When the party under investigation is a subsidiary of an international group based, for example, in the United States or the United Kingdom, then it is likely that, even if the local environment in which the subsidiary operates does not demand a high standard of accounting records, the needs of group reporting will dictate a minimum level of quality which should be helpful to the expert accountant.

Few if any countries have as many financial accounting standards as are now established in the United States. If he encounters a singular lack of local accounting standards, the expert may be able to draw comfort from international accounting standards, particularly when reviewing accounts that have been audited by public accountants.

There is an obvious danger that the expert coming from a country where the quality and standards of accounting are high may be prone to regard financial evidence in countries with lower standards as being inadequate. It is inadvisable for the expert to be the judge of this. Seek advice from an accountant or lawyer used to working in the jurisdiction concerned. They will usually be better able to indicate how a local court may react to local evidence.

§ 4.23   Verification of Claims

Again, whichever jurisdiction is involved, the basic tools that the expert brings to bear on evaluating claims are common throughout the world. He must look at the reasonability of the amounts being claimed in the overall context of the claimant's financial statements. When possible he must verify specific items in the claim with the documents supporting the amounts concerned. He must look at the application of formulas and apportionment to establish that they produce results that are sensible. Sometimes, for example, a contractor involved in several claims may recover overhead or head office expenditure several times over because of the way apportionment formulas are being applied. The expert may need to establish whether the concept of betterment—that is, new for old—has been applied so as to give an advantage to a claimant. The expert must be on the lookout for examples of double counting within claims.

There are, however, a number of areas of the expert's work that will require some tailoring according to local conditions, and these subjects are examined in §§ 4.24 through 4.29.

§ 4.24   The Global Approach to Damages

It is not uncommon to find a construction project plagued by a variety of distinct but interwoven factors. Delays can be caused by bad weather, late receipt of architect's drawings, lack of supply of components, strikes by the work force, and bureaucratic holdups of work permits or the import of vital materials. The claimants may have little difficulty in demonstrating that all of these factors in some way gave rise to delay. They may also have little difficulty in demonstrating that their costs were increased considerably above those originally expected. The difficulty may come in linking the cause and the financial effect at an individual level. In these circumstances, some jurisdictions are prepared to consider a concept known as the *global approach* to damages (see § 4.18). In this case, it can be the accountant's job on behalf of the claimant to show in broad terms that the claimant has suffered a certain total loss for which he deserves to recoup all or perhaps a proportion in damages. It is not necessary to allocate or prove these damages to specific items of claim.

The accountant should be aware that the global approach may be subject to important qualifications such as:

1. The events that are the subject of the claim must be complex and interact so that it is difficult, if not impossible, to make an accurate apportionment of the total extra cost between the various causative events
2. There must be no duplication within the amounts claimed
3. Any financial claim must exclude profit, if profit is unrecoverable under one or more of the underlying claims.

§ 4.25 Foreign Exchange Rate

Difficulties can often arise when one party to the dispute operates within a hard currency area and the other party within a soft currency region. The currency in which the award is given can in reality have a substantial effect on the true benefit of the award to the claimant (see **Chapter 3**). An award granted in a soft currency may be almost worthless unless the claimant can use that soft currency to some effect within the soft currency area. Attempts to convert into a hard currency may either be (1) impossible because of exchange control or (2) of little benefit because the true exchange rate is very disadvantageous to the claimant. The currency of the claim therefore can be a major point of contention in an international case, and the accounting expert may be called upon to apply his skills to argue the appropriate currency to be used.

§ 4.26 Purely Economic Loss

Loss of profits is a common item of damage in a variety of claims, especially in construction claims. The attitude of courts to loss of profits can vary both from country to country and case by case. It is therefore important that due account be taken of the local attitude to loss of profits claims.

Loss of profits is normally regarded in the United Kingdom, for example, as a form of purely economic loss. Generally speaking there is a reluctance to allow recovery of damages in respect to pure economic loss arising on claims in tort, unless the claims arise from physical damage to property. The notable exception to this is a claim arising from negligence, when economic loss does feature in awards of damages. On the other hand, there would appear to be no bar to claims of loss of profits flowing from contractual disputes.

In some parts of the world, for example in eastern Europe, the concept of "profit" is unknown. In such circumstances it may be difficult to prove loss of profit. In a similar manner, loss of interest or claims for financing charges may be difficult to substantiate in Muslim countries where the concept of "interest" is unacceptable. It is therefore important not to assume that, when an item of damage is not for a direct loss but for a loss like loss of profit or loss of opportunity, it will be acceptable in all jurisdictions.

§ 4.27 Proving Future Losses

The methods employed to prove and value future loss of profits claims are also a potential minefield for the unwary accounting expert. The natural concern of the court in awarding damages for loss of future profits is the subjectivity of such awards. Giving compensation now for a loss that is expected to be suffered in the future is fraught with difficulty. The only certainty, many would say, is that the award is bound to be wrong. So how do the courts in each case overcome this problem? It is important to consider the following questions:

1. Do the courts put a bar on such claims?
2. Do they admit only losses incurred and provable up to the date of trial?
3. Do they contrive to limit claims of future loss to perhaps only a year or two after the date of the trial?
4. Are they prepared to accept that future loss can be given a present value?

The subject of valuation, therefore, deserves some consideration at this point. It is of course a subject which would justify a whole book to describe the theory and practice. This chapter will discuss only some of the points of principle which the expert accountant should consider regarding the valuation of future economic loss.

The two most common approaches to valuing future streams of revenues or costs are (1) the discounted net present value approach, and (2) the multiplier approach.

What factors might influence which method is applied? The *discounted net present value* approach, or discounted cash flow approach as it is sometimes called, seeks to place a present value on expected future profits

§ 4.27   PROVING FUTURE LOSSES

or losses. It follows the principle that a dollar today is worth more than a dollar tomorrow which itself is worth more than a dollar on the next day and so on. The first step therefore is to project what these future amounts will be. The second step is to discount them back to present value, perhaps to the date when the damage arose, or perhaps to the date of the trial. This is done by applying a discount factor, which will have to be selected, according to the circumstances, by the expert. This method is used regularly in the United States for performing valuations both in and out of court. Can it be applied outside the United States? The answer in our experience is yes, but a heavily qualified yes. Although the discounted cash flow technique is based on academic study, and thus theoretically rigorous, it is not universally persuasive to a judge. It is often considered to be too subjective. First, the forecasts of cash flows are prone to error, particularly the further into the future the projections go. Second, there is considerable judgment exercised in choosing an appropriate discount rate, and a valuation can be very sensitive to the choice of this rate. For these reasons the discounted cash flow technique should be used sparingly.

The discounted cash flow technique has been accepted as appropriate for computing the value of an incomplete construction project that was expropriated by a revolutionary government in connection with a claim by the original owners for compensation. The main reasons the court accepted the method were twofold: first, the project had a finite length and only required the forecasting of costs and revenues for some two to three years into the future. Second, there was little argument between the experts on the appropriate discount rate to use for this type of project at the relevant time.

The second principal valuation route is the use of a multiplier. When this involves valuing a business, the multiplier used is known as the *price earnings ratio*. The valuer has to determine the relevant annual earnings (or losses) which are expected to arise in the future. This may be the latest available year and thus can be presented as a known historic figure to the court. Secondly, an appropriate multiplier must be selected. Again, if valuing a business, the task is to find a comparable quoted business and take its own price earnings multiple, which is typically quoted in the financial press as a starting point for computing a suitable multiplier.

A benefit of the multiplier approach is its approved objectivity, although it is perhaps less rigorous as a concept than the discounted cash flow approach. If time permits, the expert may well prefer to present valuations using both methods. Although they are unlikely to give precisely the same answer, they should produce a figure of similar magnitude and thus give extra confidence to the court.

In some jurisdictions the court will seek to avoid what it sees to be crystal ball gazing. In doing so, some rather hybrid valuation methods can emerge. In an attempt to avoid awarding damages which relate to future loss, the court may link its award to a valuation of the assets of the business. The assets could comprise the tangible fixed assets, the net financial assets (for example, receivables less payables), prospects, goodwill, and so on. Here, the courts can intend a clear distinction between goodwill or future prospects, for example, and the concept of future economic loss.

Again, no hard and fast rules can be set for every case in the international environment. Local practice and attitudes towards future economic loss must be identified and applied in a manner to suit the local conditions.

§ 4.28 Unjust Enrichment

It is a general principle of contract law that a successful claimant in a breach of contract case is entitled to be put back in the same position as he would have been had the breach not occurred. For example, an owner may have terminated a contract before completion and be found to be in breach of contract, but at the same time it could be shown that the contractor would have suffered a substantial loss if he had had to complete the contract. Following strict application of the principle of contract damages, there would in these circumstances be no loss to the contractor. It would seem, however, in international disputes, particularly those which go before arbitrators, that the contractual damage rules are sometimes set aside and an approach founded more in equity, rather than contract, applied. The rationale is that to follow the strict contractual route would leave the defendant enjoying what is sometimes referred to as *unjust enrichment*. The owner might, for example, be in possession of a building which is substantially complete for which it has paid no money across to the contractor. This is clearly unequitable and the court may apply what amounts to a quantum meruit approach to determine the damages to be awarded to the contractor. In these circumstances, the expert accountant can disregard the specific terms of the contract and look to the value of the work done.

§ 4.29 Interest

Delay in bringing a case to trial is a common feature in almost all judicial systems. The claimant may therefore seek additional damages to cover this delay, usually by way of interest on the basic damages. The way in which courts view interest varies from country to country. Local advice must be sought on this subject. In the United Kingdom, for example, interest on damages is generally awarded at the discretion of the trial judge (although there are certain particular exceptions which are covered by statute). When interest is awarded, it tends to be based on the application of simple rather than compound interest. The rate of interest applied is usually set at a point or two above base rate. In certain Muslim-based countries, the concept of certain types of interest is unacceptable and any claim for compensation in the form of interest may need to be presented in a way which does not conflict with local custom.

Interest rates vary considerably from country to country. These rates are influenced by factors such as the rate of inflation in the country concerned and the relative strength of its domestic currency. Generally speaking, a claimant should expect to receive interest on damages using rates that are pertinent to the currency in which the claim is being brought. There may be circumstances in which a claim might be brought in one currency and interest claimed at a rate applying to a different currency. This might be a result of the nature of the original contract under which the claim is brought or perhaps local court rules about the awards of interest. Anyone entering into unfamiliar jurisdictions must be alert to the possible complexities of interest on damage claims.

PART III
CONTRACTOR CLAIMS AGAINST OWNERS

CHAPTER 5

DELAY CLAIMS

Robert F. Cushman, Esquire
James D. Hollyday, Esquire
Frederic R. Miller, CPA
Vincent J. Kiernan

Robert F. Cushman is a partner in the national law firm of Pepper, Hamilton & Scheetz and a recognized specialist and lecturer on all phases of construction and real estate law. He serves as legal counsel to numerous trade associations and construction, development, and bonding companies. Mr. Cushman is editor and coauthor of several books about construction, including the following John Wiley Publications: *Construction Litigation: Representing the Contractor; Architect and Engineer Liability: Claims Against Design Professionals; Construction Defaults: Rights, Duties, and Liabilities; Construction Failures;* and *Construction Bidding Law.* A member of the Pennsylvania bar, he is admitted to practice before the United States Supreme Court and the Court of Appeals for the Federal Circuit. Mr. Cushman has served as executive vice-president and general counsel of the Construction Industry Foundation, as counsel to the American Construction Owners Association, and as regional chairman of the Public Contract Law Section of the American Bar Association. He is permanent chairman of the Andrews Conference Group's Construction Litigation and Hazardous Waste superconferences. Mr. Cushman is a charter member of the American College of Construction lawyers.

James D. Hollyday is an associate in the Philadelphia office of Pepper, Hamilton & Scheetz. Mr. Hollyday is a *magna cum laude* graduate of Temple University School of Law. His primary area of concentration is the field of construction litigation, in which he represents or counsels owners, engineers, construction managers, contractors, and sureties in a wide variety of matters, including contract drafting, contract modification, change related disputes, delay and disruption disputes, and construction failures. Mr. Hollyday holds a B.S. degree in mechanical engineering from Lafayette College and has been a registered professional engineer in Pennsylvania since 1976.

Frederic R. Miller is the director-in-charge of the Business Investigation Services Group of the Washington, D.C., office of Coopers & Lybrand and is also a leader of the firm's nationwide Task Force on Construction Claims. He has a wide spectrum of experience in construction, having worked for owners, contractors, sureties, and architect/engineers on a wide variety of construction projects including commercial and public buildings, water and waste plants, nuclear power plants, and hazardous waste sites. Mr. Miller has served as an expert in federal and state courts and before arbitration panels and board of contract appeals. He has also taught a number of professional education courses in litigation consulting and construction claims analysis. He is a Certified Public Accountant in the District of Columbia and a member of the D.C. and American Institutes of Certified Public Accountants and the Association of Insolvency Accountants. Mr. Miller obtained his undergraduate degree from Rutgers University and his M.B.A. degree from the Johnson Graduate School of Management of Cornell University.

Vincent J. Kiernan is a member of Coopers & Lybrand's Business Investigation Services Group and specializes in litigation and claims services in the construction industry. He has a wide variety of experience working for owners, contractors, and engineers on wastewater treatment plants, utilities, and commercial construction. Mr. Kiernan received his undergraduate degree from Wake Forest University and has an M.B.A. in finance from the College of William & Mary.

---

§ 5.1 Introduction

FROM THE ATTORNEY'S PERSPECTIVE

§ 5.2 Delays and Cause of Delay

§ 5.3 Excusable and Nonexcusable Delay

§ 5.4 Compensable and Noncompensable Delay

§ 5.5 Concurrent Delay

§ 5.6 Burden of Proof and Apportionment in Delay Cases

§ 5.7 Apportionment in Liquidated Damages Cases

§ 5.8 Computing the Delay

§ 5.9 Contractor's Right to Finish Early

§ 5.10 Pricing the Claim

§ 5.11 Increased Material and Labor Costs (Escalation)

§ 5.12 Costs of Idle Construction Equipment
§ 5.13 Home Office Overhead
§ 5.14 Prejudgment Interest and Financing Costs
§ 5.15 Loss of Profits

FROM THE ACCOUNTANT'S PERSPECTIVE

§ 5.16 Cost Escalation
§ 5.17 General Conditions Costs
§ 5.18 Equipment Costs
§ 5.19 Unabsorbed Home Office Overhead
§ 5.20 Interest/Cost of Capital
§ 5.21 Summary
§ 5.22 Bibliography

§ 5.1 Introduction

The modern construction project requires the coordinated efforts of the owner, the architect and its consultants, the general contractor or the construction manager, numerous contractors or subcontractors, and material suppliers all working together in close proximity of time and space to complete the work while optimizing the quality of the work, the cost of the project, and the overall time required for design and construction. The reality of the marketplace requires that in order to be the successful bidder, the contractor must compute its bid price assuming that all aspects of the work can be completed in an orderly, unhindered way without delays because of changes in the work, defects in the plans and specifications, or any of a host of other factors. If the contractor is delayed in completing the work, its cost of performance increases simply because those elements of its costs that are dependent on time require an extended period of time. For example, the contractor will likely have field overhead costs for its field office, telephones, and field supervision—costs which are directly time-related and which represent "pure" delay costs. In addition to the purely time-related delay costs, the contractor's cost of performance may increase because delayed work itself is completed in an unproductive manner or may cause subsequent related work to be done out of sequence or on a piecemeal basis instead of in an uninterrupted sequence as planned. Labor productivity rates may suffer as a result, causing the contractor's costs to increase. Although these so-called disruption costs may, in the proper circumstances, be compensable elements of delay damages, in that they are incurred as the result of delay, they may be caused by factors unrelated to delay. The discussion in this chapter is limited to time related damages. For a complete discussion of disruption damages see **Chapter 6**.

However, in order to recover its additional costs, it is not enough for the contractor to show that work was completed later than planned and that the contractor experienced coincident cost increases. To demonstrate its entitlement to compensation for delay damages, the contractor must demonstrate that under the governing contractual provisions the delay is excusable—that is, the delay was of a type for which the contractor is not contractually liable—and that the delay is also compensable—that is, the delay was of a type which entitles the contractor to compensation and not just an extension of time to perform the work. Having established its entitlement to compensation, the contractor must then demonstrate the quantum of its resulting damages.

FROM THE ATTORNEY'S PERSPECTIVE

§ 5.2 Delays and Cause of Delay

The parties to a contract may by agreement apportion the risk of delay in any way they choose. Today it is rare that a construction contract does not provide in some detail for the apportionment of the risk of delay between the parties.[1] Thus, when confronted with the question of determining liability for a project delay, the first inquiry must be to ascertain the applicable contractual provisions. Depending upon the apparent cause of any specific delay, it may be necessary to examine the changes clause, the differing site conditions clause, the termination clause, the suspension of work clause,[2] the liquidated damages clause,[3] the no damages for delay clause, and the "time is of the essence" provision, if any.

---

[1] For example, AIA A201-1987 ¶ 8.3 provides for the extension of the contract time for certain types of delay. In similar language, § 63.93(a) of the Standard General Conditions used by the Pennsylvania Department of General Services includes the following provision:

> If the Contractor is delayed at any time in the progress of the Work by any act or neglect of the Department or the Professional, or by any employee of either, or by changes ordered in the work, by labor disputes, fire, unavoidable casualties, or by delay due to suspension of work, as provided in Sections 63.161 [suspension of work because of unfavorable conditions] and 63.162 [suspension of work for the convenience of the Department], or by any cause which the Department determines may justify the delay, then the Contract Time may be extended by the approval of the Department for such reasonable time as the Department may determine.

[2] *See* the federal government's Suspension of Work clause, F.A.R. § 52.212-12.

[3] *See, e.g.,* F.A.R. § 52.212-5, Liquidated Damages—Construction.

It is well established that in a construction contract time is not generally of the essence unless it is expressly provided,[4] and a contractor's failure to complete its work in accordance with the time requirements of the contract does not entitle the owner to terminate the contract or excuse nonpayment, but may expose the contractor to liability for delay damages.[5] In *Carter v. Sherburne Corp.,* the court summarized as follows:

> Where time is of the essence, performance on time is a constructive condition of the other party's duty, usually the duty to pay for the performance rendered. *Jones v. United States,* 96 U.S. 24, 24 L.Ed. 644 (1877). Time may be made the essence of a contract by a stipulation to that effect, *Cheney v. Libby,* 134 U.S. 68, 10 S.Ct. 498, 33 L.Ed. 810 (1890) or by any language that expressly provides that the contract will be void if performance is not within a specified time. *Sowles v. Hall,* 62 Vt. 247, 20 A. 810 (1890). Where the parties have not expressly declared their intention, the determination as to whether time is of the essence depends on the intention of the parties, the circumstances surrounding the transaction, and the subject matter of the contract. *Kennedy v. Rutter,* 110 Vt. 332, 6 A.2d 17 (1939).[6]

Even when time is of the essence, the contractor is not liable to the owner in damages if delay was caused by an act or default of the owner or a person for whom the owner is responsible.[7]

§ 5.3 Excusable and Nonexcusable Delay

Stated simply, *excusable delays* are those delays from which the contractor is "excused" from liability. As a general rule, a contractor is excused from liability for delays that are the result of causes beyond the contractor's control and delays which are the result of causes that were not foreseeable.

---

[4] 13 Am. Jur. 2d *Building and Construction Contracts* § 47 (1964 and Supp. 1989). *See also* 3A Corbin on Contracts § 720.

[5] *See generally* Carter v. Sherburne Corp., 132 Vt. 88, 315 A.2d 870 (1974) and Kingery Constr. Co. v. Scherbarth Welding, Inc., 186 Neb. 653, 185 N.W.2d 857 (1971). In *Carter* the court concluded that time was not of the essence of the contracts at issue because (1) none of the contracts contained express language making time of the essence and (2) nothing in the circumstances surrounding the contracts would take them out of the general rule. Furthermore, the court noted that, even though two of the contracts included specific schedule provisions and set out "forfeitures" in the event of noncompliance, that was insufficient to make time of the essence. Finally, the provision of a forfeiture in the event of noncompliance with the schedule requirements was evidence that time was not of the essence. 315 A.2d at 874.

[6] Carter v. Sherburne Corp., 132 Vt. 88, 315 A.2d 870, 873 (1974).

[7] 315 A.2d at 874.

Thus, subject to the caveat that each contract is governed by its express provisions, delays resulting from the following causes will generally be excusable delays, entitling the contractor to an extension of time to perform the contract: an act or neglect of the owner or the architect or a separate contractor of the owner, changes, labor disputes, fire, unusual delay in deliveries, or any other causes beyond the contractor's control. Likewise, it is common for a liquidated damages provision to excuse the contractor from liability for liquidated damages for similar reasons, if the cause of the delay in completing the work is both unforeseeable and beyond the control of the contractor.[8] If the contractor demonstrates that the entire delay was excusable under the governing contractual provisions, he or she would not be liable for liquidated damages and would be entitled to an extension of time to complete the work, if the contract provides for extensions of time. Whether or not the contractor is entitled to additional compensation as a result of the delay depends upon how the contract deals with compensation for delay.

§ 5.4 Compensable and Noncompensable Delay

Although many private contracts incorporate a provision barring the recovery of compensation or money damages for delay, the current AIA documents include no such limitation, and neither does the federal government form. In the absence of a so-called no damage for delay clause, or if the clause has been rendered unenforceable because one of the numerous exceptions applies, the contractor is entitled to compensation if it can show that it did not concurrently cause the delay and if it can quantify its

---

[8] *See* J.D. Hedin Constr. Co. v. United States, 408 F.2d 424, 187 Ct. Cl. 45 (1969), in which the contractor was terminated and charged with liquidated damages when it was delayed significantly by a shortage of cement. The Court of Claims, reversing the board of contract appeals that had awarded liquidated damages to the government, held that the contractor was not required to "have prophetic insight and take extraordinary preventative action which is simply not reasonable to ask of the normal contractor." The court continued:

> Unforeseen delays excusable under [the contract] can not be arbitrarily limited to delays which subsequent to the event might within the range of possibility have been avoided. Where a contractor resorts to the usual and long-established methods employed by the commercial work in general to obtain material to perform the contract he has no cause to suspect that such instrumentalities will fail to discharge their duties.

See also Morris Mechanical Enters., Inc. v. United States, 554 F. Supp 433, 1 Cl. Ct. 50 (1982) (contractor not liable for liquidated damages because, although contractor was late in delivering a chiller machine, other delays beyond contractor's control, including unknown subsurface conditions, prevented timely completion of the room in which the chiller was to be installed).

§ 5.4 COMPENSABLE/NONCOMPENSABLE

damages with reasonable certainty. The federal government suspension of work clause[9] expressly provides for the compensation of "the increased cost of performance (excluding profit)" caused by the unreasonable period of such suspension or delay. The threshold issue is then to determine what portion of the delay was reasonable, because only the unreasonable delay is compensable. Although on its face this requirement would appear to limit the compensation for delay to the unreasonable period, regardless of the cause of the delay, the Court of Claims has held in *Chaney & James Construction Co. v. United States*[10] that a primary function of the clause is to provide an administrative substitute for what would otherwise be a breach of contract claim.[11] In *Chaney* the contractor sought compensation for delays resulting from changes required to correct for defective plans and specifications. Because the government warrants the accuracy of plans and specifications,[12] delays resulting from defects in the plans and specifications are per se unreasonable. All of the increased costs incurred by the contractor were therefore compensable.

It is probably safe to say that, although no damage for delay clauses are generally enforceable, the courts have developed so many exceptions to the rule that in practice the exceptions may have swallowed the rule. Generally speaking, if the delay for which compensation is sought was beyond the contemplation of the parties at the time of contracting,[13] or resulted from the active interference of the other party[14] or from the fraud or bad

---

[9] F.A.R. § 52.212-5.

[10] 421 F.2d 728, 190 Ct. Cl. 699 (1970).

[11] The court also noted that the wording of the suspension of work clause also makes it clear that there is no intent for an extension of time to perform to be an exclusive remedy. 421 F.2d at 732.

[12] *United States v. Spearin,* 248 U.S. 132 (1918).

[13] *See, e.g.,* Blake Constr. Co. v. Coakley Co., 431 A.2d 569 (D.C. 1981); Sheehan v. Pittsburgh, 213 Pa. 133, 62 A.2d 642 (1905); Franklin Contracting v. State, 134 N.J. Super. 198, 338 A.2d 875 (Law Div. 1975); George A. Fuller Co. v. United States, 69 F. Supp. 409 (Ct. Cl. 1947); Commonwealth Highway & Bridge Auth. v. General Asphalt Paving Co., 46 Pa. Commw. 114, 405 A.2d 1138 (1979); E.C. Ernst, Inc. v. Manhattan Constr. Co., 551 F.2d 1026 (5th Cir. 1977); Peter Kiewit Sons' Co. v. Iowa S. Utils. Co., 355 F. Supp. 376 (S.D. Iowa 1973); Phoenix Contracting Corp. v. New York City Health, 118 A.D.2d 477, 499 N.Y.S.2d 953 (1986); Lichter v. Mellon Stuart Co., 193 F. Supp. 216, *aff'd,* 305 F.2d 216 (3d Cir. 1962); Hawley v. Orange County Flood Control Dist., 211 Cal. App. 2d 708, 27 Cal. Rptr. 478 (1963); Owen Constr. Co. v. Iowa State Dep't of Transp., 274 N.W.2d 304 (Iowa 1979).

[14] *See, e.g.,* E.C. Ernst, Inc. v. Manhattan Constr. Co., 551 F.2d 1026 (5th Cir. 1977); G.C.S., Inc. v. Foster Wheeler Corp., 437 F. Supp. 757 (W.D. Pa. 1975); Peter Kiewit Sons' Co. v. Iowa S. Utils. Co., 355 F. Supp. 376 (S.D. Iowa 1973); Owen Constr. Co. v. Iowa State Dep't of Transp., 274 N.W.2d 304 (Iowa 1979); Gherdi v. Board of Educ., 53 N.J. Super. 349, 147 A.2d 535 (App. Div. 1979); Henry Shenk Co. v. Erie County, 319 Pa. 100, 178 A. 662 (1935); Allen-Howe Specialties v. U.S. Constr., Inc., 611 P.2d 705 (Utah 1980).

faith[15] or willful conduct[16] of the other party, or if the duration of the delay is unreasonable,[17] the provision may not be enforceable.[18] Although these principles appear with some consistency in the reported decisions in various jurisdictions, some depart from the general principles a great deal. At one extreme are those jurisdictions, like Washington, which say that a no damage for delay clause is unenforceable as against public policy if the delay is unreasonable or is caused by the contractee,[19] and at the other extreme are jurisdictions, like New York, which say that the clause is subject only to certain limited exceptions,[20] thus protecting the public weal from excessive claims.

§ 5.5 Concurrent Delay

Once the contractor has established that the individual delay for which an extension of time is sought is excusable and, if compensation is

---

[15] *See, e.g.,* E.C. Ernst, Inc. v. Manhattan Constr. Co., 551 F.2d 1026 (5th Cir. 1977); W.C. James, Inc. v. Phillips Petroleum Co., 485 F.2d 22 (10th Cir. 1973); Unicorn Management Corp. v. City of Chicago, 404 F.2d 627 (7th Cir. 1968); G.C.S., Inc. v. Foster Wheeler Corp., 437 F. Supp. 757 (W.D. Pa. 1975); Gherdi v. Board of Educ., 53 N.J. Super. 349, 147 A.2d 535 (N.J. Super. Ct. App. Div. 1979); Gust K. Newberg, Inc. v. Illinois State Toll Highway Auth., 153 Ill. App. 3d 918, 506 N.E.2d 658 (1987).

[16] *See, e.g.,* Northeast Clackamas County Elec. Co-op., Inc. v. Continental Casualty Co., 221 F.2d 329 (9th Cir. 1955); Ozark Dam Constructors v. United States, 127 F. Supp. 187 (Ct. Cl. 1955); Housing Auth. v. Hubbell, 325 S.W.2d 880 (Tex. Civ. App. 1959); Blake Constr. Co. v. C.J. Coakley Co., 431 A.2d 569 (D.D.C. 1981); J.R. Stevenson Corp. v. County of Westchester, 113 A.D.2d 918, 493 N.Y.S.2d 819 (1985).

[17] *See, e.g.,* E.C. Nolan v. State, 58 Mich. App. 294, 227 N.W.2d 323 (1975); American Bridge Co. v. State, 245 App. Div. 535, 283 N.Y.S. 577 (1935); Nelse Mortensen & Co. v. Group Health Coop. of Puget Sound, 17 Wash. App. 703, 566 P.2d 560 (1977); Peter Kiewit Sons' Co. v. Iowa S. Utils. Co., 355 F. Supp. 376 (S.D. Iowa 1973); Phoenix Contractors v. General Motors Corp., 135 Mich. App. 787, 355 N.W.2d 673 (1984).

[18] For an in-depth analysis of the validity and enforceability of no damage for delay clauses *see* Annotation, *Validity and construction of "no damage" clause with respect to delay in building or construction contract,* 74 A.L.R.3d 187 (1976) (Supp. 1989).

[19] Wash. Rev. Code §§ 4.24.360, .370 (1988). *See also* Cal. Pub. Cont. Code § 7102 (West 1989).

[20] In New York, on public work contracts, a no damages for delay clause is only unenforceable if one of the following exceptions applies: (1) the delay was caused by the owner's bad faith, or its willful, malicious, or grossly negligent conduct, (2) uncontemplated delays, (3) delays so unreasonable that they raise the level of intentional abandonment of the contract by the owner, or (4) delays resulting from the owner's breach of a fundamental obligation of the contract. Corrino Civetta Constr. Corp. v. City of New York, 67 N.Y.2d 297, 493 N.E.2d 905, 502 N.Y.S.2d 681 (1986). Every possible cause of delay has been made foreseeable under the current New York City no damage for delay clause and brought within the contemplation of the parties. *See, e.g.,* Castagna & Son, Inc. v. City of New York, Index No. 3803/81 (N.Y. Sup. Ct. 1989).

sought, compensable as well, it is necessary to determine whether or not the contractor was independently delaying the work. If the contractor would have been delayed in any event by causes within its control, that is, if there was a concurrent nonexcusable delay, the general rule is that it would be inequitable to grant the contractor either an extension of time or additional compensation, unless the contractor can segregate the portion of the delay which is excusable and/or compensable from that which is not. The Court of Appeals for the Federal Court articulated the rule succinctly in *Klingensmith, Inc. v. United States*.[21] The contractor sought to recover its delay costs incurred when the project was delayed by a foundation design change from footings to caissons, but there may have been concurrent delay for other causes within the control of the contractor. The court, in remanding for a determination of causes of delay, said:

> The general rule is that '[w]here both parties contribute to the delay neither can recover damage[s], unless there is in the proof a clear apportionment of the delay and the expense attributable to each party.' *Blinderman [Construction Co. v. United States,]* 695 F.2d [552 (Fed. Cir. 1982)] at 559, quoting *Coath & Gross, Inc. v. United States*, 101 Ct. Cl. 702, 714–715 (1944). Courts will deny recovery where the delays are concurrent and the contractor has not established its delayed apart from that attributable to the government. Therefore [contractor] can only recover if it can establish that the government delayed the work by requiring that the footings be changed to caissons and if it can prove how much of the delay was chargeable to the government.[22]

On the other hand, when the owner and contractor concurrently delay the work, and the responsibility for the delay cannot be apportioned, the contractor is generally not liable for liquidated damages. For example, in *V.L. Nicholson Co. v. Transcon Investment & Financing, Ltd.*,[23] the Supreme Court of Tennessee held that

> no matter how one juggles the dates and time allowances the courts below found, and we agree, that both [owner] and [contractor] contributed to the delay. While the liquidated damages clause is otherwise enforceable, as it contemplated an amount reasonably related to damages [the owner] would have suffered if the completion date were not met, we cannot enforce such a provision where both parties have mutually caused the delay.[24]

---

[21] 731 F.2d 805 (Fed. Cir. 1984).

[22] *Id.* at 809.

[23] 595 S.W.2d 474 (Tenn. 1980) (citing Glassman Constr. Co. v. Maryland City Plaza, Inc., 371 F. Supp. 1154 (D. Md. 1974)).

[24] *Id.* at 484.

§ 5.6 Burden of Proof and Apportionment in Delay Cases

The contractor bears the burden of proving the extent of the delays for which it seeks compensation and, in addition, the burden of proving damages incurred as a result of such delays. The general rule is that the contractor cannot recover for delays caused by the owner if the contractor would have been delayed in any event by causes within its control. The rule was summarized in *Pathman Construction Co. v. Hi-Way Electric Co.*:[25]

> [T]he issue of apportionment of damages in case of mutual delay is a question of fact. [citation omitted] The burden of proof is on the party claiming such damages to prove the damages were caused by default of the party to be charged, separate from any damages that may have resulted from the acts of the claimant. The amount of delay attributable to each party is a question which must be resolved by the trier of fact.
>
> Where there is sufficient evidence to allow the court to make a reasonably certain division of responsibility for delay, the assessment of damages may be allocated among several parties. Although the task is particularly difficult when, as here, the performance of the work is sequential and the delay the result of multiple causes, it is not impossible. We note that technological advances and the use of computers to devise work schedules and chart progress on a particular project have facilitated the court's ability to allocate damages.[26] Although the record does reveal incident of delay occasioned by Pathman or other subcontractors under its control we do not believe the able and experienced trial court erred in apportioning the damages to arrive at Hi-Way's aliquot share.[27]

If the claimant fails to meet this burden, its claim for damages will be denied. For example, in *Lichter v. Mellon-Stuart Co.*,[28] the court held that the claimant subcontractor's claim for delay damages was insufficient because the claimant offered no proof of allocation of damages resulting from the breaches of Mellon-Stuart (the defendant general contractor) from those resulting from other factors.[29] Similarly, in *Arntz Contracting*

---

[25] 65 Ill. App. 3d 480, 382 N.E.2d 453 (1978).

[26] A computer generated schedule or schedule analysis is not necessarily the be-all and end-all of delay apportionment. As is discussed in § 5.8, caution must be exercised when relying on a computer based scheduling system to prove delay and responsibility for delay.

[27] Pathman Constr. Co. v. Hi-Way Elec. Co., 65 Ill. App. 3d 480, 382 N.E.2d 453 (1978).

[28] 305 F.2d 216 (3d Cir. 1962).

[29] The court below observed that "'[N]o basis appears for even an educated guess as to the increased costs suffered by plaintiff due to that particular breach or breaches as distinguished from those causes from which defendant is contractually exempt from responding in damages.'" *See* 196 F. Supp. 149, 151 (W.D. Pa. 1961).

§ 5.6 PROOF AND APPORTIONMENT

Co.,[30] the Department of Energy Board of Contract Appeals articulated the rule in federal contracts:

> [E]ven assuming a Government caused delay, in order for Appellant to prevail on this issue, it must demonstrate that any such Government caused delays were not concurrent or intertwined with other delays, for which the Government was not responsible. Thus a contractor asserting a delay claim against the Government must prove not only that it incurred additional costs making up its claim, but also that such costs would not have been incurred but for some Government action. *Commerce International Co., Inc. v. U.S.*, 338 F.2d 81 (1964); *Fishbach & Moore International Corp.*, ASBCA No. 18146, 77-1 BCA ¶ 12,300.

Although the government had delayed the work, these delays were intertwined and concurrent with those of the contractor.[31] The board concluded:

> It is [contractor's] burden to show that the specifications were defective[32] or that any delays caused by Government action were not concurrent or intertwined with delays caused by [contractor's] own actions. We find that [contractor's] efforts to carry this burden are not persuasive. Accordingly, [contractor's] claim stands denied.

For purposes of determining whether the project has been delayed and for purposes of apportioning delay, only delays on the critical path of the project figure in the analysis because, by definition, delays not on the critical path will not delay the completion of the project.[33] For example, in *Fishbach & Moore Corp.*,[34] the government argued that, because the contractor, Fishbach & Moore, was itself responsible for delays concurrent with delays caused by the government, its delay claim should be denied. The board responded to the government's argument this way:

> We take no issue with the application of the *Commerce* rule[35] to the facts of this case insofar as the concurrent delay for which [contractor] is

---

[30] EBCA No. 187-12-81, 84-3 B.C.A. (CCH) ¶ 17,604.

[31] The board at this point was looking at delays associated with certain metal siding work. This work and the delays associated with it were important to the project delay because this activity was on the "as-built" critical path of the project. *See* Fishbach & Moore Corp., ASBCA No. 18146, 77-1 B.C.A. (CCH) ¶ 12,300.

[32] Defective specifications take the claim out of the suspension of work clause and the entire delay is compensable, rather than just the unreasonable part of the delay.

[33] *See* G.M. Shupe, Inc. v. United States, 5 Cl. Ct. 662, 728 (1984).

[34] ASBCA No. 18146, 77-1 B.C.A. (CCH) ¶ 12,300 (1977).

[35] Commerce Int'l Co. v. United States, 338 F.2d 81, 167 Ct. Cl. 529 (1964).

responsible affected the work in the critical path to completion of the contract. If the concurrent delays affected only work that was not in the critical path, however, they are not delays within the meaning of the rule since timely completion of the project was not thereby prevented.[36]

Thus, if a contractor seeks compensation for government-caused delays to the critical path of the project, this claim will not be barred by concurrent noncritical delays caused by the contractor.[37] Nor will the claim be barred if the contractor has eliminated from its claim critical path delays that were concurrently caused.

§ 5.7 Apportionment in Liquidated Damages Cases

Although liquidated damages are most often a claim asserted by an owner against a contractor rather than a claim asserted by a contractor, a discussion of liquidated damages is included here to clarify the differences between the courts' approach to delay claims and liquidated damages claims and because, not infrequently, a contractor's delay claim is met by an owner's liquidated damages claim.

Aside from any issue of apportionment of responsibility for delay, when faced with a claim for liquidated damages, a contractor should always determine whether the underlying liquidated damages provision is enforceable.[38] Assuming that the provision is enforceable, what is the contractor's liability when both the contractor and the owner have contributed to the delay in completing the work?

The Superior Court of New Jersey stated the general rule in terms of the presence or absence in the contract of an extension of time provision, in *Buckley & Co. v. State:*[39]

> The general rule appears to be that, in the absence of an extension of time provision in the contract, liquidated damages will be denied in their

---

[36] 77-1 B.C.A. (CCH) ¶ 12,300 at 59,224.

[37] Does this rule mean that the contractor cannot recover for noncritical path delays? The general rule in federal government contracts is that the party who gets to the float first receives the benefit of the float (until the float is actually used up and the project is delayed). *See* J. Wickwire, S. Hurlburt, & L. Lerman, *Use of Critical Path Method Techniques in Contract Claims: Issues and Developments,* 18 Pub. Conts. L.J. 338, n.39 (1974 to 1988).

[38] A discussion of the enforceability of liquidated damage provisions is beyond the scope of this chapter. For a complete discussion *see* Annotation, *Contractual provision for per diem payments for delay in performance as one for liquidated damages or penalty.* 12 A.L.R.4th 884 (1982 and Supp. 1989).

[39] 140 N.J. Super. 289, 356 A.2d 56 (Law Div. 1975).

§ 5.7 LIQUIDATED DAMAGES CASES

entirety where the contractee is responsible for any delay, even though some or most of the delay is attributable to the contractor.[40]

Where the contract includes a provision for the grant of extensions of time, the authorities permit the contractee to recover liquidated damages but only to the extent that the delay is not caused by the contractee himself. This is the rule stated in 5 Williston on Contracts (3 ed. 1961, 765, 766):

> In building contracts, there is often inserted a provision giving the architect power to certify an extension of time in certain cases, by virtue of which the effect of a delay caused by the owner operates merely as an extension of time for performance, and a new time is substituted for the old. In that event though the owner causes the delay the builder is liable in liquidated damages, but the period of delay caused by the owner is deducted from the total delay. Unless the contract contains such a provision the delay due to each party will not generally be apportioned.[41]

Because neither Buckley nor the state could prove how much of the delay was attributable to either party, the court held that neither party should recover from the other for losses resulting from the delay.

Other courts have held that, regardless of whether the contract provides for the granting of an extension of time, the contractee is not entitled to rely on a liquidated damages provision of the contractee caused all or any part of the delay.[42]

A look at two often-cited Supreme Court cases is instructive. In *United States v. United Engineering Co.*,[43] the contractor was obligated to construct a dry dock at the New York Navy yard within seven months from the date of the contract, subject to liquidated damages of 25 dollars per day if the work was not completed within the contact time. The government issued three supplements to the contract. The first supplement increased the scope of the project and included a six-month extension of the contract time to October 15, 1901. The second and third supplements, however, did not say anything as to time of completion or as to delays under the original contract or the first supplement. The contractor was delayed, solely by the government, in completing the work until May 1, 1903. Thereafter, the contractor further delayed the completion of the

---

[40] Citing Gogo v. Los Angeles County Flood Control Dist., 45 Cal. App. 2d 334, 114 P.2d 65 (1941); Nomellini Constr. Co. v. Department of Water Resources, 19 Cal. App. 3d 240, 96 Cal. Rptr. 682 (1971); and Glassman Constr. Co. v. Maryland City Plaza, Inc., 371 F. Supp. 1154 (D. Md. 1974).

[41] Buckley & Co. v. State, 356 A.2d 56 at 70.

[42] *See* Acme Process Equip. Co. v. United States, 347 F.2d 509, 535 (Ct. Cl. 1965); Gogo v. Los Angeles County Flood Control Dist., 45 Cal. App. 2d 334, 339, 114 P.2d 65, 70 (1941).

[43] 234 U.S. 236 (1914).

work. The Court affirmed the decision of the Court of Claims, denying recovery of liquidated damages, stating the applicable rule this way:

> We think the better rule is that when the contractor has agreed to do a piece of work within a given time and the parties have stipulated a fixed sum as liquidated damages not wholly disproportionate to the loss for each day's delay, in order to enforce such payment the other party must not prevent the performance of the contract within the specified time, and that where such is the case, and thereafter the work is completed though delayed by the fault of the contractor, the rule of the original contract cannot be insisted upon, and liquidated damages measured thereby are waived. Under the original and first supplemental agreements, the claimant knew definitely that he was required to complete the work by a fixed date. Presumably the claimant had made its arrangements for completion within the time named. Certainly the other contracting party ought not to be permitted to insist upon liquidated damages when it is responsible for the failure to complete by the stipulated date[;] to do this would permit it to recover damages for delay caused by its own conduct.
>
> * * *
>
> Under such circumstances we think [the government] must be content to recover such damages as it is able to prove were actually suffered.[44]

Compare that case with *Robinson v. United States*,[45] in which the Court allowed the government to recover liquidated damages for delay notwithstanding the fact that some of the delay was caused by the government. In *Robinson*, the contract had a no damage for delay clause, a time extension clause granting the contractor a day-for-day extension of time for government caused delays, and a liquidated damages provision pursuant to which the contractor was obligated to pay to the government "$420 for each and every day's delay not caused by [the government]." The parties agreed to one supplemental contract which granted the contractor an extension of time in lieu of all delays chargeable to the government which had accrued as of that point. Thereafter the completion of the work was delayed by both the contractor and the government. Because the liquidated damages clause expressly differentiated between delays caused by the government and delays caused by the contractor, and because the Court of Claims was able to determine how much of the total delay was attributable to each, the Court held that the government was entitled to liquidated damages for the delay caused by the contractor. The Court distinguished *United Engineering Co.* on the basis that there the lower court found as a matter of fact that but for the government delays the contractor would have completed the work within the originally agreed contract time.

---

[44] *Id.* at 242.
[45] 261 U.S. 486 (1923).

§ 5.8 Computing the Delay

The development of computer technology for project scheduling in recent years has made it possible to apply critical path method (CPM) analysis to extremely large and complex projects with large networks of activities. The concept of critical path analysis has not changed, but it has become relatively easy to create and update schedules because of the speed with which the computer can accomplish the necessary mathematical calculations. And as the use of this technology has become more common in the construction industry, courts and boards have recognized the value of CPM analysis in the proof and defense of delay claims.[46]

The Claims Court recently explained the critical path method this way:

> The CPM breaks down the entire project into individual tasks and assigns a number of days anticipated to perform each task. In *Haney v. United States,* 230 Ct. Cl. 148, 167–68, this Court's predecessor, the United States Court of Claims, described the CPM in the following manner:
>
>> Essentially, the critical path method is an efficient way of organizing and scheduling a complex project which consists of numerous interrelated separate small projects. Each subproject is identified and classified as to the duration and precedence of the work. (E.g., one could not carpet an area until the flooring is down and the flooring cannot be completed until the underlying electrical and telephone conduits are installed.) The data is then analyzed, usually by computer, to determine the most efficient schedule for the entire project. Many subprojects may be performed at any time within a given period without any effect on the completion of the entire project. However, some items of work are given no leeway and must be performed on schedule; otherwise, the entire project will be delayed. These latter items of work are on the "critical path." A delay or acceleration of work along the critical path will affect the entire project.[47]

Each task or activity (or subproject in the language of the *Haney* court) has an early start date, an early finish date, a late start date, and a late finish date. As these terms suggest, the *early start date,* for example, is the earliest possible date that the task can start. Adding the task duration to the early start date determines the *early finish date,* the earliest possible date on which the task could be completed. If a task is not on the "critical path," that is, if delaying the completion of that task will not immediately

---

[46] For a complete discussion of the critical path method as it applies to construction delay claims, *see* J. Wickwire & R. Smith, *The Use of Critical Path Method Techniques in Contract Claims,* 7 Pub. Cont. L.J. 1 (1974); J. Wickwire, S. Hurlburt, & L. Lerman, *Use of Critical Path Method Techniques in Contract Claims: Issues and Developments, 1974 to 1988,* 18 Pub. Cont. L.J. 338 (1989).

[47] Fortec Constructors v. United States, 8 Cl. Ct. 490 (1985).

affect the completion date of the project, the task is noncritical and is said to have "float." Every task, if delayed long enough, however, will at some point become critical to completion of the project. The length of the delay allowable before the completion date is affected is the *float*. Looking again at the *Haney* example, if the task to install the telephone conduits and the task to install the electrical conduits can both begin the same day, and if the telephone conduits will take two days to install and the electrical conduits will take seven days to install, and both must be completed before the flooring can be installed, then there are five days of float on the telephone conduit task. Looking at it another way, the float is the difference between the early start and late start dates and similarly the difference between the early finish and late finish dates for any individual task. If the task is critical, then the early start and late start dates are the same, as are the early finish and late finish.

In using project schedules to prove or disprove the existence of and liability for delays, there are a few fundamental caveats to be observed. First, the apparent mathematical certainty of the CPM schedule analysis method might suggest that there is a single unalterable critical path schedule for any given project and that this schedule can be discovered before, during or after the completion of the project if the established techniques are properly applied. However, this is not the case. Rather, there are many possible critical path schedules for any particular project, depending upon variables such as projected or actual material, labor and equipment resources and depending upon the sequence of construction preferred by the superintendent in charge. In addition, the as-planned schedule almost always differs from the as-built schedule, because as the project moves forward contingencies arise that delay some activities and accelerate others so that the critical path changes, and what was critical in the planning stages may not be critical in the actual performance of the project.[48] Thus, as the *Fortec* court noted, "if the CPM is to be used to evaluate delay on the project, it must be kept current and must reflect delays as they occur."[49]

The situation in *Fortec* is instructive on this point. There, the original CPM was only updated once during the performance of the work, and both the contractor and the government acknowledged that neither had used the CPM in evaluating contract performance. Although the government admitted that it delayed the contractor in completing certain activities, it argued that because these activities were not critical on the original or on the only update of the schedule the contractor was not

---

[48] *See* Morris Mechanical Enters., Inc. v. United States, 554 F. Supp. 433, 1 Ct. Cl. 50 (1982).

[49] Fortec Contractors v. United States, 8 Cl. Ct. 490, 505 (1985).

§ 5.8 COMPUTING THE DELAY

entitled to delay damages or an extension of time. In refusing this defense, the court said:

> Despite the obvious failure of both parties to use the CPM for scheduling purposes during construction, the Government now claims that the additional work Fortec was required to do does not justify any contract time extensions, since the CPM does not show that any of the additional work was on the project's critical path. In support of its positions, the Government relies entirely upon the once revised CPM, which does not reflect the critical path actually followed during construction.
>
> * * *
>
> [The government's resident engineer] not only admitted that the CPM in evidence did not reflect actual performance, but he also admitted that the critical path can and does change during performance. Indeed, [he] acknowledged that delay encountered in completion of a noncritical item may make that item critical so that "every month, conceivably, the critical path would change," which is precisely what happened in the instant case.
>
> * * *
>
> As a result, it is impossible to determine from the CPM diagram whether a particular activity was critical or noncritical, on schedule or behind schedule.[50]

The obvious solution would seem to be to create an "after the fact" CPM to show what the correct CPM should have been, which leads to the second caveat. Hypothetical CPM schedules generated after the fact are easily attacked. For instance, in *Fortec* the government attempted to establish the project critical path using a CPM analysis not based on the actual project schedule, with disastrous results. The government introduced into evidence CPM schedules which showed an activity called "removal of telephone poles" to be on the project critical path. Unfortunately, no such telephone poles ever existed and the validity of the entire analysis was called into question and rejected.

Similarly, in *Pathman Construction Co.*,[51] an after-the-fact CPM analysis was rejected. There, in the absence of a contemporaneous project schedule, both the contractor and the government retained experts to prepare schedules demonstrating how the contractor would have planned the project some 10 years earlier. In rejecting both of these analyses, the board observed:

> [The Contractor's expert] admitted that "the creation of an as-planned schedule 10 years after the fact, is a very messy proposition." The government's scheduling expert went further, describing [the contractor's] CPM as

---

[50] *Id.*
[51] ASBCA No. 23392, 85-2 B.C.A. (CCH) ¶ 18,096.

an "academic exercise . . . [which] wouldn't have any meaning . . . [because] there may be an unlimited number of combinations of critical paths or networks one could draw." [He] also referred to a negative factor he called "built in bias", meaning that a person who knows what has already happened on a project will be influenced, either consciously or unconsciously, by that knowledge in preparing an after the fact schedule.[52]

Because of the length of time that had passed between the actual performance of the project and the generation of the CPM, and because the people who prepared the CPM were not the same people who would have done so originally, the board concluded:

> Under all of the circumstances, we conclude that [the contractor's] as-planned CPM's are distorted [by including a change that was not known at the outset of the project], because they were not created in a timely manner, and do not represent a reasonable schedule for this project. We find them and the documents based on them unreliable.[53]

The contractor was able to prove compensable delays nonetheless, on the basis of the actual as-built schedule as explained by witnesses with firsthand knowledge of the causes and effects of the various delays.

In summary, proof of the extent and liability for delay is most convincing when that proof is based on actual contemporaneous project schedules, frequently updated at the time the delays were occurring. In the absence of such schedules, the as-built schedule for the project, supplemented by other project records and witnesses with firsthand knowledge of the delays and an expert witness's evaluation of the as-built schedule, is often preferable to an as-planned schedule generated after the fact.[54]

§ 5.9 Contractor's Right to Finish Early

Suppose that a contractor completes its work within the agreed contract time, but later than it would have been completed but for delays caused by the owner. From the contractor's perspective, because its costs would have been correspondingly lower and its profits correspondingly higher, the increased costs or lost profits should be a compensable element of the contractor's delay damages. From the owner's perspective, on the other hand,

---

[52] *Id.* at 90,837.

[53] *Id.*

[54] Note that in *Fishbach & Moore Corp.*, ASBCA No. 18146, 77-1 B.C.A. (CCH) ¶ 12,300, at 59,224, the CPM was generated after the fact and was accepted by the board, because it had a "ready and reasonable basis for segregating the delays," and because it was based "on application of the construction expertise of an expert to depict the orderly sequence of events that must be followed to accomplish a complex project such as [this one]."

the contractor should not be able to claim delay damages until the agreed contract time has been exceeded. The general rule is, however, that unless the contract provides otherwise, a contractor is entitled to finish early, and compute delay damages from the early completion date, if it can demonstrate that it reasonably could have achieved the early completion, but for an owner-caused delay. For example, when a contractor would have completed the construction of certain work 15 days before the contract deadline but was unable to do so because of owner initiated change orders, delay damages were properly calculated from the date 15 days before the contract completion date.[55] The court observed that "[the contractor's] damages were caused directly by [the owner's] actions. . . . Costs due to delay are no less damaging merely because they occur fortuitously before a contract deadline rather than after."[56] In the federal contracts contest, the Armed Services Board of Contract Appeals has summarized the rationale for this view in *CWC, Inc.*:[57]

> Stemming from the general principle that the Government is under an obligation not to hinder performance, the corollary follows, barring express restrictions in the contract to the contrary, 'a [construction] contractor has the right to proceed according to his own job capabilities at a better rate of progress than represented by his own schedule.' The Government may not hinder or prevent earlier completion without incurring liability. *John F. Burke Engineering and Construction,* ASBCA No. 8182, 1963 BCA ¶ 3713 at 18,559. *See Metropolitan Paving Company v. United States,* [9 CCF ¶ 72,357], 325 F.2d 241, 242 (Ct. Cl. 1963). A failure on the Government's part to fulfill its obligation not to interfere with or unreasonably delay the contractor's performance is remediable as a constructive change. *Johnson & Son Erectors,* ASBCA No. 24564, 81-1 BCA ¶ 15,082 at 74,599.[58]

Similarly, the Supreme Court of Arkansas has held an owner liable to the contractor in damages for delays resulting from defective plans and specifications.[59] The owner-housing authority's defense was that, notwithstanding the admitted delays, the contractor completed its work within the 350 days allotted in the contract. The court cited *Metropolitan Paving,*[60] with approval:

> In order to recover, plaintiff must surmount three hurdles. First, it must prove that defendant's employees did unnecessarily cause delay. It must then

---

[55] Sun Shipbuilding & Dry Dock Co. v. United States Lines, Inc., 439 F. Supp. 671 (E.D. Pa. 1977).

[56] *Id.* at 682.

[57] ASBCA No. 26432, 82-1 B.C.A. (CCH) ¶ 15,907 (1982).

[58] *Id.* at ¶ 78,838.

[59] Housing Auth. v. E.W. Johnson Constr. Co., 264 Ark. 523, 573 S.W.2d 316 (1978).

[60] 325 F.2d 241, 163 Ct. Cl. 420 (1963).

establish that such delays constituted a breach of contract. Finally, it must establish that it suffered damages due to the breach.

* * *

While it is true that there is not an 'obligation' or 'duty' of defendant [owner] to aid a contractor to complete prior to the completion date, from this it does not follow that defendant may hinder and prevent a contractor's early completion without incurring liability. It would seem to make little difference whether or not the parties contemplated an early completion, or even whether the contractor contemplated an early completion. Where the defendant [owner] is guilty of 'deliberate harassment and dilatory tactics' and a contractor suffers as a result of such action, we think that defendant is liable.[61]

The contractor must demonstrate that its planned schedule for the early completion of its work was both reasonable and attainable,[62] but it is not necessary for the contractor to communicate its intent to finish early to the owner.[63]

United States v. Blair[64] is often cited for the proposition that the government cannot be liable for delay damage unless the contractor's work is delayed beyond the contract completion date. There the contractor would have been able to complete its work earlier than the contract completion date but for delays caused by another contractor hired by the government. The Court held that, in the absence of an express contractual obligation, the government had no duty to aid the delayed contractor in completing its work early. *Blair* held only that delays caused by an independent third party contractor which prevent the delayed contractor from completing its

---

[61] Housing Auth. v. E.W. Johnson Constr. Co., 573 S.W.2d at 323.

[62] Owen L. Schwam Constr. Co., ASBCA No. 22407, 79-2 B.C.A. (CCH) ¶ 13,919, at 68,330 (citing Barton & Son Co., 65-2 B.C.A. (CCH) ¶ 4,874, Coley Properties Corp. v. United States, 593 F.2d 380 (Ct. Cl. 1979); Sydney Constr. Co., Inc., ASBCA No. 21377, 77-2 B.C.A. (CCH) ¶ 12,719 (1977); Green Builders, Inc., ASBCA No. 35518, 88-2 B.C.A. (CCH) ¶ 20,734 (1988).

[63] In Sydney Contrs. Co. ASBCA No. 21377, 77-2 B.C.A. (CCH) ¶ 12,719 (1977) and Barton & Son Co., 65-2 B.C.A. (CCH) ¶ 4,874 (1965) recovery of delay damages was allowed, even though the contractor had never informed the government that it intended to finish its work early. *See also* Montgomery-Ross-Fisher, Inc., PSBCA Nos. 1033 and 1096, 84-2 B.C.A. (CCH) ¶ 17,492 (1984), where the board said: "Recovery has been allowed even though the contractor never informed the Government of its intent to finish early" [citing Coley Properties Corp. v. United States, 593 F.2d 380 (Ct. Cl. 1979) and Owen L. Schwam Constr. Co., 79-2 B.C.A. (CCH) ¶ 13,919]. Absent a contract prohibition, a contractor has a right to better his progress and the Government has an implied obligation to cooperate and not to impede or delay the contractor's performance. Eickhoff Constr. Co., ASBCA No. 20049, 77-1 B.C.A. (CCH) ¶ 12,398 (1977).

[64] 321 U.S. 730 (1943).

§ 5.10 Pricing the Claim

Having established entitlement to a delay claim, the contractor is faced with a computation of the quantum of the claim. Often, there is relatively little detailed cost data in a form that is suitable for calculating damages for discrete occurrences of delay; the contractor may only have the original bid and its actual cost records, and they may not correspond directly to one another. For example, the estimator may have prepared the bid on a breakdown of work unrelated to the cost accounting system in which the actual cost of performance was recorded. When this is the case, one way to calculate the quantum of damages is to subtract the total estimated cost of performance from the total actual cost of performance, which is the so-called *total cost method* of computing damages. However, the total cost method is not favored by the courts and boards because of its inherent limitations: the loss may have been caused by a bad original estimate or because the contractor was inefficient for reasons unrelated to delays caused by the other party. In addition, in its pure form, the total cost method makes no allowance for delays caused by the contractor itself. Thus the almost universal rule is that a total cost proof of damages will only be accepted in certain limited circumstances. The Pennsylvania Superior Court has held that the "total cost method of determining damages is imprecise and frequently an inaccurate measure of damages. It should not be used unless no other method of determining damages is available and, then, only when its reliability has been established by supporting evidence."[65] To establish the reliability of the method, the court adopted the federal rule:

> The acceptability of the method hinges on proof that (1) the nature of the particular losses make it impossible or highly impracticable to determine them with a reasonable degree of accuracy; (2) the plaintiff's bid or estimate was realistic; (3) its actual costs were reasonable; and (4) it was not responsible for the added expenses. *Boyajian v. United States,* 423 F.2d 1231, 1243 (Ct. Cl. 1970) quoting *W.R.B. Corp. v. United States,* 183 Ct. Cl. 409, 426 (1968). The necessity for showing an accurate bid estimate is basic.[66]

The *Harkins* court denied recovery on the basis of the total cost method because it found that there was insufficient evidence that the original

---

[65] John F. Harkins Co. v. School Dist. of Philadelphia, 313 Pa. Super. 425, 433, 460 A.2d 260, 264 (1983).
[66] *Id.* at 431, 460 A.2d at 263.

estimate was realistic or that the damages resulted solely from the school board's actions.

Other cases have allowed contractors to recover delay damages using the "pure" total cost method.[67] Some cases have allowed a total cost recovery once the contractor adjusted the original bid to correct errors in it or adjusted the actual costs to eliminate costs for which the owner is not responsible—the *modified total cost* or *should have cost* approach.[68]

Notwithstanding these cases, a contractor is well advised to attempt to prove individual elements of damage, if possible, instead of relying on the less favored total cost method.[69]

---

[67] *See* Nebraska Pub. Power Dist. v. Austin Power Co., 773 F.2d 960 (8th Cir. 1985); Moorhead Constr. Co. v. City of Grand Forks, 508 F.2d 1008 (8th Cir. 1975) (total cost recovery allowed when contractor was faced with a situation not of his own making but induced by the city's delay, and when the court determined that no excess charges were run up by the contractor and that its bid was more realistic than the estimate prepared by the city's own engineer); Robert McMullan & Sons, Inc., ASBCA No. 19129, 76-2 B.C.A. (CCH) ¶ 12,072 (1976) (total cost recovery allowed when contractor was required to do additional work almost from the beginning and throughout the entire duration of the project and when the original bid was fair, and there was no managerial ineptitude, no financial difficulties, no contractor contribution to damages, and no evidence that prices or costs were unreasonable).

[68] *See, e.g.,* Great Lakes Dredge & Dock Co. v. United States, 96 F. Supp. 923 (Ct. Cl. 1951) in which the contractor sought an equitable adjustment for unanticipated subterranean water. The contractor adjusted both its estimate, to correct certain acknowledged errors, and its actual costs, to eliminate certain costs not attributable to the water. The court concluded that the original estimate was not reasonable and was in fact underestimated and that the contractor's calculation of damages was therefore improper. Instead, the court determined a realistic estimate of what the work "should have cost" was the average of the government's estimate of the cost of the work and the four other bids. *See also* Sovereign Constr. Co., ASBCA No. 17792, 75-1 B.C.A. (CCH) ¶ 11,251 (1975).

[69] For the pure delay elements of the claim, that is, those elements that are purely time related, the contractor is often able to prove individual elements of damage with some accuracy. When inefficiencies result from delay, the cause and effect proof is often extremely difficult. In that event, the courts and boards are more willing to accept a total cost proof of damages. For example, in Sovereign Constr. Co., 75-1 B.C.A. (CCH) ¶ 12,251 at 53,606 (1975), the board said:

> [W]e agree that the situation presented here is that type of situation contemplated by the courts and this Board in which a total cost approach is an acceptable approach provided the supporting documentation concerning the expected and actual costs is reasonably accurate. Inefficient work is an intangible commodity. It is not possible to make a direct measurement of it. In identifying and measuring inefficiency a comparison must be made to some accepted standard. In this appeal we are unaware of any better method by which the [contractor] could establish additional costs attributable to inefficiency than by establishing what the work should have cost and what the work did in fact cost. Of course, the actual costs are subject to adjustment if they include elements for which the government is not responsible.

§ 5.11 Increased Material and Labor Costs (Escalation)

If the contractor's work, or a part of its work, is delayed by the owner, the increased direct costs of material and labor resulting from the delay are a compensable element of the contractor's delay claim.[70] In *J.D. Hedin Construction Co. v. United States,* the government argued that on a fixed price contract increased wage costs were a business risk assumed by the contractor. The court observed that such is the case only if the wage increase occurs during the originally scheduled time of performance, but if the time of performance is extended because of government delays beyond the originally scheduled time, the risk of increases in wage rates does not fall on the contractor. Although this principle is straightforward, the proof of the compensable portion of increased wages is often complex. For example, the original schedule may call for some work to be performed in a period of escalated wages, the original schedule for the work may have float, and both the owner and the contractor may be responsible for parts of the delay. If any of these factors comes into play, an appropriate adjustment must be made.

The decision in *Sovereign Construction Ltd.*[71] is instructive on this point. In this case, the contractor was delayed when changed conditions were discovered for which the contractor was entitled to an adjustment. When the changed conditions were discovered, the contractor was then 30 days behind schedule for reasons not attributable to the government. In addition, after the discovery of the changed conditions, the contractor caused two months of the total additional 18-month delay. The board instructed the parties to calculate the wage escalation portion of the claim in the following manner:

1. Plot the time span during which the delayed work should have been completed but for the changed conditions, using the contractor's initial critical path schedule.

2. If any of the affected work activities had float in the original schedule, assume that the work would have been completed at the midpoint of the float period.

---

[70] J.D. Hedin Constr. Co. v. United States, 347 F.2d 235, 171 Ct. Cl. 70 (1965); United States Steel Corp. v. Missouri Pac. R.R., 668 F.2d 435 (8th Cir. 1982); Paccon, Inc., ASBCA No. 7890, 65-2 B.C.A. (CCH) ¶ 4996 (1965); Sydney Constr. Co., ASBCA No. 21377, B.C.A. (CCH) ¶ 12,719 (1977); Luria Bros. & Co. v. United States, 369 F.2d 701, 177 Ct. Cl. 676 (1966); Sovereign Constr. Ltd., ASBCA No. 17792, 75-1 B.C.A. (CCH) ¶ 11,251 (1975); D.H. Dave, Inc. v. Gerben Contracting Co., ASBCA No. 13005, 73-2 B.C.A. (CCH) ¶ 10,191 (1973).

[71] ASBCA No. 17792, 75-1 B.C.A. (CCH) ¶ 11,251 (1975).

3. Because, when the changed condition[s] giving rise to the delay were discovered, the contractor was then 30 days behind schedule for reasons attributable to the contractor and not the owner, 30 days should be added to the completion dates determined in step 2.

4. In a manner similar to step 1, plot the time spans during which the work was actually completed.

5. Compute the increased wage costs incurred by the contractor by multiplying the manhours expended after the completion dates established in step 3 by the amount of the increase in the wage rate.

6. Delete from this calculation any wage increased for personnel for which the contractor has already been compensated in prior overhead agreements between the parties.

7. Finally, because the contractor was responsible for 2/18 or 1/9 of the delay after the changed conditions were discovered, the calculation shall be reduced by 1/9. This result is the wage increase to which the contractor is entitled.[72]

With respect to increases in material costs, the basic rationale is the same as that for wage escalation. Increased costs resulting from delays attributable to the other party are compensable.[73] The threshold inquiry when increased material costs are sought is whether or not the delays attributable to the owner affected material purchase dates or whether only field labor activities were affected:

> A mere showing of the increased cost resulting from paying a higher price for material purchased during a stretched out period would not be enough, as [contractor] would have to show that the delay in purchase of the materials until the time of the higher price was caused by the suspensions of work. A suspension in the performance of work at the project site would not necessarily delay the procurement of materials.[74]

In addition to the increased costs of the materials themselves, the contractor may also be entitled to other costs related to the materials, such as the loss of benefit for purchasing large quantities,[75] storage costs,[76] or additional material handling and insurance costs.[77]

---

[72] *Id.* at ¶ 53,609.

[73] Luria Bros. & Co. v. United States, 369 F.2d 701, 177 Ct. Cl. 676 (1966); Paccon, Inc., ASBCA No. 7890, 65-2 B.C.A. (CCH) ¶ 4996 (1965); Canon Constr. Corp., ASBCA No. 16142, 72-1 B.C.A. (CCH) ¶ 9404 (1972).

[74] Paccon, Inc., ASBCA No. 7890, 65-2 B.C.A. (CCH) ¶ 4996 at 23,579 (1965).

[75] *See* Samuel N. Zarpas, Inc., ASBCA No. 4722, 59-1 B.C.A. (CCH) ¶ 2170 (1961).

[76] American Bridge Co. v. State, 245 A.D. 535, 283 N.Y.S. 577 (1935) (contractor entitled to costs of storing and repainting steel that deteriorated in storage when owner imposed delays prevented the contractor from accepting the delivery according to schedule).

[77] Stapleton Constr. Co. v. United States, 92 Ct. Cl. 551 (1941) (additional material handling and insurance costs).

§ 5.12 Costs of Idle Construction Equipment

When a project delay attributable to the owner idles the contractor's construction equipment, the contractor is entitled to compensation for the costs incurred while the equipment was idle. The dispute in computing idle equipment costs centers, in almost every case, on the rate at which the equipment should be charged. There are basically three possible sources of this cost data. The most obvious source is the contractor's actual cost to own or rent the equipment. The other frequently encountered sources are the "AGC Rates," published by the Associated General Contractors of America in the *Contractor's Equipment Ownership Manual,* which reflect the average cost of ownership of specific pieces of construction equipment, and the "AED Rates" published in the *Associated Equipment Distributor's Manual,* which reflect the average third-party rental rates for specific pieces of equipment.

The contractor's actual cost of idle equipment is, of course, the true measure of the loss incurred as a result of the owner's delay,[78] and a contractor cannot arbitrarily elect to rely on published rates instead unless the contract so provides. It would be absurd, the court noted in *L.L. Hall Construction Co. v. United States,* if a contractor could ignore actual equipment operating costs and rely on published rates, if it had or could have reasonably maintained records of such costs. "[S]ome showing must be made that secondary evidence is appropriate because the primary evidence (actual costs) is nonexistent or unavailable for good reason."[79] For example, a small contractor may not maintain detailed equipment cost records at all, or a larger contractor may not keep such records for small jobs of short duration. The general rule for contractor owned equipment is:

> The fair and reasonable measure of damages for plaintiff's equipment in this case, for contractor-owned equipment, lacking actual cost records for the delay period, is the acquisition cost of each piece of equipment involved applied to the formula set forth in the A.G.C. ownership expense manual and reduced by 50% for idle time, during which time the equipment suffered no wear and tear.[80]

---

[78] The cases recognize that in some instances the measure of the loss for contractor owned equipment is the rental rate, if the contractor can show that (1) if the equipment had not been tied up by the owner, the contractor could and would have rented it to others or (2) the contractor had other work on which the idle equipment would have been used if not for the owner's delay. It is not enough to show only that the equipment had rental value. L.L. Hall Constr. Co. v. United States, 379 F.2d 559, 177 Ct. Cl. 870 (1966).

[79] 379 F.2d at 567.

[80] *Id.* at 568.

For the contractor to avoid the AGC rates and apply instead the AED rates the contractor must show that the equipment was actually rented.[81] In *Degenarrs Co. v. United States,* the claimant "rented" equipment from a related entity. The court held that the claimant was not entitled to apply AED rated because it "failed to prove a bona fide, arm's length rental occurred."[82]

§ 5.13 Home Office Overhead

It is now well established that an allocable portion of a delayed contractor's home office overhead expenses is a compensable element of damage.[83] When a construction project is delayed, the contractor does not realize direct billings in the time period during which they were anticipated. As a result, home office overhead is said to be unabsorbed. Unabsorbed overhead is to be contrasted with extended overhead costs incurred when additional work is performed.[84] The generally accepted method for calculating unabsorbed home office overhead is the *Eichleay* formula:[85]

$$\frac{\text{Contract Billings}}{\text{Total Billings for Contract Period}} \times \text{Total Overhead for Contract Period} = \text{Overhead Allocable to the Contract}$$

$$\frac{\text{Allocable Overhead}}{\text{Days of Performance}} = \text{Daily Contract Overhead}$$

$$\text{Daily Contract Overhead} \times \text{Days of Delay} = \text{Amount Claimed}$$

Although the *Eichleay* approach was subject to intense scrutiny and met with some disfavor in the late 1970s, it is clear today that its acceptance in the federal and state courts has been reaffirmed.[86] It is also now generally

---

[81] Degenarrs Co. v. United States, 2 Cl. Ct. 482, 31 CCF ¶ 71186 (1983) (citing Cornell Wrecking v. United States, 184 Ct. Cl. 289, 291 (1968).

[82] 2 Ct. Cl. at 492.

[83] *See* 2 McBride & Wachtel, Government Contracts § 23, at 100[8]; Annotation, *Overhead expense as recoverable element of damages,* 3 A.L.R.3d 689, 706–07 (1965).

[84] *See* W. Sneed, *Proving Overhead Claims,* in Construction Contracts and Litigation 439–89 (1987).

[85] Eichleay Corp., 60-2 B.C.A. (CCH) ¶ 2,688 (1960).

[86] Capital Elec. Co. v. United States, 729 F.2d 743 (Fed. Cir. 1984); George Hyman Constr. Co. v. Washington Metro. Area Transit Auth., 816 F.2d 753 (D.C. Cir. 1987); Nebraska Pub. Power Dist. v. Austin Power, Inc., 773 F.2d 960 (8th Cir. 1985); Excavation Constr. Inc. v. WMATA, 624 F. Supp. 582 (D.D.C. 1984); Golf Landscaping Inc. v. Century Constr. Co., 39 Wash. App. 895, 696 P.2d 590 (1985).

§ 5.13 HOME OFFICE OVERHEAD

recognized that the use of the *Eichleay* formula does not automatically flow from the event of delay.[87]

> In order to recover, the contractor must show that he necessarily suffered actual damage because the nature of the delay made it impractical for him either "to undertake the performance of other work," . . . or "to [cut back on] Home Office personnel or facilities," *Eichleay Corp.*, 61-1 BCA at 15,117. A contractor generally meets this requirement by demonstrating that the delay was sudden and of unpredictable duration. *See Capital Electric Co. v. United States*, 729 F.2d 743, 745–46 & 746 nn. 4–5 (Fed. Cir. 1984).[88]

In addition, one court rejected application of *Eichleay* in a case in which even though completion of the project was significantly delayed, most of the project's billing occurred during the originally scheduled time of performance.[89]

When the contractor has shown that it has suffered a loss resulting from unabsorbed home office overhead, the application of the formula is relatively straightforward. The determination of contract billings and total billings are not problematic, nor should the determination of the total days of performance or the days of delay be difficult at this stage. The critical issue and the source of frequent challenges to the use of the formula is the makeup of the total overhead figure. The issue arises in the form of questions about the allowability of elements of overhead. And even if a given element is fundamentally allowable, it is sometimes argued that only fixed overhead can be used in the *Eichleay* calculation, after eliminating "variable" overhead. These issues will be dealt with in turn.

In federal contracts, the Federal Acquisition Regulations (FAR) specifically provide that certain elements of overhead are not allowable costs, and these should be eliminated from the overhead pool.[90] The Veterans Administration Board of Contract Appeals addressed these issues in *Salt City Contractors, Ltd.*[91] There the delayed contractor claimed the following items as elements of home office overhead: plans and blueprints, tools

---

[87] George Hyman Const. Co. v. Washington Metro. Area Transit Auth., 816 F.2d 753 (D.C. Cir. 1987).

[88] *Id.* at 757.

[89] Berley Indus., Inc. v. City of New York, 45 N.Y.2d 683, 385 N.E.2d 281, 412 N.Y.S.2d 589 (1978).

[90] The following kinds of costs are unallowable on federal contracts and should be deducted from the overhead pool: bad debts (F.A.R. § 31.205-3); contributions and donations (F.A.R. § 31.205-8); entertainment costs (F.A.R. § 31.205-14); fines and penalties (F.A.R. § 31.205-15); interest and finance costs (F.A.R. § 31.205-20); lobbying costs (F.A.R. § 31.205-22); losses on other contracts (F.A.R. § 31.205-23); organization costs (F.A.R. § 31.205-27); costs in defense of fraud proceedings (F.A.R. § 31.205-47).

[91] VABCA No. 1362, 80-2 B.C.A. (CCH) ¶ 14,713 (1980).

and supplies, workers' travel, insurance, bid bonds, depreciation expense, auto expense, advertising and promotion, cleaning and maintenance, dues and subscriptions, light and heat, penalties, interest, legal and accounting, office expense, office wages, officers salaries, rent, telephone, travel and entertainment, FICA expense, state unemployment, federal unemployment, state disability, and sales tax expense. The government took the position that only fixed overheads were properly considered and sought to have *Eichleay* applied only to the following elements of overhead: depreciation, utilities, officers salaries, and rent. The board only disallowed the inclusion of workers' travel, travel and entertainment, advertising and promotion, penalties, interest, and donations, noting:

> The other items listed as home office overhead expenses are generally allowable when reasonable. The burden of establishing that these are not reasonable expenses is on the Government (*Bruce Construction Corp. et al. v. U.S.* [9 CCF ¶ 72,235], 163 Ct. Cl. 97 (1963)). No evidence has been presented by the Government, nor does the record otherwise contain evidence, which would persuade us that these costs were not reasonable expenses incurred in the normal course of overall administration of [the contractor's] business.[92]

With respect to the issue of fixed versus variable overheads, the board had this to say:

> The Eichleay formula, in determining an average daily rate of home office expense, uses the total home office expense incurred during the period of performance. This necessarily includes some costs which may vary during the period. Even those costs which the Government defines as "fixed" costs may vary. For example, the rent for office space may increase or decrease, and utility bills certainly vary, but these are without question, allocable overhead items. It is generally accepted that the Eichleay formula is used primarily for construction contracts, where there is an assumption that almost all overhead is fixed, rather than variable [citing R. Nash, Jr., *Government Contract Changes* (1st ed. 1975) at 394.], but this is not to say that overhead costs which do not remain constant are to be excluded solely on that basis.[93]

§ 5.14 Prejudgment Interest and Financing Costs

The various courts which have dealt with the issue of the allowability of recovery of prejudgment interest have applied at least three different

---

[92] *Id.* at ¶ 72,561.
[93] *Id.* at ¶ 72,559.

theories.[94] One approach is that prejudgment interest is awarded only if damages are liquidated.[95] A second approach is that prejudgment interest may be awarded in the discretion of the court by applying equitable principles.[96] A third approach is to treat prejudgment interest as an element of damage representing the loss of the use of earning power resulting from an owner's breach of contract.[97]

§ 5.15 Loss of Profits

Generally, if the contractor's claim is held to arise under the contract and falls within the category of a change order, profit is a compensable element of damage.[98] If, on the other hand, the contractor's claim arises from an owner's breach and is therefore outside the provisions of the contract, courts have refused to grant a recovery for profits.[99]

FROM THE ACCOUNTANT'S PERSPECTIVE

§ 5.16 Cost Escalation

Sections 5.16 through **5.20** present an overview of some specific pricing issues facing contractors and owners in the development or critical analysis

---

[94] For a thorough analysis, *see* Annotation, *Award of Pre-judgment Interest to Contractor,* 60 A.L.R.3d 487 (1974 and Supp. 1987).

[95] Bartlow-Hope Elec. Corp. v. Herzog, 692 S.W.2d 404 (Mo. Ct. App. 1985).

[96] Blake Constr. Co. v. C.J. Coakley Co., 431 A.2d 569 (D.C. 1981); E.C. Ernst Co. v. Koppers Co., 626 F.2d 324 (3d Cir. 1980), *on remand, aff'd in part and rev'd in part,* 520 F. Supp. 830 (W.D. Pa. 1981); Fattore Co. v. Metropolitan Sewerage Co., 505 F.2d 1 (7th Cir. 1974); General Ins. Co. of Am. v. Hercules Constr. Co., 385 F.2d 13 (8th Cir. 1967).

[97] *See* J.M. Hollis Constr. Co., 641 S.W.2d 354, 357 (Tex. Ct. App. 1982); Shook & Fletcher Insulation Co. v. Central Rigging & Contracting Corp., 684 F.2d 1383, 1386–89 (11th Cir. 1982) (Alabama law); Nebraska Pub. Power Dist. v. Austin Power, Inc., 773 F.2d 960 (8th Cir. 1985) (Nebraska law).

[98] Peter Kiewit Sons' Co. v. Summit Constr. Co., 422 F.2d 242, 265 (8th Cir. 1969); Bennett v. United States, 371 F.2d 859, 864, 178 Ct. Cl. 61 (1967); State v. Northwestern Constr., Inc., 741 P.2d 235 at 240 (Alaska 1987); Paccon, Inc., 65-2 B.C.A. (CCH) ¶ 5,227, at 24,544 (1965).

[99] Lass v. Montana State Highway Comm'n, 157 Mont. 121, 483 P.2d 699 (1971); J.D. Hedin Constr. Co. v. United States, 347 F.2d 235, 259, 171 Ct. Cl. 70 (1965); Laburnum Constr. Corp. v. United States, 325 F.2d 451, 163 Ct. Cl. 339 (1963); Oliver-Finnie Co. v. United States, 279 F.2d 498, 150 Ct. Cl. 189 (1960).

of delay claim cost aspects. Accounting experts are increasingly involved in the cost aspects of these claims, and these sections of this chapter are devoted to analysis of major delay cost claims issues. The authors have attempted to highlight key pricing issues which in practice occur with regularity in contractors' claims.

In addition to providing some perspectives on pricing issues, these sections will also discuss some "tried and true" but not necessarily fool-proof ideas on improving the litigator's analysis of delay claims. Their aim, of course, is to ensure that delay damages are presented in a reasonable, well documented, and, above all, solidly credible fashion to the ultimate decider of the facts.

One typical delay claim presented in construction disputes is escalation costs. The standard escalation argument usually presented is that the actions of owner X forced contractor Y to perform his or her contractual obligations in a time period later than expected. Additionally, the prices of labor, materials, subcontractors, and other necessary cost elements were higher in this later period than in the original period. Therefore, contractor Y is entitled to the escalated costs.

Using a very simplistic numerical example, if an electrical contractor was supposed to complete its work by June 30, but instead had to work three weeks into July because of a delay imposed by the owner, and the International Brotherhood of Electrical Workers' wage rate jumped 10 cents per hour on July 1, then the contractor should be entitled to recover an additional 10 cents per hour for each hour worked in the three week period in July.

Such a simplistic example hardly ever occurs in real-life construction disputes. Because of the tremendous number of activities that occur in a typical construction project, most escalation claims do not occur in a vacuum. The causation for the delay claims, therefore, must be analyzed, because a key element in pricing such a claim should be: but for the action (delay) of the owner, would electrical contractor Y have been able to complete its assigned tasks by June 30? The answer to this question is typically determined by a time impact analysis performed by a retained CPM scheduling expert. However, specific pricing issues also tend to be raised in such situations. These issues include:

1. Has the impact of additional fringe benefits also been calculated; for example, markups over and above the new wage rates?
2. For materials costs, when were all the materials actually purchased? A review of actual purchase dates might indicate that all of the actual material necessary for the project was purchased in the period prior to the escalation impact.
3. For equipment costs, to what extent, if any, have equipment costs been considered in the escalation calculations, and has only the

incremental escalation impact been measured in this claim, and the delay effect elsewhere?

Thus, even relatively simple claims areas such as escalation can throw curve balls.

§ 5.17 General Conditions Costs

Another typical major cost category typically claimed when delays occur is extended general conditions costs, or jobsite support costs. These costs may be charged directly to the job yet are not identifiable to specific tasks or activities. In addition, they may be incurred throughout the entire project period. Included in these costs may be the salaries for the project's manager(s), superintendent(s), engineer(s), and accountant/clerk(s). These personnel work at the site and are typically salaried individuals. In addition, there are other nonpersonnel-related costs unique to the site, including items like the field trailer, site vehicles, jobsite radios, temporary toilets, and others.

General conditions costs are typically viewed as being primarily time-related. In other words, their incurrence and growth are directly impacted by the project's duration. Claims typically include all of these costs for the entire delay period. However, a review of these costs, on a detailed basis, can help determine whether they are totally time-related and thus impacted by the delay.

One of the most effective methods used in determining the time versus activity relationship of general conditions cost is the analysis of the actual general conditions costs incurred in given time intervals, usually on a monthly basis. This requires going into the job or project cost report for each account each month. If, indeed, the costs were fully time-related, one would expect to see approximately the same amount being incurred each month. If so, it can generally be determined that the costs were a function of the project schedule and not the amount of activity occurring at the site. For example, a rented jobsite trailer is kept at the site for the project's entire duration. The impact of a month's delay forces the contractor to incur one extra month of rent for that trailer. As such, the cost is probably time-related and the cost is likely recoverable, assuming delay causation is established (see **Figure 5–1**).

Conversely, there are other costs that by name one would think were time-related. However, a more detailed review of these costs, as actually incurred, can reveal otherwise. For example, the salaries of most project superintendents are typically thought of as a fixed type of cost. However, a monthly review of such costs might show that, in fact, the costs were incurred on a step-like function. See **Figure 5–2**. In this example, from the

130 DELAY CLAIMS

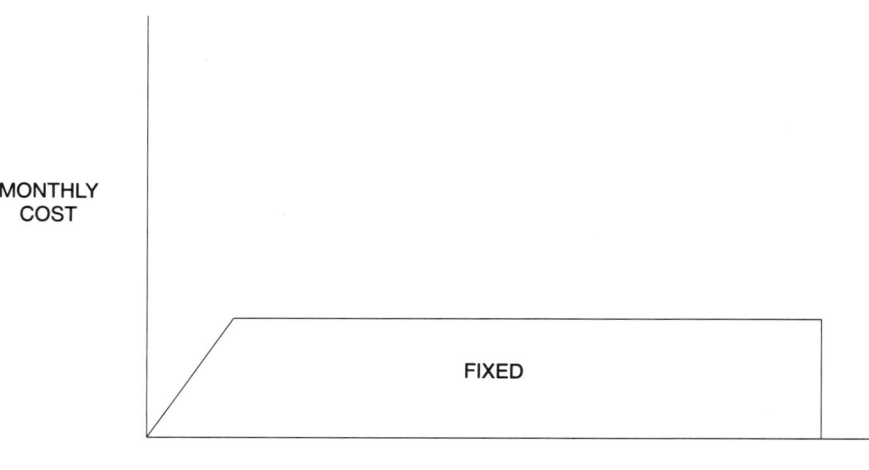

Figure 5–1. General conditions—trailer rental.

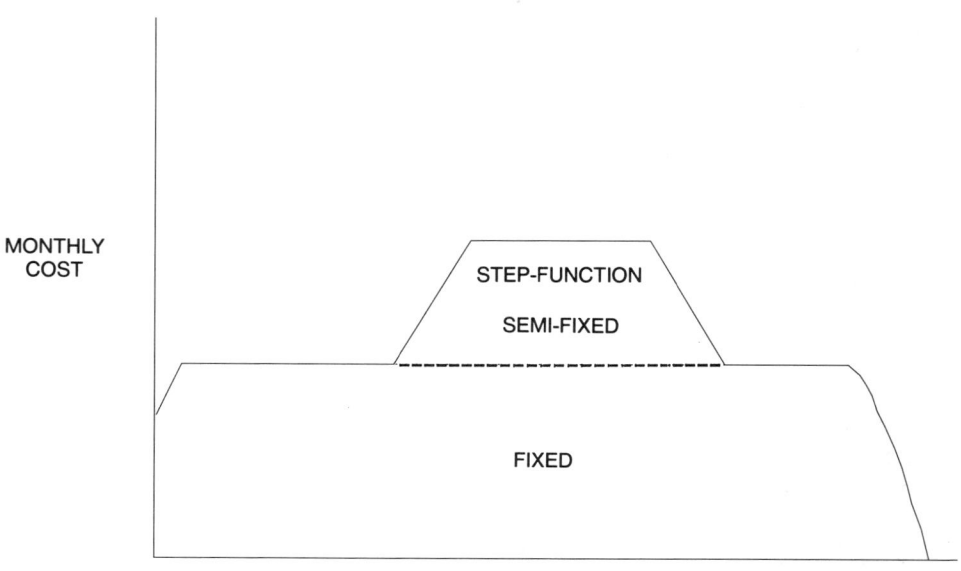

Figure 5–2. General conditions—superintendent.

cost records it appears that the contractor assigned additional supervision to the project. The claimability of this cost may be questioned for two reasons. First, supervisory costs do not appear to be totally time-related. As project activity increased, the contractor added additional superintendents. Second, if indeed there was a delay, there arguably should not have been an increase in superintendent costs during the delay period. In fact, the contractor would have had a responsibility to downscale its operations during a delay (that is, there is a duty to mitigate). These points are accentuated if there are several delays during the project. In reality, general conditions costs can be semivariable. A portion of the costs will indeed be fixed. However, based on the level of activity at a site, a certain portion of the costs may tend to increase as activity increases.

Another form of the general conditions claim is for the contractor to claim extended months of delay. This claim basically says that the contractor was forced to stay on the jobsite for an extra period of time at the end of the job. Therefore, the delay costs were incurred at the end of the job. Often, the calculation of the delay costs is then performed by subtracting budgeted costs from actual costs (that is, a total cost claim approach). This calculation can prove to be deceiving and can be an inaccurate reflection of the true cost impact to the contractor. A more accurate method might be to review the costs actually incurred during the alleged delay period or month(s). This analysis will more accurately reflect how the delay impacted the contractor and eliminate unjustified reimbursement for items such as underestimated bids and poor project controls (see **Table 5-1**).

It is important to remember that general conditions costs are overhead. Therefore, they should be reviewed for some of the same considerations as extended home office overhead costs (for example, one-time costs and allowable costs.) The intent of a general conditions claim is to reimburse the contractor for those costs that were added at the site because of the delay. It is not to reimburse the contractor for costs that would have been incurred whether the delay occurred or not.

Table 5-1

General Conditions 1 Month Delay

Cost	Budget	Actual	Claimed Overrun	Recorded Last Month	Non-Claimable
Project management	$15,000	$20,000	$ 5,000	$1,250	$3,750
Site supervision	8,000	12,000	4,000	667	3,333
Engineering	10,000	12,000	2,000	833	1,167
Trailers	3,000	3,250	250	250	0
	$36,000	$47,250	$11,250	$3,000	$8,250

§ 5.18 Equipment Costs

One of the most frequently disputed cost claim issues seen in today's construction environment is equipment cost claims. These claims can represent significant portions of contractor claims and, therefore, should and must receive a great deal of attention. Equipment inefficiency claims are sometimes presented, as well, but typically take the form of labor inefficiency claims, which are not discussed in this section. The following discussion of equipment delay claims makes one important assumption: the contractor and owner have no agreement that any equipment cost claims or charges will be priced at predetermined rates. This discussion assumes the contract was silent on this issue.

Equipment delay claims are most often presented as a contractor seeking additional reimbursement for equipment cost (for example, bulldozers) which had to be put on a jobsite for three weeks longer than it had planned because of owner-imposed or caused project delays. This type of claim is typically presented as:

$$5 \text{ bulldozers @ } \$125/\text{hour} \times 8 \text{ hours use/day} \\ \times 3 \text{ weeks delay} = \$7,500$$

This typical type of presentation is subject to scrutiny on several fronts:

1. Were all five bulldozers onsite during the delay, and were they idle throughout the entire delay period?
2. Is the appropriate hourly rate really $125/hour?
3. Did the contractor plan to use five bulldozers and are they shown in the bid?
4. Could the contractor have used the bulldozers at another site and mitigated the cost impact?
5. Would the dozers have been productive for eight hours a day and was the owner's delay really three weeks?

Thus, one can see that there are several different avenues of attack that an owner's attorney and/or claims consultant can take on such equipment cost/claim presentations. However, the well-prepared equipment cost claim should not be vulnerable to attacks on any of these fronts.

In struggling with the equipment cost issue, one of the first things that should be done is to document what portion, if any, of the claimed equipment was rented from third-party vendors. To the extent that contractors can demonstrate that (a) costs were expended to arms-length third parties to rent pieces of equipment for the jobsite and the project in dispute, and (b) the rented pieces of equipment were, in fact, idle on that project jobsite

§ 5.18 EQUIPMENT COSTS

and not used elsewhere during the delay period, then little, if any, difficulty should be encountered in claiming such costs. The contractor should analyze its own invoices and equipment use job tickets (especially those indicating "to-from" moves) as aids in documenting this important claim component.

If, on the other hand, some or all of the equipment expected to become part of a claim is owned by the contractor (or by a related company or subsidiary) and is being charged to the project at a certain rate per hour or day, then a different pricing approach may be necessary. In this situation, there is a variety of pricing alternatives that a contractor could use:

1. The internally determined charge rate.
2. A widely accepted published daily rate from a standard rate book (for example, published by the AGC, California Department of Transportation, Army Corps of Engineers, and others).
3. The true ownership cost rate; that is, the actual out-of-pocket cost of the equipment to the contractor (be it from repairs/maintenance, fuels and lubricants, and so on).

The authors have seen all of these alternatives used and, although the third typically results in the most conservative (that is, lowest) daily or hourly rate, it is also one of the easiest to defend in terms of trial credibility standards.

The widely used and accepted daily rate sources typically include costs which may not be judged as truly incremental in claims situations. For example, these published rates may contain a factor for regular preventative repairs and maintenance which will have to be performed regardless of the alleged claim impact and may not occur during the relevant claim period. Also, these rates may include depreciation charges which are arguably not a true out-of-pocket cost and, therefore, could be subject for pointed and critical expert cross-examination.

Therefore in analyzing equipment cost claims for owned equipment, the rate used in the pricing methodology should be carefully considered, and the possible impacts of other rates should be explored.

Interval rental rates might also be challenged by comparing the internal rates and actual outside rental costs available for similar equipment (see **Table 5-2**).

Additionally, analysis should be made of whether or not the equipment was originally expected to be used for eight hours per day during the alleged delay period. If, in fact, it can be shown that if not delayed, the equipment would only have been used and charged to a project for a certain number of days and hours, and the equipment was left on a delayed project site because it could not be used elsewhere, then problems in supporting such a cost would likely occur.

Table 5-2
Comparison of Equipment Rental Rates

Equipment Rate (Monthly Rate)	1988 Means Building & Construction Data	Rental Co. X	Rental Co. Y	Rental Co. Z
Backhoe-45-60 H.P.	$1,750	$2,100	$1,925	$1,900
Front end loader (1–1.5 Cy capacity)	$2,150	$3,400	$3,500	$4,400
Roller (8–12 ton)	$1,680	$2,600	$2,600	$4,500

Source: Rates are based on 1988 quotes received from rental companies in the same region and do not include delivery or operation charges.

Yet another consideration for equipment costs is to determine whether the cost behavior of the type of equipment contained in a claim is activity-related as opposed to time-related. Such an analysis is performed by comparing the actual or projected charging of equipment costs (by type) against time and direct labor costs. If a higher correlation of equipment costs to time than to direct labor exists, then the contractor should be reasonably satisfied with approaching its claim using a delay claim or time-related methodology. See **Figures 5-3** and **5-4**. However, if its equipment costs prove to be more activity related, then they may be more appropriately claimed as an add-on percentage to direct labor inefficiencies claimed (if any) than in a delay claim methodology.

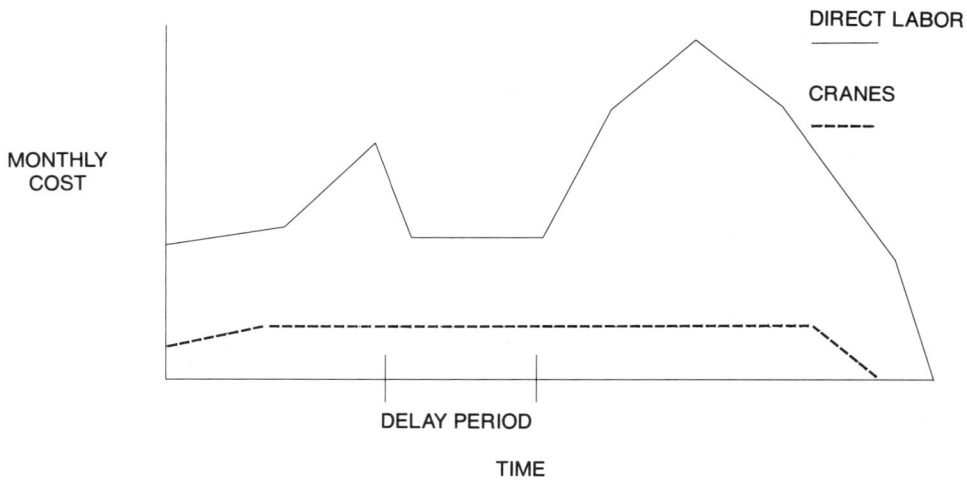

Figure 5-3. Comparison of direct labor costs to crane costs (time related).

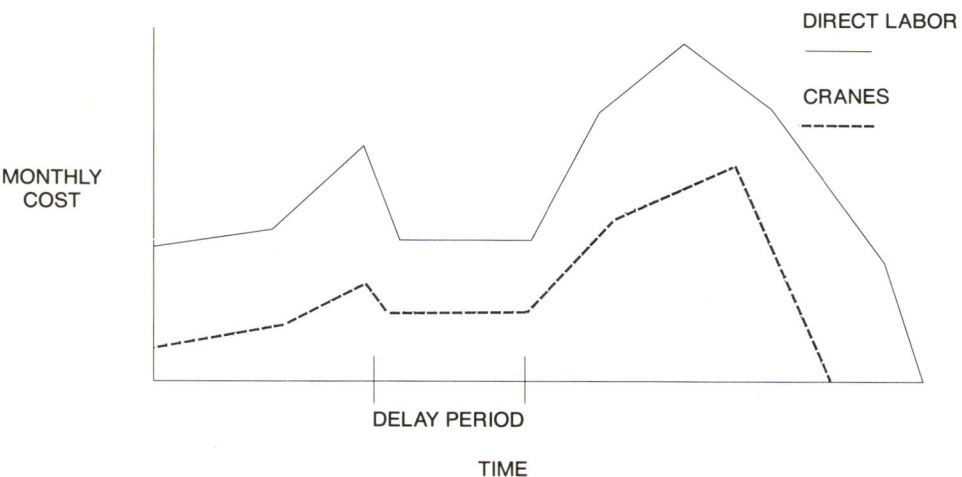

Figure 5–4. Comparison of direct labor costs to crane costs (activity related).

§ 5.19 Unabsorbed Home Office Overhead

Another major claim component frequently claimed is unabsorbed home office overhead. Although the ability to pursue such a claim at all is sometimes hotly contested in construction litigation, the focus of this section is on some analytical challenges in presenting such overhead claims as opposed to discussing their appropriateness. However, it should be noted that hope springs eternal, and the authors recognize that certain construction litigators, especially those representing owners, still are searching for the case that slays the *Eichleay* monster once and for all.

In presenting unabsorbed overhead costs, the pricing methodology one uses must be, of course, one of the first things decided. Often, the *Eichleay* method is used to calculate the contractor's overhead related delay damages. However, this method may be chosen by some contractors as an automatic response rather than as a result of careful deliberation and analysis. The *Eichleay* formula appears relatively simple and straightforward and is believed to be easy to calculate. See **Table 5–3**.

Before blithely using the *Eichleay* formula, the contractor should analyze its delay claim situation and determine certain facts which can alternately support or not support the choice of *Eichleay* as an appropriate methodology.

The contractor should consider issues such as:

1. Whether or not a true suspension in work occurred
2. What supporting documentation can be provided to show that overhead costs increased because of the delay at issue

Table 5-3
Eichleay Example

Contract amount	= $500,000	
Period of delay	= 30 days	
Contract period (including delay)	= 395 days	
Revenues during contract period	= $5,000,000	
Overhead during contract period	= $1,000,000	
1. $\dfrac{\text{Contract amount}}{\text{Revenue during contract period}}$	$= \dfrac{\$500,000}{\$5,000,000}$	= 10%
2. 10% × overhead during contract period = 10% × $1,000,000 = $100,000		
3. Overhead per day	$= \dfrac{\$100,000}{395 \text{ days}}$	= $253/day
4. 30 days of delay × $253/day	= $7,590	= Unabsorbed overhead

3. What supporting documentation can be provided to demonstrate that the contractor attempted to, but was unable to, replace the delayed project with other work.

Only after it knows the results of its exploration of these types of issues should the contractor determine that *Eichleay* is the best methodology to use, as opposed to an alternative method such as percentage markup, Carteret, Allegheny, or others. The *percentage markup method* works similarly to standard change order pricing. A contractually binding overhead rate (for example, 10 percent) is applied to the direct costs incurred during the delay. The Allegheny and Carteret methods are both based on cases tried in the 1950s. Simply put, they calculate the variance between the actual overhead rate incurred during the delay period and the "projected" or "normal" overhead rate. See **Table 5-4**. Although they have been around for many years, they have taken a back seat to the more frequently used *Eichleay* method and are not commonly seen in claims. However, that is not to say that such methods are not valuable, given specific sets of facts and circumstances.

In computing its overhead delay claim, the contractor should also be cognizant of the in-depth critical analyses that can be made of the *Eichleay* calculation. Often in claims situations, the *Eichleay* calculation is performed incorrectly or not performed completely correctly and yet its calculation and methodological problems are not probed because much

§ 5.19 UNABSORBED OVERHEAD

Table 5–4

Alternative Extended Home Office Overhead Calculations

Allegheny[a]

(1) Overhead rate during delay[c] − Projected Overhead rate = Excess overhead rate

(2) Excess overhead rate × Contract base for delay = Claimable overhead

Carteret[b]

(1) Overhead rate during delay[c] − Normal overhead rate = Excess rate

(2) Excess rate × Contract base for delay = Claimable overhead

[a] ASBCA 4163, 58-1 BCA Section 1684-1958
[b] ASBCA 1647, 6 ccf Section 61,651-1951
[c] For example, overhead rate is based on direct labor dollars.

cross-examination fury is vented on the issue of whether or not the formula is applicable and little or no attention is paid to how the calculation was actually performed.

In the experience of the authors, careful analysis of how the calculation was performed can make for withering cross-examination. Challenges such as the following can and should be made to contractors using the *Eichleay* calculation:

1. Are the cost and billing periods used comparable?
2. Has the entire period of contract performance been used?
3. Has the overhead that is clearly inapplicable been removed from the overhead pool (for example, the cost of a jet fleet for an aviation subsidiary of a construction company)?
4. Has the overhead pool been analyzed to remove items such as:
   a. One-time costs?
   b. Nontime-related costs?
   c. Unallowable items?
5. Has any overhead calculated in other claim areas been properly deducted from the *Eichleay* calculation to avoid double counting?

If such issues have not been considered, they should be and, if they have, then the contractor should feel much more comfortable as to the solid ground on which its overhead claim sits.

§ 5.20 Interest/Cost of Capital

The topic of interest or cost of capital claims is another issue that time and again develops as a theory in claims litigation. The applicability of such claims can differ based on where a case is tried. For example, some state courts allow for prejudgment interest, some do not, and the federal boards recognize interest claims in terms of Cost Accounting Standards Board Standard 414 (Facilities Cost of Capital) and costs "on" claims versus "in" claims. Additionally, the rates of interest that a contractor can recover can differ based on:

1. What it says in the contract
2. If the case is in state court and that state has a statutory rate of interest
3. If the case is a government contracts matter, and the Renegotiation Rate applies.

Yet, even outside of these issues, some other major methodological issues typically arise in developing an interest claim.

Typically, when less sophisticated interest claims are prepared, such claims will start interest charges running at the point in time from which the contractor states that additional delay, disruption, and/or other costs were incurred. However, this may not truly reflect the actual interest or financing costs incurred by the contractor. The authors believe that interest is a measure of a true economic cost that the contractor is forced to bear when it finances unexpected costs on a construction project which the owner or other responsible parties should justifiably reimburse to the contractor. However, the issue of expectation becomes the important point. A more sophisticated interest calculation can be constructed by analyzing the contractor's expected cash flow from a project versus its actual cash flow. The true burden of unexpectedly having to finance a project occurs for a contractor when (a) its expected cash flow from a project is less than its actual cash flow and (b) it must finance the "gap" in cash flow itself. For example, if a contractor did have some justifiably claimable unreimbursed cost overruns early in a project, when its actual cash flow from the project was extremely favorable (from the front-end loading phenomenon, general conditions payments, mobilization costs, and so on), then the contractor may not have truly had to finance these costs from the earliest point of time in the project (see **Figure 5–5**).

Presentation of an analysis of the amounts that the contractor actually had to finance versus what it reasonably expected to finance can make powerful damages testimony. It is also typically difficult to refute unless it is proven that the financing was obtained for another purpose.

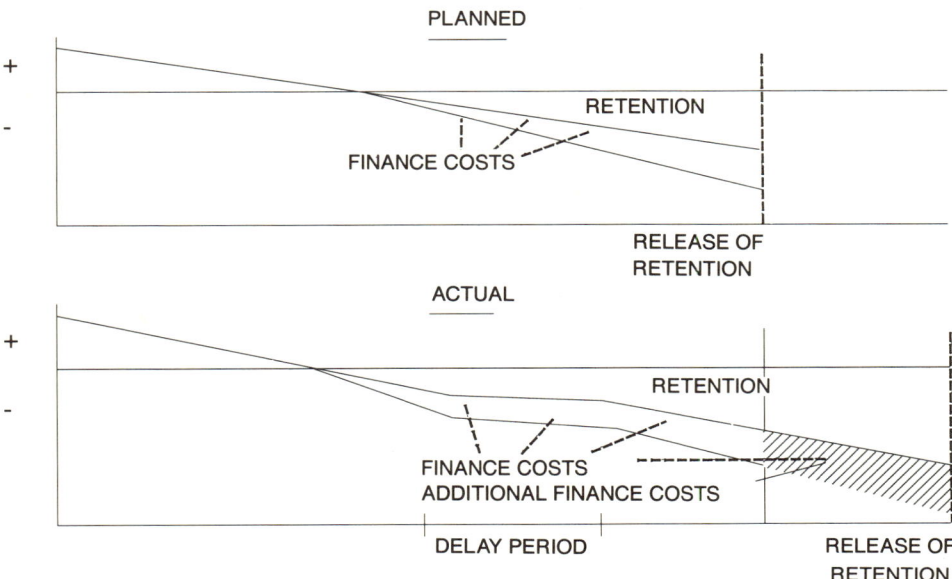

Figure 5–5. Cumulative cash flows.

§ 5.21 Summary

It is hoped that this presentation of analytical tools litigators frequently use to probe disputed cost issues in delay claims has been valuable to the reader. Because the cost issues discussed herein typically represent the larger dollar portions of claims and are less directly linked to the alleged acts or omissions, careful review of the merits of such claims can be quite beneficial to the litigator.

§ 5.22 Bibliography

F. Miller & P. Pocalyko, *Establishing Damages* in Construction Litigation: Strategies and Techniques (John Wiley & Sons 1989).

F. Miller, *Role of the Accounting Firm* in Construction Defaults: Rights, Duties and Liabilities (John Wiley & Sons 1989).

E.H. Riper, *Pricing Construction Claims* (American Construction Owner's Association Publication 1986).

R. Dunn, *Recovery of Damages for Lost Profits* (3d ed., Law Press 1987).

Appeal of Eichleay Corporation, ASBCA No. 5183, 60-2 B.C.A. (CCH) ¶ 2688 (1960).

Appeal of Allegheny Sportswear Co., Division of N.Y. Pants Co., 58-1 B.C.A. (CCH) ¶ 1684, at 6361–66 (ASBCA 1958).

Carteret Work Uniforms; Cont. Cas. Fed. (CCH) ¶ 61,561 (ASBCA 1954).

Building and Construction Cost Data 1988 (R.S. Means Co. 1988).

CHAPTER 6

DISRUPTION CLAIMS

Donald G. Gavin, Esquire
Frederic R. Miller, CPA
Daniel E. Toomey, Esquire
Robert J. Smith, Esquire

Donald G. Gavin is a partner in the law firm of Wickwire Gavin, P.C., in Washington, D.C., Vienna, Virginia, and Madison, Wisconsin. Mr. Gavin's practice primarily emphasizes public contract law, federal assistance law, and construction and surety matters, and he lectures frequently throughout the country on subjects dealing with grants, construction, and government contracts. He has authored and coauthored numerous books, articles, and other publications and has served on advisory panels to several federal grantor agencies concerned with construction under federal assistance. Mr. Gavin has been elected a fellow of the American Bar Association and is the chairman of the Public Contract Law Section of the American Bar Association. He has an economics degree from the Wharton School, a Juris Doctorate from the University of Pennsylvania School of Law, and a Masters of Law degree in Government Procurement Law from George Washington University.

Frederic R. Miller is the director-in-charge of the Business Investigation Services Group of the Washington, D.C., office of Coopers & Lybrand. He has a wide spectrum of experience in construction, having worked for owners, contractors, sureties, and architect/engineers, and has worked on a wide variety of construction projects including commercial and public buildings, water and waste plants, nuclear power plants, hazardous waste sites, and others. Mr. Miller has served as an expert witness before courts, arbitration panels, and boards of contract appeals, and has taught a number of professional education courses in litigation consulting and construction claims analysis. He is a member of the D.C. and American Institutes of Certified Public Accountants and the Association of Insolvency Accountants. Mr. Miller obtained his undergraduate degree from Rutgers University and his M.B.A. degree from Cornell University.

Daniel E. Toomey is a partner in the law firm of Wickwire Gavin, P.C., in Washington, D.C., Vienna, Virginia, and Madison, Wisconsin. He previously served as a law clerk to the chief judge of the District of Columbia Court of Appeals and as an assistant United States attorney for the District of Columbia. Mr. Toomey has been engaged in a wide-ranging civil and criminal trial practice with numerous trials and appeals in federal and state courts relating to construction contract litigation, surety law, suspension and debarment issues, and Environmental Protection Agency enforcement proceedings. He is presently an adjunct professor of construction contract law at Georgetown University, where he earned a Juris Doctorate in 1967. Mr. Toomey is vice chairman of the Architect/Engineer and Professional Services Committee of the American Bar Association's Public Contract Law Section. Along with Messrs. Gavin and Smith, he coauthored a chapter in *Construction Failures* (Wiley 1989).

Robert J. Smith is a partner in the law firm of Wickwire Gavin, P.C., in Washington, D.C., Vienna, Virginia, and Madison, Wisconsin. His practice emphasizes construction contract law, engineering issues, federal assistance law, and surety matters. He is a frequent lecturer throughout the country on subjects dealing with construction contract administration, contract specifications, construction grants, and architect/engineer professional responsibility. Prior to entering into private practice, Mr. Smith was an associate professor of engineering at the University of Wisconsin and served as chairman of the Wisconsin Transportation Commission. A member of the American Society of Civil Engineers, Mr. Smith is past chairman of its National Committee on Construction Contract Administration, and chairman of the Engineers-Joint Contract Documents Committee. He is a fellow of the American College of Construction Lawyers. A Registered Professional Engineer, Mr. Smith earned a civil engineering degree as well as a Juris Doctorate from the University of Wisconsin.

---

§ 6.1 Introduction
§ 6.2 **Delay versus Disruption**
§ 6.3 **Types of Disruption**
§ 6.4 —Noncompensable Disruptions
§ 6.5 —Compensable Disruptions
§ 6.6 **Effects of Disruption**
§ 6.7 **Nature of Disruption Damages**
§ 6.8 **Considerations in Proving Cause and Effect of Disruptions**
§ 6.9 **Considerations in Pricing Disruption Claims**
§ 6.10 —Should Cost Estimates

§ 6.11 —Industry Standards and Handbooks
§ 6.12 —Time and Motion Studies
§ 6.13 —Expert Opinion
§ 6.14 —Total Cost and Modified Total Cost Methods
§ 6.15 —Jury Verdict Method
§ 6.16 Defenses
§ 6.17 Summary

§ 6.1 Introduction

Disruption can be defined as any change in the method of performance or planned work sequence contemplated by the contractor at the time the job was bid which prevents the contractor from actually performing in that manner. In other words, disruption is a material alteration in the performance conditions that were expected at the time of bid from those actually encountered, resulting in increased difficulty and cost of performance.[1]

Disruption encompasses three general principles with respect to contract performance. First, when a contractor bids on a contract, it is entitled to schedule its performance in a series of economical operations, with each stage of performance dependent upon a previous stage. Any disruption to one stage, therefore, may have a potentially disruptive impact upon the subsequent stages.

Second, parties to contract are expected to cooperate with one another and not hinder the performance of each other. A contractor plans to perform its work in a certain manner and sequence, and the owner has an implied duty not to hinder, interfere, or disrupt the contractor's planned performance.

Third, when a contractor plans its contract performance, it must do so reasonably. It may not make unrealistic assumptions about contract performance. For example, a contractor cannot make a valid disruption claim if it has assumed that it would have sole access to the site when the contract documents indicated that other contractors would be simultaneously on-site.

---

[1] Though not the subject of this chapter, disruption claims can be asserted by subcontractors against contractors, contractors against subcontractors, and contractors against design professionals (when permitted by law). In the main, the principles enunciated herein apply equally. *See generally* Toomey & Shean, Design Professional's Liability During Construction, Construction Briefings No. 87-4.

§ 6.2 Delay versus Disruption

It is important to distinguish between a claim for delay to contract performance and a claim for disruption to contract performance. A *claim for delay* necessarily involves the assertion that contract performance was extended and the completion date affected as the result of some event. A contractor need not prove, however, that the completion date was affected in order to assert a *disruption claim.* The impact of disruption of performance is often a loss of efficiency in performance and increased costs, although the time of performance itself may not necessarily be extended as a result of the disruption.

For example, if a contractor is required by its agreement to complete its work within 250 calendar days and it suffers disruption to one of its operations, there may be an increase in the costs of performance, but the disruption does not automatically mean that its work will go beyond the 250 day mark.

To illustrate the nondelay impact of disruption, assume a general contractor was disrupted in its construction work because of a design change imposed by federal regulations, requiring that the piping work in one section of the project be ripped out and replaced. This change would affect the piping contractor by forcing it to incur more welder man-hours and to use more materials. It would also affect the heating/ventilation/air conditioning (HVAC) and electrical contractors because they would have to work together in the affected space while the piping was being ripped out, whereas under the original plans the electrical contractor would have preceded the HVAC contractor. If, in order to maintain compliance with the completion schedule, the trade contractors had to place additional workers on the job and work additional shifts, the costs of performance would increase without necessarily exceeding the performance period set forth in the contract. Moreover, the workers performing in the cramped quarters and, perhaps, working overtime would likely experience a loss of efficiency in their productivity.

This example illustrates the distinction between disruption and delay. The distinction is crucial because the damages recoverable are different for each. Typical damages associated with disruption claims are increased labor costs as the result of an increase in the number of employees placed on the job to perform additional work or to work extra shifts to meet the schedule, and/or increased equipment and material costs. In contrast, delay damages encompass those costs associated with an extended performance period and typically include increased overhead and jobsite costs, standard equipment costs, and financing costs. The other important distinction between disruption and delay claims is the impact of so-called "no damages for delay" clauses, which generally foreclose any recovery for delay damages but normally will not preclude recovery for disruption damages.[2]

§ 6.3 Types of Disruption

A contractor may experience various types of disruption. Among these are disruptions resulting from changes in the location for performance, inadequate specifications, incorrect contract drawings, differing site conditions, severe weather, strikes, unavailability of materials, failure to schedule and coordinate the work, failure to respond to requests for information in a timely manner, and so on. These disruptions may be further categorized as noncompensable and compensable.

§ 6.4 —Noncompensable Disruptions

The mere occurrence of a disruption to a contractor's planned performance does not automatically entitle the contractor to compensation. An obvious example of a noncompensable disruption is when the event should have been anticipated by the contractor, as, for example, when the contract documents indicate the presence of certain adverse soil conditions or that the contractor will have to work in conjunction with another contractor. Such events will not support a claim for disruption because the contractor should have anticipated them in light of contract indications. In other words, these types of disruptions are noncompensable because the contractor was not reasonable in planning its performance. Also included within this type of noncompensable disruption are those caused by the contractor's own actions, such as improper scheduling or inefficient material expediting. The contractor cannot complain when its disrupted performance is based on unreasonable assumptions or its own poor planning or performance.

Another type of noncompensable disruption is that which has been specifically excluded by the contract. Such contract provisions are a method of risk allocation by which each party attempts to shift to the other the risk of disruptions that are beyond the control of either party. The risk of such events is typically shared between owner and contractor, by means of contractual language that grants the contractor a time

---

[2] Buckley & Co. v. State, 140 N.J. Super. 289, 356 A.2d 56, 60–62 (1975). *See also* John L. Gregory & Sons v. A. Guenther & Sons Co., 147 Wis. 2d 298, 432 N.W.2d 584 (1986); Crawford Painting & Drywall Co. v. J.W. Bateson Co., 857 F.2d 981 (5th Cir. 1988).

extension equivalent to the time impact of the disruption but puts the burden of the costs of disruption on the contractor. Typical of this type of noncompensable disruption are those disruptions caused by "acts of God,"[3] unusually severe weather,[4] strikes,[5] and an inability to obtain materials.[6] It should be noted that contractual language as well as general legal principles require that such disruptions, even though noncompensable, be unforeseeable and beyond the contractor's control in order for a time extension to be granted.[7] Moreover, if the contract fails to provide that such events will entitle a contractor to a time extension, one will not usually be granted.[8]

This chapter focuses on compensable disruptions; however, a contractor should keep in mind that, although a disruption may not be monetarily compensable, it may entitle the contractor to a time extension when a disruption has lengthened the time necessary to perform a particular portion of the contract, thus eliminating the prospect of being assessed liquidated damages under the contract.

§ 6.5 —Compensable Disruptions

Contractors proceeding under a written contract generally have two basic means of recovery when asserting a claim for increased costs because of disruptions: (1) recovery under a specific contract provision and (2) recovery under general principles of contract law.

Specific contractual language very often provides a contractor with a remedy when its performance is adversely affected by disruptions. A

---

[3] *See* Nogler Tree Farm, AGBCA No. 81-104-1, 81-2 B.C.A. (CCH) ¶ 15,315 (1981).

[4] *See* Essential Constr. Co., ASBCA No. 18491, 78-2 B.C.A. (CCH) ¶ 13,314 (1978); J & M Lumber, Inc., ASBCA No. 25951, 82-1 B.C.A. (CCH) ¶ 15,500 (1982); R&R Constr. Co., VABCA No. 1101, 74-2 B.C.A. (CCH) ¶ 10,857 (1974); Lambert Constr. Co., DOTBCA No. 77-9, 78-1 B.C.A. (CCH) ¶ 13,221 (1978).

[5] *See* Santa Fe Eng'rs, Inc., PSBCA No. 902, 84-2 B.C.A. (CCH) ¶ 17,377 (1984); Bill's Janitor Serv., ASBCA No. 10345, 65-2 B.C.A. (CCH) ¶ 4916 (1965); Capitol Coal Sales Corp., ASBCA No. 16551, 73-1 B.C.A. (CCH) ¶ 9779 (1973).

[6] This is a very rarely invoked basis for a time extension because only infrequently would a contractor find that its materials literally cannot be obtained from another source. A time extension will not normally be granted when a contractor can obtain the materials but only at a cost substantially in excess of that anticipated. *See* Betsy Ross Flag Co., ASBCA No. 12124, 67-2 B.C.A. (CCH) ¶ 6688 (1967); Free-Flow Packaging Corp., GSBCA No. 3992, 75-1 B.C.A. (CCH) ¶ 11,105 (1975).

[7] *See* S. Head Painting Contractor, ASBCA No. 26249, 82-1 B.C.A. (CCH) ¶ 15,886 (1982); Prestex, Inc. v. United States, 3 Cl. Ct. 373 (1983); Automated Extruding & Packaging, GSBCA No. 4036, 74-2 B.C.A. (CCH) ¶ 10,949 (1974).

[8] Prather v. Latshaw, 188 Ind. 204, 122 N.E. 721 (1919).

typical example of such language is the widely used Changes clause.⁹ Although the Changes clause is typically used when the owner orders additional work, it can be applied to situations in which, although no extra work was required, the disruption has increased the contractor's cost of performance.¹⁰

Other compensable disruptions that can be adequately addressed by the Changes clause or similar contractual language are changes to the drawings or specifications,¹¹ differing site conditions,¹² and actual or constructive suspensions of work.¹³ Although these clauses do provide for compensation, they are more administrative in nature because they typically require the contractor to present its claim to the owner for a decision. Even when such clauses are present in the contract, a contractor may wish or be required to proceed by way of an action at law.

When the owner refuses to compensate the contractor for such disruptions, in accordance with a Changes clause or similar contractual stipulation, the contractor may be able to recover its increased costs in a breach of contract action by asserting that the owner breached either an express or implied obligation of the contract upon which the contractor's

---

⁹ A typical example of a changes clause can be found in the Federal Acquisition Regulations (F.A.R.). F.A.R. § 52.243-4 is the clause normally used in fixed price construction contracts. *See* 48 C.F.R. § 52.342-4.

¹⁰ *See, e.g.,* Pan-Artic Corp., ASBCA No. 20133, 77-1 B.C.A. (CCH) ¶ 12,514 (1977). In this board case, the government awarded a follow-on contract to Pan-Artic. Both the government and Pan-Artic assumed that work on the subsequent contract would be performed during the slack time on the existing contract Pan-Artic had with the government. However, the government delayed in issuing the notice to proceed, which meant that Pan-Artic could no longer use the excess labor from its existing contract. As a result, Pan-Artic's costs of performance increased and it sought compensation for the disruption caused by the government's failure to issue expeditiously the notice to proceed. Despite the fact that no additional work had been ordered, the board awarded Pan-Artic its additional costs of performance, holding that the government's delay in issuing the notice to proceed constituted a compensable disruption.

¹¹ *See* J.B. Williams Co. v. United States, 450 F.2d 1379 (Ct. Cl. 1979); Vi-Mil, Inc., ASBCA No. 25111, 82-2 B.C.A. (CCH) ¶ 15,840 (1982).

¹² A typical differing site condition clause can be found in F.A.R. § 52.236-2, which defines the rights and obligations of a contractor who encounters a differing site condition. For examples of instances in which a contractor was entitled to additional compensation as a result of a differing site condition, *see* J.F. Shea Co. v. United States, 4 Cl. Ct. 46 (1983); North Slope Technical Ltd. v. United States, 14 Cl. Ct. 242 (1988).

¹³ For a typical example of a suspension of work clause, *see* F.A.R. § 12.504. For examples of contractors who have been compensated under this clause, *see* Pathman Constr. Co., ASBCA No. 22003, 82-1 B.C.A. (CCH) ¶ 15,790 (1982); A.C.E.S., Inc., ASBCA No. 21417, 79-1 B.C.A. (CCH) ¶ 13,809 (1979). A contractor may also be compensated when the owner effectively stops contract performance without explicitly ordering that the work be stopped. This is known as a *constructive suspension. See, e.g.,* CRF v. United States, 624 F.2d 1054 (Ct. Cl. 1980).

performance depended. An example of an *express* obligation would be a specific clause that states that it is the owner's duty to make the site available to the contractor at a certain date. If the owner subsequently were to fail to provide site access to the contractor, the contractor would have the basis for a breach of contract action against the owner for that failure to make the site available as required in the contract.

In addition to construing parties' rights under the express provisions of a contract, courts frequently supplement these express rights with *implied* rights and obligations.[14] For example, one frequently implied obligation is the duty to not interfere with the contractual performance of another.[15] "[There is] an implied provision of every contract . . . that neither party to the contract will do anything to prevent performance thereof by the other party or commit any act that will hinder or delay performance."[16] There are numerous cases describing this implied duty and setting forth the acts that constitute interference with another's contract performance such that the contractor is entitled to damages. The following have been held to constitute interference with the contractor's work: an owner's failure to make the work site available when this is within its control[17]; the owner's failure to furnish materials, labor, or information in a timely manner when the owner had agreed to do so[18]; and failure to permit a contractor access to the site.[19]

Another implied duty that may form the basis for a disruption claim is the duty to schedule and coordinate the work.[20] For example, if an owner

---

[14] For a more detailed discussion of the implied obligations that courts frequently read into contracts, *see* Wickwire, Hurlburt, & Shapiro, Rights and Obligations in Scheduling, Construction Briefings No. 88-13.

[15] Williston on Contracts § 1296 (3d ed. 1968). *See also* George A. Fuller Co. v. United States, 69 F. Supp. 409 (Ct. Cl. 1947); Northeast Clackamas County Elec. Coop. v. Continental Casualty Co., 221 F.2d 329 (9th Cir. 1955); City of Dallas v. Hubbell, 325 S.W.2d 880 (Tex. Civ. App. 1959).

[16] Peter Kiewit Sons' Co. v. Summit Constr. Co., 422 F.2d 242 (8th Cir. 1969).

[17] *See, e.g.,* Continental Consol. Corp., ASBCA No. 10662, 67-1 B.C.A. (CCH) ¶ 6127 (1967); R.S. Noonan, Inc. v. Morrison-Knudsen Co., 522 F. Supp. 1186 (E.D. La. 1981); Hansel Phelps Constr. Co., ENGBCA No. 3368, 74-2 B.C.A. (CCH) ¶ 10,728 (1974).

[18] *See* Havens Steel Co. v. Randolph Eng'g Co., 613 F. Supp. 514 (W.D. Mo. 1985), *aff'd,* 813 F.2d 186 (8th Cir. 1987); Lewis-Nicholson, Inc. v. United States, 550 F.2d 26 (Ct. Cl. 1977); Fred A. Arnold, Inc., ASBCA No. 18915, 75-2 B.C.A. (CCH) ¶ 11,496 (1975); Bryant & Bryant, ASBCA No. 27910, 88-3 B.C.A. (CCH) ¶ 20,923; Tribble & Stephens Co. v. Consolidated Serv., 744 S.W.2d 945 (Tex. Ct. App. 1987).

[19] *See, e.g.,* Reliance Enters., ASBCA No. 20808, 76-1 B.C.A. (CCH) ¶ 11,831; Coatesville Contractors & Eng'rs v. Borough of Ridley Park, 509 Pa. 553, 506 A.2d 862 (1986).

[20] U.S. Indus. v. Blake Constr. Co., 671 F.2d 539 (D.C. Cir. 1982); E.C. Ernst, Inc. v. Manhattan Constr. Co., 387 F. Supp. 1001 (S.D. Ala. 1974), *aff'd in part, rev'd in part,*

favors one contractor over another[21] or permits one contractor to work in an area where another contractor is supposed to be working,[22] it may find itself faced with a claim for additional compensation as a result of the disruption caused by the owner's failure to coordinate the work.

It is extremely important that a contractor whose performance has been disrupted keep accurate records of its increased costs or time for performance.[23] It is easy for a contractor to overlook the more obvious disruption costs of a differing site condition or change order because the focus is on additional quantities of work or duration, rather than the impact the disruption has on the overall costs of performance. Whether it is a matter of the contractor's having to deal with more rock or water, or material which behaves with more difficulty than expected, there may be hidden disruption costs such as those associated with shutting down operations (*demobilization*), giving notice, analyzing the problem, labor and equipment "downtime," and starting up again (*remobilization*).

§ 6.6 Effects of Disruption

The impact of disruption on the contractor's operations usually results in a loss of efficiency, and a loss of momentum in the work will almost inevitably result. Lost productivity is a classic result of disruption, because in the end more labor and equipment hours will be required to do the same work. To illustrate, consider a simple and common disruption such as unknown and unanticipated underground utilities that were either wrongly placed on or not shown on contract drawings. A pipelaying crew consisting of a certain number of laborers, pipefitters, and equipment operators plus their assigned spread of equipment is moving along smartly and consistently with the contractor's estimated progress rates. When they encounter the unanticipated underground utilities, the crew and equipment do not stop working. However, because of the need to exercise additional care in machine excavation, to utilize additional hand excavation, and to expend additional effort in placing pipes under or around existing utility pipes, conduit, ducts, or cables lying in the path of the work in the

---

551 F.2d 1026 (5th Cir. 1977); United States v. Citizens & S. Nat'l Bank, 367 F.2d 473 (4th Cir. 1966).

[21] *See* E.C. Ernst, Inc. v. Manhattan Constr. Co., 387 F. Supp. 1001 (S.D. Ala. 1974), *aff'd in part, rev'd in part*, 551 F.2d 1026 (5th Cir. 1977).

[22] *See, e.g.,* R.S. Noonan, Inc. v. Morrison-Knudsen Co., 522 F. Supp. 1186 (E.D. La. 1981).

[23] Even though a contractor may have lost his records, if he has actual knowledge of such records, a recovery may be had based on his oral testimony of such costs. Reed Paving, Inc. v. Glen Ave. Builders, 148 A.D.2d 934, 539 N.Y.S.2d 173 (1989).

trench, it may take as much time to lay 50 feet of pipe today as it did to lay 200 feet of pipe yesterday.

Disruptions have a profound effect on the learning curve of a contractor. Each contractor's bid price contains certain productivity assumptions. Typically, these assumptions include a *learning curve* effect, that is, as the trade crews become more and more involved in a particular task, the productivity of that crew becomes better. An example of this effect is shown in **Table 6–1**. When developing its bid price, the contractor assumes production will flow, roughly, in a continuous sequence, and its price is based upon this continuous sequence or flow assumption. The economy of continual, repetitive progress, that is, the learning curve effect, results in the continued reduction in workers' hours required to perform identical tasks as workers become more skilled and efficient as the job progresses. When a disruption arises, this continuous flow is interrupted, thereby reducing or halting the contractor's planned progress and causing the learning curve effect to diminish or even disappear. Such a disruption exposes that contractor to additional costs of performance beyond the obvious costs of delay because the contractor will be deprived of the benefit of a labor force which is continually proceeding in singular, methodical steps.

Similarly, because of disruptions and changes in the crew size and composition, the learning curve is greatly diminished as a result of the necessity of going back to the same work location. In *Paccon, Inc.*,[24] a housing contractor's use of specialized crews that were highly efficient at single, repetitive tasks was disrupted because of government suspensions of work. These suspensions forced the contractor to train its crews for

Table 6–1

Example of Learning in Projected Productivity

	Quantity	Workers' Hours	Hours/ Quantity
1st 2 weeks	500 yd^3	5,000	10.0 hrs/yd^3
2nd 2 weeks	2,500 yd^3	21,000	8.4 hrs/yd^3
3rd 2 weeks	5,000 yd^3	25,000	5.0 hrs/yd^3
Last week	500 yd^3	2,000	7.0 hrs/yd^3
Overall (average) productivity:			
	8,500 yd^3	53,000	6.24 hrs/yd^3

---

[24] ASBCA No. 7890, 65-2 B.C.A. (CCH) ¶ 4996 (1965).

new tasks that they otherwise would not have performed in order to maintain progress on the job and mitigate the impact of the suspensions. As a result, the contractor was deprived of the expected increase in efficiency normally resulting from the use of a specialized crew that is continuously repeating a single function.

Disruption can also occur by the trades' having to perform work out of sequence.[25] In *Fred A. Arnold, Inc.*,[26] the government's delay in responding to a contractor's inquiry regarding the proper placement of dryer vent holes in housing units forced the contractor to install the vent holes at a later date and in a more costly manner. At the same time, the contractor was also required to remove its labor force from the work site to a nearby field where the government had accelerated a completely different aspect of the work, the relocation of a baseball field. Consequently, the Board of Contract Appeals awarded the additional costs incurred by the contractor as the result of performing work out of sequence.

A disruptive act can also cause incremental demobilization and remobilization and result in "stop and restart" costs. Typically, once a trade crew has been mobilized and is proceeding to work, that crew will not fully stop until the scheduled activity is completed. However, disruptive acts can cause multiple work stoppages and restarts, aggravating demobilization and remobilization costs.[27] Moreover, the disruption can also increase setup time for activities like moving scaffolding.

Again, accurate records must be maintained to demonstrate the incremental effect these disruptions have on costs such as labor efficiency. For example, in *Sovereign Construction Co.,*[28] the contractor encountered unanticipated unstable rock that required him to reschedule certain portions of his trade work. The contractor attempted to calculate his damages for repeated demobilization and remobilization by taking his total cost and subtracting from that figure the labor costs of all trades unaffected by the differing site condition, thereby leaving the masonry, stone, and concrete work costs. The court found, however, that, because the cost of each trade was segregable, the proper method of calculating the contractor's damages would be to segregate the masonry, stone, and concrete work individually and determine the additional costs caused by each of the inefficiencies suffered. In so finding, the court remanded the case to the trier of fact to determine the relevant specific cost figures and ordered an 11 percent

---

[25] Paccon, Inc., ASBCA No. 7890, 1963 B.C.A. (CCH) ¶ 3659 (1963).

[26] ASBCA No. 18915, 75-2 B.C.A. (CCH) ¶ 11,496 (1975).

[27] Marlin Assocs., Inc., GSBCA No. 5663, 82-1 B.C.A. (CCH) ¶ 15,739 (1982), (board held that the construction company was entitled to recover costs of repeated demobilizations and remobilizations caused by change order delays).

[28] ASBCA No. 17792, 75-1 B.C.A. (CCH) ¶ 11,251 (1975).

reduction in any final figure. This 11 percent reduction was reached by use of the so-called jury verdict method discussed in § **6.15**, and constituted the court's assessment of the percentage of the inefficiency which was caused by the contractor's own fault.

A further disruption can be caused by overcrowding. This disruptive impact can result from poor directions or attempts to accelerate to make up for lost schedule time, thereby requiring several trades (for example, welders and carpenters) to work in small confined areas simultaneously. The space needed to work effectively is understandably diminished in these situations, and this overcrowding usually results in lower productivities for all trades involved.[29]

The contractor's required reliance on the owner's scheduling subjects the owner to the potential for disruption claims if scheduling is not properly managed. A common example is the contractor's reliance on the owner to coordinate the relocation and protection of utilities traversing the jobsite. Failure to prepare the jobsite for construction by removing utility obstacles may result in major disruption claims. For instance, in *Time Contractors, J.V.,*[30] a bridge contractor was prevented from proceeding with the construction of a bridge span because of the telephone company's late and inefficient relocation of a telephone cable running through the heart of the project. Faulting the government for failing to properly manage the jobsite, the Department of Transportation Board of Contract Appeals held that the contractor was entitled to an equitable adjustment for the extra cost of performance. Utilities may further disrupt planned progress if they are unexpectedly encountered during excavation, in which case the contractor must either halt excavation or proceed in a more cautious manner.

§ 6.7 Nature of Disruption Damages

There are many types of costs that can be incurred as a result of disruptions. Because this chapter focuses on nonschedule- or nontime-related damages caused by disruptions, the damages to be described here are activity-related, that is, they are incrementally caused by, or associated with, the disruptive act.

In most disruption situations, the contractor incurs additional *direct costs*. These additional direct costs may consist of increased labor and

---

[29] *See, e.g.,* S. Leo Harmony, Inc. v. Binks Mfg. Co., 597 F. Supp. 1014 (S.D.N.Y. 1984) (acceleration resulted in double-sized crews and stacking of trades in confined work area); Continental Consol. Corp., ASBCA No. 10662, 67-1 B.C.A. (CCH) ¶ 6127 (1967) (out of sequence construction caused congestion of workers in buildings and tunnels).

[30] DOTBCA Nos. 1669, 1691, 87-1 B.C.A. (CCH) ¶ 19,582 (1987).

§ 6.7 NATURE OF DISRUPTION CLAIMS

equipment expenses, or, if necessitated by the disruptive act, the cost of additional quantities of material.

In addition to direct costs, *indirect* or *burden* costs should also be normally included in a contractor's disruption claim, if such costs are pertinent to the disruption. These indirect or burden costs could include such items as payroll taxes and fringe benefits, the cost of front-line supervisors, and the costs of materials-handling departments for materials shipped to the site. Claims for these indirect costs are typically based on rates or percentages added on to the directly claimed costs. Such costs, however, should be determined to be activity-related, because time-related costs in these categories, though probably claimable, should not be priced with the activity-related claim components.

Another loss resulting from disruption can be *increased equipment costs* caused by the additional utilization of equipment, either because the equipment had to be used for longer periods of time or because more equipment was needed in order to maintain schedules impacted by the disruptive act.[31] For example, the disruptive act may lead to increased operational and maintenance charges as well as greater idle equipment time than was originally anticipated because work must be done out of sequence or interrupted by suspension.[32]

Yet another type of loss that can be incurred as a result of disruptions is the additional *loss of profits* related to the costs. These costs should be clearly defined, not merely as "lost profits" or profits which may be speculative in nature, but rather as the justifiable and compensable rate of return on labor hours and other costs incurred as a result of the disruptive act. Although this particular type of damage is usually hotly contested in construction litigation, there is good precedent to support compensation for such losses.[33] The rationale for the recovery of profit on extra work performed because of disruption was explained by the Armed Services Board of Contract Appeals in *New York Shipbuilding Co.:*[34] "Without the payment of profit which is fair under the circumstances, the government would be getting something for nothing and the contractor would not truly be made whole." In federal government construction contracts, however, recovery of profit may be barred by the suspension of work clause,

---

[31] Mobil Chem. Co. v. Blount Bros. Corp., 809 F.2d 1175 (5th Cir. 1987).

[32] In examining the damages resulting from idle equipment, the court in Luria Bros. & Co. v. United States, 177 Ct. Cl. 676, 369 F.2d 701 (1966), ruled that idle contractor-owned equipment should be compensated at one-half AGC rates. In Folk Constr. Co. v. United States, 2 Cl. Ct. 681 (1983), the court granted recovery of the full rental rate because the contractor was unable to mitigate its damages by using its equipment elsewhere.

[33] *See, e.g.,* United States v. Callahan Walker Constr. Co., 317 U.S. 56 (1942); General Builders Supply Co. v. United States, 187 Ct. Cl. 477, 409 F.2d 246 (1969).

[34] ASBCA No. 16164, 76-2 B.C.A. (CCH) ¶ 11,979 at 57,427 (1976).

which limits recovery to costs incurred because of government-caused delay.[35]

The damages discussed in this section are not the only types that the contractor might incur. Depending upon the facts and circumstances of each case, additional, or fewer, damage categories may be relevant. It should also be noted that these damage elements are not only valid for the contractor who is directly impacted by the disruptive act but are also valid claim components for contractors indirectly affected by the ripple of the disruptive act.

§ 6.8 Considerations in Proving Cause and Effect of Disruptions

An important element of a successful disruption claim is, of course, the establishment of the causal relationship of the disruption to the damage. The nature and occurrence of the disruptive event are normally obvious. However, establishing that the contractor suffered a monetary loss as a result of the event and that the contractor is legally entitled to compensation for that loss is more difficult. In *Fischbach & Moore*,[36] a contractor sought delay, disruption, and acceleration damages resulting from government actions. The government argued that (1) the appellant failed to prove causation and (2) the government-caused delays were not segregable. The Board of Contract Appeals rejected the government's arguments, accepting the contractor's right to segregate delays to prove causation and finding that the critical path method used by the contractor in illustrating its damages was a "reasonable and ready basis for segregating delays."

The use of more precise pricing techniques, such as the so-called "modified total cost" or "specific identification" methods, rather than less precise ones like the "total cost" method, can considerably enhance the credibility of the contractor's argument that the damages being claimed were caused by the disruptive acts being alleged. To the extent that a contractor is unable to provide actual cost data for the additional work, it is important that all extra costs not resulting from the disruption be segregated and excluded from the pricing claim in order to establish valid proof of disruption damages. Although in practice it is rare for a contractor's cost records to be organized and segregated for claims purposes versus normal operating purposes, to the extent that the contractor's own contemporaneously prepared records do separately identify claimable costs, the cause and effect argument can be even more strongly supported.

---

[35] *See* Big 4 Constr. Co., DOTBCA No. 75-18, 76-2 B.C.A. (CCH) ¶ 12,029 (1976).
[36] ASBCA No. 18146, 77-1 B.C.A. (CCH) ¶ 12,300 (1977).

§ 6.9 Considerations in Pricing Disruption Claims

Given their nature, disruption damages have two principal components. One component consists of the incremental costs that would not have occurred but for the disruption. See **Table 6–2**. These include the costs of any labor or equipment demobilization and remobilization. The other component of disruption damages consists of lost productivity damages. These are the additional costs incurred by the contractor because of decreased productivity resulting from the disruption.

The first component, the incremental costs incurred as a direct result of the disruptive act, is generally more susceptible to proof by a segregated or actual cost method. That is because of the "but for" test, shown in **Table 6–3**.

As a general proposition, proving damages for lost efficiency is an inexact science. Fortunately, the courts are fairly liberal in the standards of proof and measures of damages required to support recovery, requiring only that the method chosen be fair and calculated to determine damages

Table 6–2

Example of Components of Disruption

		Demob./ Remob.	Loss of Prod.
Total cost incurred for activity X	$150,000		
Total cost estimated for this activity (assuming no disruption)	–$ 80,000		
Net overrun	$ 70,000		
Less estimate of contractor's own problems	–$ 20,000		
Claimable portion of disruption	$ 50,000		
Components:			
Incremental start and stop costs		$10,000	
Productivity impact (Loss of productivity: 4 employee hours per quantity installed × originally estimated quantity to be installed × originally estimated cost/employee hour)			$40,000

Note: This example assumes no additional quantities are made necessary by the disruptive action.

Table 6–3
"But For" Methodology

	Before Disruptive Act	After Disruptive Act	"But For" Differential
Quantity to be installed	100	100	0
Rate of installation	4 hrs./unit	8 hrs./unit	4 hrs./unit
Number of employees	4	4	0
Cost per employee hour	$10.00	$10.00	0
Actual total cost	$8,000		
Estimated cost but for disruptive act	– 4,000		
Calculated damages:	$4,000		

with reasonable certainty.[37] There are a number of methodologies. Among the more widely used are:

1. "Should Cost" estimates
2. Empirically derived indices, industry standards, and handbooks
3. Sampling (time and motion studies)
4. Expert testimony
5. Total cost
6. Modified total cost
7. Jury verdict.

§ 6.10 —Should Cost Estimates

The so-called "should cost" estimate methodology is an increasingly accepted and reasonably persuasive methodology for supporting a claim for inefficiency or lost productivity. The *should cost estimate* is one that compares actual labor costs incurred with an accurate estimate for what the labor costs should have been if no disruption had occurred. A case that involved work on the Walter Reed General Hospital in Washington, D.C., is instructive of the "should cost" method. In *U.S. Industries, Inc. v. Blake Construction Co.*,[38] the mechanical subcontractor was disrupted both by the prime contractor's failure to schedule and coordinate the work and by the need to repair completed work damaged by the prime contractor.

---

[37] Peter Kiewit Sons' Co. v. Summit Constr. Co., 422 F.2d 242 (8th Cir. 1969).
[38] 671 F.2d 539 (D.C. Cir. 1982).

§ 6.10 SHOULD COST ESTIMATES

Having sufficiently proven causation, the subcontractor offered as proof of damages, through financial data and expert testimony, the comparison of labor costs incurred before and after the disruption. The court distinguished this calculation from the total cost method, because use of the should cost method requires proof of causation and does not rely on a pre-contract bid estimate of labor costs but uses actual cost data generated from job performance determined before or after delay as compared with actual costs during the period of delay for the same or similar types of activities.[39] The mechanical subcontractor ultimately recovered a $4.8 million award for disruption damages based on the general contractor's breach of its duty not to interfere, hinder, or unreasonably disrupt the mechanical subcontractor, which resulted in the inefficient and costly performance of the contract.

In calculating and presenting damages in such situations, the opponent's strongest cross-examination efforts will focus on the following issues relative to the actual rate used:

1. Whether the claimed actual undisrupted productivity rate represents actual performance rates for a long enough period. The actual rate used should not be susceptible to allegations that it is unreasonable and was obtained only in isolated instances and therefore is not representative of an attainable actual rate.
2. Whether the actual unaffected productivity rate used to compare with the affected period represents a period of time which was truly unaffected by the disruptive act, or whether there existed other factors which would indicate that it is not representative.
3. Whether the nature of the work being performed, which was the source of the actual unaffected rate used, is, in fact, comparable to the allegedly disrupted work being performed. For example, the productivity rate of HVAC work for a standard commercial building might not be at all representative of the actual productivity for HVAC work at a resource recovery plant.

Table 6–4 is an example of how actual productivity rates might be used from different time periods on the same job to calculate a disruption impact. Naturally, the "should cost" estimate is particularly persuasive when it is based on similar work on the same contract performed in the same time frame.[40] However, when such information is not available, it might

---

[39] *Id.* at 547.

[40] *See* Time Contractors, J.V., DOTBCA No. 1669, 87-1 B.C.A. (CCH) ¶ 19,582 (1987) (contractor compared the effect of disrupted construction of a bridge span with a parallel, unaffected bridge span constructed during the same time period). *See also* Flex-Y-Plan Indus., GSBCA No. 4117, 76-1 B.C.A. (CCH) ¶ 11,713 (1976) (contractor who erected partitions in barracks compared work in occupied barracks with identical work

Table 6-4

Calculation of Disruption Impact Using Actual Productivity Rate from the Same Job

Claimed period of disruption:			June 1–June 15, 1989	
Trade:			Structural concrete on project Alpha	
Start of concrete placement:			March 15, 1989	
Completion of concrete placement:			August 31, 1989	
Actual concrete placement productivity rate:				
April 15–May 15			10 hr/yd^3	
June 1–June 25 (disruption)			12.5 hr/yd^3	
July 1–July 31			10 hr/yd^3	
Quantity installed:			4,000 yd^3	
June 1–June 25 (disruption)				
Cost/employee hour:			$15/hr	
Cost calculations for disruption period:				

	Productivity	Quantity	Cost/hr	Total
Actual	12.5 ×	4,000 ×	15 =	$750,000
Should have been	10.0 ×	4,000 ×	15 =	$600,000
Claim (actual minus should have been)				$150,000

be possible to show attainable production rates on other jobs when the nature of the work, climatic conditions, site conditions, labor skill level, and other factors are equal or very similar.

§ 6.11 —Industry Standards and Handbooks

In certain circumstances, it may not be practical, possible, or even reasonable for the contractor's disruption claim to be calculated using the should cost method based on actual costs discussed in § **6.10**. For instance, such a situation might occur when the contractor has been impacted throughout the project by a serious disruptive action and, therefore, an unimpacted period of sufficient duration does not exist, and the contractor is not currently performing this type of construction activity elsewhere. In this case, the should cost calculation may employ productivity rates that are contained in published industry sources such as MCAA, Business

---

in unoccupied barracks to demonstrate effect of government failure to provide an unoccupied work site. The actual cost of disruption was calculated by comparing installation costs per linear foot of partition between the occupied and unoccupied buildings.).

§ 6.12   TIME AND MOTION STUDIES

Roundtable, R.S. Means, and others. However, caution should be exercised in using these manuals. One must be careful to note situations that are inherently job-specific and make sure that the activities identified in the industry sources are truly comparable to the work actually performed.

An example of the successful use of industry productivity rates was the case of *Pebble Building Co. v. G.J. Hopkins, Inc.*[41] In that case, a subcontractor claimed that the prime contractor's poor supervision and failure to coordinate work caused delays that increased the subcontractor's labor costs. The court accepted an industry estimating manual as a basis for determining the effect on productivity.[42] In calculating the damages, the court subtracted the original estimate of direct labor costs from the actual costs incurred. Recognizing that the total cost method was disfavored, the court examined whether the original estimates were accurate and reasonable. The estimates used by the subcontractor were based on an electrical contracting industry manual for estimating. The subcontractor had offered evidence that past experience had proven the manual estimates to be "accurate and reasonable" predictors of labor ultimately required. Finding that the subcontractor had demonstrated that the estimates were accurate, the Supreme Court of Virginia upheld the trial court's award of delay damages. Thus, in addition to demonstrating that the estimates reflect the work actually performed, the contractor is well advised to demonstrate the reliability of the estimates themselves.[43]

§ 6.12   —Time and Motion Studies

When a contractor is unable to provide actual unaffected performance periods from the current job or other contemporaneous jobs to compare to the affected performance resulting from disruptions, the contractor may consider setting up time and motion studies. These studies would also attempt to measure losses in productivity and increases in labor costs and equipment inefficiencies by approximating both an unaffected sequence

---

[41] 223 Va. 188, 288 S.E.2d 437 (1982).

[42] 288 S.E.2d at 438. *See also* Nelson v. Commonwealth, 235 Va. 228, 368 S.E.2d 239 (1988).

[43] *See, e.g.,* Arthur Painting Co., ASBCA No. 20267, 76-1 B.C.A. (CCH) ¶ 11,894 (1976). In this case, the painting contractor encountered unusual surface conditions in preparing the surface to be painted. In seeking an adjustment for excess preparation, the contractor estimated normal preparation time. The board rejected the contractor's estimates because they were substantially lower than indicated in the estimating guide published by the Painting & Decorating Contractors of America. *See also* Ed Goetz Painting Co., DOTBCA No. 1168, 83-1 B.C.A. (CCH) ¶ 16,134 (1983), in which the contractor's expert, in calculating damages incurred by the contractor, relied on three building industry publications comparing the additional costs required to attain a white blast compared to a near white blast. The board accepted this figure as fairly representing the cost at which the contractor could have performed.

and an affected sequence of performance. To the extent possible, every effort should be made to simulate *all* conditions experienced by the contractor on the job, not just the conditions attributable to disruption. Clearly, the best place to do this would be on the actual jobsite if possible, and an effort should be made to replicate the same circumstances of the disruptive condition. A number of variances ought to be considered so as to insure fairness and to screen out or minimize the effect of other concurrent disruptive activities. In *Peter Kiewit Sons' Co. v. Summit Construction Co.*,[44] an effort by an excavation subcontractor to reflect the affected disruptions by means of a time and motion study was employed effectively and approved by the court of appeals. The subcontractor was able to demonstrate the incremental loss of efficiency on a job which was attributable to the impact of the actual disruption.

§ 6.13 —Expert Opinion

A true expert in construction productivity and efficiency can be used to document "should have been" rates, provided he has the qualifications to do so. Moreover, even when the expert is so qualified, strong challenges can be raised to his testimony if the rates developed or employed by the expert bear no relationship to those actually experienced by the contractor in unaffected periods and/or at other sites in similar, but unaffected, situations on other jobs.

In one case, a contractor attempted to prove loss of efficiency damages arising from overtime hours worked by offering the testimony of a professional engineer who stated that, through his personal observation, overtime hours generally resulted in decreased worker efficiency.[45] The engineer was presented as an expert for the purposes of calculating the labor inefficiency damages of the contractor. The court found that because the engineer was not trained in calculating labor inefficiency, he did not qualify as an expert for purposes of calculating such damages. Furthermore, his reliance on a labor inefficiency chart could not be presented in evidence absent an analysis of the reliability of the chart's calculations. Therefore, in dismissing the claim for lack of evidence, the court stated that "the very nature of labor inefficiency claims requires expert testimony," but that any figure reached as a result of this purported expert's testimony was based on completely uninformed speculation.[46]

A critical role of the expert is to apply the appropriate damage measurement methodology to the specific facts of the case and demonstrate

---

[44] 422 F.2d 242 (8th Cir. 1969).

[45] Havens Steel Co. v. Rudolph Eng'g Co., 613 F. Supp. 514 (W.D. Mo. 1985), *aff'd*, 813 F.2d 186 (8th Cir. 1987).

[46] 613 F. Supp. at 540.

that any data relied upon is supported by the contractor's actual costs. In *Cosmic Construction Co.*,[47] expert testimony on the calculation of damages was faulted for not being based on the actual experience of the contractor on the job and because the expert made assumptions not borne out by the facts. The contractor's expert developed a should cost estimate which was compared to the contractor's actual labor costs for a delay period. The difference was presented as the cost of labor inefficiency. One of the expert's assumptions was, however, that the appellant had performed all work properly and was in no way inefficient. The Board of Contract Appeals disallowed this method, discounting the expert testimony for (1) not demonstrating that the contractor's actual costs were reasonable and, (2) not demonstrating that the should cost estimates had a reasonable and ascertainable basis. These two fatal criticisms were based on the fact that, in failing to relate the should cost figure to the contractor's method of estimating his performance, the expert had employed a theoretical and unrealistic cost figure that consequently did not reflect any of the facts of the case.

A similar result was reached in *Luria Bros. & Co. v. United States.*[48] In this case, the government's breach required the contractor to perform during severe winter weather, resulting in reduced labor efficiency. The contractor attempted to prove its damages through the testimony of its chief of construction who testified, without corroboration, that labor productivity was reduced 25 to 30 percent as a result of performing in the winter weather. In rejecting the contractor's estimate, the court stated:

> It is a rare case where loss of productivity can be proven by books and records; almost always it has to be proven by the opinion of expert witnesses. However, the mere expression of an estimate as to the amount of productivity loss by an expert witness with nothing to support it will not establish the fundamental fact of resultant injury nor provide a sufficient basis for making a reasonably correct approximate of damages.[49]

Noting the witness's unsubstantiated estimate and inherent bias, the court stated: "Proof of damage is essential before estimates can be received of the amount thereof. Plaintiff's claim for loss of productivity during that period must therefore be rejected."[50] The court recognized, however, the contractor's entitlement to damages, and, based on its knowledge and experience in such cases, applied its own rates of inefficiency to the winter work delay periods. The *Cosmic Construction* and *Luria Bros.* cases demonstrate the crucial requirement that an expert

---

[47] ASBCA Nos. 24041, 24036, 88-2 B.C.A. (CCH) ¶ 20,623 (1988).
[48] 177 Ct. Cl. 676, 369 F.2d 701 (1966).
[49] 369 F.2d at 713 n.31.
[50] *Id.* at 714.

substantiate his opinion with hard data that has a direct relationship to the contractor's own experience on the job.[51]

Although the failure of an expert to prove his asserted method of calculating damages might not be fatal to a claim, the Boards of Contract Appeals do require experts to provide a minimal degree of support for their testimony. For example, in *Zinger Construction Co.*,[52] the contractor's performance on a construction contract was delayed when the government did not approve a modification order until 43 days after the modification was requested. The board ruled that the delay was unreasonable and the contractor was entitled to damages. The contractor failed, however, to provide the board with sufficient evidence upon which to establish the quantum of the damages incurred. Estimates of its delay costs were based solely on the testimony of the contractor, who was found to be an unreliable witness. Furthermore, the estimate submitted was an overall estimate, did not include a breakdown of jobs, and was a substantially higher estimate than that set forth in the complaint. Although the board recognized that a lack of direct and specific proof alone does not bar recovery of damages, because the award may be made on a jury verdict basis (see § **6.15**), the board stated that "[t]here must be evidence adduced that is sufficient to arrive at a result that is a fair and reasonably precise approximation, and the reliability of that evidence must be fully substantiated."[53] The board found that the contractor had submitted "no credible evidence" that would permit a determination of the adjustment, and denied the contractor's claim.

§ 6.14 —Total Cost and Modified Total Cost Methods

The use of the *total cost* method of calculating damages—that is, the calculation of damages by simply subtracting the bid cost from the actual cost—has continuously been criticized by boards of contract appeals and courts throughout the country. Despite the fact that it is still accepted by some courts,[54] this method is extremely difficult to protect from competent,

---

[51] *But see* Brick's, Inc., DOTBCA No. 1906, 89-1 B.C.A. (CCH) ¶ 21,381 (1989) (board rejected contractor's estimation of damages, noting its failure to substantiate the estimate with actual proof; however, because clear evidence was presented that the contractor was entitled to damages, the award was determined by the jury verdict method) (see § **6.15**).

[52] GSBCA No. 6568, 84-3 B.C.A. (CCH) ¶ 17,537 (1984).

[53] *Id.* at ¶ 87,349.

[54] Prichard Bros. v. Grady Co., 436 N.W.2d 460 (Minn. Ct. App. 1989) (allowing contractor damages on total cost basis for errors in interpreting shop drawings); Glasgow v. Pennsylvania Dep't of Transp., 529 A.2d 576 (Pa. Commw. Ct. 1987) (use of the total

Table 6–5

Example of Modified Total Cost Calculation

Total cost:	$350,000
Less bid cost:	– 215,000
Claim under total cost method:	$135,000
Less bidding errors on contractor's part:	– 42,000*
	$ 93,000
Less productivity impacts caused by contractor:	– 43,000*
Modified total cost claim:	$ 50,000

* Specifically identified and calculated.

aggressive cross-examination. One need only point to a few instances in which additional costs were incurred as a result of the contractor's own problems or because the contractor's original bid was unreasonable, in order to do damage to this method. Because of its ease in calculation, however, this method tends to show up in many disruption claim situations.

To enhance the prospects of recovery for disruption, a better approach might be to employ the *modified total cost method* of damage calculation. In this method, any and all contractor-caused productivity impacts are quantified and deducted from the total cost claim amount, so that the remaining differential, if any, is representative of the noncontractor-caused disruptive action. An example of such a calculation is shown in **Table 6–5**.

Although still facing some challenges, use of the modified total cost method can provide a much more defensible damages position for disruption impact and, at the same time, it still is far less complex and costly to develop than if a specific identification, that is, "bottom-up" approach, is used.

§ 6.15 —Jury Verdict Method

The *jury verdict method* provides the courts and Boards of Contract Appeals with a way of determining claims for disruption when damages are not readily ascertainable from a contractor's evidence.[55] The method is employed by triers of fact when they weigh the various elements of a

---

cost method permissible when no other way to frame damages exists and contractor demonstrates that its estimates were accurate and its actual costs were reasonable).

[55] Tattore Co. v. Metropolitan Sewerage Comm'n, 505 F.2d 1 (7th Cir. 1974).

contractor's claim and derive a total value for the claim, often expressed as a lump sum. From this lump sum, the court determines, as matter of fact, a percentage of damages sought which are attributable to the contractor's own inefficiencies or to causes not chargeable to the owner. The remainder constitutes the contractor's compensable loss. Use of the jury verdict is proper when sufficient proof of damages is unavailable because of conflicting evidence,[56] lack of available evidence,[57] or lack of proof of causation.[58] However, in awarding damages on a jury verdict basis, the board or court must be persuaded that:

1. Entitlement exists
2. There is no more reliable method for computing damages and
3. The record affords a basis for a fair and reasonable approximation of damages.[59]

A good example of the application of the jury verdict to a contractor's claim was found in the case of *David J. Tiernay*.[60] In this case, the contractor sought disruption damages, including loss of efficiency, resulting from the cumulative effect of a large number of change orders issued during a construction project. The board declined to accept the contractor's total cost figure, faulting the contractor for failing to prove the government's culpability for all claimed items of disruption. Resorting to the jury verdict to arrive at a fair and reasonable approximation of the contractor's damages, the board added the contractor's net operating losses to his contract price. This figure was multiplied by the 10 percent profit allowed under the changes clause of the contract. By adding this profit amount to the net operating loss, the contractor's actual loss was determined. To this figure the board applied a ratio representing the board's estimation of the contractor's meritorious and nonmeritorious claims to reach the final award.

The jury verdict may provide hope to the contractor who can demonstrate entitlement but cannot provide sufficient proof of damages. However, the jury verdict is not the method of first choice when pricing claims, and it will not be applied when the record provides an adequate basis for determining actual damages.

---

[56] Delco Elecs. Corp. v. United States, 17 Cl. Ct. 302 (1989).

[57] Algernon Blair, Inc., GSBCA No. 4072, 76-2 B.C.A. (CCH) ¶ 12,073 (1976); S.W. Elecs. & Mfg. v. United States, 228 Ct. Cl. 333, 655 F.2d 1078 (1981).

[58] Stephenson Assocs., GSBCA Nos. 6513, 6815, 86-3 B.C.A. (CCH) ¶ 19,071 (1986).

[59] Tutor Saliba-Perini, PSBCA No. 1201, 87-2 B.C.A. (CCH) ¶ 19,755 (1987); Dawco Constr., Inc. v. United States, No. 450-86C, 1989 U.S. Cl. Ct. (LEXIS 238) (Nov. 17, 1989).

[60] GSBCA Nos. 7107, 6198, 88-2 B.C.A. (CCH) ¶ 20,806 (1988).

§ 6.16 Defenses

This chapter has alluded to a number of defenses to disruption claims, such as the failure to prove causation and the contractor's own unreasonable or poor planning and/or performance. Other defenses available to the owner may include the failure of the contractor to give adequate notice required by the contract to the owner of changed conditions. Typically, under the changes clause in government contracts, the contractor is required to give notice to the owner of changes resulting from disruptions within 20 days of their occurrence. A similar requirement is contained in the suspension of work government contract clause.

When owners employ American Institute of Architects (AIA) Document A201, entitled "General Conditions of the Contract for Construction," the contractor claiming disruption claims must also give the architect notice within 20 days of the occurrence giving rise to the claim. There have been instances in which contractors' claims for loss of productivity have been barred for failure to give timely notice.[61] In the federal arena, the courts are reluctant to bar such claims for failure to give notice except when the government can show that it was prejudiced or disadvantaged by the failure to give such notice.[62] When the government has actual knowledge or oral notice of the changed condition, the requirement to give written notice is often not strictly enforced.[63]

Although this chapter does not deal with delay claims, the effect of the exculpatory no damages for delay clause requires discussion. In the context of disruption claims, there may be production costs that may in some instances be barred by a broad reading of specific language in an exculpatory clause dealing with production. However, exculpatory clauses such as the no damages for delay clause are generally construed narrowly against the person seeking to enforce them. Courts have construed no damages for delay clauses as not barring claims for loss of efficiency.[64]

Courts have generally distinguished the types of delay and attending costs that are covered by a no damages for delay clause.[65] A distinction

---

[61] *See, e.g.,* Tuttle/Whit Constr., Inc. v. State, 371 So. 2d 1096 (Fla. 1979).

[62] *Compare* Cosmo Constr. Co. v. United States, 194 Ct. Cl. 559, 439 F.2d 160 (1971) with Eggers & Higgins v. United States, 185 Ct. Cl. 765, 403 F.2d 225 (1968).

[63] *See* Appeal of Human Advancement, Inc., HUDBCA 77-215, 81-2 B.C.A. (CCH) 25,317 (1981).

[64] *See, e.g.,* John E. Green Plumbing & Heating v. Turner Constr., 742 F.2d 965 (6th Cir. 1984), *cert. denied,* 105 S. Ct. 2328 (1985).

[65] *See, e.g., id.* (no damages for delay clause in plumbing and heating contractor's contract with city building authority did not preclude a cause of action against construction manager for extra manpower costs incurred as a result of his hindering performance). *See also* Buckley & Co. v. State, 140 N.J. Super. 289, 356 A.2d 56 (1975) (a no damages for delay clause does not exculpate active interference with a contractor's performance). *But*

has been made between increased costs resulting from pure delay and those caused by disruption of the work. In the case of a pure delay, the contractor suffers a delay which causes the project to be completed after the scheduled contract completion date. In the case of disruption, the contractor may indeed finish the overall contract on schedule but may experience delays and disruption to its work prior to the completion date, resulting in increased costs. Generally, the courts have held that the no damages for delay clause will apply only in the instance of pure delay—that is, when the contractor has been delayed in its performance—and will bar a contractor's claim for increased costs only when the contractor is unable to complete the contract on time. The no damages for delay clause will not apply to disruption and impact costs, and the contractor may collect its increased costs incurred as a result of such disruptions, even in the face of a no damages for delay clause.

§ 6.17 Summary

Disruptions are an important source of potential claims for a construction contractor whose plan of work is adversely impacted by actions of the owner. The advantage of such disruption claims is that they may fall outside the limitations of a no damages for delay clause. The difficulty with such claims, however, is developing the casual connection between the act of disruption and the injury or damages to the contractor. Careful attention should be given by the contractor to give timely notice of these claims, and to endeavor to maintain accurate and contemporaneous records of impact to enhance the prospects of recovery. Fortunately, when accurate records have not been kept or the nature of the disruption is such that quantification is difficult, the courts have recognized alternative measures of damages, provided that such measurements are fair and calculate damages with reasonable certainty.

---

see Kalisch-Jarcho, Inc. v. City of N.Y., 58 N.Y.2d 377, 448 N.E.2d 413, 461 N.Y.S.2d 746 (1983) (when active interference was contemplated to be included in the no damages for delay clause, only gross negligence or bad faith may be the basis for a delay claim).

CHAPTER 7

ACCELERATION CLAIMS

Michael F. Albers, Esquire
Mark J. Hosfield, CPA
Robert L. Meyers III, Esquire*

Michael F. Albers is an attorney with Jones, Day, Reavis & Pogue, having joined the firm's Dallas office in 1981. He concentrates his practice in the areas of construction law and commercial real estate development and has represented owners, developers, lenders, and contractors in construction documentation and dispute resolution procedures, as well as project acquisition, financing, and development activities. Mr. Albers has written for and participated in the presentation of a number of programs concerning construction law, including the Practicing Law Institute's construction contracts seminars, the Texas Bar Advanced Real Estate Program, and the ABA/Joint Program on Bankruptcy in the Construction Industry. He received both his B.A. and J.D. from Southern Methodist University.

Mark J. Hosfield is a manager of Litigation & Claims Services for the national accounting and consulting firm of Coopers & Lybrand. Experienced in the analysis of project schedules, costs, and claims, he has testified as an expert witness on both scheduling and damages issues. Mr. Hosfield has assisted trial lawyers, contractors, and owners on matters involving construction delays, acceleration, and productivity loss due to disruption.

A graduate of the University of Illinois, Mr. Hosfield received a masters degree in Management from Northwestern University. He has taught college-level accounting and has been a featured speaker on the topics of design and use of automated project management systems, and construction claims analysis and presentation. He is a Certified Public Accountant, a Certified Management Accountant, and a member of the Illinois CPA Society.

---

* The authors wish to thank Mark Kopidlansky, an associate with Jones, Day, Reavis & Pogue, for his assistance in preparing this chapter.

Robert L. Meyers III is a partner in the Dallas firm of Jones, Day, Reavis & Pogue, where his practice concentrates on construction law, construction documentation, and construction litigation and arbitration. He has represented all parties to the development/construction process including owners, architects, general contractors, trade contractors, suppliers, and sureties.

A graduate of Southern Methodist University, Mr. Meyers is a member of the American Bar Association, the State Bar of Texas (secretary of the Construction Law Section), and the Dallas Bar Association (co-founder of the Construction Law Section). He has written and spoken extensively on the subject of construction law, and has served as a faculty member and contributing author for the Practicing Law Institute's construction contracts seminars for the last 15 years. Mr. Meyers has contributed to several law reviews and construction law publications including *A Businessman's Guide to Construction, Hazardous Waste Disposal and Underground Construction Law* (John Wiley & Sons), and *The Construction Lawyer*.

INTRODUCTION

§ 7.1 Acceleration in General
§ 7.2 Acceleration Claims
§ 7.3 Acceleration in Fact
§ 7.4 Constructive Acceleration

PROVING ACCELERATION

§ 7.5 Identifying Acceleration in the Project
§ 7.6 Proof of Acceleration
§ 7.7 Elements of an Acceleration Claim
§ 7.8 —Acceleration Orders
§ 7.9 —Acceleration Costs
§ 7.10 —Time Extensions
§ 7.11 Acceleration Provisions
§ 7.12 Documentation

PROVING DAMAGES

§ 7.13 Acceleration Damages
§ 7.14 Legally Recognized Measures of Damages
§ 7.15 Using Available Information
§ 7.16 Analysis of Actual Costs
§ 7.17 Reasonableness of Damage Figures
§ 7.18 Profit and Overhead Calculations
§ 7.19 Mitigation and Cost Savings
§ 7.20 Presentation of Evidence

INTRODUCTION

§ 7.1 Acceleration in General

Every owner or developer knows or soon learns that a delay in the completion of a construction project can be a very damaging and costly event. For that reason, the owner frequently tries to pressure the contractor to overcome any delays encountered during construction and may even insist that the contractor perform ahead of schedule, regardless of any delays. But the contractor is entitled to the entire contract time, plus any justifiable extensions, within which to complete performance. If the owner takes away a portion of the contractor's time by forcing the contractor to complete its performance in advance of the date to which it is entitled, the owner has accelerated the contractor and is subject to liability for any increased costs incurred by the contractor.

Acceleration is the process by which the ordinary and expected progress of events in a construction contract is quickened. Acceleration occurs when the contractor performs its work at a faster rate than required by the original contract.[1] When a contractor accelerates its performance rate voluntarily for its own purposes, it will not receive any additional compensation from the owner.[2] However, when the contractor is ordered by the owner to speed up its construction performance, the contractor may have a compensable claim against the owner for acceleration damages.[3]

There are two types of compensable acceleration: directed and constructive. *Directed acceleration* occurs when the contractor is ordered by the owner to complete the construction project ahead of the contract completion date. *Constructive acceleration,* on the other hand, occurs when the owner denies the contractor's claim for a justified time extension and requires the contractor to complete the project by the contract completion date.[4] See § 7.4.

Acceleration is prevalent in construction projects because time is money. The fact that time schedules, completion dates, milestones, and critical path charts play a significant role in the bidding and awarding of construction contracts indicates that a major concern of the owner is the time of performance of the project.[5] The time of performance is important to the owner because delays in the completion of the project mean that the

---

[1] Contracting & Material Co. v. City of Chicago, 20 Ill. App. 3d 684, 692, 314 N.E.2d 598, 604 (1974), *rev'd on other grounds,* 64 Ill. 2d 21, 349 N.E.2d 389 (1976).

[2] Mobile Chem. Co. v. Blount Bros. Corp., 809 F.2d 1175 (5th Cir. 1987).

[3] Nat Harrison Assocs. v. Gulf States Utils. Co., 491 F.2d 578 (5th Cir. 1974).

[4] K. Gibbs & G. Hunt, Construction Claims 39 (1987).

[5] M. Simon, Construction Claims and Liability 276 (John Wiley & Sons 1989).

owner incurs additional construction costs and possibly loses the use of an income-producing property during the delay. It is the desire to reduce costs and complete construction as soon as possible that often causes the owner to accelerate the contractor.

When ordered by the owner, the contractor can accomplish or attempt acceleration with a variety of techniques. These techniques include resequencing of work activities; increasing the labor force by increasing the number of crews and crew sizes, working overtime, and adding new shifts; adding extra equipment; and expediting material and equipment deliveries. The use of these techniques can be characterized as the contractor's acceleration effort.

§ 7.2 Acceleration Claims

Acceleration in construction projects usually leads to claims by the contractor against the owner, whether justified or not. As a result of its acceleration effort, the contractor incurs additional costs that were not specified in the contract. These costs may include additional equipment costs, expedited material delivery costs, and increased labor force costs.[6] Regardless of the success of the contractor's acceleration effort, the contractor is entitled to recover these additional costs from the owner. If the owner disputes the amount of the additional costs or does not adequately compensate the contractor, the contractor will almost certainly bring a claim for acceleration damages.

§ 7.3 Acceleration in Fact

Acceleration in fact is usually easy to identify. It occurs in a construction project when one party to a construction contract increases its performance effort in order to complete its required performance either in advance of its scheduled time as specified in the contract or an attached schedule, or in advance of the time-extended completion date that circumstances may justify, or by the contract completion date even though progress has otherwise been delayed.

There are two general causes of acceleration in fact. The first and less frequently encountered cause of acceleration is voluntary acceleration. *Voluntary acceleration* occurs when a party to a construction contract, usually the contractor, performs ahead of schedule for its own purposes or motives, and not on the directives of another party. In this situation, the party does not have a claim for damages against any other party to the construction contract.[7] For example, a contractor who performs

---

[6] Natkin & Co. v. George A. Fuller Co., 347 F. Supp. 17 (W.D. Mo. 1972).
[7] McNutt Constr. Co., ENGBCA No. 4724, 85-3 B.C.A. (CCH) ¶ 18,397 (1985).

ahead of schedule to lower its labor costs or to impress the owner will not be able to recover additional costs from the owner.[8] That is because the contractor incurred the additional costs pursuant to its own directives; it cannot establish that any other party is liable for its additional costs or damages.[9]

The other type of acceleration in fact is *directed acceleration,* which occurs when one party to a construction contract is ordered by another party to complete its performance in advance of the scheduled completion date in the contract. The most common occurrence of directed acceleration is when the owner orders or directs the contractor to finish its performance early.[10] The reasons behind directed acceleration are fairly simple. The party ordering the acceleration is usually doing so to decrease its costs in the construction project. For example, an owner might want construction completed early to decrease its overhead costs, or to get an income producing tenant into the project more quickly.[11] In addition, an owner might want to take advantage of lower material prices when there is a discernible threat of rising prices.[12]

Whatever the reason, an owner who directly accelerates a contractor is liable to the contractor for its acceleration costs.[13] The contractor, in bidding on the contract, bargained for a certain amount of time to complete the project. By depriving the contractor of part of its bargained time, the owner is in effect "buying back" some time from the contractor.[14] The price of the owner's purchase is the contractor's acceleration costs.

§ 7.4 Constructive Acceleration

Constructive acceleration is more common than directed acceleration and is correspondingly more difficult to identify and prove. *Constructive acceleration* occurs when an owner refuses to adjust the completion date to take into account justified time extensions and instead requires the

---

[8] *Id.*

[9] Mobile Chem. Co. v. Blount Bros. Corp., 809 F.2d 1175 (5th Cir. 1987).

[10] Norair Eng'g Corp. v. United States, 666 F.2d 546 (Ct. Cl. 1981); Johnson Controls, Inc. v. National Valve & Mfg. Co., 569 F. Supp. 758 (E.D. Okla. 1983) (a prime contractor directly accelerated its subcontractor when it ordered the subcontractor to complete construction early); Loomis & Loomis, Inc. v. Stecker & Colavecchio Architects, Inc., 6 Conn. App. 88, 503 A.2d 181 (1986) (an architect who ordered the engineer to finish its work early was found to have directly accelerated the engineer).

[11] Natkin & Co. v. George A. Fuller Co., 347 F. Supp. 17 (W.D. Mo. 1972).

[12] Loomis & Loomis, Inc. v. Stecker & Colavecchio Architects, Inc., 6 Conn. App. 88, 503 A.2d 181 (1986).

[13] Norair Eng'g Corp. v. United States, 666 F.2d 546 (Ct. Cl. 1981).

[14] Department of the Army, Office of the Chief of Engineers, Pub. No. 415-1-2, Modifications and Claim Guide B-1 (July 1987).

contractor to complete performance by the original contract completion date.[15]

A contractor cannot establish a claim for constructive acceleration unless it has been denied a time extension to which it was entitled. However, a contractor is not entitled to a time extension unless its performance was delayed by a compensable or excusable delay.[16] Therefore, whether a contractor has been constructively accelerated will depend on whether the delay in the contractor's performance was compensable, excusable, or nonexcusable.

Nonexcusable Delays

A *delay* is the period of time during which a construction project or a work activity in a construction project has been extended or not performed as a result of unforeseen circumstances. Delays can be classified into two categories: excusable delays and nonexcusable delays.[17] *Nonexcusable delays* are those delays which are caused by the fault or negligence of the contractor, or those delays for which the contractor assumed the risk.[18] The most common nonexcusable delays are those caused by:

1. Ordinary and foreseeable weather conditions[19]
2. A subcontractor's delays[20]
3. The contractor's failure to adequately manage and coordinate the project site
4. The contractor's financing problems[21]
5. The contractor's failure to mobilize quickly enough[22]
6. Delay by the contractor in obtaining materials[23]
7. Poor workmanship.

A contractor whose performance is delayed by a nonexcusable delay is still obligated to perform by the original contract date. For example, a contractor who negligently set fire to its plant would not be excused for a delay in

---

[15] Elte, Inc. v. S.S. Mullen, Inc., 469 F.2d 1127 (9th Cir. 1972).
[16] B. Bramble & M. Callahan, Construction Delay Claims (John Wiley & Sons 1987).
[17] *Id.* at 2.
[18] McDevitt & Street Co. v. Marriott Corp., 713 F. Supp. 906 (E.D. Va. 1989).
[19] *Id.*
[20] James Walford Constr. Co., GSBCA 6498, 83-1 B.C.A. (CCH) ¶ 16,277, 25 G. C. ¶ 196.
[21] Tucker v. Bitler Bros., 197 N.Y.S.2d 899 (1960).
[22] Burns v. Hanover Ins. Co., 454 A.2d 325 (D.C. App. 1982).
[23] Malor Constr. Corp., IBCA No. 1688-6-83, 84-1 B.C.A. (CCH) ¶ 17,023 (1984).

§ 7.4 CONSTRUCTIVE ACCELERATION 173

performance.[24] Because the contractor was responsible for the delay in its performance, it would not be entitled to a time extension or an adjusted completion date.[25] Therefore, the contractor would not be able to recover on a claim of constructive acceleration.[26]

Excusable Delays

The most common causes of constructive acceleration are excusable delays. *Excusable delays* are those that entitle the contractor to a time extension because they are beyond the control of the contractor. The most recognized excusable delays are found in the American Institute of Architects' (AIA) General Conditions and include:

1. Labor disputes
2. Fire
3. Unusual delay in deliveries
4. Unavoidable casualties
5. Compensable delays
6. Unforeseen delays in transportation
7. "Other causes beyond the contractor's control."[27]

Other common excusable delays are acts of God (such as earthquakes), severe storm damage, and unforeseeable or adverse weather conditions.[28]

Excusable delays include a specific category of delays called compensable delays. *Compensable delays* are delays caused by the negligence or fault of the owner. Common examples of compensable delays include:

1. The owner's failure to coordinate multiple prime contractors[29]
2. The owner's failure to provide adequate access to the project[30]
3. The owner's failure to provide the right of way[31]

---

[24] R. Nash, Government Contract Changes 327 (1975).

[25] Merritt-Chapman & Scott Corp. v. State, 54 A.D.2d 37, 386 N.Y.S.2d 894 (1976), *aff'd,* 43 N.Y.2d 690, 371 N.E.2d 790, 401 N.Y.S.2d 28 (1977). *See also,* Seifford v. Housing Auth., 192 Neb. 643, 223 N.W.2d 816 (1974).

[26] Tri-Cor, Inc. v. United States, 458 F.2d 112 (Ct. Cl. 1972).

[27] AIA A201-1976 ¶ 8.3.1.

[28] T.F. Scholes, Inc., ASBCA No. 5009, 59-2 B.C.A. (CCH) ¶ 2,375 (1959).

[29] Eric A. Carlstrom Constr. Co. v. Independent School Dist. No. 77, 256 N.W.2d 479 (Minn. 1977).

[30] Elte, Inc. v. S.S. Mullen, Inc., 469 F.2d 1127 (9th Cir. 1972).

[31] Anderson Dev. Corp. v. Coastal States, 543 S.W.2d 402 (Tex. Civ. App. 1976).

4. Suspension of the contractor's performance[32]
5. Change orders issued by the owner
6. Any interference by the owner with the contractor's performance[33]
7. Delays caused by a third party under the control of the owner, such as the architect or the engineer.

Because compensable delays give the contractor a cause of action for monetary damages, such delays are primarily encountered in delay damage cases. But, as a subset of excusable delays, compensable delays can also be the basis of a claim of constructive acceleration. (Whether the contractor can recover money damages may depend on whether the contract has a no damages for delay clause. However, such a clause does not affect the contractor's right to a time extension.)

The theory behind constructive acceleration is that the contractor should not be held responsible or accountable for delays that were not its fault. When the contractor's performance is delayed by an excusable event, the contractor is entitled to a new contract completion date that is adjusted to account for and negate the impact of the delay on the contractor's timely progress. For example, if the contractor was delayed by an excusable 50-day strike, the contractor is entitled to a 50-day time extension. Likewise, if an owner issues a change order or fails to perform a required contractual undertaking, such as laying the foundation for the construction project when contracted separately, the contractor is entitled to an adequate extension to its contract completion date.[34] Refusal by the owner to grant such justifiable time extensions constitutes constructive acceleration.[35]

PROVING ACCELERATION

§ 7.5 Identifying Acceleration in the Project

A contractor should not bring an acceleration claim unless it can show that it made a reasonable effort to accelerate its performance. If there has been no acceleration effort, there are no damages. Thus, a contractor must be able to show when its performance was due under the contract

---

[32] T.C. Bateson Constr. Co. v. United States, 319 F.2d 135 (Ct. Cl. 1963).
[33] Housing Auth. v. E.W. Johnson Constr. Co., 264 Ark. 523, 573 S.W.2d 316 (1978).
[34] Mobile Chem. Co. v. Blount Bros. Corp., 809 F.2d 1175 (5th Cir. 1987).
[35] Norair Eng'g Corp. v. United States, 666 F.2d 546 (Ct. Cl. 1981).

§ 7.5 IDENTIFYING ACCELERATION

and must be able to prove that it performed or attempted to perform ahead of that due date. For a claim of constructive acceleration, the contractor also must establish that the delaying event caused its performance to be delayed. One way to identify acceleration in the construction project is to analyze the project schedules.

Every construction project has a project schedule. A project schedule is an analytical and planning tool as well as a visual guide to the construction project. It is created to chart out and manage the sequential order in which the individual work activities will be performed. Project schedules provide a framework for analyzing both the total project and the interrelationships between the work activities within the project.

There are many different methods of designing project schedules. There are handwritten schedules, Gantt (bar) charts, program evaluation and review technique (PERT) charts (used primarily for research and development projects), precedence diagramming, and the critical path method (CPM).[36] CPM schedules can be created using either precedence or activity on arrow logic (representing the activities by arrows and the relationships between them by nodes or points of intersection). For purposes of identifying a contractor's acceleration efforts, the most accurate method is the critical path method.

The initial step in using the critical path method to identify acceleration effort is to prepare an as-planned (baseline) CPM schedule. This requires determining the work activities of the construction project and ascertaining the logical relationships among the activities. A *work activity* of a construction project is defined simply as a time-consuming task with a recognizable beginning and end.[37] It is up to the owner and contractor of each construction project to determine the length and size of each designated work activity. There are no set ground rules.

A difficulty in preparing a CPM schedule lies in ascertaining the logical relationship of the work activities (the CPM logic). CPM logic concerns itself mostly with the dependence among work activities. An activity is dependent on another when it cannot be performed until the other activity has been completed. Such dependence between activities often determines the duration of a construction project because a project can be constructed only as quickly as its dependent activities can be performed.

The as-planned CPM schedule can be displayed in the form of a network diagram or flowchart. The diagram is a pictorial display of the job's logic that charts from left to right the sequence and dependencies of events. When the CPM schedule has been analyzed, it yields the critical path, which is the longest chain or series of the dependent work activities

---

[36] M. Callahan & H. Hohns, Construction Schedules 9 (1983).

[37] *Id.* at 12.

through the performance of the total project.[38] Because a construction project cannot be completed until all of the critical path activities are performed, the baseline critical path by definition determines a contractor's time of performance for a total construction project. Delaying the completion of any critical path as-planned activity by any length of time delays the entire project by the same length of time.

The contractor usually constructs an as-planned or baseline CPM schedule before it begins construction. During construction, the contractor often finds it necessary to modify the baseline CPM. These modifications or updates are usually minor, such as correcting inaccurate logic ties or schedule drafting errors, and result in a schedule that depicts how the initial baseline CPM schedule would have looked had the contractor originally had all of the relevant information regarding the planned method of performance.[39] When properly modified, the as-planned CPM schedule serves as a baseline by which to measure acceleration efforts.

If the as-planned schedule is created at the beginning of the project and updated throughout, a valuable record has been created. If not, the as-planned CPM can be created after the project has been completed for purposes of damage analysis and claim preparation.

In directed acceleration claims, the baseline CPM schedule is measured against the as-built CPM schedule, which is a schedule of the project as it was actually performed. Because the as-built CPM schedule includes the actual time of the contractor's performance, the contractor is able to compare the as-planned and the as-built CPM schedules to establish whether it completed construction in advance of the original project time.[40] Of course, the CPM schedule analysis is just the first step in proving and pricing a directed acceleration claim. The contractor still has to prove that the owner ordered the acceleration and that the contractor incurred additional costs as a result.

To identify acceleration effort in a claim of constructive acceleration, the contractor has to analyze an impacted CPM schedule for comparison with the as-planned and the as-built schedules. The impacted schedule is a derivative of the as-planned schedule; it extends the as-planned schedule by taking into account all compensable and excusable delays. Because construction projects often involve excusable and nonexcusable delays and modifications, contractors should use all three schedules to identify acceleration efforts.

---

[38] United States Fidelity & Guar. v. Orlando Utils. Comm'n, 564 F. Supp. 962 (M.D. Fla. 1983).

[39] E. Barba & J. Kozek, *Schedule Delay Analysis: Passing Muster Before the Courts*, 8 Constr. Litig. Rep. 150 (1987).

[40] *Id.*

The proper analysis for constructive acceleration effort depends on the nature of the excusable delay. If a contractor experiences an excusable delay that affects the total project, like a labor strike, then the contractor should proceed as it would under directed acceleration. The contractor should compare the as-planned and the as-built schedules with the impacted schedule to show the impact of the project-wide excusable delay upon the critical path and the time for performance.[41] This enables the contractor to establish that it performed ahead of the impacted completion date, which is the first step in proving acceleration effort. The analysis, however, is more difficult when the contractor claims that there was a delay to an individual work activity, such as a delay in delivery of owner-furnished materials needed for the activity. In that case, the contractor has to prove a causal connection between the excusable delay and the contractor's acceleration effort. To establish such a connection, the contractor has to show that the delay to the individual work activity impacted or extended the baseline critical path.

Because all individual critical activities are dependent on one or more other critical activities, any delay to a critical activity (critical delay) extends the critical path and causes a delay to the contractor's timely performance.[42] Therefore, when a contractor can show a delay to an individual critical activity, and the activity was still completed earlier than the impacted schedule completion, establishing the contractor's acceleration effort is relatively easy. The contractor, once again, would proceed as it would when attempting to prove acceleration effort in a claim of directed acceleration. The problem, however, is when the contractor has experienced a delay to an individual noncritical activity.

Noncritical activities are activities not on the critical path. Noncritical activities have *float,* which is the amount of time a noncritical activity can be started late or completed late without impacting the critical path and project completion date. Most courts hold that an excusable delay to a noncritical activity that is not long enough to cause the activity to become critical cannot be the basis of a claim of constructive acceleration.[43] When a delay to an individual activity does not extend the critical path, the contractor's time to perform the total project has not been adversely impacted and the individual delay has not caused a delay to the project. Because there is no causal relationship between the noncritical delay and any delays in the contractor's construction of the total project, the

---

[41] Kenneth Reed Constr. Corp., ENGBCA Nos. 2748, 2749–50, 2861, 72-1 B.C.A. (CCH) ¶ 9,407.

[42] United States Fidelity & Guar. v. Orlando Utils. Comm'n, 564 F. Supp. 962, 968 (M.D. Fla. 1983).

[43] *See, e.g.,* Tri-Cor, Inc. v. United States, 458 F.2d 112, 131 (Ct. Cl. 1972).

contractor is not entitled to any extensions and, consequently, would not be able to prove a claim of constructive acceleration.[44]

This does not mean, however, that a contractor will never be able to prove acceleration effort when it encounters only delays to individual noncritical activities. If the noncritical delay exceeds the noncritical activity's float, then the noncritical activity in effect becomes a critical activity, and the delay to it becomes a critical delay that will impact the contractor's time for performance of the project.[45] The duration of the critical delay—not including the period during which the delay was a noncritical delay—will be reflected by the impacted schedule. A comparison of the as-planned baseline schedule with the impacted schedule enables the contractor to establish that it experienced a delay for which it did not receive a justifiable time extension. This entire analysis, however, depends on whether the contractor or the owner owns the float on noncritical activities. But a discussion of current arguments over float ownership and how it affects delays to the critical path is beyond the scope of this chapter.[46]

§ 7.6 Proof of Acceleration

A contractor bringing an acceleration claim needs to have documentary proof to substantiate its claim. The trier of fact, whether it be a judge, jury, mediator, or arbitrator, needs reliable progress schedules with which to gauge the impacts of delays and the resulting acceleration.

In certain cases, a bar chart will suffice.[47] But for acceleration claims, a contractor should be prepared to produce an as-planned (baseline) CPM schedule, an as-built CPM schedule, and, if necessary, an impacted schedule.[48]

The importance of reliability cannot be overemphasized. Because an as-built schedule reflects all changes, modifications, and delays that occurred during the construction process, the contractor should not wait until construction is complete to draft an as-built CPM schedule.[49] To ensure the accuracy and reliability of the as-built schedule, the contractor should continually update its as-planned schedule as changes, modifications, and delays occur, and it then should use the final modified

---

[44] Nello L. Teer Co. v. Washington Metro. Area Transit Auth., 695 F. Supp. 583 (D.D.C. 1988).

[45] United States Fidelity & Guar. v. Orlando Utils. Comm'n, 564 F. Supp. 962, 968 (M.D. Fla. 1983).

[46] See M. Callahan & H. Hohns, Construction Schedules 105-11 (1983).

[47] BECO Corp., ASBCA No. 27090, 82-2 B.C.A. (CCH) ¶ 16,124.

[48] Minmar Builders, Inc., ASBCA No. 3430, 72-2 B.C.A. (CCH) ¶ 9,599.

[49] Department of the Army, Office of the Chief of Engineers, Pub. No. 415-1-3, Modification Impact Evaluation Guide 3-2 (July 1979).

as-planned schedule as the as-built schedule. This is important because, if the contractor does not maintain an updated CPM schedule and the owner does, the court will most likely use the owner's schedule in determining whether the contractor was accelerated.[50]

Substantiating the as-built schedule and the impacted schedule with documentation is equally important. Any identification of an acceleration effort through an analysis of CPM schedules is meaningless unless the contractor can support and prove the schedules being analyzed. To substantiate the as-built schedule, the contractor should keep detailed project records that are periodically updated. These records should provide information on changes and modifications to the total project and to individual work activities.[51] Examples of project records are CPM schedules, print-outs of computerized window analysis, and work force charts. Documentation is discussed in § 7.12.

For the impacted schedule, the contractor should keep records of all correspondence and documents relating to excusable delays. For example, if the owner issues a change order or a suspension of work order, the contractor should keep that order on file so it can produce and prove it for the court if necessary. Likewise, if the contractor's labor force goes on strike, the contractor should keep copies of any and all correspondence that can be used to document the delay. This might be a written letter from a union representative giving notice of the strike, or it might be a written notice from the contractor to the owner of the contractor's delay as a result of the strike.

The importance of being able to document what changes or modifications caused the delay cannot be overemphasized. If there are excusable and nonexcusable delays and the contractor cannot show that the excusable delays caused the delay in its performance, then it will not be able to recover on its acceleration claim.[52] The contractor therefore should make a conscientious effort to keep accurate and detailed records. Its ability to recover on an acceleration claim will depend on how well it can substantiate that claim with documentation.

§ 7.7 Elements of an Acceleration Claim

Establishing acceleration of its performance will not by itself enable the contractor to recover on an acceleration claim. A contractor who brings a claim of directed acceleration against an owner must prove that

---

[50] Natkin & Co. v. George A. Fuller Co., 347 F. Supp. 17, 20 (W.D. Mo. 1972).
[51] E. Barba & J. Kozek, *Schedule Delay Analysis: Passing Muster Before the Courts,* 8 Constr. Litig. Rep. 150, 152 (1987).
[52] American Sanitary Sales Co. v. State, 178 N.J. Super. 429, 429 A.2d 403 (Div. 1981).

1. The owner ordered the contractor to accelerate the contractor's performance
2. The contractor reasonably attempted to accelerate its performance
3. The contractor incurred additional costs as a result of the acceleration.[53]

If the contractor is bringing a claim of constructive acceleration, it must also prove that

1. The contractor encountered one or more excusable delays that entitled the contractor to a time extension
2. The contractor requested a time extension on a timely basis
3. The owner refused or failed to grant the contractor's time extension and required the contractor to complete performance within the original contract date.[54]

§ 7.8 —Acceleration Orders

The contractor's first step in proving the liability of the owner is to establish that the owner ordered the contractor to accelerate its performance. The contractor does not have to prove a specific and direct command by the owner to accelerate. It has been held that a request to accelerate or an expression of concern over lagging progress, under certain circumstances, constitutes an order to accelerate.[55] Thus, a letter that reads, "I request that you take positive action to expedite the work by supplying the job with all materials necessary to accelerate progress," would be considered an order to accelerate.[56] Also, insistence by an owner that a contractor add employees and work overtime to complete work would be deemed an acceleration order.[57] However, to be construed as an order, the request must in some way pressure or coerce the contractor to accelerate its performance. A mere request by the owner for additional information on a claimed time extension or for an updated CPM schedule is not an order to accelerate.[58]

---

[53] Norair Eng'g Corp. v. United States, 666 F.2d 546, 548 (Ct. Cl. 1981).

[54] M.S.I. Corp., GSBCA No. 2429, 69-1 B.C.A. (CCH) ¶ 7,750 & B.C.A. (CCH) ¶ 7,377 (1968).

[55] Hyde Constr. Co., ASBCA No. 8393, 1963 B.C.A. (CCH) ¶ 3,911 at 19,391.

[56] Norair Eng'g Corp. v. United States, 666 F.2d 546 (Ct. Cl. 1981).

[57] Natkin & Co. v. George A. Fuller Co., 347 F. Supp. 17, 26 (W.D. Mo. 1972).

[58] Nello L. Teer Co. v. Washington Metro. Area Transit Auth., 695 F. Supp. 583 (D.D.C. 1988).

After the contractor proves it was ordered to accelerate, it must then prove that it reasonably attempted to accelerate. See § 7.5. The contractor, however, does not have to prove that it completed construction by the owner's accelerated date in order for its acceleration claim to stand.[59] The contractor only needs to show that it incurred additional costs in a reasonable effort to accelerate.[60] Thus, a contractor who does not perform by the owner's deadline can still recover as long as it performs ahead of either the original contract date in cases of direct acceleration, or the adjusted contract date in cases of constructive acceleration. For example, a contractor who completes construction after the original contract date but before the adjusted contract date has a valid claim for constructive acceleration.[61]

§ 7.9 —Acceleration Costs

The most important element of an acceleration claim is the additional costs incurred by the contractor. Costs that are commonly incurred as a result of acceleration are:

1. Overtime costs
2. Additional labor costs
3. Stacking of trades costs
4. Loss of labor efficiency costs
5. Additional equipment costs
6. Additional supervision costs
7. Increased material delivery costs
8. Increased overhead costs.[62]

A contractor cannot recover all its costs; it can recover only those costs that exceed the costs it would have incurred if it had not been accelerated.[63] Thus, the contractor must be able to show the difference between its costs before and after acceleration.[64]

The contractor can prove the cost difference by several different methods. With the *total cost method,* the contractor can show its additional

---

[59] M.S.I. Corp., GSBCA No. 2429, 69-1 B.C.A. (CCH) ¶ 7,750 at 36,316 (1968).

[60] Varo, Inc., ASBCA No. 15000, 72-2 B.C.A. (CCH) ¶ 9,717.

[61] Mobile Chem. Co. v. Blount Bros. Corp., 809 F.2d 1175, 1177 (5th Cir. 1987).

[62] R. Martell & M. Bash, *Delay Claims,* in 12th Annual Construction Contract Litigation Seminar § 3.9 (May 13, 1988).

[63] *Id.*

[64] *Id.*

costs by comparing its actual construction costs with its planned construction costs. Under this method, the planned costs serve as a baseline measurement of the amount of costs the contractor would have incurred if construction had gone as planned and the contractor had not been ordered to accelerate its performance effort to overcome delays to the project. Actual costs, on the other hand, indicate the contractor's total cost for the construction project, including any costs incurred by the contractor in its effort to overcome delays to the project. Assuming no nonexcusable delays or modifications, a comparison of the planned and actual costs indicates the amount of additional costs, if any, attributable to the contractor's claimed acceleration effort.

For the total cost method to be effective, the contractor must be able to accurately establish its planned costs. This requires the use and analysis of many of the contractor's preconstruction schedules and estimates. For example, to prove increased labor costs, the contractor needs to use its original labor estimates and employee hour charts as a baseline measurement. Similarly, to show increased equipment costs, the contractor probably will need to compare its actual equipment costs with its original equipment schedules and rental schedules.

Using a different method of cost calculation, the contractor can establish additional costs by comparing its actual costs prior to acceleration with its actual costs after acceleration. To show an increase in equipment costs as a result of acceleration, the contractor would compare cost figures from its preacceleration equipment schedules with cost figures from its postacceleration equipment schedules. The difference in the two would indicate the increase in costs attributable to the contractor's acceleration effort. Methods of calculating damages are discussed further in § 7.14.

Regardless of the cost method used, the contractor must document its claim of increased costs. To show increased labor hours, it should keep copies of job cost reports, time cards, and employee records. To show additional equipment costs, the contractor should keep copies of jobsite records. Similarly, subcontracting, material, and supply records should be kept to prove increased subcontract and material expediting costs. Contractors should also keep updated daily logs, bar chart schedules, and labor productivity schedules. For example, comparison of preacceleration and postacceleration productivity schedules will enable the contractor to show the court the amount of labor productivity before and after acceleration.[65] Also, this data may enable the contractor to recover increased overhead costs and profits on overtime wages.[66]

---

[65] S. Leo Harmony, Inc. v. Binks Mfg. Co., 597 F. Supp. 1014 (S.D.N.Y. 1984).
[66] *Id.*

§ 7.10 —Time Extensions

To prove directed acceleration, the contractor has to prove only that it was directed to accelerate, it attempted to accelerate, and it incurred additional costs as a result of its effort to accelerate. In claims of constructive acceleration, however, the contractor has to prove three additional elements. The first is that the delay in its performance was caused by an excusable delay.

Generally, whether a delay is an excusable or nonexcusable delay is a matter of contract interpretation.[67] Most construction contracts attempt to specifically delineate what will constitute excusable and nonexcusable delays. However, given the complexities of construction projects, it is impossible to provide for all contingencies. Therefore, almost all construction contracts have a time extension clause which sets out a general standard of excusable delays. This clause often states that excusable delays include all "causes beyond the contractor's control"[68] or "unforeseeable causes beyond the control and without the fault or negligence of the contractor."[69] As expected, this clause often leads to many disputes over what constitutes an excusable delay because of the difficulty of interpreting "unforeseeable."

After the contractor proves that it encountered an excusable delay, it must then establish that it gave the owner timely notice of the excusable delay and that it timely requested a time extension from the owner. Some cases have held that these requirements are waived when the owner has expressly ordered the contractor to accelerate.[70] However, in the absence of such circumstances, a failure to comply with contractual provisions requiring notice and request will probably preclude the contractor from recovering on its acceleration claim.[71]

The final element a contractor needs to prove for constructive acceleration is that the owner failed to grant the contractor a proper time extension. This is not difficult in cases of true constructive acceleration. True constructive acceleration occurs when the owner fails to grant any time

---

[67] B. Bramble & M. Callahan, Construction Delay Claims 25 (John Wiley & Sons 1987).

[68] *See, e.g.,* AIA A201-1976 ¶ 8.3.1.

[69] *Id.*

[70] *See, e.g.,* Norair Eng'g Corp. v. United States, 666 F.2d 546, 548 n.5 (Ct. Cl. 1981).

[71] Rogers Excavating, AGBCA 79-180-4, 7 C.C. ¶ 391 (1984) (the contractor had been accelerated, but the court held for the owner because the contractor had not given notice of the delay pursuant to the contract); Johnson Controls, Inc. v. National Valve & Mfg. Co., 569 F. Supp. 758, 761 (E.D. Okla. 1983) (court held for the owner because the contractor failed to comply with a contract clause requiring the contractor to submit a written application for any time extensions).

extensions to the contractor. Usually, the owner either denies the extension or tells the contractor it will consider the request for a time extension after construction has been completed.

In these cases, the primary issue is whether a time extension should have been granted. If the court finds that a time extension should have been granted, then it will hold that the contractor was constructively accelerated. Likewise, if the court finds that the time extension was properly denied because the delay was nonexcusable, then it will hold for the owner.[72]

The more difficult case to resolve is when the owner grants a time extension to the contractor, but the contractor claims that the time extension was insufficient to cover the excusable delay. This is known as *disputed constructive acceleration,* and the contractor must provide two levels of proof: The contractor first must prove that it was entitled to a time extension, then must prove that the extension granted by the owner was less than the extension to which the contractor was entitled.

If a contractor completes performance 500 days late, but had been granted a 500-day extension by the owner, then the owner will claim that the contractor performed on time. But, what if the contractor felt it was entitled to a 700-day extension? In that case, the contractor will claim that it was constructively accelerated because it was entitled to another 200 days within which to complete work. Whether the contractor will be able to prove its acceleration claim depends on whether it can prove not only that it was entitled to the 500-day extension it received but also that it was entitled to an additional 200-day extension beyond the 500-day extension. Also, the contractor may find that it waived its claim to the additional 200-day extension if it accepted the 500-day extension without taking exception.

A contractor who accelerates its performance without giving the owner notice of all its claimed time extensions may be deemed a volunteer. Thus, a contractor who accepts a 500-day time extension and completes performance within that 500 days but never gives the owner notice that it is entitled to an additional 200-day extension may be found to have voluntarily accelerated itself for that 200-day period.

The same result might be obtained if the contractor executes or accepts a change order for the accelerated work. Because a change order is simply a contract modification, a contractor who accepts or executes a change order may be found to have assented to being accelerated. Likewise, if the change order contains a specified time extension and the contractor does not object to its length, the contractor probably will be contractually

---

[72] Tri-Cor, Inc. v. United States, 458 F.2d 112 (Ct. Cl. 1972); *see also* McDevitt & Street Co. v. Marriott Corp., 713 F. Supp. 906 (E.D. Va. 1989) and Feuerland-Werkstatten GmbH, ASBCA No. 32,970 (1987).

precluded from later asserting that it was entitled to a lengthier time extension.

To avoid the dangers of waiving an additional time extension, the contractor should give the owner written notice of all time extensions it is claiming. If the contractor feels it is being accelerated because it has been granted an inadequate time extension, the contractor should give the owner written notice that it considers the time extension inadequate and that it will expect additional compensation for its acceleration effort. Finally, if the contractor chooses to execute any change orders, the contractor should expressly reserve in the change order its right to seek additional time extensions and to bring a claim to recover any additional costs it incurs.

Although the analysis of disputed constructive acceleration is more complex than that of true constructive acceleration, the results are the same. If the court finds either that the time extension was sufficient to compensate the contractor for the delay or that the contractor waived any additional time extensions, then the court will find no constructive acceleration.[73] On the other hand, if the court finds that the time extension was inadequate and the contractor did not waive the additional time extensions to which it was entitled, then the court will find that the contractor was constructively accelerated.[74]

§ 7.11 Acceleration Provisions

Every owner has the power to accelerate its contractors. The disputes that arise from directed and constructive acceleration question whether the owner had a contractual right to exercise that power. These disputes can be avoided by including in the construction contract acceleration provisions that govern the rights and liabilities of the parties. In general, a contractual right to acceleration is permissible and enforceable. An acceleration provision is nothing more than a contract clause and will be interpreted in accordance with contract law.

There are two basic types of acceleration provisions. The first gives the owner the right to accelerate the contractor when the contractor's performance has been delayed. The second gives the owner the right to accelerate the contractor even if the contractor is performing on schedule. The effect of these provisions is to transfer the risk of breaching the construction contract from the owner to the contractor. When the owner directs the contractor to accelerate pursuant to the provision, the contractor must comply or the contractor, not the owner, will be liable for breach of

---

[73] Essential Constr. Co., ASBCA No. 18,706 (1989).

[74] E.C. Ernst, Co. v. Koppers Co., 626 F.2d 324 (3d Cir. 1980), *on remand, aff'd in part and rev'd in part,* 520 F. Supp. 830, 832-33 (W.D. Pa. 1981).

contract damages. Thus, the acceleration provisions remove any doubt as to whether the owner has the right to accelerate the contractor. However, to be enforceable both provisions must be reasonable.

Whether an acceleration provision is reasonable and enforceable may depend on whether it provides that the contractor will be reimbursed for increased costs. For example, a provision that gives the owner the right to accelerate a contractor who is on schedule or who has encountered an excusable delay probably will not be enforced unless it stipulates that the owner will reimburse the contractor for acceleration costs.[75] But a provision that stipulates that the owner has the right to accelerate the contractor after a nonexcusable delay is reasonable and will be enforced regardless of whether it provides that the contractor will be reimbursed for increased costs.[76]

The following are examples of two enforceable acceleration provisions. The first provision entitles the owner to accelerate a contractor who has been delayed; the second provision entitles the owner to accelerate a contractor who is still on schedule.

Provision 1:

In the event of a nonexcusable delay in the performance or progress of the Work, Owner may direct that the Work be accelerated by means of overtime, additional crews or additional shifts, or resequencing of the Work. All such acceleration shall be at no cost to Owner. In the event of an excusable delay in the performance or progress of the Work, Owner may similarly direct acceleration and Contractor agrees to perform same on the basis of reimbursement of direct cost (i.e., premium portion overtime pay, additional crew, shift, or equipment cost, and such other items of cost requested in advance by Contractor and approved by Owner, which approval will not be unreasonably withheld) plus a fee of ___ percent (___%) of such cost, but expressly waives any other compensation therefor unless otherwise agreed to in writing in advance of performing the accelerated work. In the event of any acceleration requested pursuant to this paragraph, Contractor shall provide promptly a plan including its recommendations for the most effective and economical acceleration.

Provision 2:

Owner shall also have the right to direct that the Work be accelerated by means of overtime, additional crews or additional shifts, or resequencing of the Work notwithstanding that the work is progressing without delay in accordance with the established schedule. Contractor agrees to perform same on the basis of reimbursement of direct cost (i.e., premium portion of overtime pay, additional crew, shift, or equipment cost, and such other items of cost requested in advance by Contractor and approved by Owner which approval will not be unreasonably withheld) plus a fee of ___ percent (___%) of such cost, but expressly

---

[75] M. Simon, Construction Claims and Liability 306 (John Wiley & Sons 1989).

[76] Merritt-Chapman & Scott Corp. v. State, 54 A.D.2d 37, 386 N.Y.S.2d 894 (1976), aff'd, 43 N.Y.2d 690, 371 N.E.2d 790, 401 N.Y.S.2d 28 (1977).

waives any other compensation therefor unless otherwise agreed in writing in advance of performing the accelerated work. Contractor shall again provide promptly a plan including its recommendations for the most effective and economical acceleration.

Utilizing acceleration provisions benefits both the owner and the contractor. With such provisions, the owner knows that it has the legal right to order acceleration, and the contractor knows the costs to which it will be entitled should it be accelerated.

§ 7.12 Documentation

As discussed in § 7.6, documenting the acceleration claim is extremely important. Recovery of all or part of the damages suffered by a contractor on an accelerated project may hinge on the contractor's ability to document the costs. There are several categories of documentation that can prove helpful in developing the case. They are:

1. Notes and meeting minutes
2. Correspondence between parties
3. Detailed job cost information
4. Budgets and estimates
5. Change orders
6. Design changes
7. Project managers' and superintendents' daily log books
8. Job schedules.

Notes, Meeting Minutes, Correspondence, and Log Books

The contractor should maintain a complete job file including notes of conversations and telephone calls, meeting minutes, and all correspondence between the parties involved on the project. Also, the superintendent, project manager, and any other decisionmaking employee should keep a daily log. The daily log books should also be constructed so that they provide a diary of areas worked on each day and changes to the sequence or schedule with reasons for the changes. Comments regarding conversations, directives, interferences, or errors affecting the job should also be recorded. There is very little that is more effective in demonstrating acceleration than a written record of a directive from the owner or owner's representative to accelerate a portion of the work for a given reason. Likewise, a written record of the owner's denial of a justified time extension is very effective in demonstrating constructive acceleration.

The contractor should impress upon its field staff the importance of documenting conversations and telephone calls and maintaining a file of correspondence. It also should impress upon them the danger of including in logs and correspondence emotional comments and nonfactual judgments that might provide a distorted picture of actual field conditions. Whenever possible, the log books should record an estimate of the hours, cost, and delay effect on the job of any recorded condition.

Detailed Job Cost Information, Estimates, and Budgets

Job cost information should be maintained in detail for the entire project. If maintained in computer files, the analysis of acceleration may be made easier.

Typically, contractor's job cost systems accumulate costs into classifications called *cost codes*. Cost codes are broken down into logical segments of cost as opposed to logical segments of work. For example, for concrete work, the cost codes could include excavation, forming, rebar, pouring, and cleanup. Cost codes may correspond to the contract pay or bid items. Contractors usually record progress on a job by cost codes either in units installed (for example, cubic yards of concrete poured) or in percent completed.

Ideally the cost information that should be maintained includes:

1. Daily individual payroll records (by cost code and indicating schedule activity or area worked on). This information should include regular hours, regular labor cost, overtime hours, overtime labor cost, and benefits.
2. Daily, weekly, or monthly equipment charges (by cost code and schedule activity). Information on idle time or working time for equipment is helpful and supplements the documentation in the superintendent's log.
3. Progress information (units installed or percent completed by cost code and schedule activity). This is extremely helpful when measuring the effect on productivity of impacts to the project. It can be recorded at any time on the project, but gives more information when recorded on a weekly basis.
4. Weekly or monthly subcontractor costs (broken down by cost codes or activities worked on).
5. Any other costs affecting the project.
6. Estimates and budgets, especially when prepared in detail at the beginning of the project. These provide a good basis for comparing the costs to perform as well as anticipated productivity. As such, they should be maintained in the project files.

§ 7.12 DOCUMENTATION

Change Orders

Change orders should be maintained in a file in detail with notes and related correspondence as well as recorded in a log. The owner's failure to approve a change order for a condition that is holding up work, and the subsequent failure to grant a time extension, can be constructive acceleration. Proper documentation of the change order process, notification of the owner of the project delay, request for a time extension, and the ultimate approval process can be a valuable tool in development of both the liability and damage aspect of the acceleration claim.

Design Changes

Design changes can delay construction or add extra work. Most major design changes are documented by change orders. Design changes that delay the work by increasing the duration of one or several activities should result in a time extension. When no time extension is granted, the contractor may experience acceleration. The costs relating to this acceleration can be segregated in a separate cost or file. Although this accumulation of costs may not include everything, it will be a good indication of the extent and timing of the damage.

Schedules

As discussed in § 7.5, job schedules are extremely important. They establish the basis upon which acceleration can be measured. The contractor should develop a detailed and accurate schedule at the beginning of the project and update it at least monthly, based on the actual progress and changes to the schedule. Updating the schedule will provide a trail of actual project progress from which an analysis of acceleration of the total project or individual activities can be made. Reasons for changes to the schedule should be documented. Tying changes to a request for acceleration, resequencing, design change, or lack of progress in a given area caused by excusable delays will eliminate the need for interpretation later. The contractor should avoid producing schedule updates throughout the project which show an on-time completion date even after delays have occurred. This tendency to hide delays in construction progress makes the analysis of delays and resulting acceleration more difficult later on. Trying to prove that an event caused a critical delay early on in the project when the contractor's schedule updates show no effect on completion is not an easy task. Showing updates to the owner that reflect delayed project completion and discussing means of remedying them may lead to a request for and agreement to compensate for acceleration.

The contractor should document the project thoroughly, whether a claim is anticipated or not. Failure to do so, and the subsequent development of a claim, will make the calculation of damages and support of liability more time-consuming and possibly more difficult.

PROVING DAMAGES

§ 7.13 Acceleration Damages

Acceleration costs are those costs encountered as a result of a contractor's performing work faster than anticipated. Typical costs encountered in an accelerated environment are:

1. Labor inefficiencies. A loss in productivity can be caused by more or larger crews, resequencing of the work, overtime, or loss of planning time.
2. Increased material costs. Expedited material shipments can increase costs.
3. Excess equipment costs. More equipment may be needed when work is accelerated. The same piece of equipment may have been planned to service two separate activities that now have to take place concurrently.
4. Supervision. When work is accelerated, supervisory needs are increased. This is often a nonlinear or disproportionate increase in costs to properly manage the project.
5. Subcontract costs. In order to finish on time, some work may be subcontracted at a higher cost than that of the contractor's own labor force. The difference in the costs are damages.
6. Overhead. Greater home office time is typically spent on the accelerated project as its timely completion has become critical.
7. Subcontractor claims. Acceleration to a general contractor often creates extra costs for the subcontractors on the project. Claims can be expected.
8. Other costs may be incurred in individual situations and care must be made to identify the affect of the acceleration.

To simplify the analysis of acceleration claims, it is assumed that only acceleration has been encountered, although acceleration is often found in conjunction with disruption and delay on a project (see **Chapter 6**). The damages caused by the disruption and delay may be so intertwined with those caused by acceleration as to make it difficult to separate them.

§ 7.14 Legally Recognized Measures of Damages

The damages incurred by a contractor are those excess costs caused by the acceleration and not by the contractor's own inefficiencies or actions. Legally recognized damages are those which take into account the contractor's own contribution to extra costs.

Total Cost Approach. The total cost approach to measurement of damages calculates all project costs incurred on the project by the contractor and subtracts from that an estimate of the costs or a benchmark cost for the project. A typical type of estimate or benchmark would be the contractor's original planned costs for the project. Any amount in excess of the estimate or benchmark is considered the amount by which the contractor was damaged. This approach has several built-in assumptions:

1. The contractor did not contribute to the excess costs in any way
2. The contractor's costs are accurate
3. The estimate is reasonable or a benchmark of planned costs for the project costs can be determined
4. There is no other method available.

In most situations at least one of these assumptions is invalid; therefore, this approach is typically not favored by the courts.

Modified Total Cost Approach. The modified total cost approach is an alternative to the total cost approach. It assumes that the contractor did contribute to the excess costs. Using this approach, the contractor calculates damages using the total cost approach but then subtracts from the damages the excess costs caused by its own actions. Contractor-caused damages can be identified by log books, correspondence, and personal recollection. The amount of damages caused by the contractor can be calculated by pricing the time spent on the particular task at hourly rates. Any delay effect, extra direct costs, and overhead costs should also be considered. Care should be taken to identify all of the damages caused by nonexcusable delays or other contractor action and to assign them an accurate value. This approach is nearly always available.

Discrete Cost Approaches. Discrete cost approaches are those that determine and develop damages by adding up all of the costs attributable to specific incidences or problems and for which the contractor is not responsible (such as an excusable delay or an owner's order to accelerate). This approach can be extremely accurate when detailed cost information for the project is available. The contractor should be careful to identify all of the costs associated with a particular problem to properly calculate damages. In many cases, acceleration of a particular activity or project impacts costs throughout the project, both direct and indirect, rather than just the direct costs of a specific area or activity.

Quantum Meruit Recovery. When a project has changed so significantly as a result of acceleration that the original cost estimate and plan are no longer appropriate, the contractor may be entitled to a quantum meruit recovery. This type of recovery is based on the assumption that the work as performed is a different project than was anticipated because of significant changes. The contractor in this situation should be entitled to recover all reasonable costs to perform the work and a fair overhead and profit contribution. The contractor can, however, only recover reasonable costs and must segregate from the damages any excess costs caused by its own actions.

§ 7.15 Using Available Information

The contractor may be limited in the damage methodology by the information that is available to it. The documentation of the project throughout its performance may ultimately determine the damage approach used and is therefore extremely important. Modern computer-based job cost and payroll systems may allow a sophisticated discrete costs analysis to be performed relatively easily. Job cost systems that don't retain detailed costs over time may still allow a discrete cost analysis to be performed. The contractor may be limited to a modified total cost approach if there is limited documentation. This approach, when properly used, can still arrive at an accurate assessment of damages. Finally, the contractor may be limited in the damage methodology by the construction contract itself. If the contract specifies the manner and means for calculating acceleration damages, then the contract will control over personal preference and will determine the method by which acceleration damages are calculated.

§ 7.16 Analysis of Actual Costs

Acceleration damage can be determined using any of the approaches described in § 7.14. Acceleration on projects relates to the schedule, which is broken down into activities. It is the schedule that is delayed or accelerated. Damages are measured in costs detailed at a cost code level. Activities and costs codes are rarely in a one-to-one relationship. The contractor must therefore relate schedule issues to costs. This calls for interpretation in the modified total cost approach. Using the discrete cost approach, however, the contractor may be able to measure acceleration for each scheduling activity.

§ 7.16 ANALYSIS OF COSTS

In the total cost approach the contractor adds up the costs on the project in excess of a reasonable estimate or benchmark of the planned costs to determine damages. Using the modified total cost approach the contractor again adds up the costs on the project in excess of a reasonable estimate or benchmark. The contractor then determines, through an analysis of the project history and documentation, any actions or problems caused by the contractor, such as nonexcusable delays, and the costs of those problems. This may involve, for example, the cost of a piece of equipment left out of the estimate in error, excess labor costs caused by misfabrication of material, reperformance of work originally done in error, or productivity problems produced by poor coordination on the project. The total of the contractor-caused problems is then subtracted from the excess costs calculated earlier to determine an estimate of damages caused by the owner's acceleration. This method can almost always be used, if the necessary information is available, and is preferable to the total cost approach.

Calculations of damages using a discrete cost approach vary widely, based on the issues and specific problems encountered on a project as well as the documentation available. The contractor and its attorney and expert witness can develop variations that most accurately calculate the damages. When the contractor's job cost system retains on a weekly or monthly basis the job costs by cost code broken down into labor, equipment, material, subcontract, and other costs, and the labor and equipment hours are maintained at the same level of detail, the contractor may use this approach. Some measure of performance should be determined at the outset. If the contractor maintains performance information on a weekly or monthly basis in either units installed or percentage completed, this can be used. Otherwise, percentage completed by cost code can often be determined through an analysis of the pay requests. The pay requests typically identify the percent completed by pay item or cost code for each period's performance. If these are not available, the percentage of total actual cost for the pay item can be calculated by dividing the pay request amount for a cost code or pay item by the total actual cost appropriately increased by overhead and profit.

The contractor can calculate for each week or month the cost by category to perform a unit of progress. Also helpful is a calculation of labor hours needed to perform a unit of progress (that is, labor productivity). By calculating this for each time period, the contractor can determine how productivity or cost per unit of progress fluctuated over time for each cost code. By comparing this to a benchmark productivity, either from industry standards, the contractor's estimate, or the contractor's own performance during uninterrupted periods of work, the contractor can determine when its performance was affected.

An analysis of those specific time periods in which the acceleration occurred or began allows the determination of damages related to the acceleration. This can be accomplished by measuring the difference between actual performance per unit of progress and the benchmark planned costs during the affected periods and multiplying by the progress performed during those periods. In the case of acceleration that does not affect the entire project, the unaccelerated periods of time may provide a good benchmark for productivity.

The result of this analysis using the discrete cost approach is a calculation of acceleration damages by cost code. The contractor can take this a step further by attaching schedule activity numbers to detailed cost records and performing the analysis for each scheduling activity. This will produce an analysis of damages for each scheduling activity by cost code by time period. When this is done, it is usually very easy to isolate the effect on project costs of acceleration and other events.

The contractor must determine which method is most appropriate, given the documentation, the issues, and the type of damages caused by the acceleration. The focus should be on determining those excess costs caused by owner acceleration. Any method used that accurately calculates these damages is appropriate.

§ 7.17 Reasonableness of Damage Figures

After damages have been calculated, the contractor should determine whether they are reasonable. A good check of this is to determine damages using a total cost approach and then add on an overhead and profit percentage. This results in essentially a quantum meruit recovery amount. From this, the contractor should subtract the total payments received to date and the amount of the claim. If the result is a positive number, the contractor has not claimed all of the costs incurred on the project. This may be because some of the excess costs were the contractor's own fault, or it may be because the contractor failed to include some costs. This should be investigated.

If the result is a negative number, the contractor may be claiming an unreasonable amount. Its claimed damages and payments to date are in excess of the costs incurred. In this situation, the contractor should reevaluate the damage calculation.

§ 7.18 Profit and Overhead Calculations

Contractors are often entitled to recover profit and overhead on excess costs encountered as a result of acceleration. This is based on the theory that as a contractor deploys resources and incurs costs, it also incurs overhead to support those resources. Contractors are typically entitled to recover a fair profit for doing so. The measurement of profit and overhead to be recovered is usually a percentage of direct costs.

One method of determining how much overhead and profit is to be collected is to analyze the company's historical performance to determine the difference between the contract amount and the direct costs on projects. This difference, measured as a percentage of direct costs, can represent the historical profit and overhead percentage achieved on projects.

The contract may specify a percentage to be used for overhead and profit on extra work. This provides an alternative percentage to use in calculating profit and overhead on acceleration damages.

A third method is to use the profit and overhead percentage included in the bid. The reasonableness of the bid should be confirmed, however, before using this method.

In the absence of information to allow any other method, contractually specified amounts, or government agency guidelines for determination of profit and overhead percentages, industry standards may be used. The contractor may be able to determine what others in the industry on projects of this size and risk are charging and apply this percentage to acceleration damages.

§ 7.19 Mitigation and Cost Savings

When faced with acceleration on a project, the contractor is obligated to mitigate the damages whenever reasonable. This may involve using a larger labor force rather than incurring overtime, using extra equipment rather than more labor, or reducing jobsite overhead when an accelerated project is completed earlier. The contractor must use any available and reasonable means it has to reduce the damages. As a practical matter, on an accelerated project the contractor has been asked to perform work quickly and to do everything possible to meet a deadline. It is not always possible for the contractor in that situation to take the time to identify costs savings or the best approach. The contractor is focused not on identifying cost savings but on completing the construction as quickly as possible. Therefore, mitigation opportunities may be minimal.

Cost savings that the contractor achieved in the normal performance of the contract, such as in the buyout of subcontracts and material, do not need to be subtracted from the damages. The contractor would have been entitled to keep these had no acceleration occurred.

§ 7.20 Presentation of Evidence

The importance of the courtroom presentation of evidence cannot be understated. The presentation should be developed in such a way as to get the message across in a simple and understandable way. Graphic exhibits are often the best way of presenting the liability and damage aspects of the acceleration claim. Testimony by an expert witness is another effective way of presenting the case. Both the expert witness and the exhibits should (1) simplify the complex facts, and (2) persuasively present the case.

A variety of exhibits can be used to demonstrate acceleration and the resulting damages. The contractor, its attorney, and any expert witnesses should develop those exhibits that are most effective for the specific circumstances. Some examples are shown below.

Figure 7–1 is a barchart describing the planned, impacted, and actual project time for an accelerated project. The impacted bar illustrates the effect of events on the project's as-planned schedule, and depicts the length of time the construction would have taken had no acceleration occurred. The as-built schedule shows periods of disruption, indicated by the blank areas between the discrete time periods when construction was ongoing. The acceleration for the project is calculated as the length of time between the end of the impacted schedule and the end of the as-built schedule.

Figure 7–2 shows planned and actual labor hours on the project. This shows the higher than planned labor hours occurring late in the project, a result of work acceleration to catch up from earlier delays. **Figure 7–3** illustrates labor productivity over time. In this type of comparison, productivity refers to the number of employee hours per unit installed or completed (the greater the number of hours, the worse the productivity).

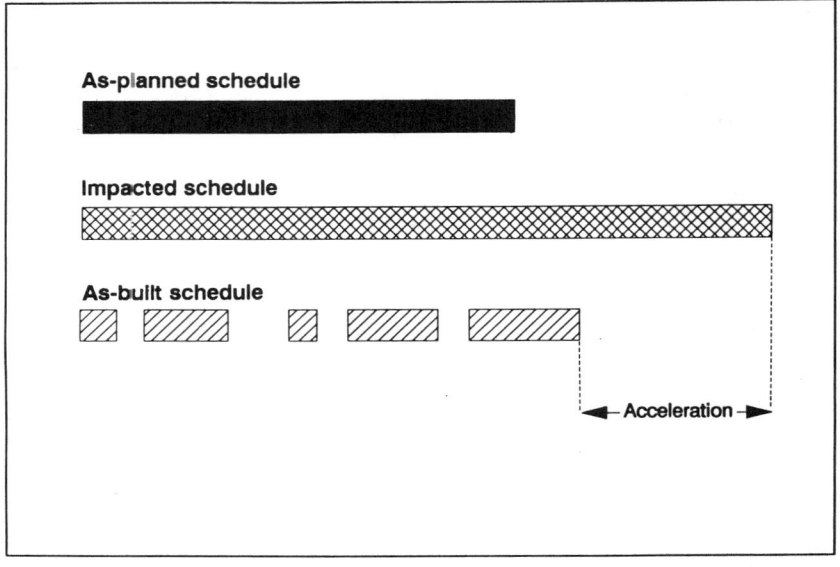

Figure 7–1. Project schedule comparison.

§ 7.20   PRESENTATION OF EVIDENCE   197

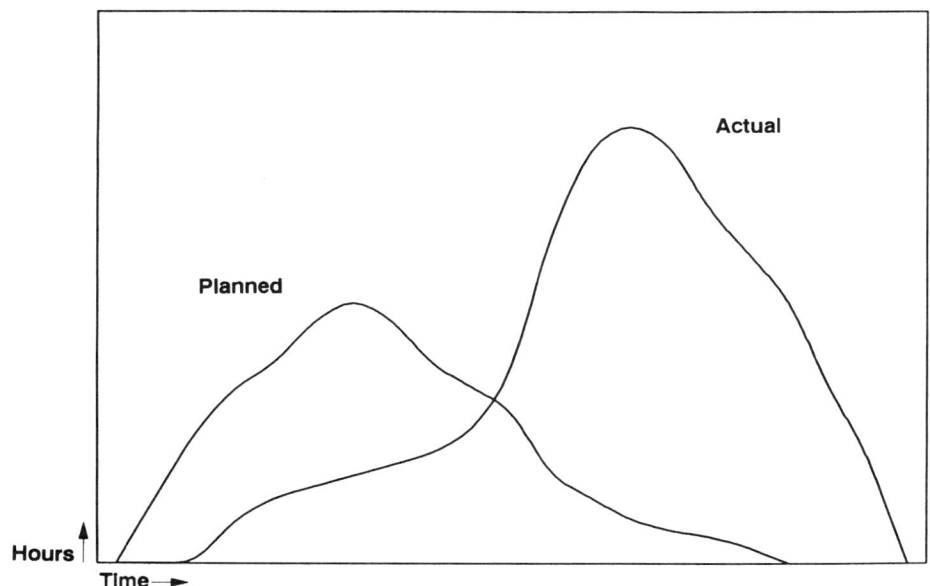

Figure 7–2. Employee hours comparison.

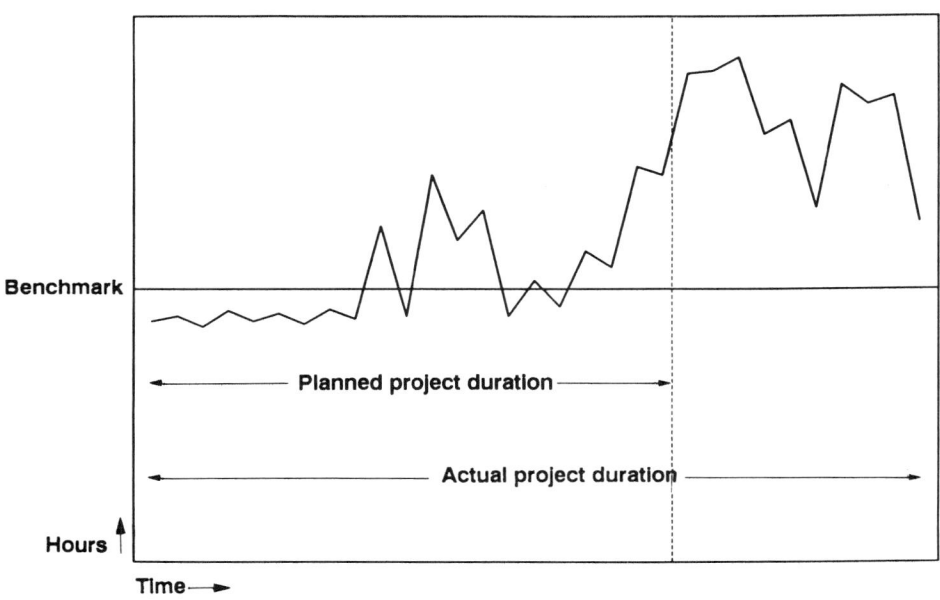

Figure 7–3. Employee hours per unit.

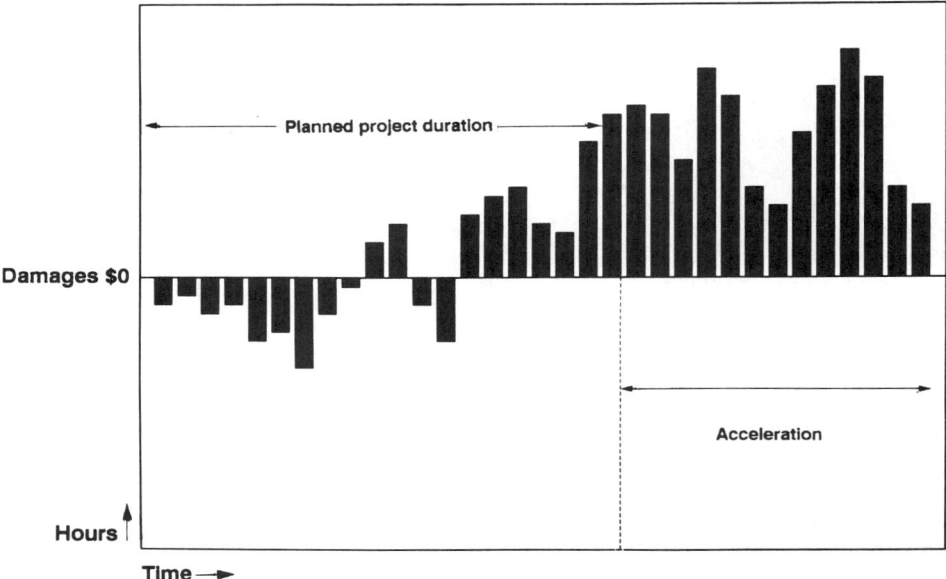

Figure 7-4. Damage summary calculated with employee hours per unit.

Thus, this graph shows that toward the end of the project, the number of labor hours per unit rises, indicating poor productivity. The benchmark productivity could represent an industry standard, an average, or a contractor's estimate (see § 7.16).

Figure 7-4 shows damages resulting from acceleration on the project. The damages are calculated by multiplying the difference between actual and benchmark productivity by the percent completed or installed units for each time period. Here, the contractor's productivity generally was better than expected (the bars below the line indicating fewer employee hours per unit than expected) until the time of acceleration or disruption.

The presentation should be clear and concise. Exhibits should illustrate simply the points in the claim. The presentation of the claim, as the final step in the claim development, is extremely important and should be done in a persuasive and well-thought-out manner.

CHAPTER 8

DIFFERING SITE CONDITIONS CLAIMS

Julian F. Hoffar, Esquire
Donna S. McCaffrey, Esquire
Mark A. Taylor, CPA
John B. Tieder, Jr., Esquire

Julian F. Hoffar, senior partner in the Washington, D.C., law firm of Watt, Tieder, Killian & Hoffar, specializes in national and international contract law. He earned his undergraduate degree from Wittenberg University and is an honor graduate of the National Law Center of George Washington University where he was a member of the Law Review. Mr. Hoffar has written several articles on construction contracting, and has lectured before several industry groups on the subject of construction law.

Donna S. McCaffrey is an associate at the firm of Watt, Tieder, Killian & Hoffar where her practice concentrates on arbitration and litigation of construction disputes. She is a 1984 graduate of the Marshall-Wythe School of Law at the College of William and Mary, and a 1981 graduate of the Catholic University of America. She is a member of the Virginia State Bar, the Federal Bar Association, and the American Bar Association's Public Contract Law Section.

Mark A. Taylor is a Certified Public Accountant and a manager in the Salt Lake City, Utah, office of Coopers & Lybrand. Mr. Taylor has extensive audit experience in a variety of industries, and has devoted substantial time servicing clients involved in the construction industry. Since 1986, he has worked full-time providing support in evaluating, pricing, and refuting complex construction claims for owners, architects, engineers, and contractors from throughout the country. He has been actively involved in assisting clients through all phases of construction claims litigation. Mr. Taylor graduated with a Masters Degree in Accounting from the University of Utah and is a member of the Construction Financial Management Association.

John B. Tieder, Jr., is a senior partner in the Washington, D.C., law firm of Watt, Tieder, Killian & Hoffar. He specializes in the representation of owners, architects, engineers, contractors, and subcontractors in construction matters. Mr. Tieder received his A.B. degree from The Johns Hopkins University in 1968 and his J.D. from the American University Washington College of Law in 1971. Mr. Tieder is the author and coauthor of a variety of publications on construction matters, and is currently a visiting lecturer in government contracts law at the Marshall-Wythe School of Law at the College of William and Mary.

§ 8.1 Introduction

CONTRACTUAL DIFFERING SITE CONDITIONS

§ 8.2 Contract Provisions

§ 8.3 Typical Clauses

§ 8.4 Type 1 Conditions

§ 8.5 —Contract Data versus Informational Data

§ 8.6 —Nature of Contract Indications

§ 8.7 —Lack of Indications

§ 8.8 —Reliance

§ 8.9 —Material Difference

§ 8.10 —Examples of Type 1 Differing Site Conditions

§ 8.11 Type 2 Conditions

§ 8.12 —Examples of Type 2 Differing Site Conditions

§ 8.13 Reverse Differing Site Conditions

§ 8.1 INTRODUCTION

§ 8.14 Notice Requirements
§ 8.15 Effect of Site Investigation and Other Disclaimer Clauses

DIFFERING SITE CONDITIONS WITHOUT CLAUSE

§ 8.16 Remedies in Absence of Contract Clause
§ 8.17 Breach of Warranty
§ 8.18 Misrepresentation and Fraud
§ 8.19 Effect of Disclaimer Clauses

FACTUAL PROOF OF DIFFERING CONDITIONS CLAIM

§ 8.20 Proving a Claim from the Prebid Stage
§ 8.21 Proving a Claim from the Bid Preparation Stage
§ 8.22 Proving a Claim from the Encountered Conditions Phase
§ 8.23 Use of Expert Testimony

PROOF OF DAMAGES

§ 8.24 Impact of a Differing Site Condition
§ 8.25 Impacted Productivity
§ 8.26 —Computation of Damages
§ 8.27 —The Productivity Benchmark
§ 8.28 —The Causal Connection
§ 8.29 —No Unimpacted Period
§ 8.30 —Selecting the Productivity Benchmark
§ 8.31 Other Approaches to Proving Damages
§ 8.32 Other Costs to Consider

§ 8.1 Introduction

This chapter deals with the proof of differing site conditions claims. **Sections 8.2** through **8.12** provide a detailed discussion of recovery under the contract clauses generally in use in the United States; they also discuss the effect of site investigation and the disclaimer-type clauses that are used in attempts to bar or limit differing site conditions claims. **Sections 8.13** through **8.16** address recovery in the absence of a contract clause under either a breach of warranty or misrepresentation/fraud theory. **Sections 8.17** through **8.21** describe how to prove the factual elements that are common to differing site conditions claims regardless of the legal theory upon which a claim is based. **Sections 8.22** through **8.30** describe the types of damages that are incurred when a differing site condition is encountered.

A *differing site condition* is a physical condition other than the weather, climate, or other act of God[1] discovered on or affecting a construction site that differs in some material respect from what reasonably was anticipated.[2] The phraseology sometimes varies; for example, early case law sometimes referred to "changed conditions,"[3] "adverse physical conditions,"[4] or concealed conditions. Regardless of the specific label, however, the condition must be physical. Thus, changes in economic,[5] labor,[6] or political[7] conditions do not constitute differing site conditions. Also, it is not necessary that the condition encountered be in existence at the time of contracting; it may come into existence after the contract award. For example, in *Lee Hoffman v. United States*,[8] when the claimant's construction site was flooded during performance because another government contractor diverted the river upstream, the court held the diversion to be a differing site condition. The project owner must either have knowledge of, be responsible for, and/or have control over the situation which gave rise to the condition.[9] The physical conditions encountered at the construction site must also differ in some "material" respect from the anticipated conditions to constitute a differing site condition. Thus, minor differences in the quantity[10] or quality[11] of a material cannot constitute a differing site condition. The same is true of differences in the character of the material.[12] Finally, a difference in weather or climate, although a physical condition, does not constitute a differing site condition except in those very rare circumstances when unexpected weather conditions interact with some other unforeseen physical condition at the site to create a material difference.[13]

---

[1] Turnkey Enters., Inc. v. United States, 597 F.2d 750, 754, 220 Ct. Cl. 179, 186 (1979).

[2] Foster Constr. C.A. & Williams Bros. Co., A Joint Venture v. United States, 435 F.2d 873, 193 Ct. Cl. 587, 613–14 (1970).

[3] 435 F.2d at 875.

[4] Jack Morehouse dba Morehouse Painting, IBCA No. 2087, 86-3 B.C.A. (CCH) ¶ 19,014.

[5] Western Contracting Corp. v. State Bd. of Equalization, 39 Cal. App. 3d 341, 114 Cal. Rptr. 227 (1974).

[6] Cross Constr. Co., ENGBCA No. 3676, 79-1 B.C.A. (CCH) ¶ 13,707 (1979).

[7] Hallman Bros. v. United States, 68 F. Supp. 204, 107 Ct. Cl. 555, 556 (1946).

[8] 340 F.2d 645, 166 Ct. Cl. 39 (1964).

[9] Frank W. Miller Constr. Co., ASBCA No. 22347, 78-1 B.C.A. (CCH) ¶ 13,039 (1978).

[10] C.E. Wylie Constr. Co., ASBCA No. 26545, 85-1 B.C.A. (CCH) ¶ 17,933 (1985).

[11] Loranger Constr., ASBCA No. 9643, 65-2 B.C.A. (CCH) ¶ 5071 (1965).

[12] Clack v. United States, 395 F.2d 773, 184 Ct. Cl. 40 (1968).

[13] Southern Paving, AGBCA No. 74-103, 77-2 B.C.A. (CCH) ¶ 12,813 (1977).

CONTRACTUAL DIFFERING SITE CONDITIONS

§ 8.2 Contract Provisions

The first and most common basis for establishing differing site condition claims is by operation of remedy-granting clauses in the applicable contract. Although there are many different forms of construction contracts currently in use in the United States, the differing site conditions clauses in standard form contracts are somewhat similar. These clauses typically provide for recovery in two circumstances: (1) when the conditions encountered are materially different from those indicated in the contract documents and (2) when the conditions encountered are materially different from those normally encountered in the type of work being performed under the contract. These two situations are typically referred to as *Type 1* and *Type 2 differing site conditions.*

§ 8.3 Typical Clauses

American Institute of Architects (AIA) Documents A201 and A201(CM) are the AIA's basic forms of fixed price construction contracts between owner and contractor. They are widely used on private construction projects and form the basis of many nonfederal public contracts. Paragraph 4.3.6 addresses differing site conditions and provides for recovery for both Type 1 and Type 2 conditions.

Article 4 of the Standard General Conditions of the construction contract prepared by the Engineers Joint Contract Documents Committee (EJCDC)[14] is more detailed than the AIA clause. It distinguishes between physical conditions in general and physical conditions in the form of underground facilities in particular. With regard to physical conditions in general, the contractor can recover only if the data provided were inaccurate or the conditions encountered differed from those indicated. This covers only a Type 1 differing site condition. With regard to underground facilities, clause 4.3.2 seems to combine the requirements of both the Type 1 and Type 2 differing site conditions for underground facilities not shown or indicated. Therefore, in order to recover, the contractor must show that the underground facility was not indicated in the contract documents and that he could not have anticipated it. Section 4.2.1 of the clause distinguishes between technical indications, which are warranted as accurate, and statement of opinion or conclusions, which are not.

---

[14] EJCDC 1910-8 (1983).

Federal Acquisition Regulations (FAR) § 52.236-2 requires the inclusion of the following clause in all fixed-price construction contracts let by United States government agencies:

Differing Site Conditions (APR 1984)

(a) The Contractor shall promptly, and before the conditions are disturbed, give a written notice to the Contracting Officer of (1) *subsurface or latent physical conditions at the site which differ materially from those indicated in this contract,* or (2) *unknown physical conditions at the site, of an unusual nature, which differ materially from those ordinarily encountered and generally recognized as inhering in work of the character provided for in the contract.*

(b) The Contracting Officer shall investigate the site conditions promptly after receiving the notice. If the conditions do materially so differ and cause an increase or decrease in the Contractor's cost of, or the time required for, performing any part of the work under this contract, whether or not changed as a result of the conditions, an equitable adjustment shall be made under this clause and the contract modified in writing accordingly.

(c) No request by the Contractor for an equitable adjustment to the contract under this clause shall be allowed, unless the Contract has given the written notice required; *provided,* that the time prescribed in (a) above for given written notice may be extended by the Contracting Officer.

(d) No request by the Contractor for an equitable adjustment to the contract for differing site conditions shall be allowed if made after final payment under this contract.

In addition to the FAR requirement, many federal agencies require the inclusion of a differing site condition clause in any contract awarded under a federal grant. For example, the Environmental Protection Agency requires all its grantees to include a clause almost identical to that required by FAR § 52.236-2 in all construction contracts.[15]

As can be seen from the emphasized language, the clauses required by FAR § 52.236-2 and the Environmental Protection Agency provide for both Type 1 and Type 2 differing site conditions claims.

§ 8.4 Type 1 Conditions

Although there may be slight differences in wording, the proof of a Type 1 differing site condition under most contracts consists of six elements:

1. The contract documents must have made express or implied indications regarding the concealed conditions that form the basis of the claim

---

[15] 40 C.F.R. p. 35, subpt. E, app. C-2.

2. The contractor must have interpreted the contract documents as would a reasonably prudent contractor
3. The contractor must have reasonably relied on the indications of concealed conditions contained in the contract when estimating the cost of performance
4. The conditions the contractor encountered at the construction site must have been materially different from the concealed conditions indicated for the same area at the site
5. The actual conditions encountered must have been reasonably unforeseeable
6. The contractor must establish increased costs arising solely from the difference between the indicated and actual conditions.[16]

§ 8.5 —Contract Data versus Informational Data

The first step in establishing a Type 1 differing site conditions claim is to determine what data are actually a part of the contract. These data must be distinguished from data simply provided for the contractor's information and not deemed part of the contract. The significance of this distinction is that a Type 1 differing site conditions claim may only be based on conditions data that are actually included in the contract.[17]

The issue of what is included as part of the contract is an issue of contract interpretation and is, therefore, a question of law to be determined by the court or other appropriate tribunal.[18] In performing that function, the court places itself in the role of a reasonably prudent contractor and decides how such a contractor would interpret the subject contract in a similar situation.[19] If the court determines that certain data are included in the contract, the contractor is bound to have knowledge of them.

A contractor is not deemed to have knowledge of data provided for "information" only. In *Gordon H. Ball, Inc.,*[20] the board of contract appeals held that the following language, included in the section of the contract containing physical data, did not incorporate the informational data referenced in the contract:

(a) Information and data referred to below are furnished for the Contractor's information and for whatever use the Contractor may find therefor.

---

[16] Weeks Dredging & Contracting v. United States, 13 Cl. Ct. 193, 218 (1987), *aff'd,* 861 F.2d 728.

[17] United Contractors, Inc. v. United States, 368 F.2d 585, 595, 177 Ct. Cl. 151, 161 (1966).

[18] P.J. Maffei Bldg. & Wrecking v. United States, 732 F.2d 913, 916 (Fed. Cir. 1984).

[19] *Id.* at 917.

[20] ENGBCA No. 3563, 78-1 B.C.A. (CCH) ¶ 13,055 (1978).

> The subsurface and other physical data such as those mentioned herein and contained in the Contract Documents, or otherwise made available to the Contractor by the Authority, are not intended as representations or warranties. It is expressly understood that the Authority will not be responsible for the completeness or accuracy thereof nor for any deductions, interpretations or conclusions drawn therefrom.[21]

Because the data were not warranted and were not made a part of the contract, the contractor was not bound to have had the knowledge he would have gleaned if he had reviewed the informational materials. It is ironic, however, that in *Gordon H. Ball, Inc.,* the contractor would have learned of the differing site condition if he had reviewed the informational data.

If a contractor does, in fact, review the informational but unincorporated data, it may not rely on an affirmative indication in the data to establish a differing site condition. Thus, in *Turnkey Enterprises, Inc. v. United States,*[22] the contractor relied on a map that was not included as part of the contract documents. The map referenced a water source for the project. The contractor asserted a differing site condition claim because the referenced water source did not exist. The claim was denied, and the court held that the map could not be viewed as a positive indication in the contract documents sufficient to support a claim for a Type 1 differing site condition. The court went on to state that the contractor's interpretation of the map as indicating a water source was unreasonable and that to the extent the contractor relied on the map, he did so at his own peril. A similar result was reached in *P.J. Maffei Building Wrecking Corp. v. United States,*[23] in which the court held that the contractor assumed the risk of inaccurate structural drawings when such drawings were merely referenced in the bid package as available from a third party and were "for information only and will not be part of the contract documents."[24]

One situation that remains troublesome is when the contractor actually reviews the referenced information and such information indicates the existence of a condition actually encountered by the contractor. Because the contractor is not entitled to rely on the data, it should follow that to disregard such information would not prejudice a differing site condition claim. If, however, the referenced data contain indications that are contrary to data set forth in the contract documents, there may be a duty on the part of the contractor to seek clarification of the contract data. This issue does not seem to have been addressed in the reported decisions.

---

[21] *Id.* at ¶ 63,746.
[22] 597 F.2d 750, 754, 220 Ct. Cl. 179, 186 (1979).
[23] 732 F.2d 913, 916 (Fed. Cir. 1984).
[24] *Id.* at 917–18.

§ 8.6 —Nature of Contract Indications

In order to constitute a contract indication, "there must be reasonably plain or positive indications in the bid information or contract documents that such subsurface conditions would be other than actually found."[25] It is not necessary, however, that the contract indications be explicit. The indication need only be enough to lull a reasonable bidder into not expecting the adverse conditions actually encountered.[26] In *Foster Construction v. United States,*[27] the court stated:

> For this part of the Changed Conditions Clause to apply, it is not necessary that the "indications" in the contract be explicit or specific; all that is required is that there be enough of an indication on the face of the contract documents for a bidder reasonably not to expect subsurface or latent physical conditions at the site differing materially from those indicated in this contract.

The standard thus encompasses not only explicit information but information and conclusions that can reasonably be drawn from all parts of the contract. For example, in *Ideker, Inc. v. Missouri State Highway Commission,*[28] the court found that the lines and drawings in the plans made positive representations of site conditions and that the "use of conventional syntax is not the only means of making a positive representation of site conditions in a contract."[29]

The types of data that can constitute contract indications are extensive. The most common are set forth below.

Data regarding subsurface conditions. The most common indications of physical conditions are boring logs and other reports of the subsurface conditions. At least one case has described the boring data as "the most reliable reflection of subsurface conditions."[30] Soils reports enjoy a similar status. Thus, a contractor was held to be bound to follow detailed data on the physical characteristics of the soil, such as "core samples" and the results of "standard penetration tests," rather than a more general description

---

[25] Pacific Alaska Contractors, Inc. v. United States, 436 F.2d 461, 469, 193 Ct. Cl. 850 (1971).

[26] Metropolitan Sewerage Comm'n v. R.W. Constr., Inc., 72 Wis. 2d 365, 241 N.W.2d 371 (1976).

[27] 435 F.2d 873, 193 Ct. Cl. 587, 613–14 (1970).

[28] 654 S.W.2d 617 (Mo. Ct. App. 1983).

[29] *Id.* at 622.

[30] Weeks Dredging & Contracting v. United States, 861 F.2d 728, 13 Cl. Ct. 193, 208–09 (1987).

such as "sand."[31] In another case, a reference to bedrock in a soils report was held to be a sufficient indication of such bedrock even though other portions of the contract did not provide such information.[32]

Owner interpretations of contract data. On some contracts, the owner goes beyond providing raw data regarding the results of prebid geotechnical investigations and provides bidders with interpretations of the data by government experts. A contractor is entitled to rely on these conclusions. For example, in *Womack & Vorhies v. United States*,[33] the contractor based its bid on government estimates of quantities. The court upheld the contractor's claim for differing site conditions and stated that:

> Intrinsically, the estimate that is made in such circumstances must be the product of such relevant underlying information as is available to the author of the invitation. If the bidder were not entitled to so regard it, its inclusion in the invitation would be surplusage at best or deception at worst. Assuming that the bidder acts reasonably, he is entitled to rely on Government estimates as representing honest and informed conclusions.[34]

It should be noted that a claim submitted under the National Society for Professional Engineers (NSPE) contract (prepared by the Engineers Joint Contract Documents Committee) cannot be based on inaccurate conclusions. Section 4.2.1 expressly disclaims any warranty of opinions or conclusions.

Inferences from design details. In addition to the data that expressly indicate the anticipated physical conditions at the site, the contract specifications indirectly may give some indication of the government's anticipation regarding site conditions. A contractor is entitled to rely on any portion of the specifications that gives some indication of the site conditions.[35] The contractor's interpretation of the contract specifications, and its reliance on that interpretation, must be reasonable under the circumstances.[36]

Thus, when the contract specifications required that certain work be performed "in the dry," the contractor was entitled to regard that as an indication that the soil to be excavated was relatively impermeable to water.[37] Similarly, the specified compaction requirements for a site could be

---

[31] Guy F. Atkinson Co., ENGBCA No. 4693, 87-3 B.C.A. (CCH) ¶ 19,971 (1987).

[32] Coastal, Inc., PSBCA No. 1728, 88-1 B.C.A. (CCH) ¶ 20,272 (1988).

[33] 389 F.2d 793, 182 Ct. Cl. 399 (1968).

[34] *Id.* at 801.

[35] Fox v. United States, 7 Cl. Ct. 60 (1984).

[36] *Id.* at 65.

[37] Foster Constr. C.A. & Williams Bros. Co., A Joint Venture v. United States, 435 F.2d 873, 193 Ct. Cl. 587 (1970).

regarded as an indication that the soil conditions were of such a nature that the compaction requirements could be met.[38] In another case, the contract specifications showed that the project was designed to be a "balanced job;" the contractor was entitled to interpret that as an indication that his bid need not include costs for the disposal of waste.[39] Finally, when the contract specifications required that rock be excavated to a close tolerance, the board held that it was an indication that performance was feasible with customary construction procedures and excessive overbreakage of the rock constituted a differing site condition.[40]

§ 8.7 —Lack of Indications

If there is no information in the contract regarding a physical condition at the construction site, then there are no indications on which a Type 1 differing site condition claim may be based.[41] Thus, in a case in which the contract specifications did not indicate the depth at which building service lateral pipe would be connected to sewer lines, the contractor was not entitled to a Type 1 differing site condition claim for the costs incurred when he was required to excavate at a greater depth than anticipated.[42]

A contractor does have an obligation to inquire about clearly conflicting or missing information. In a case in which the bid documents stated that there were concrete slabs beneath a paved area, but this information was not provided on the drawing details, nor was it ascertainable by a pre-bid site inspection, a board of contract appeals held the contractor had a duty to inquire as to whether information was missing.[43] The board found that there was an obvious discrepancy in the contract documents and that the contractor had a duty to clarify it prior to bid.[44] The claim was therefore denied.

§ 8.8 —Reliance

In order to recover for a Type 1 differing site condition, a contractor must actually rely on the information indicated in the contract. It is not enough

---

[38] Kinetic Builders, Inc., ASBCA No. 32627, 88-2 B.C.A. (CCH) ¶ 20,657 (1988).

[39] Ideker, Inc. v. Missouri State Highway Comm'n, 654 S.W.2d 617 (Mo. Ct. App. 1983).

[40] J.F. Whalen, ENGBCA No. 2859, 69-1 B.C.A. (CCH) ¶ 7519 (1969).

[41] Weeks Dredging & Contracting v. United States, 861 F.2d 728, 13 Cl. Ct. 193, 219 (1987).

[42] Sandwich Islands Constr., ASBCA No. 35244, 88-3 B.C.A. (CCH) ¶ 21,143 (1988).

[43] Giuliani Contracting Co., ASBCA No. 33341, 87-2 B.C.A. (CCH) ¶ 19,743 (1987).

[44] *Id. See also* Leavell & Co., C.H., ASBCA No. 18625, 74-2 B.C.A. (CCH) ¶ 10,885 (1974).

to prove that there is a difference between the indicated and actual conditions at the site. The contractor must show that it in some way relied to its detriment on the information about physical conditions set forth in the contract. Thus, in the *Appeal of Dravo Corp.*,[45] the contractor clearly established that there was a difference between the indicated and actual shear strengths of the soil through which he was excavating. The contractor was not able to prove, however, that he placed any reliance on the indicated shear strengths when preparing his bid. Therefore, the claim was denied. The Claims Court reached a similar result in *P.J. Maffei Building & Wrecking v. United States*,[46] because the evidence established that the contractor recognized that documents upon which he allegedly relied to prepare his bid were preliminary and might not reflect the actual conditions at the site.

In order to recover under a differing site conditions clause, the contractor must also show that its reliance on the contract indications and its interpretation of that information were reasonable. "[W]hile a contractor need not demonstrate that its interpretation is the only reasonable one, it does bear the burden of showing that its construction is at least a reasonable reading."[47] Therefore, in *Titan Midwest Construction Corp.*,[48] the board denied a differing site conditions claim because the contractor's reliance on the government's boring logs was unreasonable. The board stated that there is no authority for the contractor's assumption that boring results are representative of conditions at unsampled locations or elevations.

§ 8.9 —Material Difference

Once the contractor has shown that there is a difference between the expected conditions and those actually encountered at the site, the contractor must then prove that the difference is *material*. This means that the difference must be such that it has an impact, usually negative, on the contractor's production and costs. The determination of materiality, to a large extent, depends heavily on the particular facts and circumstances of a particular case.

Some cases have determined materiality based on the percentage difference between actual and expected quantities. For example, in *Yukon Construction Corp.*,[49] the board found that a 16 percent overrun of excavation was not a material variance because the excavation was "unclassified" to allow for certain variations, and it was done without unusual difficulty.

---

[45] ENGBCA No. 3901, 80-2 B.C.A. (CCH) ¶ 14,757 (1980).
[46] 732 F.2d 913 (Fed. Cir. 1984).
[47] *Id.* at 918.
[48] ASBCA No. 23594, 81-1 B.C.A. (CCH) ¶ 15,067 (1981).
[49] FAACAP No. 66-4, 65-2 B.C.A. (CCH) ¶ 5005 (1965).

The board also determined that a 25 percent, rather than a 16 percent, overrun is "an appropriate and reasonable indication of what is a material variation" but went on to state that

> This percentage would not, of course, necessarily apply in all cases. It is simply a rule-of-thumb, a guide to the exercise of judgment which is to be applied in the light of all the circumstances and not an arbitrary plimsoll mark which automatically applies.[50]

Materiality may sometimes be determined by analyzing the nature of the conditions. Thus, in *Bregman Construction Corp.*,[51] the board found a material variation when there were differences in the behavior and workability of the soils anticipated and those encountered. The board determined that the economic burden of working with a more difficult soil was material.[52]

Finally, the physical condition may be stated as an exact quantity. In *State Road Department v. Houdaille Industries*,[53] the court found a clearly material variance when the contract documents indicated only 408 cubic yards of muck to be removed and actual conditions necessitated the removal of 278,392 cubic yards of muck.

§ 8.10 Examples of Type 1 Differing Site Conditions

There are innumerable examples of Type 1 differing site conditions. Some of the more common are described below.

Rock.

1. The encountering of rock in an area where no rock was indicated by the contract documents is a differing site condition.[54]
2. The encountering of more and different rock than was indicated in the contract documents is a differing site condition.[55]
3. Encountering of rock that is substantially harder to drill than the contract indications would suggest is a differing site condition.[56]

---

[50] *Id.* at ¶ 23,601.
[51] ASBCA No. 9000, 64 B.C.A. (CCH) ¶ 4426 (1964).
[52] *Id.* at ¶ 21,335.
[53] 237 So. 2d 270 (Fla. Dist. Ct. App. 1970).
[54] Bernard McMenamy Contractor, Inc., ENGBCA No. 3413, 77-1 B.C.A. (CCH) ¶ 12,335 (1977).
[55] Jackson-Swindell-Dressler, A Joint Venture, ENGBCA No. 3614, 76-2 B.C.A. (CCH) ¶ 12,222 (1976).
[56] American Dredging Co. v. United States, 107 Ct. Cl. 1010 (1975).

4. Encountering rock in a different location than that indicated in the contract borings has been held to constitute a differing site condition.[57]
5. The board found a differing site condition when the contractor relied on government boring logs that indicated the presence of crushed limestone rather than the rock that the contractor actually encountered.[58]

Water.

1. The board found a differing site condition when a sudden flow of groundwater and sewage flowed into an excavation that had been open three or four months.[59]
2. Encountering subsurface mud when the state's soil boring logs indicated that no subsurface water was present and the prebid site inspection revealed only a dry, cracked surface is a differing site condition.[60]
3. Encountering artesian water, which is under pressure and therefore required a greater pumping capacity to reduce the water table than the static water indicated in the contract documents, is a differing site condition.[61]
4. The board held that, although a careful prebid review of test boring and visible evidence would put the bidder on notice that some moisture might be encountered, the presence of a water aquifer was a differing site condition.[62]
5. Encountering groundwater at an unforeseeably high level is a differing site condition.[63]

Soil conditions.

1. Although the "running sand" contractor encountered was of the same classification and strata as that indicated in the contract documents, the board found an equitable adjustment because the soil did

---

[57] Accent General, Inc., ASBCA No. 28813, 87-2 B.C.A. (CCH) ¶ 19,689 (1987).
[58] Francisco Levy Hijo, Inc., ASBCA No. 15144, 73-1 B.C.A. (CCH) ¶ 9,817 (1973).
[59] Norair Eng'g Corp., ENGBCA No. 3568, 77-1 B.C.A. (CCH) ¶ 12,225 (1977).
[60] Beco Corp. v. Roberts & Sons Constr. Co., 114 Idaho 704, 760 P.2d 1120 (1988).
[61] Metropolitan Sewerage Comm'n v. R.W. Constr., Inc., 72 Wis. 2d 365, 241 N.W.2d 371 (1976).
[62] Louis M. McMasters, Inc., ASBCA No. 80-159-4, 86-3 B.C.A. (CCH) ¶ 19,067 (1986).
[63] Town of Longboat Key v. Carl E. Widell & Son, 362 So. 2d 719 (Fla. Dist. Ct. App. 1978).

not behave as indicated and necessitated the use of a more costly excavation technique.[64]

2. Encountering soil that is much more difficult to compact than that indicated in the bearing ratios is a differing site condition.[65]

3. A contractor encountered a differing site condition when the contract documents indicated dry sand at the site and silt that resisted standard industry drying techniques was present.[66]

Foreign obstructions.

1. During the excavation and installation of a storm drain, the contractor encountered old railroad tracks that were not indicated in the contract documents. The court awarded the contractor his costs for removal of the tracks.[67]

2. The Supreme Court of Pennsylvania found a "differing condition" when the contractor encountered a high pressure gasoline line at an excavation site that was not indicated in the contract documents.[68]

3. The Maryland changed conditions clause applied when a contractor encountered parts of a buried wharf during performance of a contract to lay sewer pipe. The wharf had been covered by natural or manufactured fill.[69]

4. A contractor that encountered large concrete beams, automobile parts, and railroad ties when the boring logs indicated compact rubble recovered for a differing site condition.[70]

Concrete.

1. The court found a differing site condition when during the performance of a contract to build two earth-filled wharves, the contractor encountered concrete pilings. The authority did not disclose an engineering report that would have alerted the contractor to the condition.[71]

---

[64] S&M Traylor Bros., ENGBCA Nos. 3878, 3904, 3943, 82-1 B.C.A. (CCH) ¶ 15,484 (1982).

[65] Titan Atl. Constr. Corp., ASBCA No. 23588, 82-2 B.C.A. (CCH) ¶ 15,808 (1982).

[66] J&T Constr. Co., DOTCAB Nos. 73-4, 75-2 B.C.A. (CCH) ¶ 11,398 (1975).

[67] Haener v. Ada County Highway Dist., 108 Idaho 170, 697 P.2d 1184 (1985).

[68] Teodori v. Penn Hills School Dist. Auth., 413 Pa. 127, 196 A.2d 306 (1964).

[69] James Julian, Inc. v. Town of Elkton, 341 F.2d 205 (4th Cir. 1965).

[70] Maverick Diversified, Inc., ASBCA No. 19838, 76-2 B.C.A. (CCH) ¶ 12,104 (1976).

[71] D. Federico Co. v. New Bedford Redev. Auth., 723 F.2d 122 (1st Cir. 1983).

2. Encountering a wall of rotted wood rather than the buried concrete retaining wall indicated in the contract is a differing site condition.[72]

§ 8.11 Type 2 Conditions

In order to establish a Type 2 differing site condition, the contractor must prove that the conditions encountered were not the type of conditions normally inherent in work of the nature provided for under the contract.[73] Unlike Type 1 differing site conditions that are based on the difference between the contract indications and what was encountered, a *Type 2 differing site condition* requires the existence of an anomaly: a condition an experienced contractor working under similar circumstances would not normally anticipate.

Conditions Normally Inherent

In order to qualify as a Type 2 differing site condition, the condition must be both unknown and unusual. To qualify as unknown, the condition must be one that the contractor could not reasonably anticipate from a study of the contract documents[74] or a reasonable site investigation.[75] To qualify as unusual, the condition must also be something that would be unexpected based on the contractor's background and experience.[76] A frequently cited case is *Community Power Suction Furnace Cleaning Co.*[77] The contractor in that case was awarded a contract to clean out the heating ducts in the barracks at Fort Bragg, North Carolina. In cleaning out the ducts, it found beer cans, jars of foodstuffs, gunpowder, live ammunition, and undergarments. The Armed Services Board of Contract Appeals held that such a condition was unusual.

Experience of Contractor

In a Type 2 differing site condition, the experience of the contractor plays a role.[78] Thus, the fact that a contractor may or may not have had experience with a particular type of work in a particular geographical area is of

---

[72] Coath & Goss, Inc., ASBCA No. 20949, 76-1 B.C.A. (CCH) ¶ 11,887 (1976).

[73] Shank-Artukovich v. United States, 13 Cl. Ct. 346, 350 (1987), *aff'd without opinion*, 848 F.2d 1245 (1988).

[74] B&M Roofing & Painting Co., ASBCA No. 26998, 86-1 B.C.A. (CCH) ¶ 18,833 (1986).

[75] Kahaluu Constr. Co., ASBCA No. 31187, 89-1 B.C.A. (CCH) ¶ 21,308 (1989).

[76] Fort Sill Assocs. v. United States, 183 Ct. Cl. 301 (1968).

[77] ASBCA No. 13803, 69-2 B.C.A. (CCH) ¶ 7963 (1969).

[78] Hoyer Constr. Co., ASBCA No. 21616, 84-2 B.C.A. (CCH) ¶ 17,249 (1984).

significance.[79] The contractor is not held, however, to the level of experience of an expert, but only to the level of a general contractor familiar with the type of work provided for in the contract.[80]

§ 8.12 —Examples of Type 2 Differing Site Conditions

There are also many examples of Type 2 differing site conditions. Some of them are set forth below:

Anomalous geological conditions in the area.

1. The Armed Services Board found that the contractor was entitled to an equitable adjustment because loose cobbles are not normally found in the waters off Bangor, Washington.[81]
2. The tremendous amount of muck encountered at a site was considered unusual because muck is generally found at low elevations and the site was at a high elevation. The site inspection and survey data also did not indicate muck in the area.[82]

Conditions not normally found in similar structures.

1. The board found a Type 2 differing site condition because it is unusual to encounter an interior wall composed of four layers of brick in a hospital constructed in the 1940s.[83]
2. During the performance of a contract requiring the removal of carpet, the contractor encountered a rough and uneven concrete floor that far exceeded the allowable tolerance for laying tiles. This condition is not usually encountered in commercial or government buildings.[84]
3. A contractor was entitled to compensation for a Type 2 differing site condition because he encountered sewer pipes on a runway. This is an unusual condition.[85]
4. A Type 2 differing site condition claim was upheld because the contractor encountered a soft coating beneath four coats of hard paint.

---

[79] *Id.*
[80] Blake Constr. Co., ASBCA No. 20747, 83-1 B.C.A. (CCH) ¶ 16,410 (1983).
[81] Hurlen Constr. Co., ASBCA No. 31069, 86-1 B.C.A. (CCH) ¶ 18,690 (1986).
[82] Kinetic Builders, Inc., ASBCA No. 32627, 88-2 B.C.A. (CCH) ¶ 20,657 (1988).
[83] Hercules Constr. Co., VACAB No. 2508, 88-2 B.C.A. (CCH) ¶ 20,527 (1988).
[84] Yamas Constr. Co., ASBCA No. 27366, 86-3 B.C.A. (CCH) ¶ 19,090 (1986).
[85] Unitec, Inc., ASBCA No. 22025, 79-2 B.C.A. (CCH) ¶ 13,923 (1979).

The recognized custom in the trade is that the undercoats will be harder than the outer coats.[86]

Magnitude of the conditions was more than could be expected.

1. Although weather conditions are not normally considered differing site conditions, a contractor was allowed to recover when excessive rain mixed with the soil conditions at the site and caused a much stickier and harder-to-excavate material than normally expected.[87]
2. Although the contractor was aware of the high moisture content of the soil, the speed at which it caused the soil to deteriorate was unusual and constituted a differing site condition.[88]
3. Although a reasonable contractor would anticipate a certain deviation in the size of the floor joists, it was very unusual for 80 percent of the joists to be severely undersized.[89]
4. Water is encountered on most excavation projects; however, the presence of a sanitary sewer line and a large amount of sewage was considered a Type 2 differing site condition.[90]

§ 8.13 Reverse Differing Site Conditions

The vast majority of differing site condition claims are made as a result of the contractor's seeking an upward equitable adjustment of the contract price because of the difference between the conditions anticipated and those actually encountered. The clauses, however, also cover conditions that are more favorable than those indicated or anticipated. Thus, an owner is entitled to recover from the contractor when conditions are, in fact, better.

For example, in the appeal of *Afgo Engineering Corp.*,[91] the contractor used the government's estimate of 270 cubic yards of rock to be excavated to complete its bid. The contractor actually excavated 102 cubic yards of rock. The differing site conditions clause of the contract provided:

> [I]f . . . such conditions do materially so differ and cause an *increase or decrease in the Contractor's cost* . . . an equitable adjustment shall be made and the contract modified in writing accordingly. [emphasis supplied]

---

[86] Kahaluu Constr. Co., ASBCA No. 31187, 89-1 B.C.A. (CCH) ¶ 21,308 (1989).

[87] Southern Paving, AGBCA 74-103, 77-2 B.C.A. (CCH) ¶ 12,813 (1977).

[88] Mo Co., ASBCA No. 21403, 78-2 B.C.A. (CCH) ¶ 13,313 (1978).

[89] Kos Kam, Inc., ASBCA No. 34684, 88-1 B.C.A. (CCH) ¶ 20,246 (1988).

[90] Baltimore Contractors v. United States, 12 Cl. Ct. 328 (1987).

[91] VACAB No. 1236, 79-2 B.C.A. (CCH) ¶ 13,900 (1979).

§ 8.14 NOTICE REQUIREMENTS 217

In a counterclaim, the government sought a downward equitable adjustment to recover the costs the contractor saved as a result of the 62 percent underrun in rock removal. The board of contract appeals sustained the government's counterclaim, finding that an accurate estimate of rock quantity was not feasible, and there was a differing site condition because the quantity of rock indicated by the site borings was materially different from that actually encountered by the contractor.

§ 8.14 Notice Requirements

Virtually all differing site condition clauses require that notice be given as soon as the condition is encountered or recognized. The purpose of a notice provision is to allow the owner or its representative to investigate the condition and determine if it is in fact different from those indicated or anticipated. When the contractor fails to give notice of the condition, it runs the risk that the claim will be denied, even if it may otherwise have had merit. For example, in *M.D. Activities*,[92] the Interior Board of Contract Appeals denied a contractor's claim because the contractor had not given timely written notice to the contracting officer. The decision seemed to turn, however, on the fact that the government had no actual knowledge of the claim and therefore was unable to investigate it at the time it occurred. The lack of notice thereby prejudiced the owner, and the claim was denied.

When the owner has actual knowledge and is not prejudiced by the lack of notice, the result is frequently different. For example, in *Roger J. Au & Son v. Northeastern Ohio Regional Sewer District*,[93] the court allowed a differing site condition claim even though the contractor did not notify the owner of the condition. The court held that the owner had actual knowledge of the claim and had an opportunity to investigate it, and that there was no prejudice because of the lack of formal notice. The court held: "There is no reason to deny the claim for lack of written notice if the District was aware of differing site conditions throughout the job and had a proper opportunity to investigate and act on its knowledge, as the purpose of the formal notice would thereby have been fulfilled."[94]

Other courts and bodies are not as liberal as the Ohio Appellate Court. The contractor who fails to give notice does so at its peril. For example, in *Emerald Forest Utility District v. Simonsen Construction Co.*,[95] the court denied the contractor recovery when it encountered wet sand

---

[92] IBCA No. 2113, 88-1 B.C.A. (CCH) ¶ 20,328 (1988).
[93] 29 Ohio App. 3d 284, 504 N.E.2d 1209 (1986).
[94] 504 N.E.2d at 1210.
[95] 679 S.W.2d 51 (Tex. Ct. App. 1984).

during performance but failed to notify the owner of the condition as required by the contract. Likewise, in *Blankenship Construction Co. v. North Carolina State Highway Commission*,[96] the court refused to grant the contractor's claim under the differing site conditions clause because of its failure to give proper notice to the state. The court strictly interpreted the notice provisions of the contract and found that the contractor's oral notification was insufficient to apprise the state of the changed condition.

§ 8.15 Effect of Site Investigation and Other Disclaimer Clauses

Many contracts contain a clause that requires the contractor to conduct a site investigation and provides that the owner shall not be responsible for conditions encountered at the site. An example of such a clause is as follows:

SITE INVESTIGATION AND CONDITIONS AFFECTING THE WORK (APR 1984)

(a) The Contractor acknowledges that it has taken steps reasonably necessary to ascertain the nature and location of the work, and that it has investigated and satisfied itself as to the general and local conditions which can affect the work or its costs, including but not limited to (1) conditions bearing upon transportation, disposal, handling, and storage of materials; (2) the availability of labor, water, electric power, and roads; (3) uncertainties of weather, river stages, tides, or similar physical conditions at the site; (4) the conformation and conditions of the ground; and (5) the character of equipment and facilities needed preliminary to and during work performance. The Contractor also acknowledges that it has satisfied itself as to the character, quality, and quantity of surface and subsurface materials or obstacles to be encountered insofar as this information is reasonably ascertainable from an inspection of the site, including all exploratory work done by the Government, as well as from the drawings and specifications made a part of this contract. Any failure of the Contractor to take the actions described and acknowledged in this paragraph will not relieve the Contractor from responsibility for estimating properly the difficulty and cost of successfully performing the work, or for proceeding to successfully perform the work without additional expense to the Government.

(b) The Government assumes no responsibility for any conclusions or interpretations made by the Contractor based on the information made available by the Government. Nor does the Government assume responsibility for any understanding reached or representation made concerning conditions which can affect the work by any of its officers or agents before

---

[96] 28 N.C. App. 593, 222 S.E.2d 452 (1976).

the execution of this contract, unless that understanding or representation is expressly stated in this contract.[97]

In spite of the wording of the site investigation-type clauses, the general rule is that the contractor will only be held to have the knowledge or information that could have been gleaned from a reasonable site investigation.[98] The general test for a reasonable site investigation is what an experienced and prudent contractor could have discovered rather than what a trained geologist or other expert might have found.[99]

If a contractor fails to conduct a site investigation, it will not be compensated for differing site conditions that a reasonable investigation would have revealed. For example, in *Zenith Construction*,[100] the Armed Services Board of Contract Appeals refused to find a differing site condition when the window frames were higher than indicated in the drawings and were not in straight lines. The board found that the contractor would have discovered these conditions if it had attended the scheduled walk-through site investigations. In *Bo McAllister & Lloyd Thompson*,[101] a differing site condition claim was denied because the board found that rock outcroppings encountered during the performance of the project were visible to a person walking or riding a horse, even though the contractors stated that they did not conduct a site investigation because the area was dangerous and full of briars and snakes. Finally, in *AAAA Enterprises, Inc.*,[102] the contractor's claim for a differing site condition was denied because the oil-saturated soil it encountered during contract performance would have been discovered during a prebid site investigation.

A contractor's prior experience on a particular site or in a particular locality will not be a substitute for a reasonable site investigation. Thus, in *Stuyvesant Dredging v. United States*,[103] a contractor who relied on his prior experience in a particular locale was not able to assert a differing site condition claim because a reasonable investigation of the particular site would have revealed the conditions encountered. In addition, the government's failure to disclose site conditions may not protect the contractor who does not investigate the site. In *Cook v. Oklahoma Board of Public Affairs*,[104] a contractor conducted a "drive-through" investigation of the site. It later asserted a claim for the existence of underground water. The

---

[97] F.A.R. § 32.236-3.
[98] North Slope Technical Ltd. v. United States, 14 Cl. Ct. 242, 253 (1988).
[99] Stock & Grove, Inc. v. United States, 493 F.2d 629, 13 Cl. Ct. 209 (1974).
[100] ASBCA No. 33576, 89-3 B.C.A. (CCH) ¶ 21,894 (1989).
[101] IBCA No. 2144, 88-1 B.C.A. (CCH) ¶ 20,329 (1988).
[102] ASBCA No. 28172, 86-1 B.C.A. (CCH) ¶ 18,628 (1986).
[103] 834 F.2d 1576 (Fed. Cir. 1987).
[104] 736 P.2d 140 (Okla. 1987).

court held that the claim should be denied because a more diligent investigation would have indicated the presence of surface water and mud at the project site. This result was somewhat surprising in light of the fact that the owner possessed a soil study that indicated the existence of an underground aquifer that it had not revealed to the contractor.

On the other hand, conditions that could only have been revealed by a detailed subsurface investigation will not be chargeable to the contractor's knowledge. The board in *Southern California Roofing Co.*[105] determined that the additional roof encountered during the performance of the contract was not discernible by a reasonable site investigation. The contractor was not required to bore holes in the roof to locate a top roof system. Further, the Court of Special Appeals of Maryland held in *Raymond International, Inc. v. Baltimore County*[106] that to require a contractor to conduct diving tests to verify the contract indications regarding underwater conditions would be unduly burdensome. And, finally, in *CFI Construction Co.*,[107] the board refused to require the contractor to perform excavation before bidding in order to check the accuracy of the government's plans and specifications.

Broader disclaimer clauses that attempt to relieve the owner of any liability for any subsurface site conditions are treated somewhat differently. When differing site condition clauses are required by statute, as they are on federal government contracts, some courts have held that a disclaimer clause will not override the differing site condition clause.[108] Other courts have taken a different approach, especially when a differing site condition clause is not required by statute or regulation. State courts in New York and Ohio have enforced disclaimer provisions that are generally disregarded in federal tribunals.[109] In *James McHugh Construction Co.*,[110] the board upheld the language of a disclaimer that provided that the contractor, not the owner, would bear the risk of subsurface problems. With this provision, the owner exempted itself from responsibility for any unknown subsurface rock condition. The court enforced the disclaimer provision because the owner was not a federal agency and therefore was not bound by the strictures of federal law that hold that a mandatory differing site conditions clause may not be avoided with disclaimer language.

---

[105] PSBCA Nos. 1737, 2032-2035, 88-2 B.C.A. (CCH) ¶ 20,803 (1988).

[106] 45 Md. App. 247, 412 A.2d 1296, 1302 (1980).

[107] DOT No. 1782, 1801, 87-1 B.C.A. (CCH) ¶ 19,547 (1987).

[108] *See* United Contractors v. United States, 368 F.2d 585, 177 Ct. Cl. 151 (1966); Metropolitan Sewerage Comm'n v. R.W. Constr., Inc., 72 Wis. 2d 365, 241 N.W.2d 371 (1976).

[109] Costanza Constr. Corp. v. City of Rochester, 147 A.D.2d 929, 537 N.Y.S.2d 394 (1989); S&M Constructors, Inc. v. City of Columbus, 70 Ohio St. 2d 69, 434 N.E.2d 1349, 24 Ohio Op. 3d 145 (1982).

[110] ENGBCA No. 4600, 82-1 B.C.A. (CCH) ¶ 15,682 (1982).

In some cases, whether a differing site condition clause or a disclaimer clause will be given effect is a matter of contract interpretation. For example, in *Roy Storm Excavating & Grading Co. v. Miller Davis Co.,*[111] the court found that as a matter of contract interpretation the differing site condition clause had effect over an exculpatory clause. In *Zontelli & Sons, Inc. v. City of Nashwauk,*[112] the court held that a disclaimer clause would not be given effect when the conditions encountered were so different from what could possibly be anticipated that they could not possibly have been within the contemplation of the parties. A similar result was reached in *PT&L Construction Co. v. New Jersey Department of Transportation.*[113] The Supreme Court of New Jersey has also enforced disclaimer language in limited circumstances. The court held that the state remains liable to a contractor despite a general disclaimer in the contract if the state has made positive representations in the contract documents that are inaccurate. However, unambiguous language of disclaimer in the contract defeats a contractor's claim based on inferential conclusions from the contract documents.[114]

DIFFERING SITE CONDITIONS WITHOUT CLAUSE

§ 8.16 Remedies in Absence of Contract Clause

Absent a remedy-granting clause for differing site conditions, contractors must rely on the common-law theories of breach of warranty, misrepresentation, or fraud in the inducement for protection from the risk of adverse, unknown site conditions.

§ 8.17 Breach of Warranty

A *warranty* is a promise that a proposition of material fact is true.[115] A *material fact* is one that constitutes substantially the consideration of the contract, or one without which the contract would not have been made.[116]

---

[111] 149 Ill. App.3d 1093, 509 N.E.2d 105 (1986).
[112] 373 N.W.2d 744 (Minn. 1985).
[113] 108 N.J. 539, 531 A.2d 1330 (1987).
[114] 531 A.2d at 1342.
[115] Black's Law Dictionary 1,423 (5th ed. 1979).
[116] *Id.* at 881.

In *United States v. Atlantic Dredging Co.,*[117] a dredging contract's documents stated that the material to be removed was believed to be a mixture of mud and sand, when in actuality it was compacted sand, gravel, and cables. As the contractor's equipment was effective only for removal of the material described in the contract documents, and not the actual material found, the court concluded the misrepresentation was material, and that the contractor was entitled to additional compensation for the extra work resulting from this misrepresentation.

Owners impliedly warrant the accuracy of representations in the contract documents, and that the plans and specifications are adequate to produce a satisfactory project that is suitable for its intended use. Neither fraud nor negligence on the owner's part need to be proven to recover under a breach of implied warranty theory.[118]

In *United States v. Spearin,*[119] the plans and specifications for a dry dock project required relocation of a sewer, which the contractor performed in accordance with the plans and specifications. The relocated sewer subsequently collapsed because of increased pressure exerted by a dam within a connecting sewer that was not shown on the contract documents. The court stated "if the contractor is bound to build according to plans and specifications prepared by the owner, the contractor will not be responsible for the consequences of defects in the plans and specifications. . . . [T]he contractor should be relieved if he was misled by erroneous statements in the specifications."[120]

A breach of warranty action must be based on a positive or affirmative representation, rather than one which is merely suggestive. In *Hollerbach v. United States,*[121] statements in the specifications of a contract for dam repair that described the type and dimensions of material backing the dam were held to be positive representations giving rise to a warranty. The United States Supreme Court noted that "the specifications spoke with certainty as to a part of the conditions to be encountered by the claimants."[122]

In *Christie v. United States,*[123] the lock and dam construction contract specifications stated "[t]he material to be excavated, as far as known is showing [sic] by borings, drawings of which may be seen at this office, but bidders must inform and satisfy themselves as to the nature of the

---

[117] 253 U.S. 1, 40 S. Ct. 423, 64 L. Ed. 735 (1920).

[118] Christie v. United States, 237 U.S. 234, 35 S. Ct. 565, 59 L. Ed. 933 (1915); Carl M. Halvorson, Inc. v. United States, 461 F.2d 1337, 198 Ct. Cl. 882 (1972).

[119] 248 U.S. 132, 39 S. Ct. 59, 63 L. Ed. 166 (1918).

[120] 39 S. Ct. at 61.

[121] 233 U.S. 165, 34 S. Ct. 553, 58 L. Ed. 898 (1914).

[122] 34 S. Ct. at 555.

[123] 237 U.S. 234, 35 S. Ct. 565, 59 L. Ed. 933 (1915).

§ 8.17 BREACH OF WARRANTY

material." The contractor examined the borings results, which showed gravel, sand, and clay. The material to be excavated actually consisted of buried logs, cemented sand, and gravel. The court stated that the borings were positive representations, and allowed the contractor additional compensation for the increased costs of excavating.

Conversely, in *Sandy Hites Co. v. State Highway Commission,*[124] the court found the contract specifications made no positive representations. The plaintiff highway contractor sued under a breach of implied warranty to recover the value of additional concrete thickness provided over contract requirements. The contractor asserted the specifications caused the additional concrete thickness because they required the contractor's subgrade machine to be set to cut the subgrade to a certain depth, and thus prevented the contractor from adjusting it. In denying the contractor's claim, the court noted other provisions in the specifications that required the contractor to fill in areas found to be below the true subgrade and stated that no additional compensation for excess thickness would be allowed. The court further noted the specifications did not mention any particular kind of soil, weather, or unusual condition.

Although a public project owner cannot be held liable on an implied contract, implied warranties still attach to affirmative representations regarding site conditions made in public construction contracts. In *Champagne-Webber, Inc. v. City of Fort Lauderdale,*[125] bridge construction contract documents indicated sandy soil conditions throughout the site. Although the contract did not expressly warrant the accuracy of the soil data, it did not disclaim accuracy. The court held that, once an express contract has been entered, a public project owner can be held responsible for both the express and implied obligations contained in the contract.

No warranties are imputed to owners who provide no plans or specifications or give no description of the site. Under these circumstances, contractors may have an obligation to inquire whether owners have relevant site information in their possession.[126] Further, owners are not liable for breach of warranty if they have no knowledge of latent defects and make no contribution to the contractor's mistaken belief.[127]

To obtain the protection of the owner's warranties, the contractor must demonstrate its reliance on, and compliance with, the owner's defective contract documents in bid preparation and work performance. In *Al Johnson Construction Co. v. United States,*[128] the court denied a contractor's

---

[124] 347 Mo. 954, 149 S.W.2d 828 (1941).
[125] 519 So. 2d 696 (Fla. Dist. Ct. App. 1988).
[126] Flippen Materials Co. v. United States, 312 F.2d 408, 160 Ct. Cl. 357 (1963).
[127] Helene Curtis Indus., Inc. v. United States, 312 F.2d 774, 160 Ct. Cl. 437 (1963).
[128] 854 F.2d 467 (Fed. Cir. 1988).

breach of implied warranty claim because it did not construct a berm to the height required by the contract specifications.

Reliance cannot be demonstrated if the site's or specifications' defects are so obvious that the contractor should have known of them. In *J.A. Jones Construction Co. v. United States*,[129] the court denied a contractor's claim for additional compensation to replace defective roofing and siding sheet metal installed pursuant to an ambiguous specification in an Air Force hangar construction contract. The court noted that the contract's sheet metal specification was inconsistent with the basic specification governing all military contracts involving galvanized steel sheeting, that this inconsistency was "obvious" and "major,"[130] and that the contractor failed to bring this discrepancy to the attention of the contracting officer. Nor can reliance be proven when contractors know of deficiencies in the plans or specifications and undertake the contract anyway.[131]

§ 8.18 Misrepresentation and Fraud

Misrepresentation and fraud are two related legal theories that allow a contractor to recover if the owner (1) actively concealed information from the contractor, (2) made a representation but failed to disclose facts which materially qualify those disclosed, or (3) had possession of facts that were not known nor reasonably discoverable by the contractor.[132]

Elements

To recover for fraud a contractor must establish that:

1. The other party knowingly misrepresented or concealed information
2. The information was material to performance
3. The other party intended the contractor to rely on the information
4. The contractor reasonably relied on the information
5. Damages occurred as a result of the contractor's reliance.[133]

---

[129] 395 F.2d 783, 184 Ct. Cl. 1 (1968).

[130] 395 F.2d at 790.

[131] Consolidated Diesel Elec. Corp., ASBCA No. 10486, 67-2 B.C.A. (CCH) ¶ 6669 (1967).

[132] Warner Constr. Corp. v. City of Los Angeles, 2 Cal. 3d 285, 466 P.2d 996, 85 Cal. Rptr. 444 (1970); Weichman Eng'rs v. State, 31 Cal. App. 3d 741, 107 Cal. Rptr. 529 (1973).

[133] Daniel Hamm Drayage Co. v. Waldinger Corp., 508 F. Supp. 390, 395 (E.D. Mo.), *aff'd in part and modified in part,* 666 F.2d 1213 (8th Cir. 1981) (Illinois law applied).

A misrepresentation is material if it influenced the mind of the party to which it was made in the contract formation or negotiation of terms.[134]

Affirmative statements that mislead a contractor are actionable. In *Daniel Hamm Drayage Co. v. Waldinger Corp.*,[135] the defendant hired Hamm to rig and assemble pollution control equipment. Hamm's bid was based in part on the defendant's representation regarding the delivery date of the material to be rigged and assembled. Because delivery of this equipment was delayed six months, Hamm sued for misrepresentation. The court found that the delivery schedule was a material consideration in arriving at the bid, that the defendant failed to disclose the actual delivery schedule, that Hamm reasonably had relied on the schedule, and that Hamm suffered monetary damages.

Misrepresentations regarding the amount or character of the work to be performed under the contract, or the cost of its performance, may give rise to a fraud action if these statements purport to be factual and based upon superior knowledge. However, if these statements are merely estimates or approximations, they cannot be the basis of a fraud action.[136]

Failure to disclose information may also give rise to a fraud or misrepresentation action. An owner is obligated to disclose any information which may effect the contractor's performance and which is not equally available to the contractor.[137] *PT&L Construction Co. v. New Jersey Department of Transportation*[138] was a highway contractor's suit for extra excavation and delay costs incurred after encountering unanticipated subsurface conditions that required almost ten times the amount of stripping called for in the contract. The contractor proved that the state failed to disclose knowledge obtained during the early planning stages of the project that soggy soil conditions would make the work difficult. The court held the state's withholding of this information constituted an actionable misrepresentation of site conditions.

A similar result was reached in *Pinkerton & Laws Co. v. Roadway Express, Inc.*,[139] in which Roadway Express awarded a contract to Pinkerton to construct a freight terminal. The contract required all backfill to be mechanically compacted to a minimum density of 95 percent maximum dry weight. The contract contained soil boring logs identified as being

---

[134] McClung Constr. Co. v. Muncy, 65 S.W.2d 786 (Tex. Civ. App. 1933).

[135] 508 F. Supp. 390 (E.D. Mo.), *aff'd in part and modified in part,* 666 F.2d 1213 (8th Cir. 1981).

[136] Busch v. Wilcox, 82 Mich. 315, 46 N.W. 940 (1915).

[137] Warner Constr. Corp. v. City of Los Angeles, 2 Cal. 3d 285, 466 P.2d 996, 1011, 85 Cal. Rptr. 444 (1970).

[138] 108 N.J. 539, 531 A.2d 1330 (1987).

[139] 650 F. Supp. 1138 (N.D. Ga. 1986).

part of subsurface investigations performed by a consulting engineer. No other results of the subsurface investigations were included. The full engineering report stated laboratory results indicated the soils at the site were generally wetter than the optimum moisture content, and that the site would require drying out to achieve 95 percent compaction. Pinkerton encountered soggy soil conditions which had not been indicated in the soil borings and which made achievement of the required soil compaction very difficult. The court ruled that a question of fact existed regarding whether the owner was obligated to include the full engineering report in the contract documents, and it thus denied Roadway's summary judgment motion on the fraudulent concealment claim.

Reasonable Reliance

A misrepresentation or concealment is actionable only if the contractor reasonably relied upon it. Contractors must anticipate difficulties that are inherent in the work. Furthermore, if the contractor might have otherwise obtained the undisclosed information, there can be no recovery under a fraud or misrepresentation theory.[140] If a site condition could be discovered during a reasonable site inspection, a claim for extra compensation based on this condition will be denied.[141] The standard for judging the adequacy of a site inspection is what the reasonable, experienced contractor would discover.[142]

The nature of the misrepresentation, the time constraints involved, and the owner's expectations regarding the contractor's use of the misrepresented information are the factors to be considered in evaluating whether the contractor can establish reasonable reliance upon a misrepresentation.[143] *Joseph F. Trionfo & Sons v. Board of Education*[144] was a general contractor's misrepresentation action to recover extra compensation for additional expenses allegedly incurred as a result of misleading representations made by the Board regarding the site's subsurface. The contract documents made no representation regarding subsoil conditions, and the contract stated that the test boring data had been obtained by the architect and owner for their own use and were specifically excluded from the contract. Moreover, the contractor could obtain the test boring data only after releasing the owner from any responsibility or obligation as to its

---

[140] H.N. Bailey & Assocs. v. United States, 449 F.2d 376, 196 Ct. Cl. 166 (1971).

[141] Powell's Gen. Contracting Co., DOTCBA No. 1088, 80-2 B.C.A. (CCH) ¶ 14,680 (1980).

[142] Kaiser Indus. Co. v. United States, 340 F.2d 322, 169 Ct. Cl. 310 (1965).

[143] Al Johnson Constr. Co. v. Missouri Pac. R.R., 426 F. Supp. 639 (E.D. Ark. 1976), *aff'd,* 553 F.2d 103 (8th Cir. 1979).

[144] 41 Md. App. 103, 395 A.2d 1207 (1979).

accuracy or completeness. The contract documents also specified that the contractor would assume all responsibilities for excavation, and that the contractor was not to rely on subsurface information obtained from the owner. Additionally, nothing in the record before the court indicated the contractor was unable to obtain independent test boring data during the bidding period. Based on these facts, the court concluded the contractor was not entitled to rely on these borings.[145]

Detrimental Reliance and Damages

To prevail under a fraud or misrepresentation theory, the contractor must have been harmed by its reliance on the misrepresented or concealed fact and must have suffered damages. In most jurisdictions, an action for fraud provides two options for recovery. The first allows for rescission of the contract and returns the parties to their precontract positions. Once a party has elected to rescind the contract, it cannot rely on the contract terms in making its case, except to prove the fraudulent misrepresentation.[146]

A second option provides for the recovery of damages incurred as a direct result of the fraud. When a contractor substantially completes a project despite a misrepresentation, some courts deem the contract rescinded and award damages measured by the injury to the defrauded party. For example, in *Daniel Hamm Drayage Co.*,[147] damages were awarded for the amount of expenses in excess of the amount of compensation received, but no recovery was allowed for lost profits or overhead.

§ 8.19 Effect of Disclaimer Clauses

Implied warranties are legal presumptions and thus may be obviated by an express contractual provision.[148] Construction contracts often contain disclaimer clauses by which the owner seeks to place the risk of differing site conditions upon the contractor. Courts are generally hostile to owners' attempts to disclaim the implied warranties of accuracy and suitability.[149] However, disclaimer clauses are enforceable if there is full disclosure of a specified risk, clear allocation of the risk, and express contractual

---

[145] 395 A.2d at 1212.
[146] Davidson v. McKown, 157 Kan. 217, 139 P.2d 421 (1943).
[147] Daniel Hamm Drayage Co. v. Waldinger Corp., 508 F. Supp. 390 (E.D. Mo.), *aff'd in part and modified in part*, 666 F.2d 1213 (8th Cir. 1981).
[148] Flippen Materials Co. v. United States, 312 F.2d 408, 160 Ct. Cl. 357 (1963).
[149] Hollerbach v. United States, 233 U.S. 165, 34 S. Ct. 553, 58 L. Ed. 898 (1914); Fanning & Doorley Constr. Co. v. Geigy Chem. Corp., 305 F. Supp. 650 (D.R.I. 1969).

language evidencing the parties knowingly and consciously intended to so allocate the risk.[150]

A party cannot avoid liability for fraud or bad faith by means of a disclaimer clause,[151] nor will a disclaimer be given effect when soil boring information was incorrectly represented or withheld from the contractor. For instance, in *Fehlhaber Corp. v. United States,*[152] a disclaimer by which the government sought to deny liability for the accuracy of soil borings it furnished was disregarded. In *Baltimore Contractors, Inc. v. United States,*[153] the court refused to enforce a clause disclaiming liability for the accuracy of information regarding soil and subsurface information because the state withheld a relevant soils report it had obtained prior to bid solicitation which indicated conditions inconsistent with the borings. Similarly, in *PT&L Construction Co. v. New Jersey Department of Transportation,*[154] the court refused to enforce a clause disclaiming liability for inaccuracies in site condition data because the state failed to disclose a consultant's report indicating that soggy soil conditions would make achievement of the required soil compaction difficult.

FACTUAL PROOF OF DIFFERING CONDITIONS CLAIM

§ 8.20 Proving a Claim from the Prebid Stage

The factual proof of a differing site condition requires the presentation of evidence of the contractor's actions during three distinct periods: (1) the prebid site investigation period, (2) the bid preparation period, and (3) the period during which the condition was encountered.

Virtually every major construction contract contains a clause that requires a prospective bidder to examine the site prior to bid to satisfy the bidder as to the character and nature of the work. As a first step in the proof of claim process the contractor must present evidence as to the following.

---

[150] Flippen Materials Co. v. United States, 312 F.2d 408, 160 Ct. Cl. 357 (1963); Dravo Corp. v. Commonwealth, 564 S.W.2d 16 (Ky. Ct. App. 1977); Ell-Dorer Contracting Co. v. State, 197 N.J. Super. 175, 484 A.2d 356 (1984); Sasso Contracting Co. v. State, 173 N.J. Super. 486, 414 A.2d 603 (1980).

[151] Commonwealth v. Acchioni & Canuso, Inc., 14 Pa. Commw. 596, 324 A.2d 828 (1974).

[152] 151 F. Supp. 817, 138 Ct. Cl. 571 (1957).

[153] 12 Cl. Ct. 328 (1987).

[154] 108 N.J. 539, 531 A.2d 1330 (1987).

Length of the Bid Period

A contractor is judged by what a reasonably prudent contractor should have done prior to bid under the same circumstance.[155] When faced with a site examination clause, evidence concerning the length of the bid period becomes relevant and significant, particularly when the available time for such investigations or access to the site is limited. For example, a New York court found that the time available to a bidder to confirm the accuracy of an owner's representation directly influenced its determination of the efficacy of a site investigation and exculpatory clause. In *Public Constructors, Inc. v. State*,[156] a case involving a highway construction contract, the state had indicated to bidders that the subsurface soil consisted primarily of coarse-grained material that had the capacity to shed moisture. In reality the soil consisted primarily of fine-grained materials which lacked the capacity to permit compaction in moist conditions. The true nature of the subsurface soil necessitated a timely and costly change in excavation procedures, for which the contractor sought additional compensation. The state defended itself by asserting the existence of a site investigation clause as well as a provision which advised bidders that they were not entitled to rely upon the accuracy of descriptions of subsurface conditions contained in the bidding documents. The court rejected the state's position and concluded that the documents furnished to the bidders contained misrepresentations. Furthermore, the court held that the bidders could not reasonably have been expected to discover in a bid period of three-and-a-half weeks the kind of information that the state possessed as a result of testing the site for several years. In other words, the court found that the time for investigation did not allow for the verification of extensive or complex data.

Nature of the Site Investigation

The steps taken by the contractor during the prebid site investigation must be entered into evidence for the court to scrutinize. The failure not only to conduct a site investigation but to conduct the investigation reasonably will undermine any differing site condition claim.[157] The court must make its own independent assessment of the reasonableness of the contractor's conduct during the site investigation, whether or not the owner

---

[155] Archie & Allen Spiers Inc. v. United States, 296 F.2d 757, 759, 155 Ct. Cl. 614 (1961).
[156] 55 A.D.2d 368, 390 N.Y.S.2d 481 (1977).
[157] Weeks Dredging & Contracting v. United States, 13 Cl. Ct. 193 (1987), *aff'd,* 861 F.2d 728.

presents any evidence as to the unreasonable nature of the investigation.[158] In claims in which the site has been investigated yet the alleged changed condition remains undetected, an additional inquiry is required into the thoroughness and reasonableness of the plaintiff's actual inspection. Such an issue arose in *James Julian, Inc. v. Town of Elkton*.[159] The plaintiff contracted to lay two sewer lines for the defendant. The plaintiff presented evidence that prior to bid submission it inspected the jobsite, pursuant to its customary practice, by walking the site for the sewer and taking test borings at various designated manhole sites. During this investigation, the contractor failed to discover an old subsurface wharf, which was not shown on the contract drawings and which ultimately caused increased performance costs during construction. The contract contained a changed conditions clause. The primary defense of the owner was that the contractor had conducted the prebid site inspection negligently. The appellate court rejected the trial court's conclusion that the contractor should have discovered the wharf, instead concluding that "[the] record . . . as a whole . . . [lacked] substantial evidence to support a finding that the prebid examination of the plaintiff was careless or negligent or that it was below the standard prevalent in the industry."[160]

Location of Data in the Contract

Under some circumstances, evidence concerning the physical position of the contract indications in the contract documents may prove critical to the success of a differing site conditions claim. For example, in *E.H. Morrell Co. v. State*,[161] the contractor presented evidence that the indications relied on were set forth prominently in the special conditions of the contract. The contract also included in fine print the following general condition clause:

> The bidder shall examine carefully the site of the work and the plans and specifications therefore and shall satisfy himself as to the character, quality and quantity of surface and subsurface materials or obstacles to be encountered. . . . He shall receive no additional compensation for any obstacles or difficulties due to surface or subsurface conditions actually encountered.

Although the court determined that the owner had placed on the contractor responsibility for incorrect representations, it nonetheless found for

---

[158] 13 Cl. Ct. at 237 n. 317.
[159] 341 F.2d 205 (4th Cir. 1965).
[160] *Id.* at 209.
[161] 65 Cal. 2d 787, 423 P.2d 551, 56 Cal. Rptr. 479 (1967).

the contractor, relying on the fact that the owner had buried this exculpatory clause in the fine print of the general conditions, while making the misrepresentation in the special conditions. In doing so, the court recognized the realities of a contractor's standard bidding procedure, which is to carefully scrutinize special and technical provisions and generally review standard form provisions.

§ 8.21 Proving a Claim from the Bid Preparation Stage

Did the claimant read and use the contract data in preparing its bid? This question will be asked by the court. Therefore, it is absolutely essential to the success of a differing site condition claim that evidence concerning the methodology used by the contractor to bid and/or estimate the cost of the work be presented. The contractor must establish that its estimate was based on facts set forth in the contract documents.[162] There must be detailed testimony from the contractor's estimator as to the specific influence the contract indications had on its calculations.

It has been said that the underlying purpose of a differing site condition clause is to allow bidders to reduce their bid by eliminating contingencies for unforeseeable conditions.[163] At the very least, there must be testimony of the claimant's estimators that they read and relied on preaward data and did not factor in contingencies when formulating the bid. Mere proof of misrepresentation without the estimator's testimony will result in the claim's failure.[164]

§ 8.22 Proving a Claim from the Encountered Conditions Phase

It is basic and fundamental that the claimant's proof must encompass a comparison of the conditions as estimated to the actual conditions encountered. Thus, credible evidence of the actual conditions encountered must be produced.[165] In the event the differing nature of the conditions is established solely through the observations made by the contractor's

---

[162] Weeks Dredging & Contracting v. United States, 13 Cl. Ct. 193 (1987), *aff'd,* 861 F.2d 728.

[163] *See* Nash & Cibinic, II Federal Procurement Law 1260 (3d ed. 1980).

[164] Morrison-Knudsen Co. v. United States, 345 F.2d 535, 170 Ct. Cl. 712, 719 (1965); Frank R. Ragonese v. United States, 120 F. Supp. 768 (Ct. Cl. 1954); Dravo Corp., ENGBCA No. 3901, 80-2 B.C.A. (CCH) ¶ 14,757 (1980).

[165] Contract Master Servs., Inc. v. United States, 225 Ct. Cl. 735 (1980).

personnel at the site, it is essential to establish the pertinent qualifications and credibility of these observers.[166] In the event the qualifications of the lay witnesses to identify the materially different conditions cannot be established, expert testimony on the credibility of the documentation of the condition must be introduced.[167]

The Type 2 condition, as a general rule, carries a far more difficult burden of proof than a Type 1 condition. This difference in proof was described by the then Court of Claims:

> In the case of a 'category one' changed condition, the Government has, with relative precision, represented the subsurface or latent physical conditions to be encountered, and if it turns out that they have been materially misrepresented, a claim has been established. Under 'category two,' in contrast, the Government has elected not to presurvey and represent the subsurface conditions with the result that a claimant must demonstrate that he has encountered something materially different from the 'known' and the 'usual.' This is necessarily a stiffer test because of the wide variety of materials ordinarily encountered.[168]

In meeting this proof requirement, some of the factors that a contractor can use to establish that a site condition is not one that could reasonably have been anticipated include: (1) the customs of the trade, (2) the common knowledge within the industry, (3) a manufacturer's instructions and recommendations, and (4) the contractor's prior experience in the area.[169]

The contractor's experience and business judgment are critical factors in determining whether an encountered condition is in fact unusual within the definition of the differing site condition clause. If the condition is one that the contractor reasonably should have foreseen, it will not be regarded as an unknown or unusual condition. For example, a contractor who was familiar with road construction work in Alaska was denied his claim for a Type 2 changed site condition based on his encountering continuous landslides in the course of another road project in that state. In denying his claim, the court found that the contractor had in fact anticipated landslides in the course of his work but had used the wrong type of equipment to remove the slide material. The court found that the contractor's failure to correctly estimate the difficulty of coping with the landslides did not entitle him to relief.

---

[166] Weeks Dredging & Contracting Inc. v. United States, 13 Cl. Ct. 193, 233 (1987), *aff'd*, 861 F.2d 728.
[167] *Id.* at 229.
[168] Charles T. Parker Constr. Co. v. United States, 433 F.2d 77, 193 Ct. Cl. 320 (1970).
[169] Clack v. United States, 395 F.2d 773, 184 Ct. Cl. 40 (1968).

If the contractor knew of or had actually anticipated an unusual site condition at the time it entered into the contract, it cannot recover for a Type 1 changed site condition.[170] In this respect, knowledge of an unusual site condition is not limited to those facts actually known by the contractor but encompasses what the contractor should have known about conditions at the work site.

The contractor must then establish its specific adherence to the notice requirements of the clause. Absent the existence of a specific communiqué serving the owner with such notice, the contractor must establish through credible evidence (such as meeting minutes, daily logs, and related correspondence) that the owner was actually aware of the condition and had clear opportunity to inspect the condition prior to its disturbance.[171]

§ 8.23 Use of Expert Testimony

It is an extremely rare case in which the proof of a differing site condition claim does not require some form of expert testimony. Expert testimony is necessary regardless of whether the claim is a Type 1 or Type 2 condition or is a claim asserted in the absence of a remedy-granting clause. Its proof may very well require expert testimony in all three of the factual phases discussed in §§ **8.20** through **8.22**.

A soils expert, structural engineer, or hydrologist may be required to ascertain the nature of the indicated condition purportedly set forth in the contract documents. The expert certainly would be necessary to establish the material difference between the indication in the contract documents and the conditions actually encountered. Because the defending party will raise questions concerning the foreseeability of the actual condition, expert testimony becomes necessary to refute such allegations. For example, in *Paul N. Howard v. Puerto Rico Aqueduct Sewer*,[172] the Court of Appeals for the First Circuit relied very heavily on the apparent credibility of a soils expert's opinion regarding foreseeability:

> As noted by the trial judge in his findings, there was expert testimony to the effect that the instability of the highway embankment was unusual and unexpected.

* * *

---

[170] Vann v. United States, 420 F.2d 968, 982 (Ct. Cl. 1970); Yamas Constr., ASBCA No. 27366, 86-3 B.C.A. (CCH) ¶ 19,090 (1986).
[171] Brinderson Corp. v. Hampton Rds. Sanitation Dist., 825 F.2d 41, 45 (4th Cir. 1987).
[172] 744 F.2d 880 (1st Cir. 1984).

> It is my opinion that the designers of this project did not anticipate the events which we have observed . . . . [T]he condition which we find is not common. I do not think that the contractor anticipated the occurrences which we observed. I do not think that the soil consultant anticipated the conditions which we observed. In short, I don't [think] anybody anticipated what we observed.
>
> In October of 1979, we all suddenly learned a lesson.[173]

The claimant must establish that it relied on the contract indication and made reasonably prudent estimating decisions based on that. Thus, equipment experts, in conjunction with soils consultants, may verify that the estimator's selection of dredging, excavation, or tunneling equipment was appropriate and based on a reasonable interpretation of soil reports, boring logs, and prebid jobsite observations. A major factor in most differing site condition claims for damages involves a comparison of the estimated productivity for certain pieces of equipment to their actual productivity when working in the differing condition. The credibility of an expert, or lack thereof, can be the single most important link to recovery. In one case, the United States Court of Claims relied on its experience with expert testimony in previous cases to reject an efficiency expert's testimony and deny recovery:

> Notwithstanding the fact that Crawford's estimates regarding the other three periods are unrelated, we cannot ignore the fact that the percentages testified to were merely estimates based upon his observations and experience. Furthermore, his estimates are much higher than those testified to in other cases in which the conditions were not materially different from those present here. Taking these things into consideration and in view of the fact that no comparative data, no standards, and no corroboration support his testimony, we are constrained to reduce his estimates based on the record as a whole and the court's knowledge and experience in such cases.[174]

The absence of testimony from an estimating expert can result in a tribunal's drawing a negative inference therefrom. As one board observed:

> It is significant to the Board that none of Appellant's estimators testified. . . . [W]e cannot conclude that appellant was misled by this inaccuracy because there was no evidence that the inaccurate data was used in calculating crew size, tunneling rate or design of the shield.[175]

The reasonableness of the site investigation may be scrutinized by the court. Hence, expert testimony regarding the customs of the industry

---

[173] *Id.* at 883.
[174] Luria Bros. & Co. v. United States, 369 F.2d 701, 177 Ct. Cl. 676 (1966).
[175] Dravo Corp., ENGBCA No. 3901, 80-2 B.C.A. (CCH) ¶ 14,757 (1980).

concerning the conduct of the investigation is often necessary.[176] As with any expert, the ultimate credibility of the witness weighs heavily on the court. However, a defendant's failure to proffer its own expert testimony as to industry custom does not necessarily require the court to accept the plaintiff's version. The United States Claims Court, when faced with only one party's expert opinion as to the reasonableness of a site investigation, did not feel constrained from disregarding that expert's testimony:

> Naturally we may accept testimony as to what an alleged expert would or would not have done—but such testimony is not unassailable by the forces of objective reason, logic, credibility, as well as other material evidence which may come in through the mouth of a nonexpert.[177]

Thus, although expert testimony is persuasive to a court, it will not replace firsthand evidence of the conditions encountered or their foreseeability at bid time.

PROOF OF DAMAGES

§ 8.24 Impact of a Differing Site Condition

Damages resulting from differing site conditions are some of the most complex types of damages to compute and prove. The cost impacts from the differing condition are rarely confined to the direct costs of working through the condition. A differing condition results primarily in losses of labor and equipment productivity. Material, subcontract, and other costs may also increase as a consequence of the differing condition. Contractor damages may extend beyond these more obvious cost impacts.

Few owners initially acknowledge that a differing site condition can have impacts on costs other than direct performance costs. The differing condition may result in a delay and subsequent acceleration in the approved construction schedule. The delay and acceleration may result in crowding in the workplace and disrupting the sequence of construction activities. The potential delay, acceleration, and disruption may cause costs other than direct performance costs to increase.

Sections 8.25 through **8.32** are devoted to pricing and proving damages that result from increased direct performance costs.

---

[176] McCormick Constr. v. United States, 13 Cl. Ct. 496 (1987).
[177] Weeks Dredging & Contracting v. United States, 861 F.2d 728, 13 Cl. Ct. 193, 237 n. 317 (1987); *see also* Shank-Artukovich v. United States, 13 Cl. Ct. 346 (1987), *aff'd without opinion,* 848 F.2d 1245 (1988).

§ 8.25 Impacted Productivity

The most widely accepted approach to proving lost productivity damages resulting from differing site conditions is to compare the contractor's productivity during an unimpacted portion of a project to its productivity once the differing condition was encountered. The approach is widely accepted because it objectively displays the effect of the differing condition on a contractor's ability to perform. The approach can also show an important time-related connection between lower productivity (higher costs) and the dates a differing condition was encountered. This connection can be important in substantiating the cause and effect relationship between liability and damages.

This approach is most often used for labor equipment costs because these costs are most often captured via daily timecards, which are recorded in time sequence. Subcontract and material costs are often recorded as the invoices are received or paid. The invoice recording date may not correspond to the days the subcontractor actually incurred the costs, or when the materials were actually consumed on the project.

Productivity can be computed as cost per unit; hours per unit; units per shift, per day, or per week; or by using many other methods. Damages can be computed using any of these productivity measures. Deterioration of productivity is the key to the damage calculation. Cost per unit is often used in damage calculations because productivity is expressed in dollars, which most people understand. Also, damages are a natural result of using cost-per-unit productivity.

§ 8.26 —Computation of Damages

The computation of productivity losses requires a clear understanding of the activities impacted by the changed conditions, which should include the following: knowledge of labor crews and equipment fleets required during the unimpacted period; changes in labor and equipment requirements during the differing condition period; additional work required as a result of the differing condition; knowledge of the contractor's quantities records; and knowledge of the cost accounting systems and cost coding practices. This understanding will permit the proper matching of costs and quantities for effective productivity computation.

Once there is an understanding of the impacted activities, appropriate cost and quantity information for the unimpacted period can be accumulated and the cost per unit can be calculated (costs divided by quantities). The unimpacted cost per unit is often called the *benchmark, baseline,* or *"should have"* cost. Similarly, the costs and quantities for the impacted period can be accumulated and the cost per unit calculated. The potential loss of productivity damage equals the difference between impacted cost per unit and the benchmark cost per unit, times the number of units completed in the impacted period.

§ 8.27 —The Productivity Benchmark

One of the owner's defenses against a productivity claim in which an unimpacted period is used as the benchmark period is to claim the unimpacted period is not representative of anticipated conditions, or in other words, the unimpacted period represents only near-perfect conditions. The unimpacted period should be representative of anticipated project conditions and should include normal contract inefficiencies, such as bad weather time and low productivity elements of the job.

A danger in using an unimpacted benchmark period is that the period may not be representative and, therefore, not a fair standard for damage computations. The period used to compute the benchmark should be analyzed to determine whether it includes a complete representation of anticipated conditions. Questions similar to the following should be considered for project conditions anticipated by the contract plans and specifications:

1. Were all expected soil types (easy to difficult) excavated during the unimpacted period, or was the majority of excavation in easy soils?
2. Did the benchmark period include a proper blend of anticipated low productivity elements of the job, or was it all high productivity work?
3. Did the benchmark period include a representative amount of bad weather days, or was it always 80 degrees and sunny?
4. Did the benchmark period include a representative amount of equipment downtime?
5. Did the benchmark period consider learning curves, finishing costs, and other normal contractor inefficiencies?

If the answer to any of these types of questions indicates the unimpacted period is not representative of the contract conditions, the benchmark should be adjusted to reflect representative overall costs. The adjustment can be made by analyzing costs in other periods to demonstrate representative data, or by using estimates prepared by engineers, estimators, or other persons knowledgeable in construction methods.

§ 8.28 —The Causal Connection

The goal of the damage presentation is to complete the causal connection, the link between cause (liability) and effect (damages). The first causal connection required is the demonstration of deteriorating productivity as the contractor encountered the differing condition. Productivity fluctuations that can be corroborated by correspondence or project diary information are effective tools in proving this causal connection.

Other causal connections can be made by quantifying additional unit costs incurred because of contractor reactions to the differing conditions, such as overtime, additional labor, additional equipment, and more expensive equipment. Each reaction to the differing condition, by nature, increases construction costs (increased damages) or decreases construction costs (mitigated damages). Causal connections to specific events like removing obstructions or repairing cave-ins, for example, can be made by correlating the event dates with the higher costs. Each reaction or event should be identified and its effect on costs quantified, for use in the damage presentation.

The most persuasive damage presentation tells the story of the project by explaining the relationships between costs, reactions, and events. An effective damage presentation leaves little to the imagination.

§ 8.29 —No Unimpacted Period

The major benefit in calculating productivity damages using a representative unimpacted period as the benchmark period is simplicity. Reasonableness is inherent when the contractor's actual performance levels are used as the basis for the benchmark cost. In some cases, however, a differing condition is so pervasive in the construction activity that no unimpacted period can be isolated. Damages can then be computed and presented effectively using a productivity approach as explained in § 8.26, but the benchmark cost used must be, out of necessity, more subjective. The more subjective a part of a damage calculation is, the more susceptible it is to questions of fairness and credibility. Careful selection of the benchmark cost, in these instances, is critical in effectively proving and defending damage computations.

§ 8.30 —Selecting the Productivity Benchmark

Perhaps the most obvious starting point in selecting a productivity benchmark is the bid's productivity estimate. Using the bid's productivity figure will require testimony about the decisions, assumptions, and changes made in preparation of the estimate. All parts of the bid will be scrutinized for fairness and achievability. Bid workpapers and supporting schedules that are clear and concise and document sound estimating principles can enhance the credibility of using the bid's productivity figure as a benchmark. Also, other bidders from the project may agree to submit their bid documents or testify to corroborate the reasonableness of bid productivity computations.

There are other types of proof that can be used to select and support a productivity benchmark. The following are types of proof that may be considered, along with the related risk for each types:

1. The contractor's documented productivity on similar projects may be considered, but conditions from project to project are seldom identical.
2. Another contractor's productivity on the same project also can be a useful guide, but skills are not homogenous from contractor to contractor.
3. Productivity manuals and studies can be used in selecting a productivity benchmark but, again, the manuals and studies may not reflect project-specific conditions.
4. Expert testimony or expert reestimation after the fact are other methods that may be considered in selecting a productivity benchmark, but the expert's judgements will be susceptible to fairness and credibility questions.

The objective in selecting a benchmark when no unimpacted period exists is to create a reasonable and supportable estimate. Any of these types of proof, or a combination of them, may provide a credible productivity benchmark, depending on individual facts and circumstances.

§ 8.31 Other Approaches to Proving Damages

Another common approach to proving lost productivity damages is to segregate the costs of impacted work in extra work or force accounts. The costs in these accounts can be readily verified with timecards and rate information. However, the extra work accounts may not capture other losses in productivity from hidden impacts such as low labor morale, crowding in the workplace, stop-and-start work flow, and waiting or other unproductive time. Care must be taken to consider the ripple effects of the extra work.

Another approach is to compare the planned labor and equipment force to the actual labor and equipment force required because of the differing condition. This approach computes the difference in planned versus actual costs for impacts such as differences in labor and equipment fleet sizes, differences in crew mixes, differences in mixes or types of equipment, or differences in planned versus actual overtime. The cost differentials are multiplied by the number of impacted hours, days, or weeks to determine damages. For example, if a larger bulldozer replaces a smaller one because of the differing condition, the difference in the ownership/rental and operating rates of the two bulldozers is computed. This differential is multiplied by the number of larger bulldozer hours to quantify the impact of changing from the small bulldozer to the larger one.

§ 8.32 Other Costs to Consider

Although labor and equipment damages are the most common in differing site conditions claims, other types of costs should be considered in the damage computations.

Materials damages in differing conditions claims can be caused by the addition of new or different materials or overruns in planned materials quantities. As previously discussed, materials costs do not lend themselves to productivity analysis. New or changed materials claims are easily supported by invoices. Quantity overrun claims are easily computed as excessive quantities times price. Other costs related to materials such as tax, freight, storage, unavoidable waste or spoilage, normal overruns, and material handling should be incorporated in the materials damages.

Subcontractor costs may be included in a differing conditions claim, either as a request for reimbursement for the negotiated settlement of subcontractor claims or as a "pass-through" claim to the owner. The subcontractor will experience impacts similar to those of the general contractor when a differing condition is encountered. Because subcontractor costs are usually labor- and equipment-intensive, the subcontractor often prepares a claim against the general contractor and owner that may include productivity losses. If the general contractor negotiates and liquidates the subcontractor's claim, the amount paid in settlement of the claim should be included as part of the general contractor's claim against the owner. If negotiations are not successful, or the general contractor and subcontractor agree to pursue jointly a claim against the owner, the subcontractor's claim should be included with the general contractor's claim; in other words, it is "passed through" to the owner. In order for the subcontractor's pass through claim to be successful, damage calculations should resemble those in the general contractor's claim and should require the same proof of owner liability.

Small tools and supplies, contractually allowed markups on subcontractor costs, field and home office overhead, insurance, and bond markups should be included in the differing conditions claim. These indirect cost markups are charged as a percentage of labor or cost dollars. The markups are designed to compensate the contractor for additional indirect costs incurred related to direct performance damages. In some instances, if the markup percentages do not fully compensate the contractor for additional indirect costs, an indirect cost claim may be appropriate. For example, if additional supervisory personnel or other overhead costs were incurred to manage the changed work, the excess overhead costs might be computed and presented as overhead damages.

CHAPTER 9

CHANGES IN SCOPE CLAIMS

Louis R. Pepe, Esquire
Glenn Haese

Louis R. Pepe received his undergraduate and Master's degrees from Rensselaer Polytechnic Institute in Troy, N.Y., and his J.D. (with distinction) from Cornell Law School in 1970. He is a senior partner in the Hartford, Connecticut, law firm of Pepe & Hazard, where he specializes in construction law, usually representing the general contractor, surety, or owner. He has presented seminars in construction contract litigation throughout the Northeast and has authored numerous articles on that subject. Mr. Pepe is a member of the Advisory Committee of the American Arbitration Association and of the Construction Litigation Committee of the American Bar Association, and has been appointed chairman of the newly formed Construction Law Section of the Connecticut Bar Association.

Glenn H. Haese is director of Construction Litigation Services with the firm of Coopers & Lybrand at the firm's office in Hartford, Connecticut. Mr. Haese received a B.S. in civil engineering from the University of Arizona in 1979 and received his law degree from the University of Denver College of Law in 1982. Mr. Haese is a member of the bar of the state of Colorado and maintains memberships in the Colorado Bar Association, the American Society of Civil Engineers and the American Bar Association Forum on the Construction Industry.

§ 9.1   Introduction
§ 9.2   Formal Change Orders
§ 9.3   Constructive Change Orders

§ 9.4	Cardinal Changes	
§ 9.5	Notice and Writing Requirements	
§ 9.6	Pricing a Change Claim	
§ 9.7	Claim Preparation	
§ 9.8	Elements of Proof of a Claim	
§ 9.9	—Extra and Additional Work	
§ 9.10	—Authorization	
§ 9.11	—Reason for the Extra Work	
§ 9.12	—Value of the Extra Work	
§ 9.13	Methods of Calculating Costs	
§ 9.14	Types of Recoverable Costs	
§ 9.15	—Labor Costs	
§ 9.16	—Materials Costs	
§ 9.17	—Equipment Costs	
§ 9.18	—Bonds, Insurance, and Operational Costs	
§ 9.19	Changes and the Performance Bond	
§ 9.20	Changes and the Fast-Track Contract	

§ 9.1 Introduction

Anyone engaged in the construction industry—whether contractor, design professional, developer, surety, owner, or lawyer—asked to prepare a list of the "Top Ten" causes of construction contract disputes will invariably include changes and change orders, probably in a prominent position close to the top of the list. Indeed, most would agree that actual or constructive changes have provided the single most fertile ground for the explosive growth of litigation confronting the industry today. The tendency of changes in scope to cause serious problems in what is already a high-risk undertaking mandates a careful analysis and understanding of the legal framework surrounding the concept of "changes" to a construction contract, including the rights and remedies of the parties to that contract and the essentials of proving and pricing a change of scope claim.

Contractors generally fall into two schools of thought on the issue of changes: those who believe change orders and, more particularly, extras provide the only chance of making any profit on the job, and those who seek only to build the job as it was bid, get paid, and leave the jobsite. Although the correctness of either position can be debated, what is not subject to serious argument is the reality that, good or bad, changes in the original design are inevitable during the course of construction. In short, no construction of any significance has ever been built without changes.

§ 9.1 INTRODUCTION

The owner's admonition, therefore, often sternly and even angrily delivered at the preconstruction meeting, that budgetary requirements require that "this project will be completed with no change orders whatsoever," is as spurious as it is common. And the owner who undertakes a project not appreciating the disingenuous nature of that goal has already taken the first step down the road to litigation.

Changes are inevitable simply because they have so many sources and origins. The perfect set of plans and specifications has yet to be prepared. Design errors not discovered during the review process will always be identified during construction. And, although a drawing that shows an eight-inch riser penetrating an eight-inch concrete beam may, at a later date, be the subject of protracted and expensive litigation to determine liability, the project cannot tolerate the luxury of such a debate at the time the error is found. The impossibility of proceeding with the design as it exists must be addressed, the contractor must have a revised design, and the progress must continue. A change to the contract must be made.

Likewise, unanticipated field conditions will necessitate changes in the design and scope of work. Although subsurface site conditions or latent conditions in renovation projects are the most common source and are examined in **Chapter 7** of this book, they are not the only field conditions that precipitate changes. For example, the flooding and ponding of water on a site may so adversely affect the bearing capacity of previously acceptable soil that a redesigned footing would be required.

Advances in technology occurring after the design is complete, but before construction is done, can bring construction to a virtual standstill, while the medical staff of a hospital, for example, lobbies for the latest and the best equipment.

Redesign during construction may also result purely from financial considerations. The "softening" of the commercial real estate market after a contract for the construction of an office tower is awarded may alter the owner's lease-up and revenue projections and force it to pursue additional value engineering to reduce construction costs, with corresponding changes in the scope of the work.

Not only can changes result from a myriad of postdesign conditions and developments, they can and do affect the quantity of work,[1] the quality,[2] or the means and methods.[3]

Whatever the source, cause, or nature of the change, however, it is virtually certain that the change will have an adverse impact on the progress of the work. Such a consequence is unavoidable; it is simply a matter of

---

[1] Continental Heller Corp., 84-2 B.C.A. (CCH) ¶ 717,276 (GSBCA 1984).
[2] Green v. City of N.Y., 283 A.D.2d 485, 128 N.Y.S.2d 715 (1954).
[3] R&R Constr. Co. v. Junior College Dist. No. 529, 55 Ill. App. 3d 115, 370 N.E.2d 599 (1977).

degree. Changes in scope thus cannot be eliminated, but every reasonable effort should be made to minimize their occurrence. And when they cannot be avoided, an awareness of the steps that must be taken to preserve one's rights is essential.

§ 9.2 Formal Change Orders

The pervasiveness of changes in the construction process has not been lost on drafters of construction contracts. Virtually every such contract executed today, including the popular and commonly used general conditions document promulgated by the American Institute of Architects (AIA), contemplates changes and sets forth a procedure to implement them, as do the standard construction contracts offered by the Associated General Contractors (AGC) and the United States government.[4]

Typically the changes clause of a contract establishes a formal procedure resulting in a written amendment which describes the precise nature of the new and different work that will be performed and the compensation for it. The process is often triggered by a bulletin or similar communication from the architect, notifying the contractor of the need for a change to the design, defining, detailing, and explaining the new and different work to be performed, and requesting a cost proposal for performing that work. The contractor is expected to promptly respond with a lump-sum proposal, covering its work and the work of its subcontractors (including appropriate and applicable markups for profit and overhead), and a request for any required time extension. The change order clause probably provides for compensation on the basis of agreed-upon unit prices, if applicable to the work, or time and material, if desired by the owner, in which case no estimate or proposal is required from the contractor. However, compensable charges are ordinarily limited to the reasonable costs incurred directly in the performance of the changed work, plus allocable overhead and profit. Each contract must be carefully reviewed in order to determine if the contractor's compensation is defined and thus limited to certain authorized categories of recoverable costs, computed at defined rates.[5]

Once agreement is reached on the new scope of work and the payment for it, all is memorialized in a document, the actual change order, executed by the owner and the contractor. Then and only then may the contractor proceed to implement the change. Because a construction

---

[4] See AIA A201-1987 ¶ 7; AGC Standard Form B, art. 18; United States Standard Form 23-A.

[5] Forest Constr. Inc. v. Farrell Creek Steel Co., 484 So. 2d 40 (Fla. Dist. Ct. App. 1986).

contract, like any other contract, can be amended or modified by the mutual assent of the parties to it,[6] the rights and obligations of the owner and contractor are now defined by the original contract as amended by the change order.[7]

Those experienced in the construction process have already recognized the change procedure as occurring universally in construction contract documents but almost never in real life. They may vaguely remember actually observing the procedure once or maybe twice in the distant past, but they surely do not expect that "contractual minuet" to be played out again in their professional lifetimes.

The obvious question then is why is the formal change order clause so ubiquitous, yet so frequently ignored? The answer is equally obvious: the clause is essential because changes are unavoidable, but the overriding pressures to maintain the schedule and progress the work make adherence to it the exception rather than the rule. A change may be necessary because the work cannot proceed as designed. The work must proceed because, in construction, time is money. Accordingly, as soon as the problem is solved by redesign of the concrete spread footings, for example, the new design is immediately formed and poured, and the formal change order relegated to subsequent resolution.

But that same contract that sets forth the change order or amendment procedure almost always provides elsewhere that the contract may not be changed by oral agreement but may only be amended by a writing executed by the parties. In the hypothetical example above, the owner has the redesigned footing; it has no need for, nor interest in, a written amendment to the contract addressing that issue. The contractor has not been paid for the extra labor and materials necessitated by the larger footing—indeed, it cannot even requisition for payment without a change order—so it certainly has a desire for that document. The contractor is now confronted with the challenge of obtaining payment in the absence of the formal written modification, and whether it can is addressed in § 9.5.

The foregoing example raises, however, another subject that requires analysis before that issue is reached, and that is the concept of constructive changes.

---

[6] Carolina Metal Prods. Corp. v. Larson, 389 F.2d 490 (5th Cir. 1968); Frommeyer v. L&R Constr. Co., 261 F.2d 879 (3d Cir. 1958).

[7] This is a "bilateral" change to the contract. The owner may often have the right to make "unilateral" changes (that is, without the consent of the contractor) of a minor nature in order to progress the work. *See* AIA A201-1987 ¶¶ 7.1.3, 7.4.1; United States v. Henke Constr. Co., 157 F.2d 13 (8th Cir. 1946); White Constr. & Equip. Rental v. Rinner & Garrett, Inc., 340 So. 2d 283 (La. 1976); Albert Elia Bldg. v. New York State Urban Dev., 54 A.D.2d 337, 388 N.Y.S.2d 462 (1976).

§ 9.3 Constructive Change Orders

Because scheduling pressures, other exigencies, or simply the owner's resistance often result in the absence of a formal change order when one clearly should have been issued, the doctrine of *constructive changes* has been developed and recognized by the courts. Having its origin in government contracts, it is still most common in that field, although it has found acceptance by courts construing contracts in the private sector as well.[8] The constructive change doctrine is nothing more than a recognition that a contractor may be compelled to perform extra and additional work by a formal, written change order or by an act or omission of the owner or its agent, for which the law ought to provide a means of recovery for the contractor.[9]

Constructive changes have the same origins as their more formal counterparts, including defective plans and specifications,[10] arbitrary interpretation of the contract documents,[11] and changed site conditions.[12] Once a contractor demonstrates that it was prevented from performing the work described in the contract, or with the expected means and methods, or in the sequence planned, then it becomes entitled to compensation or to an "equitable adjustment," as it is called in government contracts.[13] That adjustment is measured by the "fair and reasonable value" of the extra work performed.[14]

Because every change to the contract potentially affects the time permitted for completion, constructive changes necessarily involve a time element. The timing aspect of constructive changes often manifests itself in the related concept of *constructive acceleration:* the owner's improper requirement, by act or omission rather than by formal directive, that the contractor adhere to a completion schedule adversely impacted by a change. In order for a contractor to recover costs incurred in meeting that schedule, it must show that: (1) it suffered an excusable delay and was, therefore,

---

[8] Siefford v. Housing Auth., 192 Neb. 643, 223 N.W.2d 816 (1974).

[9] Southern Bell Tel. & Tel. Co. v. Acme Elec. Contractors, Inc., 418 So. 2d 1187 (Fla. Dist. Ct. App. 1982).

[10] R.J. Daigle & Sons Contractors, Inc. v. Sampey Bros. Gen. Contractors, Inc., 424 So. 2d 270 (La. Ct. App. 1983); Al Johnson Constr. Co. v. United States, 854 F.2d 467 (Fed. Cir. 1988).

[11] Fox Valley Eng'g, Inc. v. United States, 151 Ct. Cl. 228 (1960).

[12] Al Johnson Constr. Co. v. Missouri Pac. R.R., 426 F. Supp. 639 (E.D. Ark. 1976).

[13] C.J. Langenfelder & Son, Inc., MDOT 1000, 1-Md. Inst. for Cont'g Prof. Ed. of Law. 97 (1980), *aff'd sub nom.* Maryland Port Admin. v. C.J. Langenfelder, 50 Md. App. 525, 438 A.2d 1374 (1982).

[14] Dale's Sand & Gravel Co. v. Westwood Constr. Co., 62 Or. App. 570, 661 P.2d 1378 (1983); Atlantic Elec. Co., 83-2 B.C.A. (CCH) ¶ 16,671 (GSBCA 1983).

entitled to an extension of time, which was improperly withheld; (2) it was under the threat of liquidated or actual damages for failure to meet the original schedule; and (3) the contractor had, in fact, accelerated its performance and thereby incurred additional costs in order to avoid those damages.[15]

If the prevalence of the changes clause in construction contracts today is a recognition of the inevitability of changes in virtually all construction undertakings, then the constructive change doctrine may be seen as an acknowledgment that frequently a change in the value of the work performed or in the time required to perform it will not be memorialized in the formal, written change order contemplated by that clause. In such a case, the law provides what the parties did not.

§ 9.4 Cardinal Changes

The foregoing sections and, indeed, even the title of this chapter are somewhat misleading in their use of the terms "changes" and "change of scope" interchangeably. That is because, technically speaking, an owner making changes to the work, whether by formal change order or constructive change, must limit those changes to the original scope of the contract if it is to avoid a material breach entitling the contractor to terminate performance. When a change made by the owner is so extensive or of such great magnitude as to be outside the scope of the original contract, the owner has made a *cardinal change,* which the contractor may refuse to perform and which will permit the contractor to abandon the project.[16]

In the leading case of *Saddler v. United States*[17] a change in the design of a levee embankment being constructed for the federal government doubled its length and more than doubled the cubic yardage of fill required. The contractor claimed damages under the theory that the government had breached the contract by ordering a change of that magnitude. The Court of Claims agreed, holding "that the nature of this particular contract was so changed by the added work, albeit the same kind of work described in the original specifications, as to amount to a cardinal alteration falling outside of the scope of the contract,"[18] thereby entitling the contractor to all the resulting costs associated with performing that work. In reaching that conclusion, however, the Court of Claims recognized the

---

[15] Norair Eng'g Corp. v. United States, 666 F.2d 546 (Ct. Cl. 1981); Tombigbee Constr. Co. v. United States, 420 F.2d 1037 (Ct. Cl. 1970); Human Advancement, Inc., 81-2 B.C.A. (CCH) ¶ 15,317 (HUDBCA 1981).

[16] Saddler v. United States, 287 F.2d 411 (Ct. Cl. 1961).

[17] *Id.*

[18] *Id.* at 414–15.

impossibility of employing a formula in determining the existence of a cardinal change:

> We think that a determination of the permissive degree of change can only be reached by considering the totality of the change and this requires recourse to its magnitude as well as its quality . . . . We do not attempt to set forth a mathematical definition by which any deviations in quantity from a contract must be measured. Obviously the differences between contract situations will admit of no such inflexible formula.[19]

In an interesting twist to this concept, a nonparty to a construction contract used the cardinal change doctrine to block circumvention of the state's competitive bidding laws. In *Alberta Elia Building Co. v. Urban Development Corp.*,[20] the contractor for the construction of a convention center was given a change order for the construction of a tunnel from the convention center to an adjacent area where the public authority was developing a "plaza." Although the contractor was pleased to accept the extra work at an increase of $428,000 to his $17 million contract, another general contractor sued to prevent the issuance of the change order on the grounds that it was outside the scope of the convention center contract and was thus a cardinal change. The court agreed, holding that, although the public authority had the right to issue change orders as to "details and minor particulars," it could not "alter the essential identity or main purpose of the contract"[21] or change the original plan so as to constitute a "new undertaking."[22] The court held the addition of the tunnel was outside the scope of the original contract, and, as such, it violated the mandatory competitive bidding statutes governing the award of such work.[23]

But cardinal changes are not limited merely to those that increase the quantity of work put in place. They may also result from the changes in the means or methods of performing the work. Accordingly, in *Peter Kiewit Sons' Co. v. Summit Construction Co.*,[24] the court found that the requirements imposed on the site work subcontractor by the general contractor with respect to the backfilling of missile sites, which resulted in a threefold increase in the cost of that work compared to what the

---

[19] *Id.* at 413, 414. The United States Supreme Court, in Freund v. United States, 260 U.S. 60, 67 (1922), provided only slightly more guidance in describing the standard in ascertaining a cardinal change as "what should be fairly and reasonably within the contemplation of the parties when the contract was entered into."

[20] 54 A.D.2d 337, 388 N.Y.S.2d 462 (1976).

[21] 388 N.Y.S.2d 462 at 467.

[22] *Id.*

[23] *Id.* at 462.

[24] 422 F.2d 242 (8th Cir. 1969).

subcontractor originally had contemplated, constituted a cardinal change. The Eighth Circuit stated that:

> In the instant case there is clearly sufficient evidence to find that the changes ordered by Kiewit went so beyond the scope of the Subcontract, including the 'changes' provisions . . . , as to breach the Subcontract. Instead of permitting Summit to backfill in a two-phase operation unimpeded by pipe in the hole as agreed . . . , Kiewit required backfilling to be done in three places *after* pipe had been laid and thereby increased immeasurably the difficulty of the backfill operation.[25]

Likewise, a cardinal change need not result only from a single massive or drastic change in the quantity or quality or sequence of work; it can be the consequence of many relatively minor changes, no one of which alone would constitute a cardinal change but the cumulative effect of which certainly does.[26]

Of course, the frustration, disruption, and delay caused by a major change or numerous minor changes do not always translate into a breach of contract.[27] But, once a contractor establishes that a cardinal change was made, it is freed from any contractual obligation to perform it. Because there is no bright-line test to separate a cardinal change from those within the original scope of the contract and within the owner's right to order, and because the abandonment of a project can have extraordinarily serious consequences, a contractor will seldom feel confident enough to abandon a project in the face of changes made by the owner. If it elects to remain and perform the changed work under protest, it is entitled to recover the "fair and reasonable value" of performing the work plus the cost of the resulting extension of time, on a theory of quantum meruit.[28]

§ 9.5 Notice and Writing Requirements

Virtually every common changes clause contains the requirement that a written, fully executed change order be obtained by the contractor before

---

[25] *Id.* at 255 (emphasis in original).

[26] Meyer Schwartz Plumbing Co. v. Shelby Constr. Co., 338 S.W.2d 781 (Mo. 1960) ("changes in plans were being made from the initiation of the job to its completion"); H.T.C. Corp. v. Olds, 486 P.2d 463 (Colo. App. 1971) ("changes made are too numerous to list, but the evidence disclosed over 40 major changes").

[27] Ferguson Contracting Co. v. State, 237 N.Y. 186, 142 N.E. 580 (1923); Wunderlich Contracting Co. v. United States, 351 F.2d 956 (Ct. Cl. 1965).

[28] Black Lake Pipe Line Co. v. Union Constr. Co., 538 S.W.2d 80, 91 (Tex. 1976); Westcott v. State, 264 A.D. 463, 36 N.Y.S.2d 23 (1942); District Concrete Co. v. Bernstein Concrete Corp., 418 F.2d 1030 (D.C. Cir. 1980); Walter Kidde Constructors, Inc. v. State of Conn., 37 Conn. Supp. 50, 434 A.2d 962 (1981).

it undertakes the performance of extra, additional, or different work. Therefore, timely notice to the owner and possibly the surety (as discussed in § 9.13), as evidenced by the written change order, is usually required prior to the owners being bound to compensate the contractor for such extra, additional, or different work. Good business practices and sound job administration mandate the necessity for such a writing and associated formal notice but, as discussed in § 9.2, compliance with either of these prerequisites is more the exception than the rule. The question then arises as to whether the contractor, who proceeded with the performance of the changed work in good faith and in the best interests of the job, will be precluded from subsequently obtaining payment for that extra or different work. Confronted with the harsh results that obtain when such rules are strictly enforced, many courts have permitted recovery in the absence of a written change order and associated formal notice to the owner and have fashioned various theories to support their efforts to do equity.

The ultimate manifestation of that judicial thinking is, of course, the constructive change discussed in § 9.3. But often the courts take a less sweeping approach. For instance, the Mississippi Supreme Court found the owner waived his right to have a written change order as a condition precedent to payment, when it was undisputed that the city had requested the alterations in question and the denial of the city's engineer to certify payment for them was bad faith.[29]

Some courts have characterized the parties' practice of ignoring the written change order requirements as a modification of that part of the contract, removing as a bar to recovery the contractor's failure to obtain the writing.[30] Others have gone so far as to treat a continuing course of conduct between the parties involving extensive nonwritten change orders to be an "abandonment" of the contractual, written change order requirement.[31] Whether the theory employed is one of waiver, modification, or estoppel,[32] the objective of the court is the same in all such cases: the prevention of an injustice to the contractor and a windfall to the owner because of a technical failure.

---

[29] City of Mound Bayon v. Ray Collins Constr. Co., 499 So. 2d 1354 (Miss. 1986). *See also* Ecko Enters., Inc. v. Remi Fortin Constr., Inc., 118 N.H. 37, 382 A.2d 368 (1978); Universal Builders, Inc. v. Moon Motor Lodge, Inc., 430 Pa. 550, 244 A.2d 10 (1968); Doral Country Club, Inc. v. Curcie Bros. Inc., 174 So. 2d 749 (Fla. Dist. Ct. App. 1965).

[30] D.K. Meyer Corp. v. Brevco, 206 Neb. 318, 292 N.W.2d 773 (1980).

[31] H.T.C. Corp. v. Olds, 486 P.2d 463 (Colo. App. 1971).

[32] Nyhus v. Travel Management Corp., 466 F.2d 440 (D.C. Cir. 1972); Reif v. Smith, 319 N.W.2d 815 (S.D. 1981); Brookhaven Landscape & Grading Co. v. J.F. Barton Contracting Co., 676 F.2d 516 (11th Cir. 1982); Huang Int'l Inc. v. Foose Constr. Co., 734 P.2d 975 (Wyo. 1987); D'Onofrio Bros. Constr. Corp. v. New York City Bd. of Educ., 72 A.D.2d 760, 421 N.Y.S.2d 377 (1979).

Yet not all courts will struggle to achieve that worthy goal, and the decisions in this area are hardly uniform. A failure to establish such a waiver of the requirement for a writing by "clear and unmistakable evidence" has been held to bar a contractor's recovery,[33] as has the failure to prove a "definite agreement to pay" by the owner.[34] Of course, such a conclusion is more likely when the owner can demonstrate that the work was done pursuant to the directives of unauthorized agents,[35] or without even its knowledge let alone its authorization,[36] thus making the contractor a "volunteer."[37]

Formal or at least actual notice to the owner is, therefore, generally a prerequisite for the contractor to recover in the government contract setting in the absence of a written change order. When the owner is responsible for extra, additional, or different work, or the events necessitating a change in the time of the performance of the same work, whether in terms of position in the project schedule or duration of the task, the owner must be given the opportunity to take corrective measures to mitigate the effects of its actions on either the project schedule or scope of the work. Otherwise, the contractor waives the right to recover for lost time or money arising from the owner's actions in requesting extra, additional, or different work.[38] Notice clauses have been held to be valid and enforceable.[39] Further, although seemingly inequitable, such clauses have been strictly enforced.[40]

Fortunately, in the government contract setting, exceptions to the written change order and formal notice requirements exist. Usually, the government has to prove that the delay in giving notice somehow operated to the detriment and prejudice of the government before such delay operates as a waiver of the contractor's right to recover.[41] If the government is unable to show that it was somehow prejudiced by the contractor's lack of strict compliance with writing and notice provisions, the claim will be decided on its merits.[42] The government usually argues that the need for early investigation to preserve the memory of witnesses, the need to compile accurate records of the contractor's actual costs in performing the

---

[33] Service Steel Erectors Co. v. SCE, Inc., 573 F. Supp. 177 (W.D. Va. 1983).

[34] Linneman Constr., Inc. v. Montana-Dakota Utils. Co., 504 F.2d 1365 (8th Cir. 1974).

[35] Hartline-Thomas, Inc. v. H.W. Ivey Constr. Co., 161 Ga. App. 91, 289 S.E.2d 196 (1982).

[36] J.A. Tobin Constr. Co. v. Kemp, 239 Kan. 240, 718 P.2d 302 (1986).

[37] Post Constr. Co., 84-1 B.C.A. (CCH) ¶ 16,959 (HUDBCA 1983).

[38] Wunderlich Contracting Co. v. United States, 351 F.2d 956 (Ct. Cl. 1965).

[39] Roberts v. Security Trust Sav. Bank, 196 Cal. 557, 238 P. 673 (1925).

[40] State Sur. Co. v. Lamb Constr. Co., 625 P.2d 184 (Wyo. 1981).

[41] Eggers & Higgins v. United States, 403 F.2d 225, 185 Ct. Cl. 765 (1968).

[42] Macri v. United States, 353 F.2d 804 (9th Cir. 1965); New Pueblo Constr., Inc. v. State of Ariz., 144 Ariz. 95, 696 P.2d 185 (1985).

changed work, and consideration of alternate construction methods in order to reduce costs, are substantial government interests which mandate that the government be given notice strictly in accordance with the contract.[43] But when a government entity has actual notice of the claim and initially takes action to permit recovery and only later raises the formal notice and writing requirements, the government waives the right to deny the claim solely on the basis of lack of formal notice.[44]

In the private sector as well, the courts have at times sought to avoid unnecessarily harsh results and have chosen not to require strict compliance with notice and writing requirements. In *Sisters of Charity v. Burke*,[45] the owner had notice of the changes, because the owner's representative had requested the changes and later observed the changed work performed. The court therefore held that any written change order requirement was waived. Further, when an owner has actual knowledge of extra work being performed and takes no steps to dispute the work as extra, the contractor is entitled to recover the reasonable value thereof.[46] Also, some courts have gone so far as to hold that, when the owner should have been aware of the changed work's being performed, strict compliance with notice and writing requirements are waived.[47]

Given the uncertainty of the legal consequences of proceeding to perform extra work without a written directive, the cautious contractor may well conclude that the only prudent course of action is to stop all progress until the written change order is forthcoming. Indeed, one court has held it was the contractor's right to do precisely that and to recover the work stoppage costs incurred while it was waiting.[48] Also, when the owner obstructs the contractor's work and refuses to issue a change order acknowledging the extra work created and the increased cost to the contractor, the owner can be held liable for breach of contract.[49]

But such hard ball is not readily played by most contractors, who instinctively want to avoid trauma to the progress of the job, finish it as quickly as possible, and trust the owner's sense of fairness to override the technical requirements of the contract. However well-intentioned, that too

---

[43] *Id.*; Appeal of Leavell & Co., ASCBCA No. 16099, 72-2 B.C.A. (CCH) ¶ 9,694 (1972).

[44] Chaney Bldg. Co., Inc. v. Sunnyside School Dist. No. 12, 147 Ariz. 270, 709 P.2d 904 (1985); Korshoj Constr. Co., IBCA No. 321, 63 B.C.A. 3848 (1963).

[45] 22 Colo. App. 230, 124 P. 472 (1912).

[46] McLaurin v. Holley, 484 So.2d 807 (La. Ct. App. 1986).

[47] Morgan v. Crowley, 91 Ga. App. 58, 85 S.E.2d 40 (1954).

[48] Dick Corp. v. State Pub. School Bldg. Auth., 27 Pa. 498, 365 A.2d 663 (1976). *See also* Ida Grove Roofing v. City of Storm Lake, 378 N.W.2d 313 (Iowa Ct. App. 1985) (precluding a contractor's recovery for removing thicker-than-expected concrete insulation on the grounds that, once the unforeseen condition was discovered, he was obligated to stop and obtain a written change order before proceeding with the extra work).

[49] Bagwell Coatings, Inc. v. Middle S. Energy, Inc., 797 F.2d 1298 (5th Cir. 1986).

can prove to be a most dangerous course of action. The better approach would appear to be to send one or more letters documenting the entire problem. The first such letter should relate the discovery of the design defect, changed condition, or other problem as soon as it is discovered and request prompt written direction. The second, if time permits, should bemoan the absence of a response and warn that commencement of work in that area is imminent and direction is required immediately. The third and final should recite the history of the problem and describe the corrective action the contractor will undertake in the interest of continuing progress, including the extra work for which it expects payment, unless it is directed in writing within 24 hours not to proceed. The onus is now clearly upon the owner to act, and if the owner still fails to do so, it is unlikely any trier of fact will subsequently deny recovery for the cost of that work.

§ 9.6 Pricing a Change Claim

Whether done prospectively in the ideal situation, in which a change order is actually issued before the extra work is undertaken, or after the fact, when a claim for an equitable adjustment is being pursued, there comes a time when the contractor must quantify the value of the extra and additional work it is being asked to perform or has, in fact, already completed. The neophyte will sadly and mistakenly perceive this as a relatively simple and straightforward undertaking. After all, the changes clause provides that unit prices will be used when applicable, and time and material will be the basis for compensation when so directed by the owner. In the event neither applies, the changes clause provides for an agreement on a lump sum, and certainly the contractor is adept at and has enough experience in estimating the cost of work and preparing a proposal. That analysis is all quite true as far as it goes, but unfortunately it stops far short of the realities. Pricing a change on the foregoing basis courts disaster by considering only the direct costs. It omits the second half of the cost element—the indirect costs—and ignores completely the second element—time.

The *indirect* or *impact costs* of a change are the hidden costs that result from every change. Many, but not all, changes extend the time for completion of the project, as will be seen in this section. When that happens, a delay claim arises and must be part of the pricing of that change. Likewise, the direct cost of extra and additional work may be dwarfed by the adverse effect upon related contract work that is required to be performed out of sequence or on a comeback basis. A simple example will illustrate the problem.

A change from eight-foot hollow metal door bucks to the 10-foot variety appears innocuous enough when issued during construction. But the contractor quickly learns the lead time for the fabrication and delivery of those new door bucks is substantial, and they are still not on site when the masons commence work on the block walls in the corridor where the bucks are to be installed. The contractor has only three choices: (1) install artificial door bucks around which the block wall can be built, (2) "tooth out" around the openings where the door bucks will ultimately be installed, or (3) stop work on the wall and move the masonry crew to another part of the job. All options are unattractive for obvious reasons of cost and time, but assume the contractor reluctantly decides to have the masons "tooth out" the wall. When the door bucks finally arrive and are installed by the carpenters, the masons are long gone from the area. They must lug by hand their mortar tubs and other equipment into a nearly complete area where finish trades are working and then fill in around the newly-installed door bucks. Not one extra square foot of block wall is ultimately installed, but no fair-minded person could argue that block wall costs the same to construct as it would have had the masons been able to proceed in that area as planned, without disruption and comeback work. But none of that extra cost is compensated by the direct cost of the taller door bucks; it is all indirect or impact cost that must be identified, quantified, and added to the direct cost if full and fair compensation is to be made.

The time aspect of every change presents a similar problem. Any extra work occurring or impacting contract work on the critical path of the project will, by definition, extend the time for completion. That is because the critical path contains no "float" in the activities which comprise it, and every day of delay to those activities delays the project completion by one day.[50] Moreover, the obvious solution of estimating the time required for the extra work or its effect on contract work usually does not work. Because the critical path of a project frequently changes during construction, and because the contractor at the time of preparing the change order proposal typically does not know when the directive to proceed will be given, it is virtually impossible to estimate the effect the change will have on the project completion time.

The costs of the delays and disruptions are as real and significant as they are unpredictable. Their quantification is addressed and explained in **Chapters 5** and **6** and is outside the scope of this chapter. But these delay and disruption costs are compensable,[51] and must be considered in pricing every change. Because it is usually impossible to measure them prospectively, however, the contractor must make clear in its proposal that it is including the direct cost of the change only and is expressly reserving the right to claim and recover those indirect costs as well as the right to have an extension of time. The owner, who wants to know the entire

---

[50] Buckley & Co. v. State of N.J., 140 N.J. 289, 356 A.2d 56 (1975).

[51] Coley Properties Corp. v. United States, 593 F.2d 380 (Ct. Cl. 1979); Regan/Nager Constr. Co., 85-1 B.C.A. (CCH) ¶ 17,778 (PSBCA 1984).

§ 9.7 Claim Preparation

A claim differs from formal litigation in that a *claim* has been defined by the federal government as a written demand or assertion by one of the parties to a construction contract seeking, as a legal right, the payment of money, adjustment or interpretation of contract terms, or other relief arising under or related to a given contract.[52] An invoice, in and of itself, is not a claim. But if the invoice is not acted upon within a reasonable time after its submission, or its liability or amount is disputed, the invoice may become a claim.[53] Once the changed work has been completed and all direct, indirect, and delay costs have been computed, using the specified procedures and limitations,[54] the contractor can then submit to the owner the total claim.

The claim should include the indirect and delay costs, that were excluded from the change order proposal submitted at the time of the contemplation or initiation of the changed work. Further, the claim should be submitted to the architect or the appropriate contracting officer, in the government setting, in order to ensure compliance with any contractual provisions requiring consideration at the administrative level prior to the initiation of formal litigation.

Prelitigation preparation and submission of the claim requires the contractor to evaluate carefully the nature and scope of the changed work to ensure that a valid, compensable claim exists. Some construction contracts require written submission of the claim and the rendering of a written decision by the architect as a precondition to the initiation of arbitration.[55]

At the point of original claim submission, the parties hope that there can be an amicable resolution to the claim submitted without resort to litigation. There can be no doubt that, if the claim has been thoroughly analyzed and realistically evaluated prior to its submission, there is a greater chance for maximum recovery in a shorter time.

---

[52] Policy Letter No. 80-3, Office of Federal Procurement Policy, 45 Fed. Reg. 31,035 (May 9, 1980); F.A.R. § 52.233-1.

[53] *Id.*

[54] *See* AIA A201-1976 ¶ 12.1.4, indicating that recoverable costs shall be limited to certain material, labor, workers' compensation insurance, bonding, equipment rental, supervision, and field office personnel costs which are directly attributable to the change.

[55] See AIA A201-1976 ¶ 12.1.4 read together with ¶ 2.2.12.

Legal counsel to the contractor must ensure the validity of the claim before proceeding with formal litigation or arbitration, or the contractor runs the risk of sanctions. The Federal Rules of Civil Procedure, as well as most state rules, have been amended in recent years to add an objective criteria to what was a subjective standard of good faith imposed on attorneys and parties to the litigation in the filing and prosecution of an action.[56] The amended Rule 11 has been the predominant basis for sanctions' being imposed against attorneys for "frivolous or harassing" proceedings and was "intended to reduce the reluctance of the courts to impose sanctions" by emphasizing the responsibilities of the attorney to ensure that a case has merit before filing of an action.[57] Significant penalties and criminal sanctions, including debarment, may be imposed for submitting an inaccurate or false claim.[58]

To strengthen the reliability of the claim and accompanying documentation, for both its benefit and that of its legal counsel, the contractor must ensure that costs incurred as a result of the changed work are calculated at the time they are incurred and not only after total contract completion. There must be a timely and consistent recognition and recording of the as-built costs of the completed, changed work, including a calculation of the impact on contract profit of the changed work.[59] If the request for the performance of the changed work results in the abandonment of further performance, the profit on the balance of the unperformed work must also be figured.[60]

In order to ensure, then, that all actual costs incurred are recorded, thus increasing the likelihood of eventual recovery, a detailed and effective system of cost accounting should be developed prior to or at the initiation of the project. Costs should be compiled regularly, at least on a monthly basis. Also, when changed work has been ordered, the contractor should compile costs on at least a biweekly basis and should isolate all associated direct, indirect, and delay costs by line item, indicating the original estimate, costs incurred prior to the initiation of the changed work, costs incurred in performance of the changed work, and the total variance from the original budget. As many separate line item cost accounts should be established as is feasible for the specific type of project involved.

The greater the detail in the compilation, the less chance there is for the exclusion of all costs as being unreasonable. More likely, only those specific line item amounts considered to be unreasonable or unrecoverable

---

[56] Fed. R. Civ. P. 11 and Advisory Committee Notes (1983).

[57] Eastway Constr. Corp. v. City of New York, 762 F.2d 243 (2d Cir. 1985).

[58] *See* Contract Disputes Act, 41 U.S.C. § 601, *et seq.;* False Claims Act, 31 U.S.C. § 231; Truth in Negotiations Act, 10 U.S.C. § 2306(f).

[59] Bennett v. United States, 371 F.2d 859 (Ct. Cl. 1967).

[60] Burnett & Doty Dev. Co. v. C.S. Phillips, 84 Cal. App. 3d 384, 148 Cal. Rptr. 569 (1978).

as a matter of law would be excluded, leaving the majority of the actual costs incurred intact and recoverable.

Equipment and labor costs should be segregated and recorded separately within the specific line item task or function to which they relate. As an example, equipment costs for soldier piles should be accumulated separately. The hours that a particular crane, front-end loader, or pile hammer are used to install the piles should be tracked separately in a function or task code. The supporting documentation should specify the duration and location of each type or piece of equipment used.[61]

After the costs have been accumulated, the auditor, legal counsel, and/or the contractor should prepare the claim for formal submission. The claim presentation should not include too much information; the inclusion of voluminous exhibits can invoke a minute and detailed review that invites squabbling over minor points. However, sufficient detail on the points of specific negotiation should be included so the opposing party can understand that a true factual basis for the claim exists.

Legal authority should probably not be cited in the claims package, but specific contractual provisions should. However, as to nonliability issues, some authority may be appropriate. As an example, a state statute governing interest recovery is probably safely included.

The audit report on the claim should probably not be submitted with the claim, because the audit report covers only limited items of the claim and usually invites a detailed response by the opposing party, thus blocking meaningful negotiations. However, an audit performed prior to the formal submission of a claim can disclose errors in the accumulation of costs to be submitted for payment in the claim package, thus avoiding possible sanctions against the attorney or contractor because of invalid claims. Also, knowledge by the opposing party that an audit has been performed in conjunction with the preparation of the claims package lends credibility to the claim when submitted.

§ 9.8 Elements of Proof of a Claim

If submission of the claim does not result in recovery from the owner, arbitration or litigation is then pursued. In order for a contractor to recover additional compensation from an owner for extra and additional work, the contractor must prove:

1. The work was outside the scope of its contractual obligation
2. The extra work was ordered or directed by the owner or someone with authority to act for the owner
3. The owner agreed by words or conduct to compensate the contractor

---

[61] J.B. Tieder, Jr. & J.F. Hoffar, Proving Construction Contract Damages (1988).

4. The work was not rendered necessary by any fault of the contractor
5. The amount being claimed is the fair and reasonable value of the extra work.[62]

§ 9.9 —Extra and Additional Work

The contractor is aided immeasurably in meeting the burden of the first element of proof by the well-settled rule of construction that any ambiguity in the plan and specifications will be construed against the drafter (owner) and in favor of the contractor.[63] Of course, the contract requirement must be reasonably susceptible of two interpretations before that axiom can be invoked; a court will not strain to find an ambiguity where none exists.[64] Furthermore, if the ambiguity is "patent," that is, so obvious that it should have been discovered during the preparation of the estimate, then the contractor must seek clarification at that time and cannot claim refuge in the ambiguity later.[65] However, once the contractor has proved its interpretation of the contract requirements to be at least as reasonable as the owner's, it has satisfied its burden of proving the work is extra.[66]

§ 9.10 —Authorization

It is fundamental that the extra work for which a contractor seeks compensation must have been requested by the owner or its authorized agent.[67] In the absence of a formal written change order (the all-too-common situation on most construction projects) the direction and authorization of the extra work in question often becomes a serious contention.

Some courts have accepted as adequate authorization the combined effect of: (1) the need for the change, (2) the owner's knowledge that the

---

[62] R&R Constr. Co. v. Junior College Dist. No. 529, 55 Ill. App. 3d 115, 370 N.E.2d 599 (1977); Savin Bros., Inc. v. State, 62 A.D.2d 511, 405 N.Y.S.2d 516 (1978), aff'd, 47 N.Y.2d 934, 393 N.E.2d 1641, 419 N.Y.S.2d 969 (1979); St. Charles Floor Co. v. Hoelzer, 565 S.W.2d 844 (Mo. Ct. App. 1978); Al Johnson Constr. Co. v. Missouri Pac. R.R., 426 F. Supp. 639 (E.D. Ark. 1976), aff'd, 553 F.2d 103 (8th Cir. 1977).

[63] United States v. Heckinger, 397 U.S. 203 (1970); Hoffman v. Pfingsten, 260 Wis. 160, 50 N.W.2d 369 (1951); Benham v. World Airways, 296 F. Supp. 813 (D. Haw. 1969); Restatement (Second) of Contracts § 206 (1981).

[64] Bruno Preski v. Warchol Constr. Co., 111 Ill. App. 3d 641, 444 N.E.2d 1105 (1985).

[65] Lane Constr. Corp., 84-2 B.C.A. (CCH) ¶ 17,490 (ENGBCA 1988); Max Drill, Inc. v. United States, 427 F.2d 1233 (Ct. Cl. 1970); Newsom v. United States, 676 F.2d 647 (Ct. Cl. 1982).

[66] Southwestern Sheet Metal Works, Inc., 79-1 B.C.A. (CCH) ¶ 13,744 (ASBCA 1979); Bennett v. United States, 371 F.2d 859 (Ct. Cl. 1967).

[67] National Bonding & Accident Ins. Co., 83-3 B.C.A. (CCH) ¶ 16,863 (ENGBCA 1983).

change was being made, and (3) the owner's failure to stop the contractor (hence, the need for the notice letter suggested in § 9.5). But others do not deem that sufficient satisfaction of the contractor's burden of proof in this regard.[68] The potential for an injustice to the contractor is greatest when the owner is a public body and authorization is questioned, as demonstrated in *Nether Providence Township School Authority v. Thomas M. Durkin & Sons.*[69] In that case, the contractor sought substantial additional compensation for clearing and grading costs incurred to correct typographical discrepancies in the site plan. Although the school authority's president asked the contractor in writing to do the corrective work subject to a subsequent resolution of the responsibility therefor, and although the school authority's architect supervised the work, the court rejected the claim because the contractor did not have written authorization to perform the work.[70] The New York Supreme Court reached a different result, however, in *D'Onofrio Bros. Construction Corp. v. New York City Board of Education,*[71] holding a change order executed by the board's area manager instead of the executive director, as required by the express terminology of the form contract providing for the original construction, was an adequate basis for recovery, because the agency representatives responsible for the project had thoroughly inspected the site and negotiated the prices, and the contractor properly believed those persons had the authority to alter the scope of work.

The question of authority most frequently arises with respect to the owner's most visible representative: his architect. The new AIA General Conditions[72] address this issue by making clear the very limited authority the architect has with respect to changes under such a contract. Paragraph 7.4 of AIA Document A201 provides:

MINOR CHANGES IN THE WORK

The Architect will have authority to order minor changes in the work *not involving adjustment in the Contract Sum or extension of the Contract Time* and not inconsistent with the intent of the Contract Documents. Such changes shall be effected by written order and shall be binding on the Owner and Contractor. The Contractor shall carry out such written orders promptly.[73]

---

[68] V.L. Nicholson Co. v. Transcom Inv. & Fin., Ltd., 595 S.W.2d 474 (Tenn. 1980); Watson Lumber Co. v. Gunnewig, 79 Ill. App. 2d 377, 226 N.E.2d 270 (1967); Universal Builders, Inc. v. Moon Motor Lodge, Inc., 430 Pa. 550, 244 A.2d 10 (1968).

[69] 505 Pa. 42, 476 A.2d 904 (1984).

[70] *Id.*

[71] 72 A.D.2d 760, 421 N.Y.S.2d 377 (1979).

[72] AIA A201-1987.

[73] *Id.* at ¶ 7.4 (emphasis supplied).

Accordingly, any contractor who, at the direction of the architect, implements a change that is likely to result in increased cost or time, does so at its peril. And that risk is not to any extent diminished by the "implied" approval of a change inherent in the architect's approval of a shop drawing, although those same general conditions vest in the architect the sole authority for review and approval of shop drawings and samples "for conformance with information given and design concept."[74] Perhaps the more prudent course of action for the contractor is not to rely on the architect at all in the event that a change is ordered that is not minor—defined as one that results in increased time or cost. Instead, direct contract with the owner, requesting a written change order for the increased time or cost is recommended. However, the architect should be kept fully apprised of the steps taken by the contractor to obtain direct owner approval of such a change order. Unfortunately, this may sour future relations with the architect; but depending on the extent of the increase in cost that may be involved in implementing the change, direct owner contact and, hopefully, intervention and discussion may be a necessary risk in face of the longer risk of non-payment for work performed. The case decisions are not uniform on the legal effect of the approval of a shop drawing that incorporates a change. That was precisely the issue in *Dehnert v. Arrow Sprinklers, Inc.,*[75] in which the contractor's shop drawings clearly showed that it intended to use plastic sprinkler heads rather than the brass heads specified. The shop drawings were approved and the plastic heads installed, but the architect, under direction from the owner, subsequently ordered their removal and replacement. The contractor refused, was terminated, and sued the architect. The Wyoming Supreme Court decided in favor of the architect, holding curiously that "the information [in the shop drawing] was inadequate for purposes of initiating the change order necessary for the substitution of plastic heads."[76]

But a diametrically opposed result was reached on similar facts in *Montgomery Ross Fisher & H.A. Lewis.*[77] The contractor's shop drawing contained "butterfly valves" in lieu of the specified "gate valves." He subsequently installed the butterfly valves, was required by the owner to remove and replace them, and sued for the extra cost. The Government Services Administration Board of Contract Appeals upheld his claim and rejected the government's position that such a shop drawing approval could not constitute authority to digress from the specifications, holding that to conclude otherwise would render the shop drawing process meaningless.

---

[74] *Id.* at ¶¶ 4.2.7 and 3.12.
[75] 705 P.2d 846 (Wyo. 1985).
[76] *Id.* at 851.
[77] 85-2 B.C.A. (CCH) ¶ 18,108 (GSBCA 1985).

§ 9.13 CALCULATING COSTS

The risk of proceeding to do extra work without clear authority is obvious. And given the economic unattractiveness of pursuing compensation for extra work after it has been performed and the job completed, the prudent contractor must establish at the preconstruction meeting precisely which of the owner's representatives has the express authority to act on behalf of and bind the owner with respect to changes in the scope of work, and that it will thereafter take directions only from that representative.

§ 9.11 —Reason for the Extra Work

Having demonstrated that the work in question was, in fact, outside the original scope of the contract and that it was authorized by the owner to perform that work, the contractor must next prove that the extra work was not the result of its failure of performance or poor workmanship. That was precisely the basis for denying the contractor's claim for an equitable adjustment in *Wrecking Corp. v. Memorial Hospital for Cancer*.[78]

§ 9.12 —Value of the Extra Work

Finally, the contractor seeking to be paid for extra work must show that the amount being claimed is the fair and reasonable value of the work.[79] It is obvious that establishing the fair market value of the materials supplied or the usual and customary rate paid for the labor required lends itself to expert testimony, perhaps by the contractor's chief estimator.[80] The overhead and profit margin to which the contractor is also entitled are usually determined by the allowance for those items set forth in the contract itself, because the agreement of the parties on such a markup is prima facie evidence of its reasonableness.

§ 9.13 Methods of Calculating Costs

The methods of establishing both direct and indirect costs, and thus quantifying the effect of the changes in the scope of the work, are varied and in almost all instances the subject of controversy. Several methods of

---

[78] 63 A.D.2d 615, 405 N.Y.S.2d 83 (1978).

[79] Black Lake Pipe Line Co. v. Union Constr. Co., 538 S.W.2d 80 (Tex. 1976); Westinghouse Elec. Supply Co. v. Fidelity & Deposit Co., Md. 560 F.2d 1109 (3d Cir. 1977).

[80] AIA A201-1987 ¶ 7.3 attempts to provide a rather specific framework for determining the compensation of the contractor subsequent to the performance of extra work pursuant to a Construction Change Directive, thereby eliminating or at least reducing the need for expert testimony.

calculating and quantifying the costs that a contractor can collect from the owner, after a breach or other justification for recovery has been established, have been developed. These methods have varying degrees of accuracy and reliability. Ordinarily, a review of the contractor's records should be undertaken prior to the selection of the method for cost calculation, and thus damage quantification, to ensure that such records provide sufficient detail and quality to allow for the application of a chosen method's procedures.

The most widely accepted method of cost calculation and damage quantification utilizes actual, historical cost records and related project documentation.[81] Further, the burden of proof shifts to the owner to prove that the costs were not reasonable if the contractor has developed its damage calculation on the basis of *actual cost* records. A presumption exists in the government contract setting that the price paid was the reasonable cost, and that presumption must be overcome by a contractor seeking to claim damages beyond the actual costs of the changed work.[82] In the case of a contractor attempting to collect the fair market value of materials incorporated into the changed work, one court has held that the actual costs to the contractor were the reasonsable costs to be considered in assessing damages.[83] The court also noted that the determination of the reasonable cost of changed work is itself objective.[84] The proper measure of an adjustment in a contract for the reduction of work has been held to be the reasonable cost of performing the work.[85] The bid pricing calculations, although useful in some instances to show a contractor's predispute anticipated costs, cannot be held as the reasonable measure of cost when they vary widely from the only estimates of the actual cost to perform the work placed into evidence.[86]

In certain cases the contractor may be able to utilize the *total cost method* to quantify the damages to be assessed against an owner. The contractor simply subtracts the estimated cost or bid of the entire project from its final actual cost of the entire project.[87] However, the total cost method is not favored by most courts and should be used only as a last resort.[88] The reason that it is not favored is that it suffers from defects as a

---

[81] Plumbers & Fitters Local 761 v. Matt J. Zaich Constr. Co., 418 F.2d 1054 (9th Cir. 1969); Charles D. Weaver v. United States, 538 F.2d 346 (Ct. Cl. 1976).

[82] Bruce Constr. Corp. v. United States, 324 F.2d 516 (Ct. Cl. 1963).

[83] *Id.* at 518.

[84] *Id.*

[85] General Ry. Signal Co. v. Washington Metro. Area Transp. Auth., 598 F. Supp. 595 (D. Colo. 1984).

[86] *Id.* at 598.

[87] Moorehead Constr. Co. v. City of Grand Forks, 508 F.2d 1008, 1016 (8th Cir. 1975).

[88] Huber, Hunt & Nichols v. Moore, 67 Cal. App. 3d 278, 136 Cal. Rptr. 603 (1977).

result of not specifically linking increased costs to specific activities. By simply subtracting the original bid estimate from the cost of the overall project, the total cost method presumes that the bid estimate was realistic. Also, it can attribute costs to the owner that might have been incurred regardless of the order for a change in the original scope of the work. Finally, it can reward a contractor's inefficiency, managerial incompetence, or other difficulties and failings by passing the additional costs generated by these inefficiencies on to the owner.[89]

Another method for establishing the cost of the changed work is the modified total cost or cost variance approach.[90] The *modified total cost method* focuses on a particular portion of the work rather than the entire project. Only that portion of the project impacted by the changed work is costed. First, the time intervals during which the extra, additional, or different work occurred are identified and segregated. Next, the employee and machinery time actually expended for the particular item of changed work is determined by reference to the superintendent's reports describing the weekly work performed, the engineer's field reports or diary, labor distribution reports, and equipment usage reports and logs. Actual costs that were for other items of work not directly related to the changed work are thereby eliminated. The actual unit costs of the overrun or increased costs resulting from the change in the original scope of the work are established during the actual time interval that such work was performed. The original estimate of the particular items of work in dispute is then reestimated based upon the original specifications without regard to the changed scope of the work. The reestimation should usually be performed by an independent third party in order to maintain the credibility of the analysis. The damages are then calculated by subtracting the independently estimated costs of performing the impacted activities from the actual cost of performing those activities.[91] The trial judge usually is careful anytime the modified total cost method is used to ensure that specific estimates have been appropriately adjusted in order to avoid recovery of unrelated costs by the contractor. Unfortunately, the contractor is unable to use the modified total cost approach when there is sufficient evidence to use the jury verdict method.[92]

The *jury verdict method* of cost and damage quantification may be used when there is clear proof that the contractor was injured and there is no

---

[89] New Pueblo Constr., Inc. v. State of Ariz., 144 Ariz. 95, 696 P.2d 185 (1985).

[90] Thorn Constr. Co. v. Utah Dep't of Transp., 598 P.2d 365 (Utah 1979).

[91] *Cf.* E.C. Ernst, Co. v. Koppers Co., 626 F.2d 324, 327 (3d Cir. 1980), *on remand,* 520 F. Supp. 830, 835 (W.D. Pa. 1981) (holding that a modified cost approach was applicable although not specifically titled as such).

[92] Specialty Assembly & Packing Co. v. United States, 355 F.2d 554, 174 Ct. Cl. 153 (1966).

more reliable method for computing damages. The court may allow the jury to make a fair and reasonable approximation of the damages once the court has determined that sufficient evidence is available.[93] When the contractor is unable to prove actual costs, evidence may be presented to the finder of fact as to the cost of the additional, different, or extra work, including actual cost data, accounting records, and testimony of expert witnesses, as well as costs experienced on similar projects.[94] In essence, the contractor offers to the trier of fact the broadest possible range of cost information and evidence that is reasonably related to the changed work. Normally, this methodology is used when liability has already been proven and the precise proof of damages is impossible. However, when there is no reasonable excuse for the failure to have kept any actual cost information, neither the jury verdict method nor the modified total cost method are available as measures of damages.[95]

In some cases, when there was no written agreement for the changed work and it appears that the change in the scope of the work rose to the level of a cardinal change, the theory of quantum meruit can be advanced. *Quantum meruit* is more accurately described as a theory or the imposition of liability than as a method of proof for cost and damage assessment. However, it is used when there is no express contract but a contract is implied in fact in order to permit recovery of a benefit conferred. This theory can be utilized in those cases in which an express contract is unenforceable or so significantly defective or so substantially breached that it is considered void. Care must be taken, however, prior to asserting the theory that no language in the original contract (when an express contract once existed) precludes recovery on implied contracts, as in the case of some AIA contracts.[96] Once successfully advanced as a theory of liability, costs and damages which usually exceed the actual costs expended in performing the changed work are recoverable as the courts focus on the value

---

[93] WRB Corp. v. United States, 183 Ct. Cl. 409 (1968).

[94] Metropolitan Sewerage Comm'n v. R.W. Constr., Inc., 78 Wis. 2d 451, 255 N.W.2d 293 (1977).

[95] Pickard's Sons Co. v. United States, 532 F.2d 739, 209 Ct. Cl. 643 (1976).

[96] Gilbert v. Powell, 165 Ga. App. 504, 301 S.E.2d 683 (1983). The court in *Gilbert v. Powell* found that the AIA contract executed between the parties had been prepared to deal with the cost of changes in the work. The theory of quantum meruit involves an implied promise to pay for work, labor, or goods sold, delivered, or incorporated into a structure. The court further stated that "if there exists a written contract that is broken, one of the remedies for the breach is quantum meruit, that is, in treating the contract as rescinded." *Id.* However, such is the situation only in cases where the original written contract does not provide for changes in the scope of work or extra work. In AIA contracts, and specifically A201-1987, Article 7, ¶¶ 7.1.1 to 7.4.1 explain in detail how changes in the scope of the work and the specific pricing are to be documented, authorized, and agreed to.

of the benefit conferred rather than on the cost of the performance to the contractor.[97] However, the use of this theory can be a two-edged sword, especially when work is completed but the work is considered to be so defective that no benefit was found to have been conferred.[98]

In the public sector, and specifically in public contracts, the term "equitable adjustment" has become the standard in quantifying the extent of damages to be awarded for additional, different, or extra work. The federal courts have recognized the existence of the *doctrine of equitable adjustment* as essentially a liability theory based on the equitable need to compensate the contractor for a change of economic position resulting from contract changes or conditions for which the government was responsible.[99] The scope of the costs and damages to be awarded under the equitable adjustment theory is much more narrow than the damages to be awarded in quantum meruit. Specifically, only the costs that result from the changes ordered by the government are compensable. Therefore, the costs of the performance of the additional, different, or extra work must be isolated from other tasks and limited to the time interval in which such changed work was performed. An equitable adjustment should reduce neither the profit the contractor would have received nor the loss it would have sustained under the contract absent any modification.[100] The contractor is entitled to an adjustment equal to the difference between the actual costs incurred on the changed work and the reasonable cost of performing the original work, including any price advantages obtained by the contractor with respect to the original work, plus profit and overhead on the work actually performed.[101] The owner, in order to prevent the contractor from collecting, must bear the burden of proving that the contractor did not utilize a reasonable method of performing the changed work.

§ 9.14 Types of Recoverable Costs

Having considered the methods by which costs and damages can be quantified and submitted to the court, the question then becomes which of the various categories of costs are likely to recovered. As a general rule, it is better to accumulate the entire spectrum of costs as elements of a total cost and damage award, in both the claims as well as the litigation setting first, and later, depending on the legal authority in the jurisdiction in

---

[97] Levan v. Richter, 152 Ill. App. 3d 1082, 504 N.E.2d 1373 (1987).

[98] Herbert W. Jaeger & Assoc. v. Slovak Am. Charitable Assoc., 156 Ill. App. 3d 106, 507 N.E.2d 863 (1987).

[99] Ragnar Benson, Inc. v. Bechtel Power Corp., 651 F. Supp. 962 (D. Pa. 1986).

[100] Bruce Constr. Corp. v. United States, 324 F.2d 516 (Ct. Cl. 1963).

[101] Allen Constr. Co. v. United States, 646 F.2d 487 (Ct. Cl. 1981).

which the claim action is being pursued, selectively delete categories for presentation. In the event grounds later arise for other theories of liability, the deleted categories of cost information could be resurrected and presented to the arbitrator, the court, or the jury. The categories of direct and indirect cost, including labor, materials and supplies used, equipment, workers' compensation insurance, bonding, interest, and general and site overhead, as recoverable elements of a total damage award have been accepted by the courts in varying degrees.

§ 9.15 —Labor Costs

Direct labor costs for work performed outside of the scope of the original contract can usually be easily calculated when the original records were properly segregated at the time the labor was performed. The additional hours of each laborer performing tasks within the scope of the changed work are then simply multiplied by the appropriate wage rates and markup percentages. Reference to union contracts can establish the appropriate wage rate. In addition, the contractor-employer is entitled to include other employee-related costs in the determination of an applicable wage rate to be used as the multiplier. Typical employee-related costs to be included are taxes and insurance, union fringe benefits, small tools, and established field supervision costs. Unless specifically precluded by the construction contract, these employee-related costs are almost universally recoverable with regard to increased direct labor costs.

§ 9.16 —Materials Costs

Changes in scope claims and damage quantifications typically involve *material costs* as an element of the claim because the contractor normally exceeds originally estimated material requirements in order to construct the changed work. In order to calculate the changed material quantities, reference is first made to the original drawings or estimates to determine the original quantity specified and estimated. Thereafter, the actual quantity is verified from purchase orders or delivery tickets for the materials actually used. However, the claim and damage quantification should include a quantity greater than that actually required in order to allow for spoilage and contingency overruns.

Supplies differ from materials in that *supplies* are those items which the contractor uses to accomplish the changed work but which are not actually incorporated into the completed structure. Again, there is a reasonableness requirement in terms of cost expended and quantity utilized in the performance of the work. Further, the best proof to offer is the original invoice, but the courts have allowed recovery when proof was offered of the cost of similar supplies purchased in the same or similar quantity in the same time frame.

If the actual *material quantity* necessary for the changed work is difficult to segregate from the basic contract materials, such as in the case of backfill, the unit price of the material must be established at the time the additional, different, or extra work was performed. An invoice from the supplier establishing the unit price or an affidavit as to the unit price charged by the supplier at the time the changed work was performed can also establish the appropriate unit charge. The actual quantity placed, plus spoilage, contingency overrun, or loss, is then simply multiplied by the unit charge to arrive at the claim or damage figure. In addition, a contractor can usually recover the *actual cost of transporting the additional material* to the site by the means normally utilized by the contractor, but not expedited or special handling costs, unless specifically requested by the owner. *Additional material handling costs* or costs generated by materials requiring different handling than originally contracted and anticipated would be recoverable. *Sales and use tax* on the cost of the materials in many cases is exempted by the government entity within whose boundaries the project is constructed and, therefore, such costs would not be recoverable.

§ 9.17 —Equipment Costs

Equipment costs are generally recoverable. However, the equipment must be of the type necessary to complete the changed work as ordered and must be in sound working condition. Normally, changed work equipment usage falls into three categories: 1) the originally intended equipment's being utilized longer than anticipated, 2) additional equipment's being required than had been originally budgeted or estimated, or 3) a combination of these two factors. A daily equipment usage log is the most preferable form of evidence to establish actual equipment usage. However, if such logs are not available (although the lack thereof had best be explainable), the equipment usage can be quantified based on the quantity of additional, different, or extra work performed and standard productivity rates for the particular type of equipment used. The manufacturer's rates for productivity should not be used to generate the claimable equipment usage as such rates are very rarely achievable by the contractor under actual field conditions.

A review of the construction contract may identify the equipment rate guide required to be used in an equipment usage claims calculation. The most widely recognized equipment rate book in the private sector is the *Rental Rate Blue Book for Construction Equipment,* commonly known as the "Blue Book" published by the Machinery Information Division of

Dataquest Incorporated. In the federal contract setting, the *Contractor's Equipment Manual* and the *Contractor's Equipment Cost Guide,* both published by the Associated General Contractors of America, are the most widely accepted sources for rate information. Also, with regard to projects which are earth-moving intensive, such as United States Army Corps of Engineers projects, the Corps publishes a *Construction Equipment Ownership and Operating Expense Schedule.*

As a check on the computations, the increased labor hours expended in performing the changed work should be correlated to the equipment usage hours. In short, whatever hours the operator of a piece of equipment expended should cross-check against the equipment usage logs. Labor hours can be used to generate equipment usage hours if equipment logs are not available. However, the operator must have been assigned to only one piece of equipment as opposed to being allowed to float between different pieces of equipment. In the event that rented equipment is used to perform the changed work, the contractor should claim the actual rental amounts paid. Utilization of rented equipment is required in certain federal contract settings.

§ 9.18 —Bonds, Insurance, and Operational Costs

The costs of additional *insurance premiums* for general liability policies covering any changed work as well as workers' compensation premiums generated by higher or different labor levels, such as from supervisor to superintendent, are also generally recoverable. Self-insurance costs are generally recoverable in the federal contract setting.

In both the public and the private contract setting, bonds guaranteeing the performance of the contractor and the payment by the contractor of subcontractors and materials suppliers are the rule rather than the exception. As the scope of the work changes, the requisite amount of both the performance and payment bonds also increase as the owner attempts to minimize the risk of financial loss by any action of the contractor. *Costs of the bonding* required of the contractor are generally allowed by the courts to be recovered, to the extent that such required additional bonding is in accordance with sound and rational business practice and the rates or premiums are reasonable under the circumstances.

Recovery of the *cost of funds or interest* is usually limited statutorily from state to state. Most state statutes allow recovery for prejudgment interest, but the interest claim must be of a liquidated nature and, therefore, ascertainable with reasonable certainty from actual cost records. The rate of interest recoverable is usually also specified by state statute. But when a construction contract rate of interest is specified, the contract rate has been held recoverable as the market rate. Postjudgment interest begins to accrue as of the date a judgment is entered and is governed by state law.

§ 9.19 CHANGES AND PERFORMANCE BOND

Overhead costs generally fall into two categories: home office overhead and jobsite overhead. Percentage overhead claims are not the equivalent of extended overhead claims, which are discussed in **Chapter 5**. Jobsite overhead is calculated by dividing the total indirect cost at the jobsite by the total direct job cost. The direct job cost is calculated with subcontractors included. Thereafter, the percentage rate is applied to the direct costs of the claim. Care must be taken to ensure that charges are not included which have been previously submitted through change orders and paid or duplicated by inclusion in other sections of the claim. Home office overhead is generally recoverable and is allocated to various jobs and contracts on the basis of total job cost.

The home office overhead rate is computed by dividing the total home office overhead cost by the total company direct contract cost. The resulting percentage is then applied to the total claimed direct and jobsite overhead cost. In the federal contract setting, contractors are now required to certify under penalty of perjury that all overhead costs included in a claim are allowable as defined in the Federal Acquisition Regulations.

By this time the difficulty and expense of proving a claim for extra work should be obvious, and the value of obtaining a written, executed change order before proceeding with the work quite apparent—however unlikely the dynamics of the construction process make that.

§ 9.19 Changes and the Performance Bond

A performance bond is required by the owner and provided by the contractor in order that the owner might have the additional security of another entity, usually a licensed insurance institution, to whom the owner can look for completion of the job in the event of a default by the contractor who is the principal under the bond.[102] What an owner seldom contemplates is its loss of that security when it makes changes to the contract, yet that can very likely be the result under the *material alterations doctrine*. That doctrine is predicated upon the well-settled rule of law that the surety on a performance bond is discharged from its liability if the parties to the underlying contract materially alter that contract without the surety's consent. Stated alternatively, the bond covers only the contract referred to, not some other and different contract.[103] Because a bilateral change order is a modification to the contract, every change or at least a material one, has the potential to invoke the doctrine and release the surety, and certainly a radical or cardinal change would have that legal effect. Accordingly, absent the language found in many performance

---

[102] National Shawmut Bank v. Amsterdam Casualty Co., 411 F.2d 843 (1st Cir. 1969).

[103] United States v. Freel, 186 U.S. 309, 316 (1902); Reliance Ins. Co. v. Colbert, 365 F.2d 530, 534 (D.C. Cir. 1966).

bonds waiving the surety's right to notice of any changes, the prudent owner sends a copy of all change orders to the surety.

Unfortunately, some performance bonds issued on government projects contain express notice provisions requiring that the surety be notified of all changes and alterations above a certain percentage of the contract price. The writing requirement appears in the expense notice provisions of the performance bond. Therefore, unless the surety is notified in writing as to a change in the total contract amount, the surety can argue that the suretyship was only valid to the extent of the original contract amount. If the contractor performs work over and above the original contract amount and, thus, beyond the amount of the original suretyship, the surety can argue that this is a material breach of the suretyship as specified in the provisions of the performance bond, unless the contractor has provided the requisite written notice as to the change in the scope of the work and, thus, total payment due. At times, the notice and change order has also been required to be in writing. Typically, the prior written approval of the surety is required in the governmental contract setting when the cost of all changes and alterations to the contract exceed 10 percent of the face amount of the bond issued.[104] Of course, the parties should decide early on in the project, and preferably prior to execution of the construction contract or delivery of the notice to proceed, how the "cost" term of such bond will be defined. Further, the courts have not uniformly required that the change order be in writing, even if the bond so stated. Although the key factor appears to be notice to the surety of the change, when the bond originally required the notice of any change to be in writing, some courts have chosen to release the surety for the failure to reduce the change order to writing, thus strictly enforcing the writing requirement,[105] while others have found that the writing requirement is not material[106] or not prejudicial enough to justify the release of the surety.[107]

But what has become painfully evident in this chapter thus far is that changes to a construction contract more often than not occur in the absence of a formal change order and often without even an express oral agreement between the owner and contractor, the *obligee* and *principal* under the bond. The question then becomes whether the parties to the contract, without an express agreement for changes, can so materially alter the contract as to release the surety on the bond. The answer to that question is unequivocally in the affirmative. The basic rule was stated in *Malone v. Santora:* "An existing contract may be modified or abrogated

---

[104] *See* AIA A311-1970 for formal waiver of notice; *cf.* FHA Form 3452; J.D. Lambert & L. White, Handbook of Modern Construction Law 95 (1982).

[105] Burnes Estate v. Fidelity & Deposit Co., 96 Mo. App. 467, 70 S.W. 518 (1902).

[106] Honolulu Roofing Co. v. Felix, 49 Haw. 578, 426 P.2d 298 (1967).

[107] Roberts v. Security Trust Sav. Bank, 196 Cal. 557, 238 P. 673 (1925).

§ 9.19 CHANGES AND PERFORMANCE BOND

by a new contract *arising by implication of the parties.*"[108] The rule was particularly applied to a surety bond in *Fergerson Contracting Co. v. Charles E. Story Construction Co.:* "It is fundamental that any agreement or dealing between the creditor and the principal in an obligation of debt, which essentially varies the term of the contract, without the consent of the surety, will release the surety from liability."[109] The Connecticut Supreme Court dispelled any lingering notion that an express agreement between the principal and obligee was a prerequisite to a material alteration defense by a surety, when it held in *New Haven v. National Steam Economizer Co.* that

> having by the bond assumed a liability measured and determined in part by the terms of the contract whose performance [the surety] has guaranteed, it is important to it that this liability should not be changed by the substitution, *by whatever means,* of a different contract; and it has the clear right to insist that such a substitution should not be made, if it is the purpose of the obligee in the bond to look to its surety. This substitution may be accomplished, in legal effect, either by material changes in the terms of the contract, or by material departures from its terms in its execution and enforcement.[110]

Material changes in the work as contemplated and delineated in the original contract cannot release the surety when the bond provides for and expressly authorizes changes to the contract without notice to the surety. However, material changes that are beyond the scope of the original contract because they were not reasonably within the contemplation of the parties at the time of contract formation, are defined as *radical* or *cardinal changes.*[111] Radical or cardinal changes or additions to the original scope of work significant enough to be considered a new undertaking may operate to release the surety of its obligation even when the surety has waived the right to be notified of changes in the contract. The courts in general, however, have been reluctant to label material changes in the scope of the work as radical or cardinal changes.[112] In contrast, in one case an increase in a contract from $2.3 million to $6.2 million discharged the surety as a result of the significance of the "change of risk."[113]

---

[108] 135 Conn. 286, 292, 64 A. 51 (1949) (emphasis supplied).

[109] 417 S.W.2d 228, 239 (Ky. 1967).

[110] 79 Conn. 482, 487, 65 A. 959 (1907) (emphasis supplied).

[111] State v. Preferred Accident Ins. Co., 149 So.2d 632 (La. Ct. App. 1963).

[112] Massachusetts Bonding & Ins. Co. v. John R. Thompson Co., 88 F.2d 825 (8th Cir. 1937).

[113] Employers Ins. v. Construction Management Eng'rs, 297 S.C. 354, 377 S.E.2d 119 (Ct. App. 1989).

It is not difficult, therefore, to imagine a flurry of changes occurring on a job—some implemented by formal change order, some by express oral agreement, some by unilateral directive, and others by implied agreement—so significantly increasing the nature and value of the original work, as well as the means and methods by which it was to be performed, that the contractor can no longer perform and defaults. In that event, the hapless owner may discover its "safety net," the performance bond, is no longer in place, because the surety has been released by the material alteration to the contract which resulted from those many changes.

§ 9.20 Changes and the Fast-Track Contract

That time is money on virtually every construction project is indisputable, and a delay adversely affects the owner just as it does the contractor. To bring its project from a vague concept to the issuance of a certificate of occupancy, the owner must proceed through three distinct stages: the design phase, the bidding phase, and the construction. Each one of those phases can be exceedingly time-consuming, and the greater the inflationary pressures at the time of contract, the costlier lost time becomes. That truism, coupled with the ever-present desire of every owner to have more built for less, gave birth some years ago to the concept of *fast-tract construction.*

Conventional construction requires that the design process be complete before the solicitation of bids begins and that the job be bought-out and formally awarded before construction begins. But on a fast-track project, construction commences before the design is complete and all the subcontractors awarded in an effort to reduce the overall time period and thereby mitigate the effect of inflation.[114] But in their rush to beat the costs of inflation, owners adopting this method of construction expose themselves to a potentially far greater risk. If defective plans and specifications are prolific parents of changes, then it clearly follows that admittedly incomplete contract documents invariably give birth to many more changes. Furthermore, the more incomplete the design is at the time the owner enters the agreement with the contractor, the greater the likelihood of frequent, costly, debilitating, and even job-threatening disputes over changes during construction.

Because the fast-track contractor typically has agreed to a fixed price or guaranteed maximum price, it is required to estimate that part of the project not yet designed. Depending on the requirements of the contract,

---

[114] Meathe v. State Univ. Constr. Fund, 65 A.D.2d 49, 410 N.Y.S.2d 702 (1978); Squires & Murphy, *The Impact of Fast Track Construction and Construction Management on Subcontractors,* 46 Law & Contemp. Probs. 55 (1983).

§ 9.20 CHANGES AND FAST-TRACK CONTRACT

its price may be established when the scope of the work yet to be defined is substantial indeed, perhaps even as much as 40 percent of the project. Of course, at such an early stage of the project, optimism abounds and mutual confidence and trust run high, for the owner, architect, and contractor share the common goal of a successful project. Against that happy backdrop, the contractor believes it fully understands the intent of the owner and architect for the undesigned portion of the project and, of course, it is assured by both that they will work with it to keep construction costs within the guaranteed maximum price.

But if estimating the cost of a fully completed design is at best an art, not a science, then surely the most talented estimator cannot have great confidence in her ability to determine the cost of work that exists only in the architect's mind. It can come as no surprise therefore that, shortly after the construction begins and the remaining drawings begin to arrive, it becomes apparent that the contractor's understanding of what the previously undersigned work would be and the architect's intent as to that same work are two different things. There begins then a protracted, expensive, and sometimes bitter battle as to what work was included in the original scope of work and what is a legitimate extra.

Sophisticated participants, therefore, take steps at the outset to define in narrative terms and in as much detail as possible the nature and parameters of the undersigned work. But those efforts, however well intentioned, only reduce the confrontations, not eliminate them, as evidenced by the plethora of litigation over this exact issue.[115]

The simple inevitability of such disputes on fast-track jobs was succinctly and matter-of-factly stated by the bankruptcy court in *In re Shamrock Construction Co.:* "It is clear that various change orders were made, which is understandable, because the entire project was designated as a 'fast track job.' The construction commenced before finalization of plans and specifications."[116]

It may be that developers, owners, contractors, architects, and engineers engaged in the construction industry are consummate, incorrigible risk-takers. Perhaps because they thrive on high risk, they do not seek to avoid it, but instead they seek it out. If that is so, then their quest for the ultimate gamble may end with a fast-track project.

---

[115] *See, e.g.,* Meathe v. State Univ. Constr. Fund, 65 A.D.2d 49, 410 N.Y.S.2d 702 (1978); E.B. Jones Constr. Co. v. County of Denver, 717 P.2d 1009 (Colo. Ct. App. 1986); E.C. Ernst Co. v. Koppers Co., 476 F. Supp. 729 (W.D. Pa. 1979); Williams Eng'g, Inc. v. Goodyear, 496 So.2d 1012 (La. 1986); Northwest Ironwork, Inc. v. Rippling River Corp., 71 Or. App. 144, 691 P.2d 111 (1984).

[116] No. B-621-73 slip op. (Bankr. D.N.J. June 30, 1980).

CHAPTER 10

TERMINATION CLAIMS FROM OWNER'S AND CONTRACTOR'S PERSPECTIVE

Jesse B. Grove III, Esquire
Craig M. Jacobsen, CPA*

Jesse B. Grove III is a partner in Thelen, Marrin, Johnson & Bridges, where he serves as head of the firm's construction law department. He is a trial lawyer with substantial experience in all phases of construction contracting and claims on major projects such as dams, tunnels, powerhouses, wastewater treatment plants, waste-to-energy projects, highways, bridges, and municipal, industrial, and commercial buildings.

Craig M. Jacobsen is a partner with the international accounting and consulting firm of Coopers & Lybrand. He specializes in evaluating, presenting, and negotiating complex construction disputes on behalf of owners, architects, engineers, and contractors. Mr. Jacobsen has extensive experience auditing and evaluating construction costs. He has designed and implemented sophisticated computer-based construction cost systems for day-to-day use as well as use in a litigation environment. Mr. Jacobsen graduated from California State University at Sacramento and is a member of the Construction Financial Management Association.

*The authors gratefully acknowledge the assistance of Paul W. Berning, Esquire, of Thelen, Marrin, Johnson & Bridges in the preparation of this chapter.

§ 10.1   Introduction

TERMINATIONS FOR CONVENIENCE

§ 10.2   Overview
§ 10.3   The Federal Compensation Scheme
§ 10.4   Danger of Unit Price Settlements
§ 10.5   Termination Clauses in Private Contracts
§ 10.6   Conclusion

OVERVIEW OF TERMINATIONS FOR DEFAULT

§ 10.7   The Nonbreaching Party's Options
§ 10.8   Considerations to Weigh Before Acting
§ 10.9   The Difference Between Termination and Rescission
§ 10.10   When Termination and Rescission Are Favorable
§ 10.11   Legal Prerequisites to Termination or Rescission
§ 10.12   When Termination or Rescission Has Been Allowed or Denied
§ 10.13   Procedures for Rescission
§ 10.14   Deciding Whether to Carry On, Terminate, or Rescind

TERMINATIONS FOR DEFAULT FROM OWNER'S VIEWPOINT

§ 10.15   Basic Damage Remedy
§ 10.16   —Before Work Begins
§ 10.17   —During the Project
§ 10.18   —At Project's Conclusion
§ 10.19   Extended Duration and Overhead Costs Incurred by the Owner
§ 10.20   Delay in Use
§ 10.21   Lost Profits
§ 10.22   Liquidated Damages
§ 10.23   Special Liquidated Damages Issues
§ 10.24   —Both Liquidated and Actual Damages Can Be Recovered
§ 10.25   —Damages Continue to Accrue after Termination
§ 10.26   Rescission and Restitution
§ 10.27   Owner's Recovery When It Defaults

TERMINATIONS FOR DEFAULT FROM CONTRACTOR'S VIEWPOINT

§ 10.28   Basic Damage Remedy
§ 10.29   —Before Work Begins
§ 10.30   —During the Project
§ 10.31   —Importance of Maintaining Good Records

§ 10.32 —Loss Contracts
§ 10.33 —Upon Full Completion of the Contract
§ 10.34 Other Damages Available
§ 10.35 Offsets Against the Contractor
§ 10.36 Rescission and Restitution
§ 10.37 —When Rescission Is the Better Option
§ 10.38 —When Rescission Is Not Favorable to the Contractor
§ 10.39 —How Restitution Is Measured for Contractors
§ 10.40 Contractor's Recovery When It Is in Default
§ 10.41 Recoveries by Parties Not in Privity with the Owner

§ 10.1 Introduction

Construction project terminations arise in a variety of circumstances and are initiated by both owners and contractors. An overview of the distinctions between the types and circumstances of termination and significant terminology is helpful to understanding termination and provides a guide to this chapter.

The first distinction is whether the termination is for convenience or for default (that is, cause). The rights of the parties and the recoveries available differ significantly depending on the type of termination.

The second distinction is whether the rights and remedies relating to termination are defined by the contract itself, by a statute or regulation, or by the common law (decisional law) that is jurisdictionally applicable. Federal procurement contracts long have provided for convenience terminations,[1] and many private construction contracts now do.[2] Although terminations for convenience are purely the creation of contracts, terminations for default can be based on common law, on statutes or regulations, or on contracts. Common law traditionally has permitted one party to terminate performance of a contract upon material breach of the contract by the other party.[3] The common-law rules have been codified in some states, and grounds for default termination sometimes are set out

---

[1] 48 C.F.R. subpt. 49.2 (1988); 48 C.F.R. §§ 52.249-1 to 52.249-7 (1988).

[2] A typical clause provides: "Owner may, at any time, upon five days' written notice, terminate the contract in whole or in part, for Owner's convenience and without cause." *See also* Arc Elec. Constr. Co. v. George A. Fuller Co., 24 N.Y.2d 99, 103, 247 N.E.2d 111, 113, 299 N.Y.S.2d 129, 132 (1969).

[3] Alder v. Drudis, 30 Cal. 2d 372, 381, 182 P.2d 195, 201 (1947). *See also* Restatement (Second) of Contracts § 237 (1981), especially illustrations 1 & 2.

in regulations controlling contracts.[4] Many construction contracts now explicitly state the grounds for default termination.[5] The authority for termination—contract, statute, regulation, or common law—governs the procedures for termination and the remedies available.

The third distinction is which party initiated the termination. Under common law, either party to a contract may terminate performance upon material breach of the contract by the other party.[6] Some statutes and regulations are to the same effect.[7] But regulations and contracts are not always even-handed. Federal regulations, for example, allow grant recipients to terminate contractors for convenience or default but allow contractors to terminate their performance only for default by the government.[8] Thus, rights can vary significantly depending on which party to the contract seeks to assert them.

The fourth distinction is what kind of relief the aggrieved party seeks. Traditionally, remedies are discussed in terms of damages at law, equitable relief, and restitution.[9] Because equitable relief rarely is appropriate for construction terminations, this chapter focuses on the remedies of damages and restitution.[10] The distinction is important because, as will be shown, recoveries vary materially, depending on the theory advanced.

Care in terminology must be taken when considering termination. Some writers speak of "terminating a contract" when they really mean that one party has terminated performance of the contract. The distinction is more than semantic. When only performance is terminated, the contract remains alive to govern the rights and recoveries of the parties to it.[11] As will be discussed, the contract itself may only be unilaterally terminated when the law permits it to be rescinded and one party exercises its right to do so. The phrase "abandon the contract" has been similarly misused. Akin to that is the term "abandon performance of the contract." More precise terminology, depending on the circumstances, would be that the contractor ceased performance of the contract because of a material breach by the owner, or that the contractor materially breached the contract by ceasing performance without excuse, or both parties agreed (perhaps implicitly) to proceed without reference to contract requirements. Precision in terminology aids courts and juries in understanding

---

[4] *See, e.g.,* Cal. Civ. Code § 1439 (1989); 48 C.F.R. § 49.402-1 (1988); 48 C.F.R. § 52.249-10 (1988).

[5] *See, e.g.,* AIA A201-1987 ¶ 14.2.

[6] Restatement (Second) of Contracts § 237, especially illustrations 1 and 2.

[7] *See, e.g.,* Cal. Civ. Code § 1439 (1989).

[8] *See, e.g.,* required termination clause in 40 C.F.R. § 33.1030 (1988).

[9] *See, e.g.,* D. Dobbs, Handbook on the Law of Remedies at XIX (1973).

[10] *But see* D. Dobbs, Handbook on the Law of Remedies § 12.22, at 907–10 (1973).

[11] Alder v. Drudis, 30 Cal. 2d 372, 381, 182 P.2d 195, 201 (1947).

the arguments being made and thereby improves the chance of success in litigation.[12]

Finally, it should be noted that contractors act as owners in dealing with subcontractors, that construction managers stand in the place of owners in dealing with contractors, and that sureties can stand in the place of general contractors.

This chapter discusses terminations for convenience first and then terminations for default. Default terminations are examined from both the owner's perspective and from the contractor's perspective. Both the damage and restitution remedies are discussed, as well as the relief available to the party in default.

TERMINATIONS FOR CONVENIENCE

§ 10.2 Overview

Termination for convenience clauses must be included in federal contracts[13] and they sometimes are found in private construction contracts.[14] Terminations for convenience, at least in theory, are no-fault events exercisable at the option of the owner. Federal contracts may be terminated whenever it is in the government's interest to do so.[15] In private contracts as well, the clause often does not require cause either. However, fault can creep into the issue. If the owner is unhappy with the contractor's progress, quality of work, or personnel, it may view terminating the contractor for convenience as the quickest way of getting rid of the contractor with the least legal difficulty. In addition, federal procurement regulations convert wrongful default terminations of contractors into terminations for convenience.[16]

The rationale for granting owners such broad rights is that the compensation scheme makes the contractor whole while ensuring the owner much-needed flexibility.[17] However, the compensation scheme is not quite that generous.

---

[12] *See, e.g.,* First Nat'l Bank v. Indian Indus., Inc., 600 F.2d 702, 708 (8th Cir. 1979).

[13] 48 C.F.R. subpt. 49.5 (1988).

[14] Arc Elec. Constr. Co. v. George A. Fuller Co., 24 N.Y.2d 99, 101–03, 247 N.E.2d 111, 112–13, 299 N.Y.S.2d 129, 131–32 (1969).

[15] *See* § (a) of required termination for convenience clause at 48 C.F.R. § 52.249-2 (1988).

[16] 48 C.F.R. § 49.401(b) (1988).

[17] *See, e.g.,* American Elec., 76-2 B.C.A. (CCH) ¶ 12,151 at 58,447, 58,479 (ASBCA 1976).

§ 10.3 The Federal Compensation Scheme

Upon termination for convenience, a federal construction contractor is entitled to:

1. The reasonable cost of its work
2. A "fair and reasonable" profit on the work completed
3. Termination expenses, including accounting, legal, and clerical costs for preparing a settlement proposal; the cost of settling terminated subcontracts; and project wind-down costs.[18]

In addition, the contractor is entitled to interest on its claim at a rate set by the treasury secretary from the date of filing with the contracting officer until the date of payment.[19]

Although some practitioners have described the contractor's recovery under the federal regulations as a total cost or cost reimbursement recovery, there are significant limitations.[20]

First, the federal contractor can recover no more than the original contract amount less payments made. (This limitation does not include settlement costs but is calculated before deducting disposal and other credits due the contractor.)[21]

Second, if the federal construction contractor would have suffered cost overruns had the entire contract been completed, the settlement for actual cost of performance must be proportionately reduced to reflect the loss.[22] Thus, the contractor is spared the additional cost of completing the project but must absorb its loss to date of termination.

Third, the contractor may recover only its "reasonable" and "ordinary" costs as judged by a "prudent person in the conduct of competitive business" standard. The burden of proof is on the contractor.[23]

---

[18] *See* § (f) in Alternate I of required termination for convenience clause at 48 C.F.R. § 52.249-2 (1988); § (h) of required termination for convenience clause at 48 C.F.R. § 52.249-2 (1988); 48 C.F.R. §§ 31.201-1 (costs also must be "allowable" and "allocable"), 31.201-2, 31.201-3, 31.201-4 (1988).

[19] 41 U.S.C. § 611 (1989).

[20] *See, e.g.,* Paul E. McCollum, Sr., 81-2 B.C.A. (CCH) ¶ 15,310, at 75,810, 75,822 (ASBCA 1981).

[21] 48 C.F.R. § 49.207 (1988).

[22] *See* § (f)(1)(iii) in Alternate I of required termination for convenience clause at 48 C.F.R. § 52.249-2 (1988). *But see* required termination clause in 40 C.F.R. § 33.1030 (1988).

[23] *See* § (h) of required termination for convenience clause at 48 C.F.R. § 52.249-2 (1988); 48 C.F.R. §§ 31.201-1, 31.201-3 (1988).

§ 10.3 FEDERAL SCHEME

Thus, the contractor must justify its costs against a reasonableness standard rather than merely presenting its costs and recovering them.

Fourth, the federal contractor, upon termination, may recover for only that extra work to which it would have been entitled if the contract had been completed. If the extra work would not have been compensable upon completion, termination alone does not make it compensable.[24] If, at termination, the contractor has claims for extra work, it will be paid only for those that would have been approved in the ordinary course of business if the contract had been completed. Thus, the contractor is not certain of recovering if it was inefficient or worked beyond its scope and is terminated for convenience.

Fifth, federal contractors no longer may be sure that the federal government will be denied offsets against their recoveries for defective work. It once was a well-settled rule that offsets for defective work were not allowed in calculating recoveries for convenience terminations.[25] But, recently, the United States Court of Appeals for the Federal Circuit has refused to apply the old rule,[26] the board of contract appeals, which set down the old rule, has questioned its continued viability and twice declined to apply it,[27] and another board of contract appeals has specifically rejected the old rule.[28] Thus, contractors can expect the government to critically examine their work, with an eye toward reducing termination settlements.

Sixth, consequential damages are not allowed to the contractor.[29]

Seventh, the contractor is permitted no anticipatory profits on the uncompleted work.[30]

Eighth, if the contractor appears to have been losing money on the job, it will be denied any profit.[31]

---

[24] Lemire Contracting, 89-2 B.C.A. (CCH) ¶ 21,763, at 109,510, 109,513 (IBCA 1989).

[25] New York Shipbldg. Co., a Div. of Merritt-Chapman & Scott Corp., 73-1 B.C.A. (CCH) ¶ 9852 at 46,015, 46,018–21 (ASBCA 1973); Western States Painting Co., 69-1 B.C.A. (CCH) ¶ 7616 at 35,372, 35,379 (ASBCA 1969); J.D. Shotwell Co., 65-2 B.C.A. (CCH) ¶ 5243, at 24,661, 24,691–92 (ASBCA 1965).

[26] Lisbon Contractors, Inc. v. United States, 828 F.2d 759, 769 (Fed. Cir. 1987).

[27] Air Cool, Inc., 88-1 B.C.A. (CCH) ¶ 20,399, at 103,195, 103,198 (ASBCA 1988); Structural Painting Corp., 89-3 B.C.A. (CCH) ¶ 21,969, at 110,501, 110,512 (ASBCA 1989).

[28] Aydin Corp., 89-3 B.C.A. (CCH) ¶ 22,044, at 110,881, 110,912–13 (EBCA 1989). *See also* Arc Elec. Constr. Co. v. George A. Fuller Co., 24 N.Y.2d 99, 103, 247 N.E.2d 111, 113, 299 N.Y.S.2d 129, 132 (1969).

[29] 48 C.F.R. § 49.202(a) (1988).

[30] *Id.*

[31] § (f)(1)(iii) in Alternate I of required termination for convenience clause at 48 C.F.R. § 52.249-2 (1988).

Ninth, even if the contractor does not appear to have lost money on the job, its claim for profits on the work completed still will be examined in terms of:

1. Extent and difficulty of the work completed compared with the total project
2. Technical support needed, such as engineering and scheduling
3. The contractor's efficiency, with particular attention to quantity and quality attained, efficiency, and economy
4. Risk assumed, amount of capital required, and its source
5. Cooperation and developmental contributions
6. Complexity of the project
7. Rate of profit the contractor would have earned but for termination
8. Extent and difficulty of the subcontracting effort
9. The rate of profit contemplated by both parties at time of contracting.[32]

Factors three and seven are particularly unfavorable to contractors who are inefficient, do low-quality work, or build a slim rate of profit into their bids.

Tenth, contractors will not be allowed a profit on their settlement costs or their subcontractors' settlement costs. However, contractors are entitled to a mark-up on their subcontractors' actual cost of completing contract work.[33]

The foregoing limitations on recovery demonstrate that the contractor terminated for convenience may not be made entirely whole. Most significantly, the contractor is denied the profit it would have received if allowed to complete the project. Yet, the contractor can have a loss imposed on it, be denied a profit entirely, see its profit reduced for inefficiency, and be subjected to offsets for defective work.

However, not all contracts governed by federal regulations are so stringent. For example, the required general conditions for many Environmental Protection Agency grants do not impose a loss on contractors with losing contracts.[34]

---

[32] 48 C.F.R. § 49.202(b) (1988); Quality Seeding, Inc., 88-3 B.C.A. (CCH) ¶ 21,020, at 106,175, 106,180–83 (IBCA 1988).

[33] 48 C.F.R. § 49.202(c) (1988); § (f)(1)(iii) in Alternate I of required termination for convenience clause at 48 C.F.R. § 52.249-2 (1988).

[34] *See, e.g.,* required termination clause in 40 C.F.R. § 33.1030 (1988).

§ 10.4 Danger of Unit Price Settlements

Owners sometimes propose calculating termination settlements on a unit-price basis using unit prices in the contract or by trying to convert the original contract from a fixed-price to a unit-price contract. This could be detrimental to the contractor. Contractors incur one-time start-up costs on projects, including bid preparation, preaward proposals, equipment modification, and productivity losses caused by the initial learning curve and employee training. When quantities of work are reduced, the start-up costs to be absorbed by each of the reduced number of units is increased. In addition, learning-curve problems are most pronounced up to the halfway point of completion. Accordingly, a unit price calculated on the entire project rather than on the truncated project could be substantially low.

The Defense Contract Audit Agency publishes the *Contract Audit Manual*, which outlines procedures to be followed by its auditors in many types of audits. Chapter 12, Auditing Terminated Contracts, is useful in determining the issues involved in all types of terminated contracts. It also provides insight into the government's likely position on certain claimed costs.[35]

§ 10.5 Termination Clauses in Private Contracts

A wide range of variations on the federal termination for convenience scheme can be found in private contracts. Some lift large sections of language from the federal regulations. Other variations grant the owner fewer rights. For example, AIA Document A201 provides for suspension of projects for the owner's convenience but does not explicitly provide for termination of convenience.[36] On the other hand, private contracts can be more restrictive for the contractor than federal contracts. Some are less generous in reimbursing shut-down costs or define compensation by the value of work completed rather than the cost of work completed. If the contractor was inefficient in performing before termination, the owner could use the difference in language to impose a loss on the contractor.

§ 10.6 Conclusion

A termination for convenience, especially of a federal contract, once may have been welcomed by contractors with losing jobs, thin profit margins

---

[35] *See also* 48 C.F.R. Subpart 45.6, Reporting, Redistribution, and Disposal of Contractor Inventory (1988).

[36] AIA A201-1987 ¶ 14.3.

or quality problems. Now, however, such contractors face substantial reductions in their recoveries. Moreover, contractors doing quality work efficiently and anticipating sizable profits will find their claims carefully scrutinized and their profits proportionately reduced. Owners, though, are likely to benefit from the flexibility and security against the unexpected provided by a termination for convenience clause.

OVERVIEW OF TERMINATIONS FOR DEFAULT

§ 10.7 The Nonbreaching Party's Options

Faced with a material (that is, serious, not minor) breach of contract, the aggrieved owner and the aggrieved contractor have three options:

> [O]ne who has been injured by a breach of contract has an election to pursue any of three remedies, to wit: "He may treat the contract as rescinded and may recover upon a quantum meruit so far as he has performed; or he may keep the contract alive, for the benefit of both parties, being at all times ready and able to perform; or, third, he may treat the repudiation as putting an end to the contract for all purposes of performance, and sue for the profits he would have realized if he had not been prevented from performing."[37]

Put more plainly, the aggrieved contractor and the aggrieved owner have these options:

1. Keep performing under the contract and sue for damages upon completion[38]
2. Consider the obligation of continued performance terminated and sue on the contract
3. Rescind the contract and seek restitution.

§ 10.8 Considerations to Weigh Before Acting

Significant legal and business decisions are involved in selecting which option in § 10.7 to follow. First, the aggrieved party must learn for which remedies it qualifies legally. Second, knowing its options, the aggrieved

---

[37] Alder v. Drudis, 30 Cal. 2d 372, 381–82, 182 P.2d 195, 201 (1947) (citation omitted).
[38] W. Jaeger, 11 Williston on Contracts § 1292, at 11 (3d ed. 1957).

§ 10.10 WHEN EACH ARE FAVORABLE   287

party must decide which of the available options makes better business sense, considering the particular circumstances it faces.

As evidenced by the relative infrequency of terminations and rescissions in the construction industry, most aggrieved owners and contractors ultimately select the performance option; they suffer through the breach until the project is complete and sue for damages. Recoveries under this option are discussed in **Chapters 5** to **9** for contractors and in **Chapters 12** to **15** for owners. The remainder of this chapter discusses when the termination and rescission options are legally available to owners and contractors, when the options make practical sense for each party, and what recoveries are available under each theory to both parties.

§ 10.9 The Difference Between Termination and Rescission

In considering whether to terminate performance of a contract or to rescind it, the first step is to understand the legal difference between those terms. One court explained the distinction in this way:

> It is contemplated that upon rescission all parties are to return those things of value which they have received and they are to be put in status quo. "Termination," however, contemplates the right of one contracting party to terminate further performance because of the breach of the other. In such case the party giving notice of termination may proceed to present his claim for damages by reason of the breach up to that date.[39]

Thus, rescission and restitution (returning the things of value received or their equivalent in money) are rearward-looking, attempting to put the pieces back as they were before the contract, while termination and a suit on the contract for damages are forward-looking because the contract will continue to control as the parties' relationship enters a new phase.

§ 10.10 When Termination and Rescission Are Favorable

From the practical standpoint, the aggrieved owner or contractor can decide between rescission and terminating the contract by asking: Am I better off with the contract or without the contract?

---

[39] B.L. Metcalf Gen. Contractor, Inc. v. Earl Erne, Inc., 212 Cal. App. 2d 689, 694, 28 Cal. Rptr. 382, 386 (1963); M&W Masonry Constr., Inc. v. Head, 562 P.2d 957, 959–61 (Okla. Ct. App. 1976); United States *ex rel.* Bldg. Rentals Corp. v. Western Casualty & Sur. Co., 498 F.2d 335, 338 (9th Cir. 1974).

Aggrieved Contractor, Owner in Material Breach

If the contractor has a contract with neutral or favorable terms and upon which it is making money, the contractor should affirm the contract, terminate performance of the contract, and sue for breach of the contract. By doing so, it retains entitlement to its anticipated profit. Conversely, if the contractor is losing money on the contract—because of a bad bid or rising prices or efficiency problems—or if the contract has unfavorable terms—for example, a no damage for delay clause or a liquidated damage clause—the contractor should rescind the contract and sue for the fair value of the work and materials it provided. To rescind, it may have to forgo profit, but it may recover all costs, even if in excess of the contract rate.

Aggrieved Owner, Contractor in Material Breach

If the owner has a contract with neutral or favorable terms and favorable prices for it—for example, it knows the contractor underbid the job or knows a substitute contractor would charge far more—the owner should affirm the contract, terminate the contractor's performance of it, and sue for breach of the contract. Conversely, if the owner agreed to pay far too much for the project or if the contractor's performance is so defective as to have little practical value and if the contract has no particularly owner-favorable provisions, the owner should rescind the contract. In such circumstances, the contractor would be entitled only to the fair value of what it provided to the owner, as measured from the owner's standpoint. If the quality of the contractor's performance was low, it would have less value to the owner than its cost under the contract, thereby benefiting the owner.

§ 10.11 Legal Prerequisites to Termination or Rescission

With the distinction between rescission and terminating performance of a contract in mind, the aggrieved owner or contractor must examine the legal prerequisites to termination and rescission. When contracts are breached in only relatively minor (that is, immaterial) ways, the aggrieved party's only remedy is to recover damages.[40] However, the aggrieved party may rescind and seek restitution or terminate and seek damages if (1) the

---

[40] Restatement (Second) of Contracts § 373 illustration 2; A. Corbin, 5 Corbin on Contracts, § 1104, at 562 (1964); W. Jaeger, 11 Williston on Contracts, § 1292, at 9, 115 (3d ed. 1957); Fountain v. Semi-Tropic Land & Water Co., 99 Cal. 677, 680, 34 P. 497, 498 (1893).

§ 10.11 LEGAL PREREQUISITES   289

defaulting party repudiates the contract,[41] or (2) the defaulting party's nonperformance amounts to a "total"[42] or "vital" breach of contract or goes to the "essence" of the contract.[43] Other authorities use words less strong, such as "material breach," to describe a triggering breach.[44] Thus, although the leading commentators use different language to describe the triggering breach, they view rescission and termination of performance as alternative responses equally available to the nondefaulting party upon a serious enough breach of contract.[45]

Some states permit rescission on grounds in addition to material breach, including:

1. Consent of the rescinding party was obtained by mistake, fraud (either intentional, constructive, or negligent), duress, menace, or undue influence
2. Consideration becomes void (if the consideration is a promise to perform an existing legal duty, consists of acts or forbearances previously performed, consists of compromising or forbearing from a wholly worthless claim, or is deemed illegal)
3. Contract is unlawful or against the public interest.[46]

In addition, construction contracts often specify grounds for default termination, including: failure to pay subcontractors, materialmen, or employees; failure to provide an adequate labor force; failure to meet the construction schedule or to work with due diligence; defective performance; and disregarding laws or regulations.[47] Contractual grounds for termination are not exclusive and do not foreclose using common-law grounds for termination unless the contract specifically so provides.[48]

---

[41] Restatement (Second) of Contracts § 373(1), comment a & illustration 10; W. Jaeger, 12 Williston on Contracts, § 1455, at 14; 11 Williston § 1315, at 115 (3d ed. 1957); A. Corbin, 5 Corbin on Contracts, § 1104, at 560 (1964).

[42] Restatement (Second) of Contracts § 373(1), comment a & illustration 2.

[43] A. Corbin, 5 Corbin on Contracts § 1104, at 558, 562 (1964). *Accord* United States *ex rel.* Bldg. Rentals Corp. v. Western Casualty & Sur. Co., 498 F.2d 335, 339 (9th Cir. 1974).

[44] W. Jaeger, 12 Williston on Contracts § 1455, at 14; 11 Williston § 1315, at 115 (3d ed. 1957). See also Cal. Civ. Code § 1689(a)(2) (1989), and Economy Swimming Pool Co. v. Freeling, 236 Ark. 888, 889, 370 S.W.2d 438, 440 (1963).

[45] W. Jaeger, 12 Williston on Contracts § 1455, at 14; A. Corbin, 5 Corbin on Contracts § 1104, at 560 (1964). *But see* discussion in Integrated, Inc. v. Alec Fergusson Elec. Contractor, 250 Cal. App. 2d 287, 296–97, 58 Cal. Rptr. 503, 509–10 (1967).

[46] *See, e.g.,* Cal. Civ. Code §§ 1689, 1571–73 (1989).

[47] *See, e.g.,* AIA A201-1987 ¶ 14.2.

[48] Monmouth Pub. Schools, Dist. 38 v. D.H. Rouse Co., 153 Ill. App. 3d 901, 903–04, 506 N.E.2d 315, 317 (1987).

However, default termination is not proper when the nonperformance is excusable. The excuse can arise from the contract or from common law.[49] Even if grounds for termination exist and are not excusable, the nondefaulting party must act in good faith or the termination will be deemed wrongful.[50] Courts may examine whether the owner, saddled with an unfavorable contract, provoked the contractor's breach.[51] Other contract defenses can be interposed, undercutting the basis for termination. On the other hand, the party wishing to terminate can itself be in default, so long as the conditions precedent to termination are satisfied.[52] Moreover, the termination can eventually be justified on the basis of facts of which the terminating party was unaware at the time of termination.[53]

§ 10.12 When Termination or Rescission Has Been Allowed or Denied

Before terminating or rescinding, the aggrieved owner or contractor should make sure that the situation it faces satisfies the applicable standard for the jurisdiction under which the contract is to be enforced. The following kinds of events have been held to justify termination or rescission:

1. There was a "substantial failure" by the owner to make progress payments, not merely a "slight deviation" in time or amount paid.[54] Failure to make progress payments for three months has been held to be grounds to terminate performance.[55]
2. The owner requested such substantial changes in the project that they ran to the "entire consideration of the contract" and cost nearly

---

[49] United States *ex rel.* Susi Contracting Co. v. Zara Contracting Co., 146 F.2d 606, 608 (2d Cir. 1944). *See also* AIA A201-1987 ¶ 8.3.1.

[50] Paul Hardeman, Inc. v. Arkansas Power & Light Co., 380 F. Supp. 298, 331 (E.D. Ark. 1974).

[51] Restatement (Second) of Contracts § 373 comment d.

[52] Paul Hardeman, Inc. v. Arkansas Power & Light Co., 380 F. Supp. 298, 331 (E.D. Ark. 1974).

[53] College Point Boat Corp. v. United States, 267 U.S. 12, 15–16, 45 S. Ct. 199, 200–01 (1925); Pots Unltd., Ltd. v. United States, 600 F.2d 790, 793 (Ct. Cl. 1979); Joseph Morton Co. v. United States, 3 Cl. Ct. 120, 122 (1983), *aff'd,* 757 F.2d 1273, 1277 (Fed. Cir. 1985).

[54] Integrated, Inc. v. Alec Fergusson Elec. Contractor, 250 Cal. App. 2d 287, 296–97, 58 Cal. Rptr. 503, 510 (1967); Porter v. Arrowhead Reservoir Co., 100 Cal. 500, 504, 35 P. 146, 148 (1893); United States *ex rel.* Bldg. Rentals Corp. v. Western Casualty & Sur. Co., 498 F.2d 335, 339 (9th Cir. 1974).

[55] M&W Masonry Constr., Inc. v. Head, 562 P.2d 957, 961 (Okla. Ct. App. 1976).

§ 10.12   TERMINATION ALLOWED OR DENIED 291

half the contract price, yet the owner refused to agree on the price for a change order, offering only 20 percent of the additional cost.[56]

3. The contractor was unable to complete the project, as evidenced by its absence from the project for two months and by its completing only 6 percent of the work after half the performance period had passed.[57] However, merely experiencing high front end costs and falling behind the construction schedule do not warrant termination.[58]

4. The contractor failed to diligently perform its work with an adequate crew, thereby breaching a time is of the essence clause in the contract.[59]

5. The contractor's work was so shoddy that the project leaked water and part of it collapsed.[60]

The right to terminate or rescind has been lost under the following circumstances:

1. The contractor performed the majority of its work properly and stood ready to correct any defects. The evidence showed that the owner had orally extended the contractor's completion date, so that the termination occurred before the revised termination date.[61]

2. The subcontract contained a clause obligating the subcontractor not to delay or cease work in the event of a dispute with the contractor or another subcontractor, thereby amounting to a waiver by the subcontractor of its right to rescind.[62]

---

[56] Coleman Eng'g Co., Inc. v. North Am. Aviation, Inc., 65 Cal. 2d 396, 400–01, 407, 420 P.2d 713, 717, 720–21, 55 Cal. Rptr. 1, 5, 8–9 (1966).

[57] First Nat'l Bank v. Indian Indus., Inc., 600 F.2d 702, 707 (8th Cir. 1979); Commercial Contractors, Inc. v. United States Fidelity & Guar. Co., 524 F.2d 944, 955–56 (5th Cir. 1975); Oregon State Highway Comm'n v. DeLong Corp., 9 Or. App. 550, 569–71, 495 P.2d 1215, 1225 (1972), *cert. denied,* 411 U.S. 965, 93 S. Ct. 2142, *reh'g denied,* 412 U.S. 944, 93 S. Ct. 2771 (1973).

[58] Paul Hardeman, Inc. v. Arkansas Power & Light Co., 380 F. Supp. 298, 315 (E.D. Ark. 1974).

[59] Call v. Alcan Pac. Co., 251 Cal. App. 2d 442, 446–47, 59 Cal. Rptr. 763, 766–67 (1967) (termination proper both under a specific termination clause in the subcontract and under common law).

[60] Economy Swimming Pool Co. v. Freeling, 236 Ark. 888, 370 S.W.2d 438 (1963). *See also* Cantrell v. Woodhill Enters., Inc., 273 N.C. 490, 497, 160 S.E.2d 476, 481 (1968).

[61] Marathon Oil Co. v. Hollis, 167 Ga. App. 48, 48–49, 305 S.E.2d 864, 866–67 (1983).

[62] B.C. Richter Contracting Co. v. Continental Casualty Co., 230 Cal. App. 2d 491, 500–501, 41 Cal. Rptr. 98, 104 (1964). The clause read:

Subcontractor, in the event of any dispute or controversy with Contractor or any other subcontractor over any matter whatsoever, shall not cause any delay or

3. The owner allowed a delinquent contractor to continue performing. By doing so, the owner waived its right to terminate until a reasonable time had passed unless the contractor abandoned performance. To again make time of the essence, the owner must notify the contractor of a reasonable new deadline date.[63]

Restitution is not permitted under the following additional circumstances:

1. The contractor conferred no benefit on the owner[64]
2. The contractor unreasonably performed work after repudiation of the contract by the owner[65]
3. The contractor fully performed, and the owner simply owes the contractor a definite sum of money.[66]

§ 10.13 Procedures for Rescission

Certain procedures must be followed in order to rescind. The rescinding party must rescind promptly upon becoming aware of the grounds, must notify the other party to the contract, and must tender back the benefits it has received so far from the bargain. The rescinding party must take care not to inadvertently waive other remedies.[67]

Under some circumstances, including fraud, the rescinding party need not tender back the consideration it received or will be excused for delaying in rescinding or tendering back the consideration.[68] Some states do

---

cessation in or of Subcontractor's work or the work of any other subcontractor or of the Contractor but shall proceed under this Subcontract Agreement with the performance of the work required thereby.

[63] D. Joseph DeVito v. United States, 413 F.2d 1147, 1153–54 (Ct. Cl. 1969); Call v. Alcan Pac. Co., 251 Cal. App. 2d 442, 447–48, 59 Cal. Rptr. 763, 767 (1967); United States *ex rel.* Susi Contracting Co. v. Zara Contracting Co., 146 F.2d 606, 608 (2d Cir. 1944). *See also* Paul Hardeman, Inc. v. Arkansas Power & Light Co., 380 F. Supp. 298, 311 (E.D. Ark. 1974). *But see* the discussion of problems with overly rigid application of waiver principles in Acret, Construction Litigation Handbook § 10.03, at 172 (1986).

[64] Restatement (Second) of Contracts § 373, illustration 11 & comment e.

[65] Restatement (Second) of Contracts § 373, comment e & illustration 13.

[66] Restatement (Second) of Contracts § 373(2) & illustration 5.

[67] Cal. Civ. Code §§ 1691, 1692 (1989); Restatement (Second) of Contracts § 378.

[68] California Farm & Fruit Co. v. Schiappa-Pietra, 151 Cal. 732, 739–40, 91 P. 593, 595–96 (1907); Thackrah v. Hass, 119 U.S. 499, 502, 7 S. Ct. 311, 312 (1886); Cal. Civ. Code §§ 1691, 1693 (1989).

not permit rescission if the rescinding party itself was in default or if the defendant's nonperformance was justified by the rescinding party's own performance.[69]

§ 10.14 Deciding Whether to Carry On, Terminate, or Rescind

Owners and contractors should proceed with caution in deciding whether to continue performing, to terminate, or to rescind. If the termination is found to be wrongful, either because grounds for it did not exist or because the other party's nonperformance was excused, the terminated party can recover for wrongful termination and thereby turn the remedial tables. The decision should not be made emotionally or in anger. Rather, the decision should be coolly and rationally considered based on comprehensive legal research into the applicable law, a thorough understanding of the contract, full knowledge of the facts, and careful consideration of the business ramifications of both going forward with the contract and with halting performance.[70] The important issues are whether the owner or contractor qualifies for termination or rescission and whether the contract, on balance, is favorable and worth retaining or unfavorable and worth avoiding. Warnings should be considered rather than simply surprising the other party with termination.

TERMINATIONS FOR DEFAULT FROM OWNER'S VIEWPOINT

§ 10.15 Basic Damage Remedy

The owner is entitled to default remedies if it rightfully terminates the contractor for a material breach of contract or if the contractor wrongfully ceases performance, thereby repudiating the contract. The owner's basic damage remedy depends on the stage of the project at termination and is discussed in §§ **10.16** through **10.18**.

---

[69] Martin v. Rollins, Inc., 238 Ga. 119, 119–21, 231 S.E.2d 751, 752 (1977). *But see* American-Hawaiian Eng'g & Constr. Co. v. Butler, 165 Cal. 497, 516–17, 133 P. 280, 288 (1913); Integrated, Inc. v. Alec Fergusson Elec. Contractor, 250 Cal. App. 2d 287, 298, 58 Cal. Rptr. 503, 510 (1967).

[70] J. Sweet, Legal Aspects of Architecture, Engineering, and the Construction Process § 38.01, at 872–74 (1985).

§ 10.16 —Before Work Begins

The cost of completing the project less the amount of the contract with the terminated contractor is the owner's measure of recovery if default termination occurs before work begins.[71] Because the owner usually awards the contract to the low bidder, it is almost certain to pay the replacement contractor more for the project than it would have paid under the breached contract. The owner may recover that difference, so long as the scope of work does not change in the new contract.[72] Indeed, the owner may recover the difference between the terminated contractor's bid and the cost of completion from the contractor even though the project is not completed.[73]

§ 10.17 —During the Project

For a contract terminated during work on the project, the total cost of the completed project less the amount of the contract with the terminated contractor is the measure of recovery.[74] There is, however, a split in authority on whether the owner is entitled to the reasonable cost of completion or the actual cost of completion.[75] Some courts require a showing of reasonableness to ensure that the owner did not expand or upgrade the project at the defaulted contractor's expense.[76] Other courts believe it is unfair to place the burden of proving reasonableness on owners, who are not familiar with construction, and instead employ an actual cost measure of damages to ensure that owners get their expectancy.[77] One court resolved this dilemma by imposing a reasonableness test, but putting the

---

[71] Restatement (Second) of Contracts § 347; C. McCormick, Damages § 169, at 650 (1935); Ray v. William G. Eurice & Bros., 201 Md. 115, 129–30, 93 A.2d 272, 279–80 (1952); Carrig v. Gilbert-Varker Corp., 314 Mass. 351, 356–57, 50 N.E.2d 59, 62–63 (1943).

[72] Martin v. Phillips, 122 N.H. 34, 36–37, 440 A.2d 1124, 1125 (1982).

[73] Fairlane Estates, Inc. v. Carrico Constr. Co., 228 Cal. App. 2d 65, 72, 39 Cal. Rptr. 35, 40 (1964).

[74] Restatement (Second) of Contracts § 347, illustration 6; C. McCormick, Damages § 169, at 650 (1935); Sea Ledge Properties, Inc. v. Dodge, 283 So. 2d 55, 57–58 (Fla. Dist. Ct. App. 1973).

[75] See Kirkpatrick v. Temme, 98 Nev. 523, 525–26, 654 P.2d 1011, 1012–13 (1982) and cases cited therein; D. Dobbs, Handbook on the Law of Remedies § 12.21, at 897 (1973).

[76] See cases cited in Kirkpatrick v. Temme, 98 Nev. 523, 525–26, 654 P.2d 1011, 1012–13 (1982). See, e.g., Martin v. Phillips, 122 N.H. 34, 36–37, 440 A.2d 1124, 1125 (1982).

[77] Kirkpatrick v. Temme, 98 Nev. 523, 526, 654 P.2d 1011, 1013–14 (1982).

§ 10.18   AT PROJECT'S CONCLUSION   295

burden of proof on the party in default to show that costs incurred by the nonbreaching party were unreasonable.[78] Another court, looking to the rule of mitigation, required the nondefaulting party to mitigate its damages but put the burden on the party in breach to prove that the nondefaulting party did not take reasonable steps to mitigate.[79]

§ 10.18   —At Project's Conclusion

The cost of completing the project or repairing the defective work less any unpaid amounts due the contractor is the owner's measure of recovery when default termination occurs as the project nears completion.[80] So long as the incomplete or defective work is on the punch list, the owner should have no trouble recovering the actual cost of completion or correction. Even when repair costs approached five percent of the contract price, the contractor has been required to pay them.[81] Another court wrote that the cost measure should be used "unless the cost of completion is grossly and unfairly out of proportion to the good to be attained."[82]

However, when the cost of correction or completion would be high, a different damage measure may be used: diminution in value of the project as a result of the contractor's failure to fully or adequately perform.[83] *Diminution* is the fair market value of the project as planned less the fair market value of the project as built. Courts apply this measure of damages to prevent economic waste.[84] The diminution measure is disfavored by some commentators because it denies the owner what he bargained for and thereby defeats a basic tenet of contract law: placing the

---

[78] Parem Contracting Corp. v. Welch Constr. Co., 128 N.H. 254, 259, 512 A.2d 1104, 1107 (1986).

[79] Conner v. Southern Nev. Paving, Inc., 741 P.2d 800, 801 (Nev. 1987).

[80] Restatement (Second) of Contracts § 347; C. McCormick, Damages § 168, at 647 (1935); D. Dobbs, Handbook on the Law of Remedies § 12.21, at 897 (1973); Fidelity & Deposit Co. of Md. v. Stool, 607 S.W.2d 17, 21 (Tex. Civ. App. 1980).

[81] M.W. Goodell Constr. v. Monadnock Skating Club, 121 N.H. 320, 323, 429 A.2d 329, 331 (1981).

[82] Jacobs & Young, Inc. v. Kent, 230 N.Y. 239, 244, 129 N.E. 889, 891 (1921).

[83] Restatement (Second) of Contracts § 348(2)(b); C. McCormick, Damages § 168, at 647 (1935); D. Dobbs, Handbook on the Law of Remedies § 12.21, at 897 (1973); W. Jaeger, 11 Williston on Contracts § 1363, 344–46 (3d ed. 1957) (The late Samuel Williston was a professor at the Harvard University School of Law.); A. Corbin, 5 Corbin on Contracts 485, 491 (1964); *See* Annotation, *Modern Status of Rule as to Whether Cost of Correction or Difference in Value of Structures Is Proper Measure of Damages for Breach of Construction Contract*, 41 A.L.R.4th 131 (1985).

[84] Restatement (Second) of Contracts § 348 comment c.

nonbreaching party in as good a position as it would have been with full performance.[85] Indeed, to ensure full compensation, one court awarded a homeowner, saddled with a defective and leaky swimming pool, the purchase price paid, the cost of removing the defective pool, and the cost of filling the hole.[86] Professor Arthur Linton Corbin, the late author of the treatise on contracts and professor at the Yale University School of Law, would put the burden of proving economic waste, and thereby establishing entitlement to the diminution measure, on the defaulting contractor.[87]

Courts consider some or all of these factors in deciding whether to apply the cost or the diminution in value measure of damages:

1. How substantial was the contractor's performance?
2. Will the owner's intended use of the project be seriously impeded if it receives value-measured damages rather than cost-measured damages? Is the building safe to use?
3. Was the contractor's failure in good faith or was it a willful or intentional failure?
4. Will the owner be unjustly enriched by application of the cost measure rather than the value measure?[88]

Courts, concerned with economic waste, refused to apply the cost measure and used the diminution in value measure in these cases:

1. The contractor, over the owner's objection, built a home facing the wrong direction on the building site. The house was a mirror image of what the owner wanted, and the owner wanted the house entirely rebuilt.[89]
2. The contractor failed to use the specified brand of plumbing supplies but did use supplies of equal quality in all respects. The defect was discovered only after substantial completion, and correction would have required tearing down much of the house.[90]

---

[85] Restatement (Second) of Contracts § 347 comment a; § 348 comment c; D. Dobbs, Handbook on the Law of Remedies § 12.21, at 897 (1973).

[86] Bause v. Anthony Pools, Inc., 205 Cal. App. 2d 606, 609, 614, 23 Cal. Rptr. 265, 267, 270 (1962).

[87] A. Corbin, 5 Corbin on Contracts § 1089, at 492 (1964).

[88] C. McCormick, Damages § 168, at 649 (1935); D. Dobbs, Handbook on the Law of Remedies § 12.21, at 898–900 (1973); Fidelity & Deposit Co. of Md. v. Stool, 607 S.W.2d 17, 22–23 (Tex. Civ. App. 1980) (cost of correction rather than diminution in value measure used when contractor's breach was intentional).

[89] Grossman Holdings, Ltd. v. Hourihan, 414 So. 2d 1037, 1040 (Fla. 1982).

[90] Jacobs & Young, Inc. v. Kent, 230 N.Y. 239, 244, 129 N.E. 889, 891 (1921).

3. The contractor used ordinary brick instead of the specified "face brick veneer." The owner wanted the nonspecification brick removed and replaced with the specified veneer, but the court refused because the repair would have cost nearly as much as the contract price for the entire house.[91]

In sum, courts will award owners the cost of completion, correction, or replacement if the owner otherwise would not get substantially what it bargained for, so long as economic waste would not result.[92]

Besides the basic cost of completion, correction, or replacement, owners may seek a variety of other damages resulting from the contractor's breach. In exercising its right to terminate a contractor, an owner does not waive its other damage claims.[93] These additional damages are discussed in §§ **10.19** through **10.23** and **Chapters 12** through **15**.

§ 10.19 Extended Duration and Overhead Costs Incurred by the Owner

If delayed completion causes the owner to incur additional supervisory, co-ordination, or overhead costs directly associated with the project, it may recover those.[94] The owner may also recover premiums for bonding around liens placed on the project by the defaulting contractor's creditors.[95]

§ 10.20 Delay in Use

Owners may recover for loss of use caused by delayed completion of the project. The *delay* is the actual time to complete the project, less the contractually agreed time to complete the project, less delays caused by the owner, less excusable delays. Typically, the damages for delay are measured by the rental value of the project as completed, whether or not the owner intended to rent out the project.[96] Owners may want to calculate

---

[91] Witty v. C. Casey Homes, Inc., 102 Ill. App. 3d 619, 623, 430 N.E.2d 191, 194–95 (1981).

[92] *Cf.* illustrations 3 and 4 to Restatement (Second) of Contracts § 348; D. Dobbs, Handbook on the Law of Remedies § 12.21, at 897–99 (1973).

[93] Armor & Co. v. Nard, 463 F.2d 8, 10–11 (8th Cir. 1972) and authorities cited therein; B.L. Metcalf Gen. Contractor, Inc. v. Earl Erne, Inc., 212 Cal. App. 2d 689, 693, 28 Cal. Rptr. 382, 385 (1963).

[94] Restatement (Second) of Contracts § 347(b).

[95] C. McCormick, Damages § 169, at 651–52 (1935).

[96] Ryan v. Thurmond, 481 S.W.2d 199, 206 (Tex. Civ. App. 1972); C. McCormick, Damages § 170, at 652 (1935); D. Dobbs, Handbook on the Law of Remedies § 12.21, at 903–04 (1973). *But see* Restatement (Second) of Contracts § 348(1).

the cost of their lost use of money resulting from a delayed sale, but they cannot have a double recovery by receiving damages for rental value and for the cost of money.[97] When, because of its nature, the property has no meaningful rental value, the value of its use is the measure of damages.[98] Examples include charitable projects such as churches, and public works projects such as highways.

Other measures of delay damages are the owner's additional cost of borrowing money because of the delay or the cost of renting a substitute property—in effect, the cost of cover.[99] The owner may pay a premium for substitute property because it must rent on short notice and for only a short term.

The owner may incur additional interest costs if an advantageous financing opportunity is missed or if carrying multiple mortgages becomes necessary.[100] However, in some cases the owner actually may avoid interest expense because construction loan draws are made when needed. When the draws are made later than planned because of the delay, interest costs are reduced. **Chapter 14** contains an expanded treatment of delay damages.

§ 10.21 Lost Profits

Delayed completion of some projects can be especially damaging to the owner. For example, a fruit processing plant's timely completion is critical to profits earned only once a year, during the harvest season. Delayed completion of facilities may cause the owner to suffer a loss of profits or market position that may never be regained.

The owner may seek to recover its lost profits as an alternative or supplement to delay damages, but it faces two problems in recovering lost profits. First, it must prove that the profits were foreseeable and within the contemplation of the parties at the time of contracting.[101] Second, the owner must prove, with reasonable certainty, that it indeed would have earned profits but for the delay caused by the contractor.[102] This presents

---

[97] Ryan v. Thurmond, 481 S.W.2d 199, 208 (Tex. Civ. App. 1972).

[98] C. McCormick, Damages § 170, at 652 (1935); D. Dobbs, Handbook on the Law of Remedies § 12.21, at 903–04 (1973).

[99] C. McCormick, Damages § 170, at 652 (1935); D. Dobbs, Handbook on the Law of Remedies § 12.21, at 903–04 (1973).

[100] Conner v. Southern Nev. Paving, Inc., 741 P.2d 800, 801 (Nev. 1987).

[101] Restatement (Second) of Contracts § 351; C. McCormick, Damages § 170, at 652–53 (1935); D. Dobbs, Handbook on the Law of Remedies § 12.21, at 904 (1973).

[102] Restatement (Second) of Contracts § 352 illustrations 5 & 6, § 348; D. Dobbs, Handbook on the Law of Remedies § 12.21, at 904 (1973).

§ 10.22 **LIQUIDATED DAMAGES** 299

a special problem for new businesses or existing businesses that are expanding.[103] **Chapter 15** discusses in detail claims for lost profits.

§ 10.22 Liquidated Damages

Liquidated damages clauses, once viewed skeptically by courts as a forfeiture, now are widely accepted and are authorized by a number of statutes.[104] From the owner's standpoint, they are a way around the problems of uncertainty and unforeseeability that can hinder attempts to recover for delayed use (see § 10.20) and lost profits (see § 10.21).[105] In addition, liquidated damages clauses are particularly useful in solving the valuation problems that arise when completion of public works projects is delayed.[106] On the other hand, contractors often view a liquidated damages clause as an economic sword of Damocles, hanging over the project from the first day. Moreover, the clause denies the contractor the chance to argue that the owner's delayed-use and lost-profit damages were uncertain and unforeseeable. The clause does have the advantage of providing the contractor with a fixed point of reference in weighing whether it is more economical to accelerate performance and meet the deadline or to pay the liquidated damages amount. There are many cases in which the owner's recovery by way of liquidated damages was less than if it could have recovered if it sued for delay and lost-profit damages.[107] Contractors can attempt to minimize the disadvantage of such clauses by negotiating for lower liquidated damages amounts or for bonuses in case of early completion.

---

[103] *See* Restatement (Second) of Contracts § 349 illustration 3, § 351, § 352 illustrations 5 & 6; D. Dobbs, Handbook on the Law of Remedies § 12.21, at 904 (1973); Exton Drive-In, Inc. v. Home Indem. Co., 436 Pa. 480, 489–90, 261 A.2d 319, 324–25 (1969), *cert. denied,* 400 U.S. 819, 91 S. Ct. 36 (1970).

[104] Alpine Constr. Co. v. Water Works Bd., 377 So. 2d 954, 955 (Ala. 1979); Oregon State Highway Comm'n v. DeLong Corp., 9 Or. App. 550, 575–77, 495 P.2d 1215, 1228 (1972), *cert. denied,* 411 U.S. 965, 93 S. Ct. 2142, *reh'g denied,* 412 U.S. 944, 93 S. Ct. 2771 (1973) and authorities cited therein; Pacific Employers Ins. Co. v. City of Berkeley, 158 Cal. App. 3d 145, 152, 204 Cal. Rptr. 387, 391–92 (1984); Cal. Civ. Code § 1671 (1989). *See also* Annotation, *Per Diem Payments for Delay,* 12 A.L.R.4th 891 (1982).

[105] *See* C. McCormick, Damages § 155, at 619 (1935).

[106] *See, e.g.,* Oregon State Highway Comm'n v. DeLong Corp., 9 Or. App. 550, 556, 575–77, 495 P.2d 1215, 1219, 1228 (1972), *cert. denied,* 411 U.S. 965, 93 S. Ct. 2142, *reh'g denied,* 412 U.S. 944, 93 S. Ct. 2771 (1973).

[107] Heinkel v. City of Corvallis, 13 Or. App. 375, 381–83, 510 P.2d 579, 582 (1973); Burns v. Hanover Ins. Co., 454 A.2d 325, 326–27 (D.C. App. 1982); X.L.O. Concrete Corp. v. John T. Brady & Co., 104 A.D.2d 181, 184, 186, 482 N.Y.S.2d 476, 479–80 (1984), *aff'd,* 66 N.Y.2d 970, 489 N.E.2d 768, 498 N.Y.S.2d 799 (1985). *See, e.g.,* Industrial Indem. Co. v. Wick Constr. Co., 680 P.2d 1100, 1103–06 (Alaska 1984) (passes through benefits of general contract to subcontractor).

To withstand judicial scrutiny, liquidated damages must not be a penalty. Rather, they must be a reasonable estimate of damages that are themselves difficult or impossible to establish with exactness.[108] Even if the liquidated damages appeared to be reasonable at the time of contracting, they can be held unenforceable if, by the time of the breach, they were unreasonably large or there was no evidence to support awarding them.[109]

The owner cannot opt out of a liquidated damage clause and attempt to recover actual delay or lost-profit damages.[110] However, if the court refuses to enforce the liquidated damages clause, the owner can seek its actual delay or lost-profit damages.[111]

Liquidated damages may be collected only for unexcused delays, and the contractor's liability for them ends upon substantial completion of the project.[112] But liquidated damages continue to accrue after the target completion date even if the owner chooses to allow the contractor to continue performing rather than terminating it for material breach of contract.[113] In the termination context, the days of delay for which the terminated contractor is liable are calculated as total days to date of substantial completion, less days to reach scheduled completion, less days of excusable delay, less days of delay caused by the owner or new contractor. The resulting number of days is multiplied by the contract's daily damage rate to provide the damage total.[114]

Owners may not be able to rescind and recover liquidated damages.[115] Even if a state allows recovery of consequential damages in conjunction with rescission, it is questionable whether a contractually established liquidated damage formula should control after rescission. Owners also can lose their entitlement to liquidated damages if they fail to hold them back

---

[108] Restatement (Second) of Contracts § 356; Cal. Civ. Code § 1671 (1989).

[109] Restatement (Second) of Contracts § 356(1); Heinkel v. City of Corvallis, 13 Or. App. 375, 381–83, 510 P.2d 579, 582 (1973); Eller Bros. v. Home Fed. Sav. & Loan Ass'n, 623 S.W.2d 624, 628–29 (Tenn. Ct. App. 1981); Kenworthy v. State, 236 Cal. App. 2d 378, 401, 46 Cal. Rptr. 396, 411 (1965).

[110] Heinkel v. City of Corvallis, 13 Or. App. 375, 381–83, 510 P.2d 579, 582 (1973).

[111] United States v. United Eng'g & Constr. Co., 234 U.S. 236, 242, 34 S. Ct. 843, 845 (1913); Aetna Casualty & Sur. Co. v. Board of Trustees, 223 Cal. App. 2d 337, 341, 35 Cal. Rptr. 765, 768 (1963).

[112] Heinkel v. City of Corvallis, 13 Or. App. 375, 380–83, 510 P.2d 579, 581–82 (1973).

[113] Alpine Constr. Co. v. Water Works Bd., 377 So. 2d 954, 955 (Ala. 1979).

[114] *See, e.g.,* I. Richter & R. Mitchell, Handbook of Construction Law and Claims 305 (1982).

[115] *But see* Cal. Civ. Code § 1692 (1989) (party may rescind, receive restitution and still collect consequential damages); Lawson v. Durant, 213 Kan. 772, 774–75, 518 P.2d 549, 551 (1974) (plaintiff rescinded one part of contract and collected liquidated damages under another part).

from their final payment to the contractor.[116] If the contractor's performance bond incorporates the general contract, the surety is liable for the liquidated and actual damages caused by the contractor.[117]

When delays have been caused by both the owner and the contractor, the delays may be apportioned according to fault, and liquidated damages may be awarded to the owner for contractor-caused delay.[118] Some courts allow apportionment only when the contract clearly provides for allocating fault for delays between the owner and the contractor.[119] Courts also have apportioned responsibility for delays between general contractors and subcontractors when the subcontract incorporated the general contract.[120] But owners must take care with the language used in the liquidated damages clause in order to meet the courts' requirements for permitting apportionment.

§ 10.23 Special Liquidated Damages Issues

In the termination context, two particular issues frequently arise regarding liquidated damages:

1. May the owner recover actual damages for the cost of completing the project as well as liquidated damages for delayed use and lost profits?
2. May the owner recover liquidated delay damages from the scheduled completion date until the project is reasonably completed after terminating the contractor or after the contractor abandons the work?

The better-reasoned answer to both questions is yes. First, so long as the contract clearly specifies that the liquidated damages are to compensate the owner for delayed use and lost profits, they should be no bar to recovering the additional cost of completion. Indeed, if there were no liquidated damages clause, the owner could recover the additional cost of completing

---

[116] Chrysler Corp., 75-1 B.C.A. (CCH) ¶ 11,236 at 53,486, 53,490 (ASBCA 1975). *But see* Illinois State Toll Highway Auth. v. Gust K. Newberg, Inc., 177 Ill. App. 3d 6, 15–16, 531 N.E.2d 982, 987–88 (1989).

[117] Pacific Employers Ins. Co. v. City of Berkeley, 158 Cal. App. 3d 145, 152, 204 Cal. Rptr. 387, 391–92 (1984).

[118] Jasper Constr., Inc. v. Foothill Junior College Dist., 91 Cal. App. 3d 1, 13–14, 153 Cal. Rptr. 767, 774 (1979).

[119] *See, e.g.,* Robinson v. United States, 261 U.S. 486, 487–88, 43 S. Ct. 420, 420–21 (1923); Nomellini Constr. Co. v. State *ex rel.* Dep't of Water Resources, 19 Cal. App. 3d 240, 244–46, 96 Cal. Rptr. 682, 684–86 (1971); Jasper Constr., Inc. v. Foothill Junior College Dist., 91 Cal. App. 3d 1, 13–14, 153 Cal. Rptr. 767, 774 (1979).

[120] Conner v. Southern Nev. Paving, Inc., 741 P.2d 800, 801–02 (Nev. 1987); Ely v. Bottini, 179 Cal. App. 2d 287, 298, 3 Cal. Rptr. 756, 763 (1960).

and also could recover consequential damages for its delayed use and lost profits. See § **10.24**.

Second, a rule that entitlement to liquidated damages are cut off when the contractor has been justifiably terminated or wrongfully stops performing is both legally and practically flawed. Legally, such a rule would rest on the proposition that termination of work on the project terminates the contract. The better view in the case of abandonment or justifiable termination is that work may have ceased but the contract still should govern the rights and remedies of the parties. Practically, cutting off liquidated damages at the time of termination would reward the contractor who is in material breach and would penalize the aggrieved owner. See § **10.25**.

§ 10.24 —Both Liquidated and Actual Damages Can Be Recovered

The clear weight of modern authority is to permit recovery of actual damages for the cost of completing the project plus liquidated damages for the delay.

> A provision in a contract liquidating certain items of damage will not prevent the recovery of actual damages for other items to which the liquidation provision does not apply, unless the contract expressly provides that damages other than those enumerated shall not be recovered.[121]

The rule was followed by the Oregon Court of Appeals when a contractor building a bridge was terminated for default.[122] The court affirmed the award of $1.7 million in actual damages for the excess cost of completing the bridge project and $952,000 in liquidated damages for

---

[121] Lawson v. Durant, 213 Kan. 772, 775, 518 P.2d 549, 551 (1974). *See also* Spinella v. B-Neva, Inc., 94 Nev. 373, 376, 580 P.2d 945, 946, 947 (1978); Twin River Constr. Co. v. Public Water Dist. No. 6, 653 S.W.2d 682, 694 (Mo. Ct. App. 1983); J.E. Hathaway & Co. v. United States, 249 U.S. 460, 464, 39 S. Ct. 346, 347 (1919); Burns v. Hanover Ins. Co., 454 A.2d 325, 326–27 (D.C. App. 1982); Louis Lyster Gen. Contractor, Inc. v. City of Las Vegas, 83 N.M. 138, 146, 489 P.2d 646, 654 (1971); Clemente Constr. Corp. v. P.T. Cox Contracting Co., 172 Misc. 904, 909–10, 16 N.Y.S.2d 483, 488 (1939); Town of North Hempstead v. Sea Crest Constr. Corp., 119 A.D.2d 744, 745–46, 501 N.Y.S.2d 156, 157–58 (1986); Avey v. Leather Prods. Co., 73 Ohio App. 245, 55 N.E.2d 813 (1942). *See also* Austin-Griffith, Inc. v. Goldberg, 224 S.C. 372, 386–88, 79 S.E.2d 447, 449, 455 (1953).

[122] Oregon State Highway Comm'n v. DeLong Corp., 9 Or. App. 550, 556–57, 571–73, 578–79, 495 P.2d 1215, 1219, 1226, 1229 (1972), *cert. denied,* 411 U.S. 965, 93 S. Ct. 2142, *reh'g denied,* 412 U.S. 944, 93 S. Ct. 2771 (1973).

delayed completion of the project. The court noted that the liquidated damages clause made clear that it would compensate the state only for delays; it also recognized the difficulty in attempting to prove actual delay damages on a public works project.

The Supreme Court of Alaska, however, ruled to the contrary in a construction case, holding that recovery of liquidated damages for delay would amount to enforcement of a penalty.[123] The court's reasoning was flawed in two regards. It relies for much of its authority on Court of Claims cases interpreting federal contract provisions giving the government an option, when a contractor fails to prosecute the work with due diligence, of taking over the work *or* collecting liquidated damages. The Alaska court, however, was interpreting a contract that specifically provided for recovery of the excess cost of completion plus liquidated damages for delay. The court somehow concluded, on the basis of the facts before it, that enforcing the liquidated damages clause would constitute a penalty and violate public policy by providing "double damage[s]." The court provided no guidance on why it believed an owner had no right to be compensated for delay in completing a project. However, the court may have been influenced by its conclusion that the owner had breached the construction contract by improperly withholding a progress payment. Chief Justice Robert Boochever strongly dissented, arguing that actual damages for extra cost of completion and liquidated damages for delay did not constitute a double recovery.[124]

Owners who wish to recover liquidated damages for delayed use and lost profits and also wish to recover actual damages for the additional cost of completion should draft their construction contracts with care, explicitly stating the harm for which the liquidated damages are intended to compensate and the method of calculation.

§ 10.25 —Damages Continue to Accrue after Termination

Whether owners may recover liquidated damages for delays after contractors wrongfully stop work on a project or after they are rightfully terminated is an issue that has bedeviled courts. The better rule is that they may.

---

[123] Arctic Contractors, Inc. v. State, 564 P.2d 30, 51, 53 (Alaska 1977).

[124] *Id.* at 44, 49–51 (Alaska 1977). *See also* Industrial Indem. Co. v. Wick Constr. Co., 680 P.2d 1100, 1103–06 (Alaska 1984). *But see* United States v. American Sur. Co., 322 U.S. 96, 101, 64 S. Ct. 866, 869 (1944) (recognizing how contractors can exploit allowance of only actual or liquidated damages).

After tracing the history of the dispute over when the accrual of liquidated damages stops, the Massachusetts Supreme Judicial Council held:

> [W]here a contract provides for liquidated damages until "the entire completion of the work," the appropriate rule is to allow recovery of reasonable liquidated damages beyond the date the defaulting party abandons the work. A contrary rule would permit a party to limit his liability for liquidated damages by totally abandoning the work and would deny the injured party those damages which were agreed to as fairly measuring damage caused by delay.[125]

The court then held that the owner could recover liquidated damages for the reasonable amount of time needed to complete the project with another contractor but not for unreasonable delay in completion.[126] However, the court remanded the matter to the trial court to hear proof from the owner that the time needed to complete the project was reasonable.

Another court allowed recovery of liquidated damages from the time the owner rightfully terminated the contractor until the project was completed.[127] Indeed, the liquidated damages were calculated from the date the termination became effective, even though that was before the scheduled completion date for the project.

The United States Supreme Court, interpreting California law, held that a clause providing liquidated damages for delays was inapplicable for

---

[125] City of Boston v. New England Sales & Mfg., Inc., 386 Mass. 820, 824, 438 N.E.2d 68, 70 (1982) (and authorities cited therein). *See also* Austin-Griffith, Inc. v. Goldberg, 224 S.C. 372, 386–88, 79 S.E.2d 447, 455 (1953); Spinella v. B-Neva, Inc., 94 Nev. 373, 375–76, 580 P.2d 945, 946–47 (1978) (contractor rightfully terminated for cause); Clemente Constr. Corp. v. P.T. Cox Contracting Co., 172 Misc. 904, 910–11, 16 N.Y.S.2d 483, 488–89 (1939); Comment, *Applicability of Liquidated Damages Clause to Delays Where Contractor Abandons,* 35 Nw. U. L. Rev. 74, 76 (1941). *But see* Twin River Constr. Co. v. Public Water Dist. No. 6, 653 S.W.2d 682, 694 (Mo. Ct. App. 1983) (discusses conflicting authorities but follows 1909 Missouri Supreme Court decision denying liquidated damages after termination of contractor); Louis Lyster Gen. Contractor, Inc. v. City of Las Vegas, 83 N.M. 138, 146, 489 P.2d 646, 654 (1971) (no liquidated damages after contract terminated; the court appeared to confuse order by owner that contractor terminate performance of the contract with rescission of the contract by the owner). *See also* Annotation, *Liability of Building or Construction Contractor for Liquidated Damages for Breach of Time Limit Provision Where He Abandons Work after Time Fixed for Its Completion,* 42 A.L.R.2d 1134 (1955).

[126] City of Boston v. New England Sales & Mfg., Inc., 386 Mass. at 825, 438 N.E.2d at 70. *See also* Austin-Griffith, Inc. v. Goldberg, 224 S.C. 372, 386–88, 79 S.E.2d 447, 455 (1953).

[127] Oregon State Highway Comm'n v. DeLong Corp., 9 Or. App. 550, 559–60, 570–71, 581, 495 P.2d 1215, 1219–20, 1225, 1230 (1972), *cert. denied,* 411 U.S. 965, 93 S. Ct. 2142, *reh'g denied,* 412 U.S. 944, 93 S. Ct. 2771 (1973).

§ 10.25   DAMAGES AFTER TERMINATION   305

delays occurring after the contractor abandoned the project.[128] However, a California Court of Appeal bluntly held that the Supreme Court had incorrectly described California law:

> The *Six Companies* cases therefore present this comedy of errors: The district court correctly interpreted the limited scope of the holdings in the California cases, and therefore reached a correct result under California law. The court of appeals also reached the correct result, but stated that it was contrary to California law, which it incorrectly interpreted and then disapproved and refused to follow. The Supreme Court then held that the court of appeals should have followed California law, which it too stated incorrectly, and reversed, reaching an ultimately improper result under the law of this state.[129]

The court then stated that California law provides: "The contract measure of damages for delayed completion . . . does not limit or reduce the damages available to the owner because the contractor failed to complete the construction."[130] It then held: "[T]he contractor's abandonment of the project after the date set for completion does not alone render inapplicable the contract clause providing for liquidated damages for delayed completion."[131] The California court added:

> In cases where the contractor abandons the work and the owner takes over completion of the project, it is unclear what the appropriate measure of damages is, that is, to what extent the owner should receive actual damages (difference between contract price and cost of complete project) and/or liquidated damages.[132]

After citing conflicting authorities, the California court concluded that it did not have to resolve that issue because the contractor's surety had completed the project, and the court held that the surety stood in the shoes of the contractor.[133] The surety was required to pay the liquidated damages.

The unqualified Massachusetts rule is the better one. So long as the owner completes the project reasonably promptly, cutting off liquidated

---

[128] Six Cos. v. Joint Highway Dist. No. 13, 311 U.S. 180, 188, 61 S. Ct. 186, 188 (1940), *reh'g denied,* 311 U.S. 730, 61 S. Ct. 438 (1941).

[129] Pacific Employers Ins. Co. v. City of Berkeley, 158 Cal. App. 3d 145, 155, 204 Cal. Rptr. 387, 394 (1984) (citing Six Cos. v. Joint Highway Dist. No. 13, 24 F. Supp. 346, 348 (N.D. Cal. 1938), 110 F.2d 620, 625 (9th Cir. 1940), 311 U.S. 180, 188, 61 S. Ct. 186, 188 (1940), *reh'g denied,* 311 U.S. 730, 61 S. Ct. 438 (1941)).

[130] Pacific Employers Ins. Co. v. City of Berkeley, 158 Cal. App. 3d at 155–56, 204 Cal. Rptr. at 394.

[131] *Id.,* 158 Cal. App. 3d at 156, 204 Cal. Rptr. at 394.

[132] *Id.*

[133] *Id.*

damages upon termination of the contractor or cessation of work by the contractor serves only to deny the aggrieved owner its expectancy and to reward defaulting contractors. The Massachusetts rule should apply whether the contractor is terminated before or after the completion date so long as the termination is justifiable. The requirement that the work be completed reasonably promptly protects the contractor by requiring the owner to move expeditiously in completing the project and to carefully document its efforts in doing so.[134] If the contractor believes it could have completed the project on time, it can put on evidence to show that the owner's action was unjustified. If the contractor can prove its point, the owner thereby will be liable for wrongful termination and most likely will be precluded from recovering liquidated damages for delay.

§ 10.26 Rescission and Restitution

Owners invoke the remedy of rescission infrequently because of the requirement that each side make restitution of the benefit it has received. Typically, rescission is used up the contracting chain, by contractors against owners or by subcontractors against general contractors who are seeking payment in quantum meruit for the work they have done.[135]

When rescission is favorable to the owner. Rescission is useful when the project is excessively costly or has utterly no value as constructed. Because the owner cannot literally make restitution of the project, it must return to the contractor the fair value of the work performed.[136] Because the contract has been terminated, fair value must be determined by the marketplace and not by contract rates.[137] When there are several ways of calculating fair value, such as cost to contractor or value to owner, value will be calculated in the way less favorable to the breaching party.[138] That rule is of great benefit to an owner left with a project of little value.

A classic example of an owner-favorable use of rescission came in the case of an Arkansas man who made a down payment on a fallout shelter. The shelter leaked water because the concrete was honeycombed, and one wall eventually caved in. The homeowner successfully sued for rescission of the construction contract, restitution of his down payment,

---

[134] *See, e.g.,* the discussion of steps the owner took to complete in City of Boston v. New England Sales & Mfg., Inc., 386 Mass. 820, 824, 438 N.E.2d 68, 70 (1982).

[135] Acret, Construction Litigation Handbook § 10.02, at 166 (1986).

[136] Restatement (Second) of Contracts § 371 illustration 1, & § 384(2)(b).

[137] Restatement (Second) of Contracts § 371, and § 373 comment d.

[138] Restatement (Second) of Contracts § 371, comment b & illustration 1; Martin v. Phillips, 122 N.H. 34, 36–37, 440 A.2d 1124, 1125 (1982).

§ 10.26   RESCISSION AND RESTITUTION   307

and discharge of materialmen's liens against his home. Because the project was worthless, the homeowner was not required to make restitution to the contractor.[139]

Restitution also can be favorable if the value of the work performed is less than the owner has paid for it under the contract.[140] Thus, rescission, in particular instances, can be a powerful weapon for an aggrieved owner.

At least theoretically, rescission should be a good remedy for an owner who has agreed to pay too much for a project and whose contractor has otherwise materially breached. Because of competitive bidding, the situation is unlikely to arise on commercial or government projects.[141] It could arise in the consumer context, however, with the contractor's defective work or nonperformance alerting the homeowner to the contractor's excessively high price. The homeowner then could use the performance-related breach as grounds for rescission and thereby avoid the high contract price. In addition, consumer protection statutes may provide homeowners with rescission remedies that are more expansive than common law provides.[142]

When rescission is not beneficial.   Owners should avoid rescinding if the contractor has a loss contract, so long as the contractor's work is acceptable because, by rescinding, the owner will become liable for the value of the work, even if it exceeds the contract price. In rescinding, owners may have to forgo damages.[143] Therefore, if damages are substantial and the jurisdiction does not allow rescinding parties to recover damages, the owner should not rescind.

Because rescission will not be allowed after work has been completed and only money is left owing, the owner should not wait to rescind an unsatisfactory contract until the project is completed. In addition, owners should not rescind for trivialities, lest they be accused of wrongful termination. Because courts may look to see whether the owner provoked the contractor's breach in order to rescind, owners considering rescission should take care to act in good faith in their conduct leading up to termination.

---

[139] Economy Swimming Pool Co. v. Freeling, 236 Ark. 888, 370 S.W.2d 438 (1963). *See also* Restatement (Second) of Contracts § 384(2)(a).

[140] Cook v. Jacklitch & Sons, Inc., 315 N.W.2d 660, 663 (N.D. 1982) (no quantum meruit recovery by contractor if no net benefit to owner); First Nat'l Bank v. Indian Indus., Inc., 600 F.2d 702, 708 (8th Cir. 1979); Martin v. Phillips, 122 N.H. 34, 37–38, 440 A.2d 1124, 1125–26 (1982); Ed Hackstaff Concrete, Inc. v. Powder Condominium "A" Owners Ass'n, 679 P.2d 1112, 1114 (Colo. Ct. App. 1984).

[141] United States v. Systron-Donner Corp., 486 F.2d 249, 252 (9th Cir. 1973).

[142] *See, e.g.,* Cal. Civ. Code § 1689.8; Cal. Bus. & Prof. Code § 17203 (1989).

[143] *Cf.* Economy Swimming Pool Co. v. Freeling, 236 Ark. 888, 889, 370 S.W.2d 438, 440 (1963) (damages denied) *with* Cal. Civ. Code § 1692 (1989) (damages recovered in addition to rescission).

§ 10.27 Owner's Recovery When It Defaults

Even if the owner wrongfully terminates the contractor, thereby materially breaching their contract, or if the contractor rightfully stops performing, the owner may be able to offset its damage claims for defective work against the contractor's recovery. However, it seems unlikely that a court would award delay damages to an owner in breach because the owner either would have denied the contractor the chance to complete the project on time or the contractor would have rightfully halted performance under the contract before the completion date.

When both the owner and the contractor materially breach the contract, each may recover against the other, with the damages offset, including for delay.[144] The owner also is entitled to an offset for the cost of performance that the contractor avoids.[145]

TERMINATIONS FOR DEFAULT FROM CONTRACTOR'S VIEWPOINT

§ 10.28 Basic Damage Remedy

The contractor is entitled to default remedies if it rightfully terminates performance of the contract—that is, stops work—or if the contractor is wrongfully terminated by the owner, who thereby repudiates the contract.[146] The basic damage remedy for contractors depends on the stage of the project at the time of the termination.

§ 10.29 —Before Work Begins

The contract amount less the cost of performance is the measure of recovery for a termination at this stage.[147] If the contractor has done no work at

---

[144] United States v. William F. Klingsmith, Inc., 670 F.2d 1227, 1230–31 (D.C. Cir. 1982); Martin v. Rollins, Inc., 238 Ga. 119, 121, 231 S.E.2d 751, 753 (1977); Restatement (Second) of Contracts § 374. *But see* Paul Hardeman, Inc. v. Arkansas Power & Light Co., 380 F. Supp. 298, 339 (E.D. Ark. 1974).

[145] Restatement (Second) of Contracts § 347 and illustration 6.

[146] Paul Hardeman, Inc. v. Arkansas Power & Light Co., 380 F. Supp. 298, 337–38 (E.D. Ark. 1974).

[147] Restatement (Second) of Contracts § 344, illustration 1; C. McCormick, Damages § 164, at 640 (1935); D. Dobbs, Handbook of the Law of Remedies § 12.24, at 912 (1973); Juengel Constr. Co. v. Mt. Etna, Inc., 622 S.W.2d 510, 514 (Mo. Ct. App. 1981).

all, this measure of damages provides the contractor with its expected profit for the project, which is the only loss it suffered. If the contractor has preparatory expenses, such as bid costs or mobilization, it may recover them, as shown in § 10.30.[148] If the contractor cannot provide its profit, it can recover its preparatory costs.[149] But uncertainty about the amount of damage, in the face of certain damage, does not bar recovery of profits.[150]

§ 10.30 —During the Project

The contract price, less the cost of completion, less any progress payments is the measure of recovery at this stage.[151] This formula should yield the contractor's unpaid cost of performance until termination and its expected profit for the entire project. The remedy also has been described in the alternative as:

1. Contract price less cost of completion (as above)
2. Profit the contractor would have made on the project but for termination plus its expenditures in part performance or
3. The proportion of the contract price equal to the proportion of the project completed plus the profit that the contractor would have earned on the balance of the project.[152]

These three alternatives illustrate the difficulty in applying this basic measure of damages. Indeed, the alternatives are almost circular: How is the cost of completion to be determined?

1. Is it merely the contract amount, less progress payments made, less retentions, less profit as reflected in the contractor's estimate? This does nothing to test an estimator's overly optimistic profit forecast against the realities of performance.
2. Is it the contractor's cost of performing plus its profit, as deduced from the estimate? This again provides for an untested profit margin

---

[148] D. Dobbs, Handbook of the Law of Remedies § 12.24, at 914 (1973).

[149] Restatement (Second) of Contracts § 349; D. Dobbs, Handbook of the Law of Remedies § 12.24, at 914 (1973).

[150] Tull v. Gundersons, Inc., 709 P.2d 940, 943 (Colo. 1985); California Lettuce Growers, Inc. v. Union Sugar Co., 45 Cal. 2d 474, 487, 289 P.2d 785, 793 (1955).

[151] Restatement (Second) of Contracts § 344 illustration 2, § 347 illustration 6; D. Dobbs, Handbook of the Law of Remedies § 12.24, at 913 (1973); Tull v. Gundersons, Inc., 709 P.2d 940, 944 (Colo. 1985).

[152] C. McCormick, Damages § 164, at 640 (1935).

and also provides, in effect, a quantum meruit or restitutional recovery which would reward a contractor that was losing money on the project by paying the contractor its costs even if they exceed the contract rate.

3. Is it calculated by applying the percentage of completion to the project cost, which would be the contract amount less expected profit? This formula also provides for an untested rate of profit, to the contractor's benefit, yet would deny the contractor the benefit of the efficiencies realized as the project nears completion, to the owner's benefit.[153]

The commentator who set out the three alternative measures observed that, if costs could be proven with precision, the three measures would yield the same results. However, as the commentator himself recognized, such precision is difficult to achieve in the real world of complex construction projects.[154]

One approach in figuring the cost of completion is to base the method of calculation on the stage the project had reached when the contractor was terminated. That is, if the project were far along, the contractor's records should yield a sizable amount of cost data that can be compared with its estimate. If the quantity takeoffs in the estimate have proven to be accurate, the contractor's unit costs so far in the project can be multiplied by the units remaining in order to determine cost of completion. If the quantities in the estimate are unreliable, the contractor should look to actual quantities used and spread them across the entire project. Or, if necessary, a reestimate of all quantities for the project can be prepared, based on quantities used to date. If the cost trend reflects efficiencies resulting from the learning curve, unit costs can be modified as appropriate. If the job is on budget or nearly on budget, both quantities and costs from the estimate can be used in such a calculation.[155]

On the other hand, if the contractor was terminated early in the project before much cost data had been accumulated, the better measure is actual costs to date plus the anticipated profit reflected in the estimate.[156] Moreover, if the owner has breached, it is appropriate to determine the cost of performance by the contractor's actual costs or by the fair market of such

---

[153] See § **10.4**; Paul Hardeman, Inc. v. Arkansas Power & Light Co., 380 F. Supp. 298, 316 (E.D. Ark. 1974).

[154] C. McCormick, Damages § 165, at 642 (1935). *See also* D. Dobbs, Handbook of the Law of Remedies § 12.24, at 913–14 (1973).

[155] *See, e.g.,* New Pueblo Constructors, Inc. v. State, 144 Ariz. 95, 104–05, 696 P.2d 185, 194–95 (1985).

[156] *See* Jasken v. Sheehy Constr. Co., 642 P.2d 58, 59 (Colo. Ct. App. 1982); C. McCormick, Damages § 165, at 642 (1935).

costs rather than by the value to the owner of performance.[157] Indeed, the United States Supreme Court held that a terminated contractor should receive its cost of performance, so long as it was not extravagant or unnecessary, plus anticipated profits, but no more than the contract amount.[158]

The owner may attempt to use the replacement contractor's cost of completion to measure the terminated contractor's cost of completion and, thereby, its damages.[159] However, if the circumstances and conditions under which the replacement contractor worked are different from those faced by the terminated contractor, such a measure may not be used.[160] Otherwise, the terminated contractor could be denied its expectancy and be penalized for the inefficiencies of the replacement contractor.[161] Indeed, Corbin flatly states that the cost of completion should be the terminated contractor's and not that of the owner or replacement contractor.[162] He notes that the terminated contractor may have obtained favorable agreements with subcontractors and materialmen that its replacement cannot match.

§ 10.31 —Importance of Maintaining Good Records

To maximize their recoveries, contractors should compile estimating and cost-accounting data in as much detail as possible. They should not glibly rely on bald statements about costs and profits. As one commentator wrote:

> The difficulties of proving the prospective cost, that is, the further amount that it would cost to finish the job, cannot be adequately met in the case of a substantial building contract by merely placing the builder or his superintendent on the stand to give his lump sum opinion or estimate. This may be admissible, but, to warrant a finding, the estimate should include detailed figures as to the cost of the different materials and operations, based, if possible, on "percentage of completion" reports.[163]

---

[157] Restatement (Second) of Contracts § 371, comment b and illustration 1.

[158] United States v. Behan, 110 U.S. 338, 344–46, 4 S. Ct. 81, 83 (1883).

[159] United States ex rel. N. Maltese & Sons, Inc. v. Juno Constr. Corp., 759 F.2d 253, 256 (2d Cir. 1985).

[160] Paul Hardeman, Inc. v. Arkansas Power & Light Co., 380 F. Supp. 298, 316 (E.D. Ark. 1974).

[161] Westminster Elec. Corp. v. Salem Eng'g & Constr., 712 F.2d 720, 724 (1st Cir. 1983); United States ex rel. N. Maltese & Sons, Inc. v. Juno Constr. Corp., 759 F.2d 253, 255–56 (2d Cir. 1985).

[162] A. Corbin, 5 Corbin on Contracts § 1094, at 510 (1964).

[163] C. McCormick, Damages § 165, at 644 (1935) (footnotes omitted).

When an owner used the foregoing quotation in an attempt to deny a contractor its profit, the court rejected the owner's argument because the contractor had thorough documentation to support its damage claim.[164] The president of the contractor, as part of his testimony, presented a lump sum estimate of lost profits, but he backed it up with detailed cost estimates for each phase of building the canceled golf course project. In addition, the court held that profits the contractor earned on other projects, both for the owner and for others, could be evidence of profits on the terminated project.

Other courts have allowed contractors to put in evidence of the profits built into their estimates and to call other bidders to testify about the profit margin they built into their bids.[165]

Although looking to the estimate to determine profit has pitfalls, the owner need not be left defenseless. The court can allow the owner to challenge the reasonableness of the estimate, either in quantity errors or in unrealistic unit costs. To do so, the owner can be allowed to point to quantities actually used and to costs actually incurred on the project, on other similar projects by the contractor, or on other similar projects by other contractors.

Even if the contractor cannot prove its profit with certainty, it still can recover its cost of performance until termination.[166]

Other possible grounds for recovery available to the contractor are covered in **Chapters 8** and **9**.

§ 10.32 —Loss Contracts

The damage formulas discussed in **§ 10.30** for partially completed projects do not explicitly state the contractor's remedy if it is losing money at the time of its termination. However, the requirement in one remedial formula that progress payments and the cost to complete be deducted from the contract price to determine the contractor's recovery

---

[164] Tull v. Gundersons, Inc., 709 P.2d 940, 944–45 (Colo. 1985); Restatement (Second) of Contracts § 352 comment b.

[165] Alaska Children's Servs. v. Smart, 677 P.2d 899, 902 (Alaska 1984). *See also* Contract Audit Manual, published by the Defense Contract Audit Agency, which outlines procedures to be followed by its auditors in many types of audits. Chapter 12, Auditing Terminated Contracts, is useful in determining the issues involved in all types of terminated contracts. It also provides insight into the government's likely position on certain claimed costs.

[166] United States v. Behan, 110 U.S. 338, 344–46, 4 S. Ct. 81, 83 (1883); Restatement (Second) of Contracts § 349; D. Dobbs, Handbook of the Law of Remedies § 12.24, at 914–15 (1973).

implies that a contractor with a loss contract could be denied a recovery or see its recovery reduced. At least one state explicitly imposes a loss on contractors that are wrongfully terminated. Georgia's damage formula for loss contracts is: actual expenditures by the contractor up to termination, less progress payments, less the value of unused materials, less the contractor's projected loss to complete the project according to specifications.[167] Such a formula would allow owners to retain unpaid progress payments and retentions after wrongfully terminating contractors who were losing money on projects. It is difficult to understand how a court can impose a loss on the nonbreaching contractor, which has been denied the chance to turn the project around, and award the loss to the owner who breached the contract by wrongfully terminating the contractor.[168]

However, in most instances commercial realities dictate against terminating contractors with loss contracts. If the contractor is losing money on the project, the owner has no economic incentive to terminate the contractor unless its state uses a damage formula like Georgia's. The owner cannot be sure of finding another contractor as willing to lose money as the contractor it is contemplating terminating. Thus, the owner has every incentive to hold the incumbent contractor to its bad bargain, assuming the contractor's performance is adequate. If it is not, the owner may have grounds to terminate the contract for cause and place the excess cost of completion on the contractor.

§ 10.33 —Upon Full Completion of the Contract

The full contract price is the measure of recovery after completion of the contract.[169] Because the contractor bargained to complete the project for the contract price, the court will hold the owner to its bargain. Such a rule, however, does not preclude the contractor from seeking to recover for extra work outside the scope of the contract or for delay damages. See **Chapters 8** and **9**.

§ 10.34 Other Damages Available

The contractor may be able to recover additional damages if it is wrongfully terminated.

---

[167] Marathon Oil Co. v. Hollis, 167 Ga. App. 48, 51, 305 S.E.2d 864, 868 (1983).

[168] United States v. Behan, 110 U.S. 338, 345–46, 4 S. Ct. 81, 84–85 (1883).

[169] C. McCormick, Damages § 164, at 640 (1935); D. Dobbs, Handbook of the Law of Remedies § 12.24, at 912 (1973).

Interest. The contractor may recover postjudgment interest once a judgment is entered in its favor. In addition, the contractor may be able to recover prejudgment interest on its termination damages from the time payment for them is due, provided they are a definite or certain sum.[170] They are not definite or certain if a court or jury must first calculate them.[171] As soon as possible after termination, the contractor should determine its unpaid cost of performance, its wind-down costs, and its anticipated profit, and bill the owner. The contractor also should willingly submit to reasonable requests by the owner for back-up documentation of its final bill.

Home office overhead. Courts are divided on whether a contractor's home office overhead expenses are recoverable or must be reduced to the extent of nonperformance.[172] The better rule is that reducing a recovery is proper only when the owner can prove that overhead costs actually were saved as a result of its breach. Such a rule is supported by the contractor's argument that, but for the owner's breach, it would have used revenue from the terminated project after paying direct costs to cover its home office overhead.[173]

Reasonable posttermination costs. The contractor is entitled to compensation for reasonable costs incurred in performing on the contract after termination.[174]

Winding-up costs. Whatever formula is used for calculating the contractor's basic damages, the contractor is entitled to reasonable winding-up costs that were not included in its estimate and, therefore, are not part of the contract price. For example, the contractor's equipment rental contracts may require a premium charge for early return.[175] In addition, if the damage remedy is based on the proportion of work performed, winding-up costs—both ordinary demobilization costs and those caused by premature termination—should be added to the contractor's damage claim. When other measures are used, winding-up costs are reflected in the contractor's actual cost of performance, which is part of the basic damage remedy.

---

[170] Restatement (Second) of Contracts § 354; Cal. Civ. Code § 3287 (1989).
[171] Marathon Oil Co. v. Hollis, 167 Ga. App. 48, 51–52, 305 S.E.2d 864, 868–69 (1983).
[172] Juengel Constr. Co. v. Mt. Etna, Inc., 622 S.W.2d 510, 514–15 (Mo. Ct. App. 1981).
[173] *Id.*
[174] Restatement (Second) of Contracts § 350 illustration 18. *See also* 48 C.F.R. § 31.205-42 (1988).
[175] Tull v. Gundersons, Inc., 709 P.2d 940, 945–46 (Colo. 1985).

Delay, disruption, and acceleration damages. If the basic damage formula used does not cover all of the contractor's costs, the contractor should include its delay, disruption, and acceleration damages.[176] Such a damage gap may occur when the cost-of-completion and proportion-of-completion formulas for calculating damages are used. Although the contractor is not entitled to a double recovery, it is entitled to termination damages in addition to any normally compensable damages. Because termination typically is preceded by considerable turmoil on the project, the contractor should be particularly alert for resulting delay, disruption, and loss of efficiency damages. See **Chapters 5**, **6**, and **7**.

Damages for loss of good will and loss of future profits resulting from wrongful termination. A wrongfully terminated contractor who seeks damages for lost good will or lost future profits is certain to encounter objections from the owner that such damages were not foreseeable and are not certain enough to be compensated.[177] In one case, a wrongfully terminated trenching subcontractor was denied such damages on grounds of foreseeability.[178] Citing *Hadley v. Baxendale,*[179] the court held that such damages would not foreseeably flow from the breach of a trenching contract and found no evidence that the defendant had agreed to accept liability for such damages. Contractors, facing a possible termination situation, should attempt to lay a factual foundation supporting their contemplated claims for lost good will and lost profits.

§ 10.35 Offsets Against the Contractor

Even when a contractor is wrongfully terminated, certain offsets against its recovery may be allowed. They include:

1. Payments already made by the owner under the contract.[180]
2. Value of materials on hand that the contractor can sell or use in other projects.[181] (The contractor should be wary of the owner's

---

[176] *See* explanation of Eichleay Formula for recovering extended overhead in Richter and Mitchell, Handbook of Construction Law & Claims 312–14 (1982).

[177] Restatement (Second) of Contracts §§ 351, 352 illustrations 5 and 6, § 348; C. McCormick, Damages § 170, at 652–53 (1935); D. Dobbs, Handbook on the Law of Remedies § 12.21, at 904 (1973).

[178] Traylor v. Henkels & McCoy, Inc., 99 Idaho 560, 562–63, 585 P.2d 970, 972–73 (1978).

[179] 156 Eng. Rep. 145, 151 (Ex. 1854).

[180] D. Dobbs, Handbook of the Law of Remedies § 12.24, at 913 (1973).

[181] Restatement (Second) of Contracts § 347 illustration 7; D. Dobbs, Handbook of the Law of Remedies § 12.24, at 913 (1973). *See also* 48 C.F.R. Subpart 45.6, Reporting, Redistribution, and Disposal of Contractor Inventory (1988).

attempts to inflate the value of materials that may only prove to be scrap.)

3. Costs that the contractor unreasonably incurred after being notified of termination, that is, reasonably avoidable costs.[182] This potential offset demonstrates how important it is for a contractor, though stung by a wrongful termination, to continue to act reasonably, prudently, and in good faith in bowing out of the project.

4. The contractor's profit, on the theory that termination from the owner's project allowed the contractor to bid and win another profitable project. The contractor can counter such an argument by showing that it would have bid the second job and could have handled both. Thus, the contractor can show that, as a result of the owner's breach, the contractor suffered lost volume in business by virtue of its termination from the first contract.[183] Conversely, if the contractor seeks to mitigate its damages from termination by bidding on other projects, it may be able to recover costs expended in such bidding.[184]

§ 10.36 Rescission and Restitution

The restitutional remedy is especially favorable for a contractor saddled with a losing project.[185] When the contractor elects to rescind the construction contract, it forgoes profit, but it may recover the reasonable value of its performance—the quantum meruit—so that the owner would not be unjustly enriched.[186] The usual method of measuring quantum meruit is the actual cost of performance, without reference to the contract.[187] However, such a remedy has proven controversial.[188]

---

[182] Restatement (Second) of Contracts § 350 illustration 1.

[183] Restatement (Second) of Contracts § 347 comment f and illustration 16, § 350 illustration 10.

[184] Tull v. Gundersons, Inc., 709 P.2d 940, 946–47 (Colo. 1985).

[185] Restatement (Second) of Contracts § 344 comment d and § 373 comment d.

[186] Restatement (Second) of Contracts § 344(c) and comment a; C. McCormick, Damages § 164, at 642 (1935); United States *ex rel.* Bldg. Rentals Corp. v. Western Casualty & Sur. Co., 498 F.2d 335, 338 (9th Cir. 1974); B.C. Richter Contracting Co. v. Continental Casualty Co., 230 Cal. App. 2d 491, 499–500, 41 Cal. Rptr. 98, 104 (1964).

[187] Restatement (Second) of Contracts § 371 comments a and b, illustration 1; C. McCormick, Damages § 166, at 645 (1935); United States *ex rel.* Bldg. Rentals Corp. v. Western Casualty & Sur. Co., 498 F.2d 335, 338 (9th Cir. 1974).

[188] *See* Restatement (Second) of Contracts § 373 comment d.

§ 10.37 —When Rescission Is the Better Option

In four particular instances, contractors faced with a material breach by the owner should consider rescission:

1. When the contractor is losing money on the contract. This would include a contract payable on a unit price basis, with an unfavorable unit price.[189]
2. When the contract has unfavorable provisions, such as a no damages for delay clause or a liquidated damages clause, which have become important in the context of the particular project
3. When the contractor's expectancy damages, such as profit, will be hard to prove
4. When the owner makes such massive changes in the project that they constitute abandonment of the contract, and a total cost recovery by way of restitution would be more favorable to the contractor than attempting to prove entitlement to damages under the contract.[190]

§ 10.38 —When Rescission Is Not Favorable to the Contractor

Rescission is not favorable to the contractor when:

1. The contractor's cost of performance is more than the value conferred on the owner by the project, and benefit to the owner is based on value[191]

---

[189] Stein, Construction Law 11-74 through 11-75 (1989); Restatement (Second) of Contracts § 373 comment d; C. McCormick, Damages § 164, at 646 (1935); Paul Hardeman, Inc. v. Arkansas Power & Light Co., 380 F. Supp. 298, 315 (E.D. Ark. 1974) (owner recognizes possibility that contractor with losing contract may stop work); M&W Masonry Constr., Inc. v. Head, 562 P.2d 957, 958, 961 (Okla. Ct. App. 1976) (subcontractor terminated performance although restitution measure would have been more favorable); United States *ex rel.* Susi Contracting Co. v. Zara Contracting Co., 146 F.2d 206 (2d Cir. 1944) (contractor with losing contract is wrongfully terminated and recovers in quantum meruit).

[190] C. Norman Peterson Co. v. Container Corp., 172 Cal. App. 3d 628, 218 Cal. Rptr. 592 (1985) (and cases cited therein); State v. Guy F. Atkinson Co., 187 Cal. App. 3d 25, 32, 231 Cal. Rptr. 382, 385 (1986). *But see* WRB Corp. v. United States, 183 Ct. Cl. 409, 426 (1968); Highland Constr. Co. v. Union Pac. R.R., 683 P.2d 1042, 1045-49 (Utah 1984); Simon, Construction Law Claims & Liability, a Current Treatise §§ 15.6, 15.6.1, at 15.6-1 to 15.6-3 (1989).

[191] Restatement (Second) of Contracts § 344 comment a, § 373 comment d.

2. The contractor would have earned a profit but for the owner's breach of contract[192]
3. The contract has important terms that are favorable to the contractor.

§ 10.39 —How Restitution Is Measured for Contractors

Most commentators favor measuring the benefit to the owner in terms of the project's cost, either to the contractor or on the open market, rather than on the increase in value resulting from the project. The *Restatement (Second) of Contracts,* published by the American Law Institute, follows a fair market value approach by speaking of "what it would have cost to obtain [performance] from a person in the [contractor's] position."[193] One court looked to the estimates of other bidders on the project from which the contractor was terminated.[194] Dan B. Dobbs, author of the treatise on remedies and professor at the University of Arkansas School of Law, writes that contractors can recover their own costs of performance so long as they are not "extravagant."[195] Charles T. McCormick, the late author of the treatise on damages and professor at the University of Texas School of Law, wrote of actual cost to the contractor and did not mention limitations of reason or fair market value.[196] Whichever measure is used, contractors pursuing quantum meruit claims, especially on losing contracts, should be prepared to show that their costs were reasonable by their own standards and within the industry generally.

The *Restatement (Second) of Contracts* acknowledges that allowing a contractor to recover all of its costs on a losing contract is "controversial."[197] Dobbs has found the cases divided on this issue and provides an excellent discussion of the pros and cons, with suggestions for

---

[192] Restatement (Second) of Contracts § 344 illustrations 1, 2. *But see* Cal. Civ. Code § 1692 (1989) (consequential damages allowed); Main Cornice Works, Inc. v. National Union Fire Ins. Co., 258 F. Supp. 377, 379 (S.D. Cal. 1966) (reasonable profit allowed in addition to restitution).

[193] Restatement (Second) of Contracts § 371(a). *See also* United States *ex rel.* Susi Contracting Co. v. Zara Contracting Co., 146 F.2d 606, 611 (2d Cir. 1944); United States *ex rel.* Bldg. Rentals Corp. v. Western Casualty & Sur. Co., 498 F.2d 335, 338 (9th Cir. 1974).

[194] Paul Hardeman, Inc. v. Arkansas Power & Light Co., 380 F. Supp. 298, 340–41 (E.D. Ark. 1974).

[195] D. Dobbs, Handbook of the Law of Remedies 915 (1973).

[196] C. McCormick, Damages 643 (1935).

[197] Restatement (Second) of Contracts § 373 comment d.

middle-ground remedies.[198] However, at least two United States Courts of Appeals have held that the better rule does not limit the contractor's recovery to the contract price or to unit prices in the contract.[199] The Second Circuit would look to the contract price as evidence of the project's value to the owner or of the contractor's costs.[200] In the face of such controversy, whenever a contractor is faced with a termination situation and before taking any inalterable steps, it should learn whether the jurisdiction under whose rules it will be litigating limits quantum meruit recoveries. Such limitations can have a dramatic effect on recovery.[201]

In calculating their costs, contractors should include all direct and indirect costs, including labor, materials, supervision, overhead, administration, equipment, bonds, and insurance.[202] If the owner continues to use any of the contractor's equipment after termination, the contractor is entitled to rental payments.[203]

In addition, some states allow contractors to rescind and to recover damages, including consequential damages, so long as there is no double recovery.[204]

§ 10.40 Contractor's Recovery When It Is in Default

A contractor is entitled to compensation even when it materially breaches its contract with the owner, either when the owner rightfully terminates the contractor or when the contractor wrongfully ceases performing. The generally accepted recovery is the value of the contractor's performance less the damages to the owner. However, courts and commentators are divided on the theoretical basis for the recovery and the method of measuring the value of the performance. Because of the divergence in viewpoints,

---

[198] D. Dobbs, Handbook of the Law of Remedies 915–18 (1973).

[199] United States *ex rel.* Bldg. Rentals Corp. v. Western Casualty & Sur. Co., 498 F.2d 335, 338 (9th Cir. 1974); United States *ex rel.* Susi Contracting Co. v. Zara Contracting Co., 146 F.2d 606, 610 (2d Cir. 1944). *See also* B.C. Richter Contracting Co. v. Continental Casualty Co., 230 Cal. App. 2d 491, 500, 41 Cal. Rptr. 98, 104 (1964).

[200] United States *ex rel.* Susi Contracting Co. v. Zara Contracting Co., 146 F.2d 606, 611 (2d Cir. 1944).

[201] C. McCormick, Damages § 164, at 642 (1935).

[202] D. Dobbs, Handbook of the Law of Remedies § 12.24, at 915 (1973); C. McCormick, Damages § 165, at 644 (1935).

[203] United States *ex rel.* Susi Contracting Co. v. Zara Contracting Co., 146 F.2d 606, 611–12 (2d Cir. 1944).

[204] *See, e.g.,* Cal. Civ. Code § 1692 (1989). *See also* Main Cornice Works, Inc. v. National Union Fire Ins. Co., 258 F. Supp. 377, 379 (S.D. Cal. 1966) (subcontractor allowed to rescind and recover profit).

the contractor should determine how courts in the jurisdiction in which its contract will be enforced approach this issue.

The consensus seems to be that restitution is the theoretical basis for the recovery.[205] The *Restatement (Second) of Contracts* would measure the value of the contractor's performance as value to the owner of the project as completed, less cost of completion, less other damages to the owner, but no more to the contractor than the ratable portion of the contract when that can be determined.[206] Dobbs would measure the value of the contractor's work as contract price less cost of completion, but not to exceed the contract price and not to exceed the contractor's actual expenditures.[207] If economic waste would result from correcting the contractor's work, Dobbs would measure the value of the contractor's work as value to the owner of the completed project less the owner's loss in expectancy (for example, loss in resale value).[208] Dobbs describes a number of other methods of calculating the contractor's recovery but argues that a contract-price-less-cost-of-completion formula best protects the owner's expectancy interest while basing it on the contract, upon which both parties agreed, rather than on property values or other measures, to which the contractor was not a party.[209] The difference could be significant and should be considered in the event of termination.

Some courts deny recovery if the contractor's breach was willful.[210] This approach has been criticized for missing the point of restitution: Each party should pay for the benefits it received and should not be unjustly enriched. The *Restatement (Second) of Contracts* does not mention fault in discussing restitutionary remedies for breaching parties.[211]

When the contractor's breach is not material, the contractor may sue on the contract as if it had been fully performed.[212] The doctrines of

---

[205] Restatement (Second) of Contracts § 374; C. McCormick, Damages § 167, at 646–47 (1935); A. Corbin, 5A Corbin on Contracts §§ 1125–1126, at 18–29 (1964); D. Dobbs, Handbook of the Law of Remedies § 12.24, at 920 (1973). *But see* Dobbs at 918–19.

[206] Restatement (Second) of Contracts § 374, illustrations 2, 3, 4, and comment b. *See also* A. Corbin, 5A Corbin on Contracts § 1124 (1964); C. McCormick, Damages § 167, at 646–47 (1935).

[207] D. Dobbs, Handbook of the Law of Remedies § 12.24, at 924 (1973).

[208] *Id.* at 923.

[209] *Id.* at 922–24.

[210] C. McCormick, Damages § 167, at 646–47 (1935); D. Dobbs, Handbook of the Law of Remedies § 12.24, at 919–21 (1973); A. Corbin, 5A Corbin on Contracts § 1123 (1964). *See, e.g.,* Martin v. Rollins, Inc., 238 Ga. 119, 121, 231 S.E.2d 751, 752–53 (1977).

[211] D. Dobbs, Handbook of the Law of Remedies § 12.24, at 921 (1973); Restatement (Second) of Contracts § 374 comment a.

[212] C. McCormick, Damages § 167, at 647–48 (1935).

substantial performance and divisibility may provide protection to the contractor.[213] Also, courts may permit use of the "jury verdict" method of determining damages when proof is difficult and there is fault on both sides.[214]

In some states, a party in breach may rescind if the other party has materially breached the contract in an unrelated manner.[215] Thus, the contractor may have the right to rescind the contract so long as it pays offsetting damages to the owner. However, if the contractor may not rescind because it is in breach, the contractor may be required to recover under the contract and suffer the consequences if it was a loss contract.[216]

The foregoing remedy formulas present opportunities for owners to directly or indirectly seek offsetting damages for defective work and delays. Such offsets would be reflected in the cost of correcting defective work, the cost of completing the project, the value of the project, and the owner's loss of expectancy.

§ 10.41 Recoveries by Parties Not in Privity with the Owner

When contractors are terminated, they may fail to pay subcontractors or sub-subcontractors. Although not in privity of contract with the owner and not otherwise protected by the general contract, the unpaid subcontractors and sub-subcontractors sometimes seek to recover from the owner on such theories as third-party beneficiary, unjust enrichment, and quasi contract. The subcontractors and sub-subcontractors are likely to find courts reluctant to assist them, holding that they were unintended

---

[213] D. Dobbs, Handbook of the Law of Remedies § 12.24, at 919 (1973); Nestos Painting Co., 86-2 B.C.A. (CCH) ¶ 18,993 at 95,912, 95,916 (GSBCA 1986). *But see* W. Jaeger, Williston on Contracts 266–75, and 11 Williston on Contracts § 1363, at 341 (3d ed. 1957).

[214] State v. Guy F. Atkinson Co., 187 Cal. App. 3d 25, 32–33, 231 Cal. Rptr. 382, 385–86 (1986); Simon, Construction Law Claims & Liability, a Current Treatise § 15.6.2, at 15.6-5 (1989).

[215] *See, e.g.,* American-Hawaiian Eng'g & Constr. Co. v. Butler, 165 Cal. 497, 516–17, 133 P. 280, 288 (1913); Integrated, Inc. v. Alec Fergusson Elec. Contractor, 250 Cal. App. 2d 287, 298, 58 Cal. Rptr. 503, 510 (1966). *See also* A. Corbin, 5A Corbin on Contracts § 1122 (1964). *But see* Martin v. Rollins, Inc., 238 Ga. 119, 120–21, 231 S.E.2d 751, 752 (1977).

[216] *Cf.* Martin v. Rollins, Inc., 238 Ga. 119, 121, 231 S.E.2d 751, 752 (1977) (party in breach cannot rescind and seek restitution) and Marathon Oil Co. v. Hollis, 167 Ga. App. 48, 51, 305 S.E.2d 864, 868 (1983) (loss imposed on wrongfully terminated contractor pursuant to contract; no mention of rescission and restitution).

beneficiaries of the general contract and that they conferred a benefit on the contractor or subcontractor, not on the owner.[217]

Subcontractors and sub-subcontractors concerned about the financial soundness of their upstream business associates should seek explicit contractual protection from the owner and general contractor or protection by way of payment bonds. They also should be mindful of their mechanic's lien and stop notice rights.[218]

---

[217] Insulation Contracting & Supply v. Kravco, Inc., 209 N.J. Super. 367, 371–79, 507 A.2d 754, 756–61 (App. Div. 1986); Frank W. Whitcomb Constr. Corp. v. Cedar Constr. Co., 142 Vt. 541, 546, 459 A.2d 985, 988–89 (1983). *But see* Acret, California Construction Law Manual § 2.23, at 56–57 (3d ed. 1982); Fidelity & Deposit Co. v. Harris, 360 F.2d 402 (9th Cir. 1966); Pike Indus., Inc. v. Middlebury Assocs., 140 Vt. 67, 70–71, 436 A.2d 725, 727 (1981), *cert. denied,* 455 U.S. 947, 102 S. Ct. 1446 (1982); COAC, Inc. v. Kennedy Eng'rs, 67 Cal. App. 3d 916, 922, 136 Cal. Rptr. 890, 893 (1977).

[218] *See, e.g.,* Cal. Civ. Code §§ 3104, 3110, 3158 (1989).

CHAPTER 11

PAYMENT DELAY CLAIMS

W. Robert Buxton, Esquire
Robert L. Ivey, Esquire
James T. Schmid, CPA
Louis D. Victorino, Esquire

W. Robert Buxton is a 1960 graduate of the Harvard Law School and a partner in the San Francisco law firm of Pillsbury, Madison & Sutro who concentrates his legal practice in the area of tort, commercial, and construction litigation.

Robert L. Ivey is a partner in the Los Angeles office of Pillsbury, Madison & Sutro specializing in construction and government contracts litigation. His principal work is large and complex lawsuits involving multiple parties, experts, and computerized litigation support. He is a 1967 graduate of the University of Virginia Law School where he was Notes editor of the *Virginia Journal of International Law*. He is the coauthor of articles entitled "The Inspection Clause" (Government Contractor Briefing Papers No. 88-10, September 1988) and "Qui Tam Lawsuits" (Government Contractor Briefing Papers No. 89-10, September 1989).

James T. Schmid was formerly a director with Coopers & Lybrand in their San Francisco office responsible for providing litigation consulting services to contractors, owners, bonding companies, and other entities involved in the construction industry. He is now a principal of Hemming Morse & Co. Mr. Schmid received a B.S. in Engineering from Oakland University in Rochester, Michigan and an M.B.A. from the University of Michigan. He is a coauthor of *Construction Industry Forms, Construction Litigation: Representing the Contractor,* and *Construction Failures* (John Wiley & Sons). A frequent speaker on construction litigation, he also has acted as an arbitrator for the American Arbitration Association.

Louis D. Victorino is a 1970 graduate of the UCLA School of Law, where he was an editor of the *UCLA Law Review.* He is coauthor of Government Contractor Briefing Papers entitled "Qui Tam Lawsuits," "The Inspection Clause," "Compliance Programs," and "Bid Protest Suits in the Federal Courts." He is a frequent lecturer and speaker, including courses entitled "The Masters Institute on Construction Contracting" and "Fundamentals of Government Procurement" for the seminar division of Federal Publications, Inc. He is a partner in the Los Angeles office of Pillsbury, Madison & Sutro.

---

§ 11.1　Introduction

CONTRACT TERMS

§ 11.2　Contract Payments Provisions

§ 11.3　Important Payments Clause Issues

§ 11.4　Common Payment Terms

§ 11.5　—AIA Payments Clauses

§ 11.6　—AGC Subcontract Payments Clauses

§ 11.7　—FAR Payments Provisions

§ 11.8　—Suggested Payments Clause

INTEREST DAMAGES

§ 11.9　Interest as a Cost

§ 11.10　—Theoretical Basis for Interest Cost

§ 11.11　—Interest as a Collectible Damage

§ 11.12　Proving Interest Cost Damages

§ 11.13　—Traceability (Causation)

§ 11.14　Calculating Interest for Commercial Contracts

§ 11.1 INTRODUCTION

§ 11.15   Calculating Interest for Federal Government Contracts
§ 11.16   Determining the Proper Interest Rate

OTHER DAMAGES

§ 11.17   Foreseeability of Other Payment Delay Damages
§ 11.18   Increased Bonding Cost
§ 11.19   Loss of Other Work
§ 11.20   Collection Fees and Credit Checks
§ 11.21   Costs of Defending Subcontractors' Suits
§ 11.22   Delayed Payment Checklists

§ 11.1 Introduction

When a construction contractor is paid in an untimely fashion, bottom-line profits are directly affected. The contractor prepares a bid based upon more than an analysis of the specifications and drawings; in today's climate of high interest rates, an accurate bid estimate must also take into account expected cash flow. Direct labor and material costs must be compared to anticipated progress or milestone payments that the bid's payments clause stipulates. The cost of any shortfall and the duration of that shortfall must be financed by the contractor, which results in a cost that must be worked into the bid price. A change in the anticipated financing obligation, therefore, must be treated like any other change to the contract: if the change has resulted in increased costs (damages), the contractor must be compensated.

Anticipated contract financing can be affected in two general ways. The most common occurs when the owner fails to make timely progress or milestone payments. If the payments clause provides that payments are to be made on the 10th of each month, based on the costs incurred during the preceding month, a failure to receive payments on the 10th increases the contractor's anticipated financing obligation.

A more subtle impact on the contractor's financing obligation occurs when the owner imposes extra or changed work or the contractor encounters changed conditions or delays (referred to collectively henceforth as "changed or extra work") and no timely contract modifications are made to increase the contract price. In these instances, also, the contractor's anticipated cash flow is adversely affected. Without a timely contract price adjustment, the increased costs incurred to perform the changed or extra work must be financed by the contractor. Indeed, if the changed or extra work is disputed, payment may not be received until after completion of the project, and perhaps not until after the matter has been litigated.

This chapter discusses both types of payment delays. **Sections 11.2** through **11.8** describe measures that can be taken even before contract performance begins to either minimize the impact of payment delays or to maximize a construction contractor's ability to collect the increased costs associated with the payment delay. **Sections 11.9** through **11.16** discuss the damage most commonly incurred as a result of payment delays: interest. Included in this discussion is a delineation of the types of facts that must be proved in order to maximize the likelihood of recovery. **Sections 11.17** through **11.21** set forth additional types of damages that are often incurred as a result of payment delays and discuss whether the courts have permitted recovery for them and in what contexts. Finally, § **11.22** contains checklists that can be followed to both minimize damage and maximize the collectability of damages unavoidably incurred.

CONTRACT TERMS

§ 11.2 Contract Payments Provisions

Any strategy aimed at maximizing recovery for payment delays must begin with the contract terms. A well-crafted payments clause often eliminates debate and litigation over compensation for payment delays. Many professional or trade associations have prepared model contract terms and conditions for construction projects and each, of course, includes a suggested payments clause. Perhaps the most commonly used are those published by the American Institute of Architects (AIA) and the Associated General Contractors of America (AGC). The federal government has its own standard terms and conditions, which can be found in applicable acquisition regulations such as the Federal Acquisition Regulation (FAR). When one of these standard forms is proposed, the contractor may or may not be able to propose alternative language. If negotiation on language is permitted, there are important alternatives to be considered. When negotiation is not permitted, it is important to understand the rights and obligations imposed by these standard clauses and how the clauses affect the ability to collect payment delay damages.

§ 11.3 Important Payments Clause Issues

Whether drafting a tailored clause for a particular project or evaluating one of the standard payments clauses, there are several significant issues to be considered. Aspects of the clause relating to payment delays to which special attention must be given include the following.

§ 11.3 PAYMENTS CLAUSE ISSUES

Frequency of Invoices

Obviously, the more frequently invoices are submitted, the less work in progress will have to be financed. At the same time, payment requests that are submitted too frequently can result in an administrative burden. In general, it is reasonable to request that invoices be permitted on major projects every two weeks. On many commercial projects, only monthly billings are permitted.

In addition to the obvious financing impact of billing frequency, there is another more subtle impact. Often a payments provision is silent as to how long the owner may take to honor the payment request. In these situations the courts will usually require that payment be made within a "reasonable" time. Sometimes a payments clause states explicitly that payment will be made within a reasonable time. In either event, what is a reasonable time can be influenced by the frequency of billings provision. A reasonable time under a payments clause permitting billing every two weeks will likely be shorter than that permitted under a clause permitting billings only monthly. The more frequent billing period manifests a recognition by the parties that costs are being incurred rapidly, warranting accelerated review and payment by the owner. In such situations, it is only reasonable that review and payment be prompt.

Payment Request Substantiation

Most payments clauses require that payment requests be accompanied by supporting documentation.[1] The scope of supporting documentation, however, varies greatly; some contracts require special forms, certifications, accounting records, or even notarization. To avoid payment delays, agreement on the forms to be utilized and the documentation that must accompany a request must be negotiated at the time of contracting, prior to the submission of the first payment request. Otherwise, it is likely that at least the initial payment request will be held up until a properly supported request is received. Any resulting delay in receiving payment probably will not be compensable.[2]

Payment Request Processing Time

As already noted, standard payments clauses are often silent regarding how much time may be taken by the owner after a valid payment request is received before it must make payment. If the clause addresses the issue at all, usually a "reasonable" time is stipulated. In order to accurately

---

[1] *See, e.g.,* AIA A201-1974 ¶ 9.3.1.
[2] *In re* American Casualty Co., 851 F.2d 794 (6th Cir. 1988).

project cash flow and to maximize the likelihood of recovering payment delay damages, a definite time period for payment should be established. In most instances, seven to fourteen calendar days should be acceptable. It is important, moreover, that the contract define whether the days allowed for review are calendar or work days.

Partial Payment

It is inevitable in a long-term contractual relationship that some type of payment dispute will arise. The dispute may relate to the adequacy of the payment request, the actual progress of the construction, the quality of work, or any number of other issues. Usually, payment is withheld until the disputed item is resolved. This is so even when the dispute involves at most only a portion of the payment sought. Without contract protection, such withholdings are difficult and expensive to fight. Thus, one of the important protections that the payments clause can provide is to stipulate that undisputed portions of a payment request must be paid in a timely fashion.

Remedies for Delayed Payment

Quite often the biggest problem associated with payment delay claims is that the amount in controversy, although significant in comparison to expected profit, is too small to merit the time and expense of litigation. For example, a construction contractor may have a perfectly valid payment delay claim based on payments' being routinely made 60 days late, but the interest expense for 60 days over the life of the contract may still be less than the cost of litigation. Thus, it must be one of the objectives of the payments clause to provide a realistic remedy for nonpayment. The most effective remedy is the right to stop work. The law generally provides that a contractor has the right to stop work if not paid; a failure to make payments in accordance with the contract provisions is a breach of contract entitling the nonbreaching party to discontinue performance.[3] Exercising such a right must be done with great care, however. If it is determined later that the owner was justified in withholding payment, stopping work is unjustified and is itself a breach of contract by the construction contractor.

Some protection is provided by a payments clause that specifically reserves to the contractor the right to stop work if payments are not received

---

[3] United States *ex rel.* N. Maltese & Sons, Inc. v. Juno Constr. Corp., 759 F.2d 253 (2d Cir. 1985); Building Maintenance Co., ENGBCA No. 4115, 83-2 B.C.A. (CCH) ¶ 16,629 (1983); Zobel & Dahl Constr. v. Crotty, 356 N.W.2d 42 (Minn. 1984); Integrated, Inc. v. Alec Fergusson Elec. Contractor, 250 Cal. App. 2d 287, 58 Cal. Rptr. 503 (1967).

within the stipulated time.[4] Such provisions, however, are very difficult to obtain in the negotiation process. More common are provisions stipulating that interest will accrue at a stated amount for late payments and that attorneys' fees and costs (as well as other collection costs) will be assessed if litigation is necessary to collect money due under the contract. Such provisions are routinely upheld by the courts.[5]

§ 11.4 Common Payment Terms

The three most common payment terms are found in (1) AIA Document A201 (1974 ed.), General Conditions of the Contract for Construction, Article 9, "Payments and Completion," (2) AGC Document 600 (1984 ed.), Subcontract for Building Construction, Paragraph 5, "Payments," and (3) FAR 52.232-5, "Payments Under Fixed-Price Construction Contracts." Each of these clauses contains both beneficial and negative provisions. Unfortunately, the most onerous of these clauses—the FAR provision for federal government contracts—is also the least negotiable. The following analyses of each of these clauses utilize the criteria set forth in § 11.3.

§ 11.5 —AIA Payments Clauses

The AIA payments clause is silent regarding the frequency with which payment requests may be submitted. Paragraph 9.2.1 of Document A201, however, contemplates that this is a subject to be negotiated prior to submitting the first payment request. In order to accurately project the contract financing cost, it is best to propose the intended frequency in the bid. Paragraph 9.4.1 contemplates a scheme whereby the architect is given seven days in which to review a payment request and the owner then has three days in which to make payment. This scheme is consistent with either monthly or twice-monthly payment requests.

Paragraph 9.2.1 states simply that the form and support to be required for payment requests are to be negotiated prior to the first payment request. As already noted, this is unacceptable and usually results in the initial payment's being significantly delayed. Agreement on payment forms

---

[4] *See* Integrated, Inc. v. Alec Fergusson Elec. Contractor, 250 Cal. App. 2d 287, 58 Cal. Rptr. 503 (1967); Big Boy Drilling Corp. v. Etheridge, 44 Cal. App. 2d 114, 11 P.2d 953 (1941); J. Acret, California Construction Manual § 3.16 (3d ed. 1982).

[5] LaFarge Conseils Et Etudes, S.A. v. Kaiser Cement & Gypsum Corp., 791 F.2d 1334 (9th Cir. 1986); E.C. Ernst Co. v. Koppers Co., 626 F.2d 324 (3d Cir. 1980), *on remand, aff'd in part and rev'd in part,* 520 F. Supp. 830 (W.D. Pa. 1981); National Farm Workers Serv. Center v. Caratan, Inc., 146 Cal. App. 3d 796, 809, 194 Cal. Rptr. 617 (1983); De La Hoya v. Slim's Gun Shop, 80 Cal. App. 3d Supp. 6, 146 Cal. Rptr. 68 (1978).

and the requisite support should be made during the bidding process. It is usually best to attach the proposal forms to the bid.

Paragraph 9.3.1 of the AIA form presumes that dates for payment will be set forth in the contract. This is often left to negotiation after contract award or, in some instances, is overlooked entirely. The parties should confirm that the contract contains a finite schedule for payment or be sure to negotiate one as part of the agreements contemplated under ¶ 9.2.1. As already noted, payment dates should be finite, expressed either in terms of calendar days after the submittal of each request or by date (for example, the 15th day of each month).

AIA Document 201 ¶ 9.7.1 gives the contractor the right to stop work if payment is not received, through no fault of the contractor, within the time stipulated. This essentially restates the law that would be applied even if this provision were not included. The benefit of the clause, however, is to establish a precise date upon which a stop work action may be taken: seven days after payment was due. This provision is acceptable. Added to the remedies provided in this clause is the additional remedy established at ¶ 7.8, which provides that, as a general matter, payments due and payable under the contract, from either party, shall bear interest at a rate to be agreed upon by the parties. The rate, however, must be established during the bidding process and should be part of an offer.

§ 11.6 —AGC Subcontract Payments Clauses

The AGC provision regarding payments is based upon the AIA clause. Because of the special problems associated with the prime contractor/subcontractor relationship, certain alterations and additions have been made.

The AGC clause, ¶¶ 5.2.1 and 5.2.3 of AGC Document 600, suggests that the parties will agree on the frequency of payment requests as is done in the AIA model. However, the specific language requiring such agreement in the AIA model (AIA Document A201 ¶ 9.3.1) is not present in the AGC clause. Thus, it is imperative that this potential ambiguity be clarified at the time of bidding.

Oddly, the AGC clause contains no general requirement that supporting documentation accompany a payment request. Paragraph 5.2.4 requires supporting documentation only as it relates to billings for "stored" materials. Nevertheless, if litigated, the courts would likely uphold any reasonable request for supporting documentation. Thus, it is best to establish the form and content of payment requests and supporting documentation in the bidding process.

Paragraph 5.2.4 states that the subcontractor "may" include amounts for "stored" materials if approved in advance by the owner. This language suggests that the prime contractor could deny such payments independently from the decision by the owner to make such payments to the prime contractor. Compare this provision to AIA Document A201 ¶ 9.3.2. The word "may" should be changed to "shall" or agreement regarding billings for "stored" materials otherwise should be clarified.

The AGC clause, ¶ 5.2.5, specifically conditions payment to subcontractors upon receipt of payment by the prime contractor from the owner. This is likely to be unacceptable to many subcontractors.[6] From a prime contractor's perspective, this requirement is deemed necessary in order to spread the risk of nonpayment by the owner.

Paragraph 5.4 provides for the payment of interest on delayed payments to subcontractors only to the extent interest is obtained by the prime contractor from the owner. Like ¶ 5.2.5 which similarly conditions payment, this provision is likely to be unacceptable to most subcontractors. It would prevent the payment of interest when the late payment was due solely to actions of the prime contractor.

Because payment to the subcontractor is conditioned on payment from the owner, the right to stop work is similarly conditioned in ¶¶ 5.2.5 and 5.2.6. Subcontractors are likely to request that a right to stop work exists as long as the payment delay is not caused by the subcontractor's acts.

§ 11.7 —FAR Payments Provisions

The payments provisions contained in federal government contracts are generally the most onerous provisions to be encountered. Moreover, because most government construction contracts are awarded pursuant to formal, sealed bidding procedures (see FAR Part 14), no negotiation on language is permitted. The FAR payments provision most frequently used is FAR 52.232-5:

PAYMENTS UNDER FIXED-PRICE CONSTRUCTION CONTRACTS (APR 1989)

   (a) The Government shall pay the Contractor the contract price as provided in this contract.

   (b) The Government shall make progress payments monthly as the work proceeds, or at more frequent intervals as determined by the Contracting Officer, on estimates of work accomplished which meets the standards of quality established under the contract, as approved by the Contracting Officer. The Contractor shall furnish a breakdown of the total contract price showing the amount included therein for each principal category of the work, which shall substantiate the payment amount requested in order to provide a basis for determining progress payments, in such detail as requested by the Contracting Officer. . . .

---

[6] *See* American Subcontractors Assoc. & Associated Specialty Contractors, Commentary on AGC Subcontract Document No. 600 (1984).

(c) Along with each request for progress payments, the contractor shall furnish the following certification, or payment shall not be made:

I hereby certify, to the best of my knowledge and belief, that—

(1) The amounts requested are only for performance in accordance with the specifications, terms, and conditions of the contract;

(2) Payments to subcontractors and suppliers have been made from previous payments received under the contract, and timely payments will be made from the proceeds of the payment covered by this certification, in accordance with subcontract agreements and the requirements of chapter 39 of Title 31, United States Code; and

(3) This request for progress payments does not include any amounts which the prime contractor intends to withhold or retain from a subcontractor or supplier in accordance with the terms and conditions of the subcontract.

_____
(Name)

_____
(Title)

_____
(Date)

* * *

(e) If the Contracting Officer finds that satisfactory progress was achieved during any period for which a progress payment is to be made, the Contracting Officer shall authorize payment to be made in full. However, if satisfactory progress has not been made, the Contracting Officer may retain a maximum of 10 percent of the amount of the payment until satisfactory progress is achieved. When the work is substantially complete, the Contracting Officer may retain from previously withheld funds and future progress payments that amount the Contracting Officer considers adequate for protection of the Government and shall release to the Contractor all the remaining withheld funds. Also, on completion and acceptance of each separate building, public work, or other division of the contract, for which the price is stated separately in the contract, payment shall be made for the completed work without retention of a percentage.

* * *

(h) The Government shall pay the amount due the Contractor under this contract after—

(1) Completion and acceptance of all work;

(2) Presentation of a properly executed voucher; and

(3) Presentation of release of all claims against the Government arising by virtue of this contract, other than claims, in stated amounts, that the Contractor has specifically excepted from the operation of the release. A release may also be required of the assignee if the Contractor's claim to

amounts payable under this contract has been assigned under the Assignment of Claims Act of 1940 (31 U.S.C. 3727 and 41 U.S.C. 15).

Payment requests are usually made monthly. On major projects the contracting officer is likely to approve submissions every two weeks. This is something that can be clarified at the bidders' conference or through formal correspondence prior to bidding.

Progress payment requests under federal government contracts are subject to audit by agency accountants (for example, the Defense Contract Audit Agency). The degree of scrutiny a request receives, at least initially, far exceeds that which is experienced on other types of projects.

The federal government is notoriously slow in paying. Some protection is afforded by the relatively recently enacted Prompt Payment Act,[7] which requires that the government pay invoices within 30 days of their receipt at the government billing office designated in the contract.[8] Amounts not paid within the time permitted automatically accrue interest at the rate established by the secretary of the treasury under § 12 of the Contract Disputes Act of 1978.[9]

The federal payments provision does not expressly provide a right to stop work. Indeed, applicable law and regulations require that all federal government contracts contain a disputes clause that makes stopping work even more risky than in commercial contracts.[10] Although the right to stop work exists, it is available in only very limited situations. Mere delayed payment is not enough; a contractor would have to demonstrate that the government had withheld payment, in significant amounts, without right for a substantial time.[11]

§ 11.8 —Suggested Payments Clause

Given the significance of the payments clause, it is important to have a standard provision in the construction contract's terms and conditions. We suggest the following:

PAYMENTS

(a) Monthly progress payments shall be paid by the Owner to the Contractor on or before the tenth of each month. Payments will be made for the value

---

[7] 31 U.S.C. §§ 3901–3906 (1988).
[8] *See* F.A.R. §§ 32.900–32.909 and F.A.R. § 52.232-25, Alternate I.
[9] 41 U.S.C. § 611 (1978).
[10] *See* F.A.R. § 52.233-1.
[11] Northern Helix Co. v. United States, 455 F.2d 546, 199 Ct. Cl. 998 (1972); General Dynamics Corp., DOTCAB 1232, 83-1 B.C.A. (CCH) ¶ 16,386 (1983).

of work completed, for materials or equipment not incorporated in the work but delivered and suitably stored at the site, and for materials or equipment suitably stored at some other location agreed upon in advance. Payments for materials or equipment stored on or off the site shall be conditioned upon submission by the Contractor of bills of sale or such other procedures satisfactory to the Owner to establish the Owner's title to such materials or equipment or otherwise protect the Owner's interest, including applicable insurance and transportation to the site for those materials and equipment stored off the site.

(b) Invoices for payment shall be submitted in the form attached to this contract as Exhibit 1 and transmitted to the following:

> _____[Address]_____
>
> _____
>
> _____

(c) Invoices remaining unpaid after the tenth of each month shall bear interest from the eleventh day of the month, compounded monthly, at the nominal rate of 1 percent per month.

(d) Should either party be required to employ an attorney to enforce this payment provision, to protect its interest in any matter pertaining to this provision, or to collect damages for breach of this provision, the prevailing party shall be entitled to recover reasonable attorneys' fees, costs, charges, and expenses (including but without limitation verification of financial status) expended or incurred therein.

INTEREST DAMAGES

§ 11.9 Interest as a Cost

The payment of interest is usually considered the most straightforward way to make whole a construction contractor who has been denied timely payment of costs. Although that principle is self-evident, interest is surprisingly difficult to obtain. American courts are reluctant to award interest as an item of damages; they impose special rules and restrictions that are not required of almost any other kind of damages. The reasons for this are probably twofold. First, our law is derived from, and is still heavily influenced by, the English legal system, which early displayed hostility to interest.[12] Interest was condemned on ethical and religious grounds, and regarded as unlawful. Ever since, it has been afforded special treatment by both English and American law.

---

[12] J. Acret, Construction Litigation Handbook 186 (1986).

Second, interest has been accorded unusual treatment by the accounting profession. When important damages issues are at stake, our courts often look to accounting principles to aid in reaching decisions. Under traditional rules of cost accounting, if money is acquired by borrowing (and its use, therefore, financed by interest), the interest paid is a cost. If, on the other hand, money used in the operation of a company is equity capital, no cost is recognized.[13] As a result, it has been argued that, so as not to favor those who use borrowed capital over those who use equity capital, interest should be recognized as a cost in neither instance.[14]

§ 11.10 —Theoretical Basis for Interest Cost

Historical treatment notwithstanding, there is a cost associated with the use of money, be it borrowings or equity, and that cost is measured by interest, either actual or imputed. Many scholarly treatises have been written on the theoretical basis for treating interest as a cost, regardless of the source of funds.[15] The overriding logic supporting the thesis is summarized in Professor Charles T. Horngren's excellent and widely recognized text, *Cost Accounting: A Managerial Emphasis:*[16]

> Interest is the cost of using money. It is the rental charge for funds, just as rental charges are made for use of buildings and equipment. Whenever a time span is involved, it is necessary to recognize interest as a cost of using invested funds. This applies even if funds in use represent ownership capital and if the interest does not entail an outlay of cash. The reason why interest must be considered is that the selection of one alternative automatically commits a given amount of invested funds that otherwise could be invested in some other opportunity. The measure of the interest in such cases is the return foregone by rejecting the alternative use.

From a construction contractor's point of view, distinguishing between borrowed capital and equity capital often makes no sense. Once the contractor has acquired capital, no distinction is made in how it is used; a contractor typically does not use borrowed capital for one category of expense and equity capital for another. Bills are paid with dollars, and dollars are fungible. Thus, if a contractor must finance construction progress, it should not matter whether the dollars used were borrowed or equity capital.

---

[13] J. Booth, Interest and Federal Contracts 55 (1982).
[14] J.P. Bedingfield & L.I. Rosen, Government Contract Accounting (1979).
[15] *See, e.g.,* J. Booth, Interest and Federal Contracts (1982); F. Scovell, Interest as a Cost (1924 reprint 1976).
[16] (4th ed. 1977).

There are important advantages to the contractor in treating interest on both borrowed and equity capital as a cost. Identifying a cost associated with the use of equity capital provides recognition and traceability to its use, providing important information to shareholders, lenders, and managers. Because the use of equity capital to finance a construction contract deprives a company of an opportunity to expend those funds elsewhere, imputing an interest cost permits an informed comparison with other potential uses.[17]

§ 11.11 —Interest as a Collectible Damage

Courts do sometimes award interest as an element of damages. Because of the historical aversion to interest noted in § 11.9, however, numerous limitations and restrictions have been imposed. First, a distinction is made between interest "on" a claim and interest "in" a claim. The distinction is between interest awarded *on* an amount due under a contract payments provision (an invoice that is not timely paid) and interest as an element *in* a claim for increased costs (interest on the cost of changed or extra work). Note that a single claim can comprise both types of interest damages: interest in a claim for the cost of financing changed or extra work can be combined with a claim for interest on the total amount claimed once that amount has been determined and a demand made for payment. In this situation, the two types of interest do not run concurrently; the interest "in" the claim is replaced by the interest "on" the claim once demand for payment is made.

Interest Cost on a Claim

In general, interest is permitted on a claim only when expressly provided for in the contract or authorized by statute. Both federal and state statutes authorize interest damages after a judicial judgment has been obtained.[18] Prejudgment interest on a claim is more difficult. Absent a contract provision imposing prejudgment interest, the traditional rule is that interest will begin to accrue only at the time that the claim becomes "liquidated."[19] A debt is considered liquidated if (1) the debtor knows the amount owed or (2) the debtor could compute the figures from available information.[20]

---

[17] *See generally* R. Anthony, Accounting for the Cost of Interest (1975).

[18] *See, e.g.,* 41 U.S.C. § 611 (1978); Cal. Civ. Proc. Code §§ 685.010–685.110 (Deering 1988).

[19] Chesapeake Indus., Inc. v. Togova Enters., Inc., 149 Cal. App. 3d 901, 911, 197 Cal. Rptr. 348 (1983).

[20] *Id.,* 149 Cal. App. 3d at 911, 197 Cal. Rptr. at 354.

§ 11.11 INTEREST AS COLLECTIBLE DAMAGE

When indefinite or unliquidated damages are involved, the award of interest is usually left to the discretion of the court.[21] In such situations, the courts generally look to the following four factors:

1. Whether the claimant had been diligent in pursuing the claim
2. Whether the defendant had been unjustly enriched
3. Whether actual borrowings were used and interest costs were incurred by the claimant
4. Whether there are countervailing equitable considerations against the award.[22]

The most common manner in which an arguably unliquidated amount becomes liquidated is for a formal demand to be made. A *demand* is nothing more than a formal request for payment of a specific amount together with a clear statement of the basis for making the request.

Interest Cost in a Claim

An owner who breaches a construction contract by requiring changed or extra work, or by delaying the project, is liable to the contractor for its resulting damages. That liability is generally limited by two factors: foreseeability and certainty.

As announced in the landmark case of *Hadley v. Baxendale*,[23] a wronged party can only recover damages that were reasonably foreseeable at the time of entering into the contract. Foreseeable damages are not necessarily those that are actually contemplated by the parties but are simply those that are the natural result of the breach. Furthermore, if at the time of contracting the parties knew or should have known of special or unusual consequences that would result from a breach, the breaching party will be liable for those unusual damages. This is referred to as the *foreseeability requirement*.

Once the fact of damages has been clearly established, the amount of damage need only be proven to a reasonable degree of certainty.[24] Courts will not deny recovery solely because a claim is not capable of precise proof or mathematical certainty.[25] But the damages must be proved with

---

[21] Feather v. United Mine Workers, 711 F.2d 530 (3d Cir. 1983); National Farm Workers Serv. Center, Inc. v. Caratan, Inc., 146 Cal. App. 3d 796, 194 Cal. Rptr. 617 (1983).

[22] Feather v. United Mine Workers, 711 F.2d 530, 540 (3d Cir. 1983).

[23] 9 Ex. 341, 156 Eng. Rep. 145 (1854).

[24] E.C. Ernst Co. v. Koppers Co., 626 F.2d 324, 327 (3d Cir. 1980).

[25] Allen v. Gardner, 126 Cal. App. 2d 335, 340, 272 P.2d 99 (1954).

sufficient precision so that they cannot be categorized as "speculative."[26] This is referred to as the *certainty requirement.*

An owner who causes a construction contractor to increase its costs by performing changed or extra work, without concurrently adjusting the contract price, is effectively requiring the contractor to finance that portion of the project. In such a situation, it is not only foreseeable but likely that the contractor would incur increased financing expenses—interest. As a result, courts have recognized interest as a collectable damage, but special limitations remain.

Interest expense resulting from financing changed or extra work has been found recoverable when the contractor can demonstrate actual borrowings due to the increased financing responsibility.[27] One case has held that, in order to recover interest in these situations, the contractor must demonstrate: (1) a need to borrow money to finance the changed or extra work; (2) actual borrowings and interest expenses; and (3) actual use of the borrowed money in performing the changed or extra work.[28] Other, more recent decisions, however, have applied less rigorous requirements.[29] Except for several cases involving federal government contracts, no case has explicitly permitted interest on the use of equity capital. On the other hand, there is no clear precedent denying such relief.

§ 11.12 Proving Interest Cost Damages

To be successful in seeking interest damages, the case must be carefully prepared. Particularly when seeking interest "in" a claim, the quality of the proof of actual damage will likely determine the outcome. In addition, as a case for interest is prepared, a number of computational issues will arise. The following sections discuss the type of proof that is likely to be the most successful and give guidance on how the interest claim might be prepared. In some instances, the factors to be considered vary depending on whether interest is sought "on" or "in" the claim.

---

[26] Zirin Laboratories Int'l, Inc. v. Mead-Johnson & Co., 208 F. Supp. 633 (E.D. Mich. 1962).

[27] Nebraska Pub. Power Dist. v. Austin Power, Inc., 773 F.2d 960 (8th Cir. 1985); Havens Steel Co. v. Randolph Eng'g Co., 613 F. Supp. 514 (W.D. Mo. 1985); Joseph Bell v. United States, 404 F.2d 975, 186 Ct. Cl. 189 (1968); Dravo Corp. v. United States, 594 F.2d 842, 219 Ct. Cl. 416 (1979).

[28] Dravo Corp. v. United States, 594 F.2d 842, 219 Ct. Cl. 416 (1979); Joseph Bell v. United States, 404 F. 2d 975, 186 Ct. Cl. 189 (1968); *cf.* Keco Indus. Inc., ASBCA Nos. 15184, 15547, 72-2 B.C.A. (CCH) ¶ 9576 (1972).

[29] Singer Co. v. United States, 568 F.2d 695, 215 Ct. Cl. 281 (1977); Aerojet Gen. Corp., ASBCA No. 17171, 74-2 B.C.A. (CCH) ¶ 10,863 (1974).

§ 11.13 —Traceability (Causation)

Although some cases have permitted the recovery of interest "in" a claim based on general proof of increased financing needs, the likelihood of recovery is greatly increased with detailed evidence tying borrowings to changed or extra work. Indeed, even suits for interest "on" claims are greatly enhanced by such proof; one of the factors considered by the courts in determining whether to award prejudgment interest is whether actual borrowings were necessitated and incurred.

In proving actual interest expense, begin with a detailed chronological analysis of the increased change or extra work costs incurred. Identify the time periods (by month) during which the changed or extra work was performed and develop a cumulative total by month of the resulting increased costs. If increased costs cannot be identified to individual months, accumulate them by quarter or, if necessary, by year.

Compare this chronological accumulation of costs with a similar chart of company borrowings. Research financial records to determine individual debt and cumulative debt. For this purpose, the source of the borrowings should be ignored. The debt might arise from borrowings from a financial institution, a parent corporation, or a pension fund. In many instances, there will be a nearly direct correlation between increased contract costs on the one hand, and growth in overall debt (or growth in particular debt) on the other. Such evidence is highly persuasive in demonstrating "actual borrowings."[30] In other instances, total debt will not significantly change or may even decrease during the months analyzed. In these situations counsel should determine the reasons for the debt experience. Several factors should be considered: (1) completion of another major contract(s) with the release of large retainages; (2) adoption of alternative sources of financing, such as a stock offering; or (3) unusual sales of assets or another form of asset reduction. Concrete evidence of such influences demonstrate that, but for the added or changed work, borrowings would have been reduced even further.

There are special problems if the company had no debt during the period of performance. In that situation, a chronological chart of investments should be developed and compared with the chronological accumulation of costs. This evidence may demonstrate a decrease in investment corresponding to the increased financing requirements of the changed or extra work. Alternatively, the evidence may demonstrate a corporate policy of investing excess cash and the rates at which further investments could have been made. In this regard, contemporaneous, written corporate policies memorializing investment policies should also be identified

---

[30] Aerojet Gen. Corp., ASBCA No. 17171, 74-2 B.C.A. (CCH) ¶ 10,863 (1974).

and located. This evidence demonstrates that, if additional funds had been available, the company would have invested it at stated return.

§ 11.14 Calculating Interest for Commercial Contracts

The date on which interest damages begin depends on the nature of the claim. For delayed payment claims (interest *on* the claim), interest begins to accrue when the payment becomes overdue. Thus, evidence must be produced to demonstrate that (1) an invoice in the correct form and with requisite support (2) was submitted to the person designated in the contract for payment and (3) either the time for review and payment stipulated in the contract passed or more than a reasonable time passed without payment. If doubt exists regarding any of these facts, it is best to submit a demand letter specifying the exact amount overdue. Interest is likely to accrue no later than the demand date. As noted in § 11.11, a demand letter need be nothing more than a statement of the exact amount due, a demand that payment be made immediately, and a brief statement of the grounds supporting the demand.

The beginning point for calculating interest *in* a claim can be more difficult. If the increased costs resulting from changed or added work each month are significant and continue for several months, interest should begin with the first month in which it is estimated that the changed work began, with the principal increased each month by the cost of that month's increased costs. By utilizing the chart prepared to demonstrate traceability (see § 11.13), the monthly increases in added costs can be identified and added to the principal in the interest calculation.

In some instances, the nature of the changed work will prevent precise identification of increased cost by month. The changed work may have begun from the very first stages of construction and be so intertwined with the unchanged work that a monthly analysis is not possible. An effort should nevertheless be made to equitably distribute the increased costs over time in a manner that approximates their incurrence. One estimating tool that might be employed is to determine the ratio of total increased costs to the original contract amount. This ratio can then be applied to monthly total costs incurred to estimate the monthly change costs. The interest calculation can then be performed monthly.

A simpler approach is to identify the starting and stopping points for the increased costs and begin the interest calculation at the midpoint of the time period. Apply interest to the entire claim amount from the midpoint forward. This approach, of course, assumes that one-half of the claim dollars were incurred prior to the selected midpoint and one-half after. If facts are known that are inconsistent with this assumption, the technique should be adjusted for the known facts.

§ 11.15 Calculating Interest for Federal Government Contracts

Interest "in" a claim has been most thoroughly developed in federal government contracting.[31] Subsequent to that development, however, applicable regulations were amended to effectively eliminate a contractor's ability to recover such interest.[32] In 1978, however, Congress enacted the Contract Disputes Act of 1978,[33] which provides for payment of interest on contractors' claims from the date of certification of the claim.

Certification under the Contract Disputes Act is generally analogous to a demand letter in commercial contracting. When a contractor believes it has encountered constructive changes on a project, it may submit a claim for increased costs. If the contracting officer disagrees that additional payments are due, the matter becomes a dispute and the contractor may submit a certified claim. A *certified claim* is a request for payment that demands a specific sum, is accompanied by detailed support substantiating the demand, and includes a formal certification in the precise form prescribed by applicable regulations. If the certified claim is denied, but the contractor is successful in litigating the matter, interest will accrue from the date of the certification.

§ 11.16 Determining the Proper Interest Rate

When *prejudgment interest* is awarded, the interest rate is that rate which is statutorily set by law, unless otherwise provided by the contracting parties. (There is also a statutory limitation on the rate to which the parties may agree.[34]) Thus, counsel must refer to the state law applicable to the contract to determine the proper rate. In contracts with the federal government, prejudgment interest is governed by the Prompt Payment Act, which stipulates that interest will accrue at the rate established by the secretary of

---

[31] J. Booth, Interest and Federal Contracts 29-38 (1982).

[32] The Defense Acquisition Regulation was amended to make use of the cost principles of the regulation mandatory in the pricing of equitable adjustments. Defense Procurement Circular No. 79 (effective Jul. 1, 1970). Those regulations provide that interest is not an allowable cost and, therefore, cannot be considered in pricing contract changes. F.A.R. § 31.205-20.

[33] 41 U.S.C. § 601 (1978).

[34] Cal. Civ. Code §§ 1916.1–1917.005.

the treasury pursuant to 50 U.S.C. App. § 1215(b)(2) (1978). Similarly, if the claim for interest is based on disputed, changed, or extra work, the applicable interest rate is that set by the secretary of the treasury under the Contract Disputes Act.[35]

Interest in a claim should be calculated at the interest rate actually incurred during the performance of the changed or added work. This is not, however, always easily determined. The starting place in identifying the appropriate rate is the chronological analyses of borrowings and costs previously prepared (see § 11.13). The analysis of borrowings should be expanded to identify the nominal (or stated) rates of borrowing. If the prior analysis was able to tie increased performance costs to a single loan, the interest charges for that loan should be used. If no single loan was used, a composite rate should be developed. A *composite rate* is the weighted average of all interest expense of the company during the period analyzed. Such a composite rate might be the weighted average of only short-term borrowings or the weighted average of all debt. There is also authority to support the calculation of a marginal rate, which focuses on the incremental cost of borrowing rather than average cost of borrowing.[36]

After identifying the nominal interest rate, the loan agreements should be analyzed to identify significant terms that affect the nominal rate. Terms that affect the interest rate include requirements for compounding and compensating balances. The effect of compounding is obvious; the effect of compensating balances is more subtle. A requirement for compensating balance means, generally, that the lender requires that the borrower maintain a stated amount of funds in its account at all times. For example, if $200,000 in new funds are required to fund changed or extra work, a requirement for a compensating balance of $20,000 would require the contractor to borrow $220,000. A claim for interest on the $200,000 claim for changed or added work calculated at the nominal rate of the loan would thus understate the contractor's actual damage. The contractor was required to borrow and pay interest on $20,000 more than the claim amount.[37] The effect of the compensating balance should be considered in calculating the effective rate of interest.

When equity capital was utilized to finance changes, the same general approach as was used with borrowed funds should be followed. In this case, however, an analysis of investments made during the period that contract changes were being financed should be made to determine the rate of return obtained. Because of the courts' reluctance to award interest without borrowings, there is little guidance on how the rate should be derived. There is no reason to believe that the concepts of "average

---

[35] 41 U.S.C. § 611 (1978).
[36] Aerojet Gen. Corp., ASBCA No. 17171, 74-2 B.C.A. (CCH) ¶ 10,863 (1974).
[37] *See* Aerojet Gen. Corp., ASBCA No. 17171, 74-2 B.C.A. (CCH) ¶ 10,863 (1974).

return on investment" and "marginal return on investment" should not apply.

OTHER DAMAGES

§ 11.17 Foreseeability of Other Payment Delay Damages

As noted earlier, an owner who breaches a construction contract by failing to pay the contractor on time will be held responsible for all foreseeable direct damages. "Foreseeability" is a mercurial term and courts have struggled, and often disagreed, in attempting to establish a working definition. Any cost that is the natural and probable outgrowth of the failure to pay is a potential collectible damage. Some of the more important costs that should be considered are discussed in §§ 11.18 through 11.21.

§ 11.18 Increased Bonding Cost

The ability to obtain a bond (bonding capacity) and the cost of the bond are functions of the contractor's financial condition. As we have seen, one effect of delayed payment is that it forces the contractor to finance the work in progress. Sometimes that financing obligation is so great that it uses additional borrowings or significant equity. In these situations, the construction contractor may find it more costly, or impossible, to obtain bonds for new work. Because such an impact is often difficult to prove, however, most courts are reluctant to award damages for the increased cost of the bonds. When contractors have been able to clearly demonstrate a nexus between the delayed payment, increased debt (or decreased equity), and resulting loss of bonding capacity, however, an award has been made.[38]

There are two keys to collecting increased bonding costs. First, evidence that such damage was foreseeable must be presented. Evidence that the owner was familiar with bonding requirements and procedures is also important. For example, evidence that an owner is experienced in construction and routinely includes a bonding requirement in construction contracts would tend to confirm its knowledge of industry practices. Second, as always, detailed proof tying increased bonding costs to the

---

[38] Laas v. Montana Highway Comm., 157 Mont. 121, 483 P.2d 699 (1971); cf. United Telecommunications, Inc. v. American Television & Communications Corp., 536 F. 2d 1310 (10th Cir. 1976); Roanoke Hosp. Ass'n v. Doyle & Russell, 215 Va. 766, 215 S.E.2d 155 (1975).

increased contract financing obligation is necessary. Evidence developed in connection with the interest claim should establish the tie between the changed work and increased borrowings or reduction in equity. This evidence and testimony from a bonding agent regarding methods of arriving at bonding charges are important. Further, at the time that new bonds are sought, a letter from bonding agents regarding changes in bonding charges would be helpful. Finally, evidence of specific bonds obtained and fees paid must be adduced.

§ 11.19 Loss of Other Work

Lack of bonding capacity, or simply insufficient financing as a result of delayed payments, may cause a contractor to miss out on other profitable work. In these situations, an argument can be made that the lost profit on the other work is a damage resulting from the delayed payments. Courts, however, generally refuse to award damages for unearned profits on other work for which it is alleged a contractor was unable to bid because of the owner's breach. Courts feel that such damages are far too speculative and fail the certainty requirement because it is always uncertain that the contractor would have (1) bid on other jobs, (2) been awarded the additional contracts, or (3) earned a profit on the prospective contract.[39]

§ 11.20 Collection Fees and Credit Checks

When an owner delays making a contract payment, it is usually wise to immediately have a credit check performed on the owner. If the owner is late making payments on a number of accounts, this is a strong indication of financial jeopardy, requiring the construction contractor to take immediate action to protect against the owner's bankruptcy. Clearly, therefore, a credit check should be viewed as a natural consequence of a delayed payment. Even if the credit check demonstrates that the owner is financially sound, the contractor may wish to engage assistance in collecting unpaid invoices. This may involve assistance from an attorney or a professional collection agency. All of these measures cost money and are caused by the delayed payment.

Because these costs are rarely significant in amount by themselves, there is little case law dealing with their recovery. However, the doctrines of foreseeability and certainty make it likely that such costs would

---

[39] *See, e.g.,* Manshul Constr. v. Dormitory Auth., 111 Misc. 2d 209, 444 N.Y.S.2d 792 (1981).

be recoverable. Moreover, adoption of contract language such as that contained in the clause suggested in § 11.8 should eliminate any doubt about the collectibility of these changes.

§ 11.21 Costs of Defending Subcontractors' Suits

If a prime contractor is not paid, it is likely its subcontractors also will not be paid. Often subcontractors have less patience than the prime contractor and sue to collect amounts owed. The prime contractor necessarily incurs legal fees in defending the lawsuit, even if the defense is only to bring the owner into the litigation. Courts have allowed the contractor to recover such defense costs.[40]

In proving these costs the complaint filed by the subcontractor should be used to establish the basis for the action. If the complaint alleges a number of breaches in addition to nonpayment, the subcontractor will need to testify regarding the primary basis for the suit, if any. In addition, the prime contractor should be prepared to place into evidence detailed billings from its attorney itemizing time spent in defending the action. The attorney should also be prepared to defend billing rates based on comparison to fees charged by other attorneys possessing similar skills in the same geographic locale.

§ 11.22 Delayed Payment Checklists

When faced with nonpayment of performance costs, there are various actions that can be taken by the contractor to reduce the risk of loss and to substantiate a damages claim for the resultant losses. The following is a checklist of actions that should be considered. When an invoice is not honored within the time stipulated, consider the following:

Delay in Response to Payment Request

_____ 1. Upper management should personally investigate the status of the project to ensure that accurate information is being received about progress of the work and problems with the owner or its representative

_____ 2. Assure that billings to date on the project are current

_____ 3. Immediately perform a credit check to determine the financial health of the owner

---

[40] Lea County Constr. Co., ASBCA No. 13964, 72-1 B.C.A. (CCH) ¶ 9298 (1972).

4. If the financial health of the owner appears weak, immediately consult an attorney expert in both construction and bankruptcy law
5. At as early a date as possible, make a formal demand for payment specifying the exact amount due and advising that interest will be sought at the contractual rate or, absent a contractual provision, at the legal rate
6. Analyze lien rights that may be available under state or federal law and give necessary notices when appropriate
7. Check the company's liquidity and assess the extent to which it can sustain a drain on cash flow
8. If the project is ahead of schedule, consider whether it would be appropriate to slow the pace of work
9. When substantial amounts remain unpaid for a significant time (or in accordance with contract provision), consider stopping work. It is usually best to consult with an attorney before taking this step
10. Establish contact with the highest levels of management of the owner and make them aware of the problems being encountered
11. Follow up regularly, daily if necessary, with inquiries on the status of payment
12. If a federal contract is involved, assure that Prompt Payment Act interest is automatically awarded upon payment.

Delay in Response to Changed or Extra Work Payment Claims

1. Identify with as much precision as possible all added or changed work as the work is being performed
2. Implement document collection procedures which will ensure that all relevant correspondences, memoranda, and other documentation are identified and retained
3. Consider establishing a separate line of credit for the changed or added work if new borrowings are necessary
4. Discuss with the owner the possibility of obtaining partial payment on disputed changes such as labor only, or material only, or 75 percent of the quoted change order price
5. Document terms of credit carefully
6. Document intercompany loans, if any, identifying the source and amount of funds, purpose of the intercompany loan, and intercompany fees, if any

§ 11.22 CHECKLISTS

____ 7. Document bonding costs, obtaining letters from agents explaining any increased charges

____ 8. Make a demand for payment of claims as soon as the amount involved has been identified

____ 9. Under federal government contracts, certify a formal claim as soon as the basis for the claim and the pricing can be established.

PART IV
OWNER CLAIMS AGAINST CONTRACTORS AND DESIGN PROFESSIONALS

CHAPTER 12

COST CLAIMS

Paul W. Pocalyko, CPA
James L. Shea, Esquire

Paul W. Pocalyko is a manager in the litigation and claims practice of the Philadelphia Business Investigation Services Group of Coopers & Lybrand. He has an extensive background in financial and analytical analysis in construction disputes. He has worked in several multimillion dollar construction matters including nuclear and electric power plants, commercial buildings, highways, and hospitals. He has provided investigative and accounting expertise addressing issues of entitlement and damages. Mr. Pocalyko is a member of the American and Pennsylvania Institutes of Certified Public Accountants. He holds a B.S. degree in accounting and an M.B.A. from Lehigh University.

James L. Shea is a partner in the law firm of Venable, Baetjer & Howard. He specializes in civil litigation, representing clients in a variety of matters, including civil RICO suits, construction and real estate disputes, and products liability cases. Mr. Shea graduated from Phillips Academy, Andover, Princeton University, and University of Virginia Law School, where he was a member of the Order of the Coif. He served as law clerk to federal Judge Joseph H. Young and later served as an assistant attorney general for the State of Maryland. He has been a frequent faculty member for litigation-related programs and seminars.

§ 12.1 Introduction
§ 12.2 Methods of Investigating Contractor-Claimed Costs
§ 12.3 Contractor's Books and Records
§ 12.4 Contractor's Relationship with Subcontractors and Material Suppliers
§ 12.5 Problems with the Collection of Records

§ 12.6   How Are Contractor Costs Overstated?
§ 12.7   Was the Work Properly Billed as Cost-Plus Rather Than Under a Fixed Fee Term?
§ 12.8   Did the Contractor Perform Efficiently?
§ 12.9   Are the Claimed Material Costs Reasonable?
§ 12.10 Are the Alleged Labor Costs Excessive?
§ 12.11 Are the Office and Administrative Costs Excessive?
§ 12.12 How to Judge a Reasonable Profit Margin
§ 12.13 Were There Premature Payments?
§ 12.14 Owner's Involvement in Cost-Plus Contracting
§ 12.15 Nature and Extent of the Surety's Obligations

§ 12.1 Introduction

This chapter covers owners' claims against contractors for excessive costs. Although contractors in pursuit of additional funds from owners are the most frequent proponents of excess cost claims, the converse can be true as well. Thus, this chapter looks at ways in which an owner can claim or counterclaim against a contractor who has overbilled for work done. The price term in such a setting is usually, but not always, cost plus; the questions of importance are whether and by how much a contractor has overcharged the owner.

Several examples will more clearly define and delineate the context of this chapter. Occasionally, a prime contract between an owner and a general contractor has a price term that requires the owner to pay the reasonable cost of the contractor's work plus profit and overhead. Even under a fixed price contract there may be questions about overcharging and overpayment: did the owner pay too much, too early? The time-honored but questionable practice of front end loading requires an assessment of whether the amount billed for work completed at a particular stage was fair and reasonable given the construction activity to that point.

Changes in the scope of work also give rise to potential owner cost claims. Non-negotiated extras are usually determined on a cost-plus basis, again posing the question of whether the contractor has reasonably charged the owner for the work. Another type of owner cost claim arises when there is a failure to accord sufficient credit to the owner. Some construction contracts have a savings clause that enables an owner to reap all or a portion of the benefit when construction costs are less than originally anticipated. Accordingly, when an owner can demonstrate that a contractor has overstated the cost of construction, the owner should recover the overcharge pursuant to the savings provision. Similarly, when there are reductions in the scope of work (or some work is simply not completed) the question becomes whether the owner was accorded sufficient credit for

§ 12.1   INTRODUCTION   353

the reduction. This claim necessitates a comparison of the reasonable cost of the entire scope of work with what was actually done.

The potential of an owners' cost claim arises also in a lien action against the owner by a subcontractor whose contract was with the general contractor or, perhaps, the owner's lessee. In those cases the extent of the lien is measured in essence by a quantum meruit standard, which typically is the fair and reasonable cost of the improvement to the property. Although a fixed fee subcontract could be evidence of that fair and reasonable cost, it alone is not necessarily determinative.[1] This point would normally arise as an owner's defense to or avoidance of a lien claim brought by a subcontractor. It could also, however, be the basis for an owner's claim against a contractor for fraud or for breach of trust. It might also be the basis for a suit against the general contractor's surety, pursuant to a contractual responsibility to protect the property from liens.[2]

At issue in all of these situations is the fair and reasonable cost of work performed or to be performed. In most, if not all jurisdictions, reasonable cost is determined by an objective, fair-market standard.[3] Each of the elements of cost must be judged by that standard. Thus, the amount of labor and material utilized as well as the per unit charge for that labor and material must be reasonable. When an amount for overhead and/or profit is also at issue, that too must be assessed on an objective basis. Questions about these costs arise, for example, if the prime contract requires payment for profit and overhead but does not specifically set forth an amount. They also arise when a general contractor's claimed costs include the profit and overhead of its subcontractors. Under many circumstances, the owner should not be responsible for paying inflated or unreasonable amounts for the profit and administrative cost of the subcontractors, even if the contractor is entitled to a cost-plus price term.

It is plain that the actual cost paid by the contractor in any of these contexts is not enough by itself to establish the reasonableness of the cost. Nor is the sum for which the contractor has been billed by its subcontractors and suppliers necessarily determinative. There must be additional evidence of the reasonableness of these costs, both in terms of quantity and per unit price, in order to establish the requisite proof. **Sections 12.2** through **12.13** discuss the means by which that issue is analyzed and the areas in which it can be shown that costs claimed by contractors are excessive and unreasonable.

The investigative approach, as well as the ultimate presentation of the evidence in court, will vary depending on whether the contractor has

---

[1] *See, e.g.,* Umbaugh Builders, Inc. v. Parr Co., 86 Misc. 2d 1036, 385 N.Y.S.2d 698 (1976).

[2] See § **12.14**.

[3] Corbin, 5 Contracts § 1004 (1964 and Supp. 1980). *See also* Hunter, Damages for Breach of Construction Contracts, in Modern Law of Contracts § 8.06 (1986).

overstated costs because of an intentional design to defraud or because of incompetence and inefficiency. Naturally, the two are not exclusive; both fraud and incompetence can appear on the same project. Either will suffice to warrant a return of the overstated costs. But a coherent, logical explanation of the contractor's conduct inevitably sharpens the investigation and makes the presentation of the evidence persuasive and compelling.

Often, however, the owner's investigators, particularly at the outset, may be uncertain of the contractor's motives. The decision of which explanation for the overcharges fits cannot be made prematurely; too narrow a focus obscures otherwise helpful information. The trick then is to assess the information continuously, testing the various hypotheses thoroughly without unduly prolonging the process.

§ 12.2 Methods of Investigating Contractor-Claimed Costs

The costs claimed by many contractors should undergo an intense scrutiny to test their validity, accuracy, and, most importantly, reasonableness. The means by which an analysis and investigation can be undertaken include:

1. Assessing the contractor's accounting books and records
2. Investigating the relationships that exist between the contractor and subcontractors or material suppliers
3. Analyzing other ongoing projects of the contractor
4. When appropriate, analyzing the books and records of selected subcontractors and material suppliers.

These approaches can uncover a significant amount of information important to both understanding and investigating a claim. They can also be costly and time-consuming. A shrewd assessment of which to pursue and when is vital to success.

§ 12.3 Contractor's Books and Records

Much has been written and many hours of litigation time have been spent attempting to unravel the mysteries of the construction contractor's accounting and the related records which purport to substantiate a contractor's costs. In general terms, an investigation by an accountant of a contractor's records is required to establish an opinion as to the reasonable cost of construction. This examination cannot take place in a vacuum. If, for example, a claim is for extra work on a contract, the analysis of the books and records should extend beyond just the records supporting

§ 12.3 CONTRACTOR'S RECORDS

the claimed extras. By reviewing the entirety of cost records for the project, one may find that the costs related and billed as extras should have been included as contractually obligated work. A good example of this arose when a grading subcontractor exceeded its estimated billing for contractually obligated work as the result of poor estimating and mismanagement by the general contractor. The subcontractor and in turn the contractor proceeded to bill these charges as extras in order to recover the subcontractor's costs. Only an overall analysis of all the subcontractor's contractual billings revealed the problem.

Typically, the most fundamental analysis that provides insight into true costs is a comparison of cash received by the contractor for a project with the cash it has expended. The data for this analysis are the contractor's cash receipts and cash disbursement books and records. These may take many forms, from a manual ledger to a sophisticated, computerized system. The importance of this analysis is two-fold. It provides an understanding of what obligations are owing to subcontractors or material suppliers and to what extent claims for additional costs are substantiated by actual payments. When the contractor has actually spent money, its costs are most likely to be real. Even then, additional procedures should be employed to test further the dollar claims. See § 12.4. On the other hand, when no cash outlay has occurred, the potential for overstated costs is greater.

The contractor's invoices to the project owner also serve as an initial means by which to check the costs claimed. Often, the costs have not been previously reported to the owner and are presented only as the project is completed. In some cases, these are cost overruns uncovered in a final project accounting and in others they are mere puffery, developed with a goal to negotiate a settlement with the owner for the highest possible sum. When the contractor contends that the work was beyond the scope of original project cost estimates and billings, an analysis of those invoices is in order. A detailed comparison of scheduled values (invoices) billed and percentages of completion with the claimed extras often results in the elimination of claimed costs.

An additional means by which to investigate the contractor's books and records involves the use of analytical comparisons which can reveal what was going on in the contractor's business. Such information might well impact the claims or additional costs identified by the contractor. The owner who presumably has some familiarity with the contractor may be able to provide leads as to what analysis might be fruitful.

An analysis of the historical background of the company, for example, might identify the contractor's ability to have performed effectively for the owner. Questions that should be addressed include:

1. Whether the type of construction was within the usual expertise of the contractor. If a contractor had little experience with a given type of work, the credibility of the claims for cost overruns is reduced.

2. What was the relative size of the project in relation to the contractor's normal projects? Contractors who get involved in jobs that are their largest ever or significantly larger than their prior experience often have difficulty controlling costs and job administration.
3. What was the financial condition of the contractor throughout the construction process? Unusual stresses such as cash shortages, large payables, or debt balances can have an influence upon the actions of the contractor during construction.

Another useful analysis requires a thorough investigation of the contractor's relationships with subcontractors and material suppliers. The financial transactions between these parties should be at arm's length to ensure that claims for cost overruns are not merely cost inflation by the subcontractor or supplier. Signs of suspect relationships include frequent collaboration on projects or, more glaring, investment relationships such as partnerships among contractors and sharing of office space and personnel. Such relationships lead to practices that promote cost inflation, like early payment to subcontractors before work is completed, purchases of materials at prices higher than those quoted by the suppliers, and limited scrutiny of billings submitted for payment.

Another technique that can illuminate areas of overstated costs involves analyzing the contractor's other ongoing projects. If an owner can show that a contractor diverted labor, material, or subcontractor work charged to the disputed project to other projects, it can destroy the credibility of the contractor's alleged cost overrun. Although the benefit is enormous, the task is difficult. Only through a systematic and detailed review of time cards, daily reports, delivery tickets and the like can these types of diversions be identified and quantified.

§ 12.4 Contractor's Relationship with Subcontractors and Material Suppliers

Even if a general contractor's claim is supported by the contractor's own accounting books and records, in many cases the contractor itself may not have performed any significant amount of construction work. The typical general contractor employs a variety of subcontractors and material suppliers whose invoices supplement the costs records. The general contractor's summarization of these costs often would not reflect as accurate a picture as would a thorough and complete examination of each item of claimed costs. Red flags that should signal the need to dig beyond the general contractor's records include advance payments to or favorable payment arrangements with subcontractors and significant relationships between the contractor and subcontractors or material suppliers. Particularly with large

§ 12.4 SUBCONTRACTOR AND SUPPLIERS

and important claims, the owner must often assume the burden of showing that the claims are overstated and the contractor overpaid.

The selection of which subcontractor's books and records to examine is critical to both the outcome and the minimization of litigation costs to the owner. A shotgun approach does not readily produce viable information and is more likely to distract attention from pertinent facts and data than to uncover them. The selection of which subcontractor costs to scrutinize is influenced by several factors, including the areas of claimed cost overruns, the appearance of cost inflation, the potential for contractor inefficiencies, the relationships between parties, a variance in a contractor's claim against the owner when compared with the subcontractor's claims against the contractor, and the appearance of cash flow or other business pressures faced by the general or its subcontractors.

Often, cost overruns emanating from subcontractors are passed on to the owner in the form of a general contractor's overrun or claimed extra. General experience has revealed that the underlying subcontractor's records often reveal improper billing methods, large and unwarranted job profitability, poor bidding techniques, and an assortment of other sins that are in no way related to costs which the owner should be asked to pay.

When there is reason to believe that payments have been made in excess of the reasonable face value of the work, that subcontractor's books and records should be analyzed. In one particular case, a subcontractor was to perform a minimal amount of excavation and road work which the owner had estimated at $30,000. When the general contractor presented the subcontractor's bill for $50,000, the owner balked and began to question the charges. An analysis of the books and records supported costs of only $22,000 for the work performed and excess profits of $28,000. Further investigation uncovered a series of credits posted to another job which this subcontractor had performed for the general contractor. These credits were reflected in the subcontractor's books, but the general contractor had only posted net of credit costs on the other job. The timing of the credits plainly suggested a connection between the overcharge and the credits. This could not have been discovered if only the general contractor's books had been examined.

A subcontractor who has been inefficient is a good candidate for investigation. The first set of records that can reveal information important to this search are daily contract status reports. Clerk of the work records and other job progress reports memorialize the difficulties encountered during construction. In one case, unanticipated site conditions encountered by a grading contractor were alleged to have caused extensive cost overruns and extras. An analysis of the daily records combined with testimony showed that the site development was complicated by inefficient scheduling by the general contractor and that the alleged overruns were at least in part a result.

The types of contractor/subcontractor relationships discussed in §§ 12.2 and 12.3 also serve to identify the records which should be sought. An example of an unusual relationship that should prompt questions by an investigator is one in which a general contractor serves as a sub-subcontractor on the project. The general contractor may lend its employees and equipment to one of its subcontractors, who then includes these charges in the subcontractor costs. The abuse in these relationships through multiple markups of overhead and profit can be costly to the unsuspecting owner. These charges can be effectively challenged based on the simple reasonableness of the actual cost to perform the work versus the excessive profits engendered in the process and ultimately charged as a cost to an owner.

The claims of a general contractor should be reviewed against the balances owed and possibly claimed by subcontractors. Often, a general contractor will withhold payment to its subcontractors if it believes or contends that the owner owes it money, effectively breaching provisions of the contractor/subcontractor agreement. If the subcontractor has not either filed mechanic's liens or instituted a recovery action, there is reason to question the credibility of the general contractor's claim.

A final characteristic that can identify which records to pursue is the identification of a subcontractor with cash flow problems. Cash shortages can be accompanied by odd payment arrangements, including early payments when little work has been performed or excessive payments beyond reasonable costs incurred. There is other significance to such discoveries, but, at a minimum, they often show up in contractors' demands for additional compensation.

§ 12.5   Problems with the Collection of Records

In today's litigatory environment, the collection of necessary records is often considerably complicated, despite the fact that the discovery standard in the federal courts and most state courts is quite liberal. Under the federal rules, "any matter, not privileged, which is relevant" is subject to discovery.[4] The fact that the records and information sought may not be admissible as evidence is not grounds for nonproduction as long as the "information sought appears reasonably calculated to lead to the discovery of admissible evidence."[5]

Notwithstanding the liberality of the rules, general contractors and subcontractors (like most businesses) frequently object to the production of the records by which costs might be tested and analyzed. The second

---

[4] Fed. R. Civ. P. 26(b).

[5] *Id.*

§ 12.5 COLLECTION OF RECORDS 359

paragraph of Federal Rule of Civil Procedure 26(b)(1) (or its state analogue) is typically the express or implied basis for objection:

> The frequency or extent of use of the discovery method set forth in subdivision (a) shall be limited by the court if it determines that: (i) the discovery sought is unreasonably cumulative or duplicative, or is obtainable from some other source that is more convenient, less burdensome, or less expensive; (ii) the party seeking discovery has had ample opportunity by discovery in the action to obtain the information sought; or (iii) the discovery is unduly burdensome or expensive, taking into account the needs of the case, the amount in controversy, limitations on the parties' resources, and the importance of the issues at stake in the litigation.

In light of this provision and because the courts are increasingly wary of discovery abuse, an owner seeking the records described above must show that they are both relevant and the best, if not the only, source of the relevant information.

In order to accomplish that, it is critical to engage a qualified expert, preferably a cost accountant with a background in the construction industry. This should be done at an early point in the litigation to facilitate a determination of the following:

1. What records are needed
2. What records are likely to exist
3. What records present the most practical, efficient, and cost-effective information on the matters at issue.

Once these questions have been answered, the records should then be pursued by means of a request for production of documents or, in the case of a nonparty (such as a subcontractor or supplier), a subpoena duces tecum. Be aware, however, that very often the responses to such legal process are a few scattered records and a statement from the contractor's attorney that "everything has been produced." Experience has shown that is rarely the case.

When it is apparent that potentially important records have not been produced, the deposition of the custodian of records or other knowledgeable person can be quite helpful. From that person one can determine the particular company's cost accounting system, how its records are kept, and the specifics of what is available. Without this information the owner's counsel is often working in the dark and thus ineffectively and inefficiently.

Having determined what is needed and what is available, counsel for the owner is then in a position to move to compel the production of specific records. Motions to compel should be used sparingly and as a last resort.

They are unpopular with the courts and significantly increase the cost of litigation. With the help of an expert and with the information gained from a deposition, however, a prudent assessment of the importance of such a motion is possible. If the owner's counsel has done his homework, the motion can focus on specific records known from depositions to exist and can persuasively explain why these records are the best and often the only source of the information. It may be effective to submit with the motion an affidavit of the expert explaining the importance of the records sought. A risk of this tactic is that opposing counsel may press for an early, extra, and unwanted deposition of the owner's expert in order to inquire into the basis for the affidavit.

Once the critical records have been identified, counsel for the owner can usually, by compromise, avoid any additional objection raised by a general contractor or subcontractor seeking to avoid discovery. For example, the owner can offer to pay for the reasonable costs of copying and selecting or collating the records. It is preferable, however, to have the owner's expert look at the original books and records and choose the specific ones for copying. If the contractor protests that the records contain trade secrets or other proprietary information, a confidentiality agreement, using language similar to that in the protective order included in this section, can easily provide the needed protection. In it, the parties agree to use the records only for purposes of litigation, limiting access to them to those who need it. Such agreements might also limit the reproduction of records and require their return at the end of litigation. In the end, objections relating to burden, expense, or privilege can be overcome, once a convincing case of relevance and materiality is developed.

IN THE UNITED STATES DISTRICT COURT FOR THE DISTRICT OF

| Plaintiff, |
| v. |
| Defendants. |

CIVIL ACTION NO.

PROTECTIVE ORDER

IT IS HEREBY ORDERED by the United States District Court for the District of _____, that in connection with the Deposition Subpoena issued by this Court and the Notice of Deposition with Exhibit A to _____ (hereafter the "Deponent") for the production of documents and records on March _____, 19____ at the request of _____ (hereafter the "Defendant") in connection with that case

§ 12.5 COLLECTION OF RECORDS

entitled _____, Civil Action No. _____, in the United States District Court for the District of _____ (hereafter the "_____ Litigation"), the Motion To Quash Subpoena And Motion For Protective Order filed herein by the Deponent, and the response thereto filed by the Defendant, that the Deponent shall produce for inspection and copying by the Defendant the documents and records described in the aforesaid Exhibit A to the Notice of Deposition on the following terms and conditions:

1. Within two (2) weeks of the date of this Order the Deponent shall produce at its office, at _____, _____, _____ , its documents and records described in Exhibit A to the above described Notice Of Deposition for the period from January ____, 19 ____ to February ____, 19 ____. In addition thereto, within two (2) weeks of the date of this Order the Deponent shall produce at its office, at _____, _____, _____ , all of its documents and records (for all periods of time) related to work, labor, materials, and services provided by the Deponent for construction of the _____.

2. The Defendant shall, within 30 days after a bill is presented by Deponent, pay to the Deponent for the cost of the Deponent pulling the requested documents and records from their place of storage for inspection and copying by the Defendant, and for returning those documents and records to their place of storage within the Deponent's business records. That cost shall be calculated at the rate of $11.50 per hour for the time of Deponent's employee devoted to pulling the aforesaid documents and records from their place of storage and returning those documents and records to their place of storage. Deponent is not required to answer any questions during inspection of the documents by Defendant.

3. The Defendant shall have the opportunity to examine those documents and records produced at the Deponent's place of business, and to make photocopies of those documents and records. The documents and records of the Deponent shall not be removed from the Deponent's premises by the Defendant and the Defendant shall make a reasonable effort to maintain those documents and records in the same order and the same condition as when they were furnished to the Defendant for inspection and copying.

4. The Defendant shall be permitted to make photocopies of the documents and records produced by the Deponent for examination at the Defendants' cost as described in paragraph 2 and this paragraph. The Defendant shall furnish the labor for making such photocopies and the Defendant shall be entitled to make those photocopies on the Deponent's photocopying equipment. Defendant shall, within thirty (30) days after a bill is presented by Deponent, pay to Deponent, at the rate of $.15 per page, for photocopies made by Defendant on Deponent's photocopying equipment to cover the cost of the equipment and supplies.

5. All of the documents and records produced by the Deponent and copied by the Defendant, and the contents thereof, shall be "Confidential" information. Confidential information shall be made available only to the following persons who are involved in the _____ Litigation: the Court and

its officers; witnesses and jurors at the trial of the _____ Litigation; counsel of record for the parties to the _____ Litigation; paraprofessionals, stenographic and clerical employees assisting such counsel of record; in-house counsel, partners of, and employees for the parties who are engaged in preparation of the _____ Litigation for trial; and independent experts retained by any party. All of the above-described persons shall be informed by counsel of record for the parties in the _____ Litigation of the provisions of this Order and the confidential nature of the information.

6. Confidential information may be used solely for the purpose of the _____ Litigation and for no other purpose. Nothing contained in this Order shall prevent the use of Confidential information at trial or at depositions in the _____ Litigation. If such Confidential information is used as an exhibit or exhibits in depositions, all such exhibits shall be treated as confidential and shall be placed in an envelope marked "Confidential", and sealed, only to be opened and viewed by those persons described above.

7. All documents containing Confidential information which are filed with the Court shall be filed in sealed envelopes or other appropriate sealed containers on which shall be endorsed the title of this action, an indication of the nature of the words "Confidential Information", and a statement substantially in the following form:

> "This envelope is sealed pursuant to Order of the Court and contains information designated confidential in this case by _____ and is not to be opened and its contents are not to be displayed or revealed except for purposes of litigation."

8. Within sixty (60) days after the conclusion of the _____ Litigation all documents and testimony designated or treated as Confidential information, and all copies thereof, shall be returned to the counsel for _____.

If any Confidential information is furnished under this Order to any expert or other person, the attorneys for the party retaining such expert or furnishing such Confidential information shall be responsible to insure that all Confidential information is returned to counsel for _____.

9. Nothing contained in this Stipulation and Order shall limit or waive the rights of the parties hereto with regard to other information sought, including documents for the remainder of the five (5) year period originally requested by Defendant. Upon twenty (20) days notice by Defendant to Deponent, Deponent shall produce for inspection and copying the documents and records of Deponent as described in the aforesaid Exhibit A for the remainder of the five (5) year period which have not heretofore been produced by the Deponent, subject to the same terms and conditions (other than the time limitations set forth in paragraph 1 hereof) as are set forth in this Order.

10. The terms of this Order shall survive any final disposition of this case and the _____ Litigation.

11. The parties hereto, by their respective counsel, do hereby stipulate and consent to the foregoing Protective Order and they further stipulate to a

similar Protective Order, containing the same terms and conditions in any other existing or future litigation involving the Defendant and development and construction of the _____, provided that the Protective Order shall be binding on all parties to such future litigation.

_____
Judge

§ 12.6 How Are Contractor Costs Overstated?

Having gained access to the sources of the contractor's cost information, an owner is likely to discover numerous ways in which the true costs of the work it has paid for have been overstated by the contractor. In theory, the possibilities for cost inflation are virtually limitless. Techniques for and conditions conducive to cost overstatement are as old as the construction industry itself. **Sections 12.7** through **12.13** set forth instances of overstatement which no doubt will recur in one form or another from one project to the next. Those sections also discuss the means by which the overcharges can be proven in litigation. The ability of the owner to discover, develop, and prevail in litigation about these cost issues is dependent upon expert assistance. An effective cost accountant must examine the data underlying the contractor's claimed costs. In many cases this analysis will reveal unfounded costs. In other cases, the cost accountant can define the question to pose to an architect or engineer, cost estimator, or construction management consultant. These experts can develop the evidence and provide the testimony necessary to carry the owner's burden of proof.

§ 12.7 Was the Work Properly Billed as Cost-Plus Rather Than Under a Fixed Fee Term?

Confusion sometimes exists as to the legitimate nature of billings to the owner for extra work. A contractor or subcontractor may contend that the costs incurred are the result of work done outside the scope of its contractual obligation. Under many construction contracts, if the work were indeed an "extra," the contractor would be entitled to the payment under a cost-plus term rather than the fixed fee. Such claims are common and are the cause of heated disputes and litigation. In many cases, a contractor will have what appears to be a well-documented claim for the extra. Its invoices will typically contain explanations of why the work is beyond the scope of the contract. The owner, when faced with these claims, should perform some elementary checking to determine their accuracy and legitimacy.

Common sense is often the best guide for this inquiry. The difficulty sometimes lies in obtaining the detailed records that support the extra charges (discussed in § 12.6). The types of equipment being utilized or the descriptions of work performed, as reflected in the document supporting the invoice, may contradict the contractor's claim. In one case, for example, fine grading was to be an extra paid for on a cost-plus basis, while the rough grading was performed under a subcontract for a fixed fee. The grading contractor contended that its fine grading costs were several hundred thousand dollars. A review of the types of equipment used for the work invoiced, however, revealed inappropriate machinery for fine grading. An examination of the relevant daily reports further showed that roadways were being excavated and large amounts of dirt were being stockpiled on the days when the contractor claimed the fine grading was being done.

On such an issue, the retention of an expert in the trade involved is crucial. Some of the information can be seen and analyzed by a layperson; for example, bulldozers are not necessary for fine grading. But some of the less obvious points could be seen only by someone knowledgeable in the industry. For the full picture, expert analysis and testimony are necessary.

§ 12.8 Did the Contractor Perform Efficiently?

Inflated costs can result from the inefficiency of the contractor and an owner should not be responsible under a cost-plus term to pay for that inefficiency. One example of such inefficiency is a general contractor, entitled to a fee on a cost-plus basis, who fails to require competitive bidding on subcontracts. Although competitive bidding may not be legally required, an owner has a right to complain if a contractor fails to do so in a cost-plus setting because that would be inefficient performance on the contractor's part. In essence, a cost-plus term creates a relationship of trust between owner and contractor and requires the contractor to use due care in deriving its cost.[6]

Many contractors state that they do not usually shop quotes from their subcontractors and suppliers. This may be an accepted practice when the contract is for a fixed price, because under that term the general contractor assumes the risk of a cost overrun. In a cost-plus setting, however, that risk is the owner's and not the general contractor's.

Therefore, the owner can make an effective argument that a failure to shop subcontract prices creates a presumption that subcontract costs are inflated. This alone may not be sufficient to carry an ultimate burden of proof, but it should at least shift the burden onto the general contractor to justify its prices. Even if a court would not shift the burden of going

---

[6] *See, e.g.,* AIA A111-1978 art. 3.

forward, evidence that the general contractor failed to obtain competitive bids or otherwise compare prices would corroborate an owner's evidence that the subcontracts included costs and profits that were higher than ordinarily expected in the industry and higher than what was reasonable.

Another frequently encountered area of inefficiency that causes inflated costs and overcharges is a general contractor's failure to properly coordinate a job. Expert analysis by an experienced construction manager will uncover this. For example, in the construction of a building, poor planning by the general contractor could prevent concrete from being poured at a time when the trucks could drive to the location and pour the concrete directly rather than having to pump it. In that case, the costs associated with the pumping would be excess and something which an owner ought not to have to pay. Similarly, when a contractor stockpiles topsoil near a highway in order to advertise and sell it for its own profit, thus requiring the grading and excavating subcontractor to drive much farther to remove the topsoil, an owner should not be responsible for the extra time and equipment charges involved for that. There are innumerable additional examples illustrative of the basic point that an owner should protest an overcharge resulting from the contractor's inefficiency.

§ 12.9 Are the Claimed Material Costs Reasonable?

A component of a claimed cost overrun often relates to the material used. An owner should evaluate the materials utilized on the project in terms of the cost, the quantity, and the quality of the materials that the contractor claims to have used.

Typically, a contractor buys material, adds some form of a markup, and passes the cost on to a project owner. Things become somewhat complicated when an owner, often in an effort to reap cost savings, reduces the specification in anticipation of a credit. The contractor might not pass on the full savings it achieved from the owner's change. In one case, a roofer was asked to switch brands of roofing insulation because the owner knew it would save a base cost of 20¢ per square foot. In preparing the original roofing bid, the contractor had added a 15 percent profit margin to its material cost. When preparing the credit, the contractor failed to back out the profit on the amount saved. A comparison of the offered and appropriate credit is shown in **Table 12-1**. The failure to pass through the 15 percent profit credit is an all-too-common method by which owners are shortchanged on these reductions. On the other hand, the same contractor was very careful to add its 15 percent to all additions (change orders) that had been requested. A careful check of pricing sheets and the bid files prepared by the contractor to support its billing, when obtained, would reveal these discrepancies.

Table 12-1

An Example of a Comparison of Offered and Appropriate Credit

	Credit Billed by Contractor	Appropriate Credit
Square Feet of Roofing	100,000	100,000
Cost Difference of Products	$ .20	$ .20
	$ 20,000	$ 20,000
Reduction of Contractor Profit (15%)	0	3,000
Credit Due Owner	$ 20,000	$ 23,000

An analysis of the quantity of materials can also unveil cost savings. Materials that were never delivered, or were delivered to other sites, or appear to be excessive or unexplained quantities can all explain cost overruns. One example of quantity overstatement involved a roadway general contractor who contended that extra materials were needed to construct a suspended roadway. A comparison of the alleged overruns to actual cost information and bidding documents uncovered significant overstatement in the actual quantities of material required.

Delivery tickets and shipping records provide the primary means for checking such questionable material quantities. The form of such records varies by trade and can be either very formal or very informal. After the data is summarized, an experienced construction expert can compare the quantities used with take-offs from the construction drawings. There are many different methods by which to verify the quantities. For example, in one case a paving contractor was required to provide hot mix at a specified depth of two inches on a parking lot. When challenged by the project inspectors, the contractor claimed its yield was well within the specifications. A quantification of the hot mix delivery tickets revealed a massive shortage of material. To further support the case, the parking area was cored and samples of the depth were recorded in the one-inch to one-and-one-half-inch range.

Occasionally, a contractor simply overstates the quantities actually used. If so, the overstatement of costs is plain. However, when the materials are buried or are otherwise not visible, it may be difficult to discover the overstatement. In one case, the installation of a sewage treatment facility called for a specified amount of stone bedding in the drainage areas. Simple calculations determined the total stone necessary to fill the space. The contractor, however, produced invoices for stone well in excess of that amount. Without having to dig up the sewage beds to refute the contractor's contended overrun, the owner could prove that the additional stone was neither specified nor required and provided no benefit to the owner's operation of the facility. Under the circumstances, the owner could hardly be charged for such costs.

The owner is also entitled to material of a quality consistent with the specifications. The delivery records and supplier invoices can provide extensive information to determine whether the material was of the prescribed quality. If, for example, plumbing specs required schedule 40 PVC pipe for sewers and the contractor substituted schedule 21 PVC pipe, there would be a significant reduction in quality and a corresponding cost reduction. Similarly, a paving contractor may attempt to substitute thicker layers of cheaper base course hot mix in lieu of the specified wearing course hot mix. In both cases, the owner is left with an inferior product. If it has paid for the specified material, it has a cost claim against the contractor.

The assessment of material cost, quantity, and quality is best accomplished by a consultant intimately familiar with the trade. Such expertise is critical both for finding the overstatement of cost and for proving that it occurred.

§ 12.10 Are the Alleged Labor Costs Excessive?

Labor costs are a simple product of the hours worked times the rate charged. In one case, a masonry contractor claimed labor costs that seemed high. He supported his claim with figures that would have required crews with sizes ranging from 40 to 60 men working in very tight quarters. A comparison of those numbers with the owner's check of the work records uncovered a significant overstatement of up to 20 men daily. Expert testimony from engineers and construction management specialists about the impossibility and/or inefficiency of the contractor's labor claims corroborated the daily work records.

In another case, a painting subcontractor claimed extremely high costs per hour. A review of his wage records did not support his position. Another subcontractor had labor shortages as a result of his own inefficiency and had hired prison labor to substitute. The owner naturally took the position that it was not responsible for the consequently increased labor costs. Like the other components of a cost claim, labor costs provide multiple opportunities for overstatement.

§ 12.11 Are the Office and Administrative Costs Excessive?

A contractor's cost claims may also contain a variety of overhead costs that must be analyzed to determine their applicability and, ultimately, their reasonableness. Project overhead can usually be divided into job-specific overhead and home office overhead. In challenging these costs, several techniques help develop an accurate basis for establishing what an owner should be paying.

Typically, job-specific overhead is supported by invoices for costs such as trailer rentals and trash removal, while home office overhead is represented as a percentage of the total operating costs of the contractor's office. The job-specific overhead can be investigated by analyzing the relevant invoices, which can reveal many improper charges in areas such as equipment rentals and purchases. A contractor who purchases equipment to be used on more than one project may attempt to charge the entire cost of the equipment to one (or both) jobs. For example, a painter can reuse a paint sprayer on more than one job and a roofer can use a tar kettle to melt roof tar on more than one roof. Cost claims on one job incorporating the entire cost of the sprayer and the kettle are therefore excessive.

Equipment rentals provide another opportunity for cost overstatement. They are often included in claims based on hourly charges for the equipment. When those hourly charges in total exceed the price for which the contractor could have purchased the equipment, the owner is being asked to pay an unreasonable cost.

There are numerous methods for calculating and allocating home office overhead to a project contractor. Contractual language, agreed-upon percentages, and formulas such as the Eichleay method (in delay situations) can each be a legitimate basis for establishing a claim. An initial focus in checking such claims might be the bid information. If, for example, a contractor bid the job with a 15 percent factor for home office overhead and its claimed overhead is 25 percent of the total claim, a basis by which to reduce these costs is plain. It is difficult for the contractor to effectively link overhead cost increases to specific actions or events in the construction process because construction records are not maintained on this basis. Thus, the opinion of a construction accounting expert is often needed to argue these overhead issues effectively.

§ 12.12 How to Judge a Reasonable Profit Margin

Part of a general contractor's claimed cost may include subcontractors' profits which are unreasonably high. There are two ways to look at what could be described as excessive profits. They could be high prospectively or retrospectively. In the first situation, the subcontractor's bid sheets would show an estimate of a very high profit margin or, more likely, reflect unrealistically high figures for the quantities of labor and material or the cost per unit. In that context, an owner required to pay reasonable costs should not have to compensate the contractor for a subcontractor's excessive profits.

§ 12.14 COST-PLUS CONTRACTING   369

Alternatively, the profits might be deemed high only retrospectively. In that situation, a subcontractor's good work or good luck could explain the high profits. As such, it is reasonable for an owner to be required to pay the high profit margin.

The task in some construction litigation is to distinguish the prospectively inflated profits from those resulting from good work or good luck and visible only through hindsight. To do so, the amount of profit and an examination of the circumstances which caused the profit to accrue is the first area of inquiry. The absence of competitive bidding by the contractor will also inform this analysis. Alone, however, such an analysis does not tell the full story. The real essence of the answer lies in the bid sheets and the work-up done by the subcontractor. The prices or quotes obtained and the methodology followed must be examined by thorough questioning on deposition and distilled by expert analysis. Only by these painstaking measures can the fine line between compensable and noncompensable subcontractor profit be drawn.

§ 12.13 Were There Premature Payments?

Under most contracts a subcontractor is not paid prior to the performance of work. When early payments occur, questions should be asked about the propriety of these payments. Discovering this practice is accomplished through an analysis of subcontractor cost records, accounts receivable documentation, and job progress information.

A related practice that can cost the owner significant sums of money is *front end loading,* which occurs when the contractor submits disproportionately high early billings. In effect, the contractor receives payments in excess of costs incurred in the initial phase of the contract. In the later phases of the contract, cost overruns may occur for which the contractor then has no funds available to make payment. Front end loading can be discovered by a comparison of contractor costs incurred with the billings to the owner. In certain circumstances its occurrence can constitute a breach of contract or even fraud or misrepresentation.

§ 12.14 Owner's Involvement in Cost-Plus Contracting

The owner itself must tread a fine line during construction. It must determine whether to become involved in the cost issues as construction proceeds. If it does, it will be in a better position both to prevent a contractor from charging excessive costs and to provide litigation counsel and experts with leads on where excessive costs may have occurred.

On the other hand, an owner's involvement to a degree sufficient to accomplish either of these goals often gives rise to facts on which the contractor will later argue that the owner acquiesced in the costs incurred or that there was a waiver or an estoppel of the claim for overcharges.

There is no easy answer to this dilemma. Most useful is a clear statement at the outset of the owner's role and careful documentation of the limits on its involvement as construction progresses. In addition, many construction contracts provide an alternative to direct owner involvement through a provision enabling the owner to require the contractor to certify its costs. Such provisions can be used to manage and control these issues before problems proliferate. Ongoing investigation without the litigation or arbitration process is therefore available and frequently desirable. As in medicine, an ounce of prevention is often worth a pound of cure.

§ 12.15 Nature and Extent of the Surety's Obligations

An owner's cost claims against the contractor may face the obstacle of a contractor's insolvency. If that is a potential, it is necessary to involve the surety in the discussions about excessive costs. Under a typical performance bond, a surety obligates itself to perform the general contractor's duties under the construction contract. In most contexts this will require the surety to finish construction upon the default of the contractor, but that is not the limit of the surety's obligation. Particularly when the contract is incorporated into the bond (as most are), the construction contract and the performance bond are construed together to determine the obligations of the surety.[7] Thus, the surety must insure that the general contractor performs all the provisions of the bonded contract. That may include keeping the job free of mechanic's liens as well as discharging other contractual obligations of the general contractor. In sum, absent statutory or contractual restrictions, the liability of the surety is coextensive with that of its principal.[8]

Therefore, consideration should be given to pursuing the surety under the performance bond for overcharge claims. This would include a claim that the contractor has received too much for the contract work or extras as well as a claim for the failure of the contractor to pass along contractually required savings for either reductions in scope, incomplete work, or lower-than-projected costs. In short, the surety may well be the entity which can make pursuit of the issues described in this chapter worthwhile.

---

[7] Carrols Equities Corp. v. Villnave, 57 A.D.2d 1044, 395 N.Y.S.2d 800 (1977).

[8] General Builders Supply Co. v. MacArthur, 228 Md. 320, 179 A.2d 868 (1962).

CHAPTER 13

CAUSES OF ACTION AND POSSIBLE RECOVERIES

Steven J. Comen, Esquire
Martin E. Stauffer, CPA

Steven J. Comen, partner in the New England-based law firm of Hinckley, Allen, Snyder & Comen, is a graduate of the University of Michigan Law School and a former law clerk for a United States district court judge. He has represented clients in state and federal courts, arbitration proceedings, boards of contract appeals, and before administrative agencies throughout New England and New York, Michigan, and other states, and in Puerto Rico and the Virgin Islands. Mr. Comen has lectured extensively to professional groups of lawyers, contractors, architects, engineers, and others associated with the construction industry. He has spoken for seminars sponsored by *Engineering News Record,* Associated General Contractors, the Engineering Societies of New England, The Construction Industries of Massachusetts, and Continuing Legal Education of New England.

Martin E. Stauffer is a certified public accountant and partner-in-charge of the Construction/Real Estate Group for the Hartford office of Coopers & Lybrand. Mr. Stauffer provides consulting, auditing, and tax services exclusively for construction and real estate clients. He is also Northeast Regional Real Estate Coordinator and member of the National Real Estate Steering Committee for Coopers & Lybrand, and he has served on the American Institute of Certified Public Accountants' Committee on Real Estate Accounting. He is the author of numerous articles that have appeared in local and national publications, and he is a frequent speaker on construction and real estate matters.

CAUSES OF ACTION

§ 13.1 Introduction and Hypothetical Construction Failure

CLAIMS AGAINST THE ARCHITECT

§ 13.2 Breach of Warranties
§ 13.3 —Architect's Defenses
§ 13.4 UCC
§ 13.5 Negligence
§ 13.6 —Architect's Defenses
§ 13.7 Strict Liability
§ 13.8 Application of Liability Rules to Hypothetical

CLAIMS AGAINST THE OWNER

§ 13.9 Breach of Warranties
§ 13.10 —Owner's Defenses
§ 13.11 UCC
§ 13.12 Negligence
§ 13.13 Strict Liability
§ 13.14 —Owner's Defenses
§ 13.15 Application of Liability Rules to Hypothetical

CLAIMS AGAINST THE GENERAL CONTRACTOR

§ 13.16 Breach of Warranties
§ 13.17 —Contractor's Defenses
§ 13.18 UCC
§ 13.19 Negligence
§ 13.20 Strict Liability
§ 13.21 Application of Liability Rules to Hypothetical

CLAIMS AGAINST A MANUFACTURER/SUPPLIER

§ 13.22 Breach of Warranties and UCC
§ 13.23 —Manufacturer/Supplier Defenses
§ 13.24 Negligence
§ 13.25 Strict Liability
§ 13.26 Application of Liability Rules to Hypothetical

CLAIMS AGAINST THE SURETY, INSURER, AND LENDER

§ 13.27 Types of Bonds
§ 13.28 All Risk Policies
§ 13.29 Indemnification Clauses
§ 13.30 Subcontractor's Payment or Performance Bond Construed to Extend Coverage for Defects and Failures
§ 13.31 Claims Against the Lender

§ 13.1 INTRODUCTION

DAMAGES AVAILABLE TO OWNER AS PLAINTIFF

§ 13.32 Damages Generally
§ 13.33 Recovery of the Cost of Corrective Work
§ 13.34 Liquidated Damages Clause
§ 13.35 Recovery of the Cost to Complete
§ 13.36 Recovery of Interest Rate Differential
§ 13.37 Recovery of Land Value
§ 13.38 Recovery for Loss of Permanent Financing Commitment
§ 13.39 Recovery of Lost Revenues
§ 13.40 Recovery for Loss of Use
§ 13.41 Recovery of Additional Supervision Costs
§ 13.42 Lost Profits
§ 13.43 Recovery of Additional Professional Services
§ 13.44 Recovery of Attorneys' Fees
§ 13.45 Recovery of Depreciation
§ 13.46 Recovery for Loss of Professional Reputation

DAMAGES AVAILABLE TO CONTRACTOR AS PLAINTIFF

§ 13.47 Compliance with Contractual Notice Provisions
§ 13.48 Determining Actual Damages
§ 13.49 Recovery of Damages for Defective Plans and Specifications
§ 13.50 —Economic Loss Doctrine
§ 13.51 Recovery Despite No Damages for Delay Clause
§ 13.52 Recovery for Defective Performance Specifications
§ 13.53 Recovery of Consequential Costs and Damages

OTHER CONSIDERATIONS

§ 13.54 Statutes of Limitations and Repose
§ 13.55 Claims Preparation and Settlement Considerations

§ 13.1 Introduction and Hypothetical Construction Failure

The Mianus Bridge collapse, the John Hancock Tower window failure, and the Hartford Coliseum collapse are celebrated cases which come to mind when thinking of construction defect and failure claims. These cases are but a handful from the rapidly growing area of construction litigation in which defect and failure claims play a major role. In recent years construction defect and failure claims have been asserted in a myriad of construction disputes and on projects of every size, kind, and location.

The unprecedented rise in construction defect and failure claims is, in part, a result of the significant expansion of the scope of liability and damages for such claims, as well as the increase in ongoing construction.

In the typical construction defect and failure case, the plaintiff is an owner seeking compensation for additional costs to repair faulty work and for delays in the completion of a project. Other parties, such as the general contractor and subcontractors, may seek relief for extra work and delay. However, with the expanded scope of liability and damage claims, entities asserting such claims are frequently outside the circle of members of the traditional construction team. It is not unusual for buyers, tenants, or even users to seek relief, nor is it unusual for plaintiffs to seek damages for purely economic loss.

The issues involved with construction defect and failure claims are best illustrated by example and, therefore, reference at times may be made in this chapter to the following hypothetical situation. Take a typical, privately funded project—the construction of a commercial office building—for which the plans and specifications require foundation walls to be constructed out of high strength concrete. Halfway through the foundation work, the owner discovers that the concrete in numerous walls fails to achieve its designed compressive strength. The walls must be removed and replaced. As a result, the work of follow-on trades is delayed and the contract date for substantial completion is overrun by six months. The owner wants to sue the general contractor for delay damages and the cost of replacing the concrete. However, the more defendants there are, the more likely it is the case will settle or that a future judgment will be paid. Therefore, the owner also wants to sue the architect, the concrete supplier, and the supplier's cement manufacturer. The follow-on trades want to sue the owner as well as these same defendants for delay damages. Finally, tenants under lease for the building want to sue all these parties for lost profits.

In such scenarios, the same basic issues usually arise and must be addressed by the attorneys prosecuting or defending such claims. These are:

1. Against whom can the aggrieved party initiate an action?
2. What are the theories of liability?
3. What are the theories of defense?
4. What damages are recoverable?

This chapter addresses these issues, enumerating the various causes of action and defenses available, and emphasizing the advantages and disadvantages of each. Also embodied in the chapter is a discussion of recent developments in the case law as well as the current standard form contract language.

CLAIMS AGAINST THE ARCHITECT

§ 13.2 Breach of Warranties

Warranty is defined as an express or implied promise of "indemnity against defects in an article sold."[1] It is unusual to find express warranties relating to construction defects in contracts between the owner and the architect. For example, the current standard form agreement between owner and architect prepared by the American Institute of Architects (AIA), AIA Document B141-1987, contains no such warranties. It is equally unusual for such warranties to be implied,[2] the rationale being the long-held treatment of architects as professionals.[3]

The architect often does, however, make express warranties as to the performance of his work. For example, article 1.1.2 of AIA Document B141-1987 provides in part: "The Architect's services shall be performed as expeditiously as is consistent with professional skill and care and orderly progress of the work." The architect also impliedly warrants that he possesses and exercises "the care of those ordinarily skilled in the business."[4]

In addition to these performance warranties, the architect usually agrees in the contract with the owner to perform specific duties. These may be limited to the preparation of plans and specifications or may include supervision of construction.[5]

The standard for breach of warranty is very similar to that applied in negligence claims (see **§ 13.5**). Proof of such claims is ultimately a question of fact for which expert testimony is extremely important.

---

[1] Ballentine's Law Dictionary 1360 (3d ed. 1969).

[2] *See, e.g.,* Queensbury Union Free School Dist. v. Jim Walter Corp., 91 Misc. 2d 804, 806, 398 N.Y.S.2d 832, 833 (1977) ("in this state no cause of action is known to the law against an architect for a breach of implied warranty"); Sears, Roebuck & Co. v. Enco Assocs., Inc., 43 N.Y.2d 389, 372 N.E.2d 555, 401 N.Y.S.2d 767 (1977) (no action lies for breach of implied warranty on behalf of an owner against an architect); Looker v. Gulf Coast Fair, 203 Ala. 42, 45, 81 So. 832, 835 (1919) ("[a] competent architect, pursuing an independent profession, is not an insurer of the accuracy or perfection of his work"); Donnelly Constr. Co. v. Oberg/Hunt/Gilleland, 139 Ariz. 184, 189, 677 P.2d 1292, 1297 (1984) ("[d]esign professionals, in the absence of an express guarantee, do not 'warrant' that their work will be 'accurate'").

[3] Queensbury Union Free School Dist., 91 Misc. 2d 804, 805, 398 N.Y.S.2d 832, 833 (1977).

[4] Surf Realty Corp. v. Standing, 195 Va. 431, 443, 78 S.E.2d 901, 907 (1953).

[5] AIA B141-1987 art. 2 provides that the architect's "basic services" include preparation of drawings and specifications, inspection of the work, certification of payment to the contractor, rejection of work, review of shop drawings, preparation of change orders, and interpretation of the contract documents.

Damages for breach of warranty (or contract) are generally limited to those which were within the contemplation of the parties at the time the contract was made.[6] The reason for this is that the plaintiff is entitled to the benefit of its bargain and no more.

§ 13.3 —Architect's Defenses

Claims for breach of warranty or contract generally require that the plaintiff be in privity of contract with the breaching party.[7] Accordingly, construction failure claims against the architect are usually limited to those brought by the owner. Therefore, parties not in contract with the architect sometimes attempt to assert the cause of action under a third-party beneficiary theory. In order to be successful, the claimant must show that it was an "intended" rather than an "incidental" beneficiary.[8] However, construction contracts often expressly exclude third parties as intended beneficiaries. For example, AIA Document B141-1987 provides at ¶ 9.7 that "[n]othing . . . in this Agreement shall create a contractual relationship with or a cause of action in favor of a third party against either the Owner or the Architect." This and similar clauses effectively bar the use of breach of warranty claims by third parties against architects.

Remote parties may, of course, initiate a breach of contract action against those with whom they are in privity. The defendant may then implead the next person in the chain of privity, who in turn may implead the next, until the responsible person is made a party. However, each link in the chain of privity must be founded on an independent breach and is subject to any existing contractual defenses.

---

[6] Hadley v. Baxendale, 9 Ex. 341 (1854); *see generally,* 22 Am. Jur. 2d *Damages* § 56; 5 A. Corbin, Corbin on Contracts § 1007 (1965). A consequential damage that is often within the contemplation of the parties at the time the contract is made is damage from delay. Many contracts between owners and general contractors expressly provide for liquidated damages for delay but it is unusual for owner-architect agreements to contain such provisions.

[7] *See, e.g.,* E.C. Ernst, Inc. v. Manhattan Constr. Co., 551 F.2d 1026, 1030 (5th Cir. 1977); *but see* Donnelly Constr. Co. v. Oberg/Hunt/Gilleland, 139 Ariz. 184, 677 P.2d 1292, 1295 (1984) (implied duty to use reasonable skill and care extends to those with whom design professional is not in privity).

[8] *See* Restatement (Second) of Contracts § 302(2) (an incidental beneficiary is a beneficiary who is not an intended beneficiary); *id.* at § 315 (an incidental beneficiary acquires by virtue of the promise no right against the promisor or the promisee); *see also* Renel Constr. Inc. v. Brooklyn Coop. Meat Distrib. Center, 59 A.D.2d 391, 399 N.Y.S.2d 429 (1977) (incidental beneficiary has no right to enforce contract); E.C. Ernst, Inc. v. Manhattan Constr. Co., 551 F.2d 1026, 1031 (5th Cir. 1977) (same).

Exculpatory language in the owner-architect agreement may provide a defense for the architect. The existence of contractual duties relating to inspection and rejection of work could provide fertile ground for liability in defect and failure cases. However, the contract may expressly limit the inspection required of the architect.[9] In such instances it is more difficult to establish an architect's liability for failure to reject defective work.[10]

§ 13.4 UCC

Article 2 of the Uniform Commercial Code (UCC), adopted in one form or another by every state in the United States, provides for specific express and implied warranties in the sale of goods.[11] However, the UCC does not apply to a contract for the rendition of services.[12] It is therefore inapplicable in an action against an architect involving professional services.

§ 13.5 Negligence

Under tort law, architects have a duty to conform their conduct to a standard of reasonable care.[13] The duty is imposed by society to protect persons and their property from physical harm. The violation of this tort duty is sometimes called "professional negligence" and is very similar to the implied performance warranty discussed in § 13.2.[14] To maintain an action for professional negligence requires proof that (1) the architect breached his duty of care, (2) the breach proximately caused an injury

---

[9] AIA B141-1987 ¶ 2.6.5 expressly limits the level of inspection required:

> [T]he Architect shall not be required to make exhaustive or continuous on-site inspections to check the quality or quantity of the work. On the basis of on-site observations as an architect, the Architect shall keep the Owner informed of the progress and quality of the work, and shall endeavor to guard the Owner against defects and deficiencies in the work.

[10] *See, e.g.,* Moundsview Indep. School Dist. v. Buetow & Assoc., Inc., 253 N.W.2d 836, 839 (Minn. 1977) (plaintiff not entitled to clerk-of-works inspection when contract called only for general supervision).

[11] U.C.C. § 2-313 (1977).

[12] *Id.* § 2-102 (1977).

[13] Restatement (Second) of Torts § 324A (1965).

[14] In fact, breach of this warranty has been described as "negligent breach of contract." *See, e.g.,* Keel v. Titan Constr. Corp., 639 P.2d 1228, 1232 (Okla. 1981) (negligent failure to perform contract with reasonable care is a tort as well as a breach of contract).

resulting in actual damages, and (3) the plaintiff and the risk were reasonably foreseeable.[15] As with breach of warranty, proof of professional negligence is a question of fact for the fact finder. In proving such claims, great importance is attached to the use of experts.[16]

A defect or failure claim based on negligence has several advantages over a claim of breach of warranties (or contract). For one thing, a cause of action in negligence is not limited by the privity of contract doctrine. The abolishment of this limitation began in the area of products liability law with Judge Cardozo's decision in *MacPherson v. Buick Motor Co.*,[17] which held that a negligent manufacturer may be liable to a nonpurchaser when physical harm was foreseeable.[18] This rule was adopted throughout most of the country through UCC § 2-318.[19] The *MacPherson* rule was eventually extended to the construction industry in the case of *Inman v. Binghamton Housing Authority*,[20] which held that there was no distinction

---

[15] Donnelly Constr. Co. v. Oberg/Hunt/Gilleland, 139 Ariz. 184, 677 P.2d 1292, 1295 (1984).

[16] *See* Annotation, *Necessity of Expert Testimony to Show Malpractice of Architect*, 3 A.L.R.4th 1023 (1981). *But see* M.J. Womack, Inc. v. House of Representatives, 509 So. 2d 62, 65 (La. Ct. App. 1987) (architect's negligence may on occasion be established without need for expert testimony).

[17] 217 N.Y. 382, 111 N.E. 1050 (1916).

[18] In *MacPherson*, the plaintiff was personally injured by an automobile which he purchased from a dealer. He filed a negligence action against the manufacturer who raised the defense of lack of privity. The New York Court of Appeals held that, when a product is reasonably certain to put life and limb in peril if negligently manufactured and will probably be used by nonpurchasers, the manufacturer is under a duty to use responsible care to prevent such harm.

[19] *See, e.g.,* Mass. Gen. L., ch. 106, § 2-318 (1989), which provides:

> Lack of privity between plaintiff and defendant shall be no defense in any action brought against the manufacturer, seller, lessor or supplier of goods to recover damages for breach of warranty, express or implied, or for negligence, although the plaintiff did not purchase the goods from the defendant if the plaintiff was a person whom the manufacturer, seller, lessor or supplier might reasonably have expected to use, consume or be affected by the goods. The manufacturer, seller, lessor or supplier may not exclude or limit the operation of this section. Failure to give notice shall not bar recovery under this section unless the defendant proves that he was prejudiced thereby. All actions under this section shall be commenced within three years next after the date the injury and damage occurs.

[20] 3 N.Y.2d 137, 143 N.E.2d 895, 897, 164 N.Y.S.2d 699 (1957). In *Inman*, the architect, builder, and owner were sued for negligence by a tenant whose child sustained injuries in falling off a porch. The plaintiff alleged that the architect who designed the apartment house was negligent in failing to place a protective railing around the porch. In rejecting the privity defense raised by the architect and builder, the court declared that "there is no visible reason for any distinction between the liability of one who supplies a chattel and one who erects a structure." The court concluded that "the principle inherent in the *MacPherson* doctrine applies to determine the liability of architects or builders for their handiwork." *Id.* at 899.

§ 13.5 NEGLIGENCE

between the liability of one who supplies a chattel and one who erects a structure. Thus, a cause of action in negligence permits the plaintiff to bring in numerous defendants, whereas a cause of action for breach of contract does not.

A claim of negligence is not limited by exculpatory language in the contract, such as the inspection clause in the current AIA standard owner-architect agreement discussed in § 13.2. A duty of care may arise from the architect's affirmative undertaking or representation regardless of an absence of a contractual duty.[21]

Finally, the scope of recoverable damages may be greater for a negligence claim, because consequential damages are generally recoverable in tort.[22] However, for cases involving purely economic loss, the law is unsettled. Subsequent to the *Inman* case, a number of courts have extended the *MacPherson* rule to situations involving purely economic loss.[23] The

---

[21] The courts have upheld the principle that:

> one who undertakes to act, even though gratuitously, is required to act carefully and with the exercise of due care and will be liable for injuries proximately caused by failure to use such care.

57A Am. Jur. 2d *Negligence* § 112 at 168 (1989) (citing numerous cases). This principle is also set forth in the Restatement (Second) of Torts § 324A, which provides:

> one who undertakes, gratuitously or for consideration, to render services to another which he should recognize as necessary for the protection of the other person's person or things is subject to liability to the other for physical harm resulting from his failure to exercise reasonable care to perform his undertaking, if (a) his failure to exercise such care increases the risk of such harm, or (b) the harm is suffered because of the other's reliance upon the undertaking.

See, e.g., City of Columbus v. Clark-Dietz & Assocs. Eng'rs, 550 F. Supp. 610 (N.D. Miss. 1982), *which* involved the failure of a protective levee surrounding a construction site for a waste water treatment plant. The general contractor asserted, among others, a claim against the architect/engineer for negligent supervision of construction. The court noted that the architect's contract with the owner limited the architect's "duty for supervising construction to an obligation to observe the general progress of the work and not to make continuous and exhaustive inspection." *Id.* Despite this limited contractual obligation which ran to the owner and not to the general contractor, the court held that the architect owed a duty of care to the general contractor: "[W]e hold that Clark-Dietz owned no duty of supervision to [the general contractor] other than to *exercise reasonable care when it provided instructions and test results at the job site." Id.* at 627 (emphasis added). *See also* Shoffer Indus. v. W.B. Lloyd Constr. Co., 42 N.C. App. 259, 257 S.E.2d 50, 55 (1979) (architect's active course of conduct created a duty to the general contractor to exercise reasonable care, even though architect's contract was with owner).

[22] *See* Restatement (Second) of Torts § 917 (1979). However, in some jurisdictions damages are subject to the economic loss rule. See § 13.50.

[23] *See, e.g.,* E.C. Ernst, Inc. v. Manhattan Constr. Co., 551 F.2d 1026 (5th Cir. 1977) (electrical subcontractor entitled to recover delay damages from architect for architect's negligent refusal to approve emergency generator system); Donnelly Constr. Co. v.

extension followed the same development in the area of products liability. One of the first such cases was *A.R. Moyer, Inc. v. Graham*,[24] in which the architect negligently prepared plans and specifications, and the Florida Supreme Court permitted the general contractor to maintain a negligence action against the architect for recovery of delay damages.

However, this extension of liability to include purely economic harm has been criticized.[25] It is viewed as going beyond the purpose of tort law, which is the protection of persons and property from physical harm.[26] For this reason, courts in several jurisdictions have refused to extend negligence liability to architects for purely economic losses.[27]

§ 13.6 —Architect's Defenses

Damages must be reasonably *foreseeable* before they are recoverable under negligence.[28]

A party who pleads a negligence count also exposes itself to a defense of contributory and comparative negligence, which may constitute a partial or total bar to recovery.[29]

Oberg/Hunt/Gilleland, 139 Ariz. 184, 677 P.2d 1292 (1984) (contractor stated cause of action in claiming actual damages against architect for negligent preparation of plans and specifications); Womack, Inc. v. House of Representatives, 509 So. 2d 62 (La. Ct. App. 1987) (contractor entitled to recover delay damages from architect for negligent preparation of plans); Shoffner Indus. v. W.B. Lloyd Constr. Co., 42 N.C. App. 259, 257 S.E.2d 50 (1979) (contractor stated cause of action in seeking damages for additional labor costs resulting from architect's negligent approval of trusses); *see also* Owen v. Dodd, 431 F. Supp. 1239 (N.D. Miss. 1977); City of Columbus v. Clark-Dietz & Assocs. Eng'rs, 550 F. Supp. 610 (N.D. Miss. 1982); Davidson & Jones, Inc. v. County of New Hanover, 41 N.C. App. 661, 255 S.E.2d 580 (1979); Annotation, *Tort Liability of Project Architect for Economic Damages Suffered by Contractor*, 65 A.L.R.3d 249 (1975 & Supp. 1989).

[24] 285 So. 2d 397 (Fla. 1973).

[25] Moorman Mfg. Co. v. National Tank Co., 91 Ill. 2d 69, 435 N.E.2d 443 (1982).

[26] *Id.*, 91 Ill. 2d at 81, 435 N.E.2d at 448.

[27] *See* Bryant Elec. Co. v. City of Fredericksburg, 762 F.2d 1192 (4th Cir. 1985) (contractor not entitled to recover delay damages against architect for negligent design and supervision); Bates & Rogers Constr. Corp. v. North Shore Sanitary Dist., 128 Ill. App. 3d 962, 471 N.E.2d 915 (1984) (contractor could not recover economic damages for alleged negligence of architect/engineer in design and administration of construction). Other courts have refused to extend negligence liability to include purely economic harm on the basis that the architect's duties under its contract were limited. *See, e.g.*, Raymey Constr. Co. v. Apache Tribe, 673 F.2d 315 (10th Cir. 1982).

[28] 57 Am. Jur. 2d *Negligence* § 136 (1989).

[29] *See, e.g.*, Columbus v. Clark-Dietz & Assocs. Eng'rs, 550 F. Supp. 610, 627 (N.D. Miss. 1982) (applying comparative negligence statute to owner and architect).

§ 13.7 Strict Liability

Limitations periods for negligence actions are generally much shorter than those for contract actions. The former are usually 3–5 years,[30] while the latter are 10 years or more.[31]

§ 13.7 Strict Liability

Unlike a claim for negligence or breach of warranties, a claim of strict liability in tort does not require proof that the defendant was at fault. The elimination of the requirement began with *Greeman v. Yuba Power Products, Inc.*,[32] which permitted a user of a defective lathe to recover for personal injuries he suffered when a piece of wood flew out of the machine and struck him.[33] The doctrine of strict products liability has since been adopted in almost every jurisdiction. The elements are set forth in the Restatement (Second) of Torts, which provides:

> (1) One who sells any product in a defective condition unreasonably dangerous to the user or consumer or to his property is subject to liability for physical harm thereby caused to the ultimate user or consumer, or to his property, if
>
>   (a) the seller is engaged in the business of selling such a product, and
>
>   (b) it is expected to and does reach the user or consumer without substantial change in the condition in which it is sold.
>
> (2) The rule stated in Subsection (1) applies although
>
>   (a) the seller has exercised all possible care in the preparation and sale of his product, and
>
>   (b) the user or consumer has not bought the product from or entered into any contractual relation with the seller.[34]

Under § 402A, liability is extended to each person responsible for placing the defective good in the stream of commerce, which can be manufacturers,

---

[30] *See, e.g.,* Ark. Stat. Ann. § 37-237 (1967) (5 years); Mass. Gen. L., ch. 260, § 2A (1989 Supp.) (3 years). Such statutes are usually construed to provide that the actions do not "accrue" until the facts giving rise to them were discovered or, with due diligence, should have been discovered. *See, e.g.,* Gore v. Daniel O'Connell's Sons, Inc., 17 Mass. App. 645, 461 N.E.2d 256 (1984). However, the accrual is usually limited by statutes of repose to relatively short periods. *See, e.g.,* Mass. Gen. L., ch. 260 § 2B (1989 Supp.) (six years after opening to use or substantial completion).

[31] *See, e.g.,* Mass. Gen. L., ch. 260 § 1 (1989 Supp.) (20 years).

[32] 59 Cal. 2d 57, 377 P.2d 897 (1962).

[33] *Id.* at 63, 377 P.2d at 901.

[34] Restatement (Second) of Torts § 402A (1965).

distributors, retailers, and so on. The public policy behind this doctrine is that the industry is in a better position to spread loss resulting from defective products among all consumers.[35]

Although the doctrine is intended for defective "products," it has been applied to defective improvements to real estate. However, the courts have held that the doctrine is inapplicable to architects because they market "services" rather than "goods."[36]

§ 13.8 Application of Liability Rules to Hypothetical

In the hypothetical in § 13.1, only the owner could maintain a claim against the architect for breach of warranty because only the owner can satisfy the requirement of privity. The owner's claim would likely assert that the architect failed to exercise appropriate skill and care in inspecting the placement and curing of the concrete. However, if the owner-architect agreement contained a provision similar to article 2.6.5 of the AIA Document B141-1987, it is unlikely that the architect would be in breach of warranty for failing to reject the concrete unless the poor placement and curing practices occurred during the period of his scheduled inspection. No plaintiff could maintain a claim against the architect under the UCC or in strict liability because the architect provided only services. However, the subcontractors, owner, and tenants could maintain direct claims against the architect for negligence, particularly if the architect took any steps to approve the placement and curing of the concrete. The tenants, however, may have difficulty establishing foreseeability of harm. Moreover, depending on the jurisdiction, the owner and subcontractors may not recover delay damages, nor the tenants lost profits, as these are pure economic damages.

CLAIMS AGAINST THE OWNER

§ 13.9 Breach of Warranties

It is unusual for construction contracts to contain express warranties of the owner. The current AIA standard form contract between owner and contractor, AIA Document A201-1987, contains none, nor does the

---

[35] *See, e.g.,* Little Rock School Dist. v. Matson, Inc., 576 S.W.2d 709 (Ark. 1979).

[36] *See generally,* 57A Am. Jur. 2d *Negligence* § 845 (1989).

current AIA standard form contract between owner and architect, AIA Document B141-1987.

However, certain warranties are implied by law. The main implied warranty is that the plans and specifications are adequate. The seminal case is *United States v. Spearin*,[37] which held that the contractor, who is bound to build according to the plans and specifications, is not responsible for the consequences of defects in those plans and specifications.[38] The owner, in fact, must compensate the contractor for any additional costs caused by the defects.[39]

The owner also impliedly warrants that when it furnishes material and equipment to be used on the construction project, such material and equipment are suitable for their intended purpose.[40] The owner also has a duty to disclose superior knowledge of material matters pertaining to the contract if they are unknown or not readily available to the contractor.[41] Finally, the owner impliedly warrants the quality of the work which it performs itself.

§ 13.10 —Owner's Defenses

As with the architect's warranties (see § 13.2), the warranties of the owner are subject to the requirement of privity of contract. Thus, a claim against the owner for breach of warranty ordinarily may be asserted only by parties with whom the owner has a contract, usually the architect and general contractor.

The owner may also raise as a defense any exculpatory clause contained in its contract with the plaintiff. For example, many owner-contractor agreements contain a no damages for delay clause which insulates the

---

[37] 248 U.S. 132, 136, 39 S. Ct. 59 (1918).

[38] *See also,* J.L. Simmons Co. v. United States, 412 F.2d 1360, 188 Ct. Cl. 684 (1969); J.D. Hedin Constr. Co. v. United States, 347 F.2d 235, 171 Ct. Cl. 70 (1965). *But see* Emerald Forest Util. Dist. v. Simonsen Constr. Co., 679 S.W.2d 51 (Tex. Ct. App. 1984); Town of Urania v. M.P. Dumesnil Constr. Co., 492 So. 2d 888 (La. 1986).

[39] *See, e.g.,* Pine Bluff Hotel Co. v. Monk & Ritchie, 122 Ark. 308, 183 S.W. 761 (1916); Sandy Hites Co. v. State Highway Comm'n, 347 Mo. 954, 149 S.W.2d 828 (1941); Mayor & City Council v. Clark-Dietz & Assocs. Eng'rs, 550 F. Supp. 610, 625 (N.D. Miss. 1982).

[40] Thompson Ramo Woodridge, Inc. v. United States, 361 F.2d 222, 175 Ct. Cl. 527 (1966); Harlan Fuel Co. v. Wiggerton, 203 Ky. 546, 262 S.W. 957 (1924). However, the contractor may incur liability if it knows of the defective quality of the materials and makes no objection before using them. City of New Orleans v. Vicon, Inc., 529 F. Supp. 1234, 1243 (E.D. La. 1982).

[41] Gallegos v. Graff, 32 Colo. App. 213, 508 P.2d 798 (1973); Sponseller v. Meltebeke, 280 Or. 361, 570 P.2d 974 (1977).

owner from liability to the contractor for delay.[42] Most agreements also contain a changes clause which requires the contractor to submit to the owner its claim for additional costs within a specific time period. Such clauses may be used by the owner as a defense to breach of warranty or contract claims.[43]

§ 13.11 UCC

In most circumstances the owner is a buyer, not a seller. Therefore, article 2 of the UCC would be inapplicable. However, even if the owner were to sell the property, the UCC would not be applicable because it is applicable to the sale of goods.[44] The sale of buildings constitutes the sale of realty, to which the UCC is inapplicable.[45]

§ 13.12 Negligence

Like everyone else, owners have a duty to conform their conduct to a standard of reasonable care. Owners, however, traditionally have limited direct involvement in the planning, performance, and supervision of construction projects. Accordingly, occasions when they are directly negligent are relatively few.

§ 13.13 Strict Liability

In *Schiapper v. Levitt & Sons,*[46] strict liability in tort was extended to an owner/developer for defective improvements to real estate. In *Schiapper,* the defendant owner constructed a residential subdivision, acting as architect, contractor, and developer. The plaintiff leased one of the homes built

---

[42] *See, e.g.,* E.C. Ernst, Inc. v. Manhattan Constr. Co., 551 F.2d 1026, 1029 (5th Cir. 1977) (no damages for delay clause).

[43] *See, e.g.,* S&M Constr., Inc. v. City of Columbus, 70 Ohio St. 2d 69, 434 N.E.2d 1349 (1982); Sasso Contracting Co. v. State, 173 N.J. Super. 486, 414 A.2d 603 (1980); Eastern Tunneling Corp. v. Southgate Sanitation Dist., 487 F. Supp. 109 (D. Colo. 1980).

[44] See § **13.22**.

[45] Gable v. Silver, 258 So.2d 11 (Fla. Dist. Ct. App. 1972). *But see,* Pollard v. Saxe & Yolles Dev. Co., 12 Cal. 3d 374, 525 P.2d 88, 115 Cal. Rptr. 648 (1974) (applying implied warranty provisions of UCC art. 2 to new apartment buildings).

[46] 44 N.J. 70, 207 A.2d 314 (1965).

§ 13.13   STRICT LIABILITY

by the defendant. The plaintiff's 16-month-old infant was badly scalded by tap water as a result of the defendant's decision not to install a mixer valve on the hot water heater at the time of construction. In holding the defendant strictly liable, the court analogized the defendant's mass production of homes to the mass production of automobiles and concluded that the defendant was in a better position to bear the loss than the plaintiff. Strict liability attached even though the defect occurred in a fixture of the house and not in the structure of the house itself.[47]

The *Schiapper* doctrine has been extended in subsequent cases. In *Kriegler v. Eichler Homes, Inc.*,[48] the plaintiff bought a home from the original purchaser six years after the construction was completed. He sued the defendant, a mass producer of homes, in strict liability for damages resulting from a defective heating system installed when the home was constructed. The California Appeals Court held that the *Schiapper* doctrine was applicable. In so holding, the court made strict tort liability available to a remote buyer and for a defect in an improvement to real estate which caused only economic damages.[49]

In *Patitucci v. Drelich*,[50] the defendant, a "professional builder," developed a tract of land with 12 homes and sold one to the plaintiff. The plaintiff discovered that the sewage system was of inadequate capacity and sued the defendant under strict liability. The New Jersey court held that the doctrine was applicable even though the defendant was not a mass-producer of homes.[51]

There appear to be no cases applying the doctrine of strict liability to sellers of nonresidential construction. The reason may be that commercial and public construction projects are less apt to be mass-produced. However, if the rationale of *Patitucci* were followed, the doctrine of strict liability could extend to commercial area of the industry as well.

---

[47] *Id.* at 96-97, 207 A.2d at 329-30.

[48] 269 Cal. App. 2d 224, 74 Cal. Rptr. 749 (1969). Other courts have held that the doctrine is inapplicable to remote purchasers against the original builder. *See, e.g.,* Wright v. Creative Corp., 30 Colo. App. 575, 498 P.2d 1179 (1972).

[49] *See also* Stuart v. Crestview Mut. Water Co., 34 Cal. App. 3d 802, 110 Cal. Rptr. 543 (1973) (strict liability applied to developer of defective water system for property damage caused by system); Avner v. Longridge Estates, 272 Cal. App. 2d 607, 77 Cal. Rptr. 633 (1969) (strict liability applied to tract developer for defective work in lot manufacturing).

[50] 153 N.J. Super. 177, 379 A.2d 297 (1977).

[51] 153 N.J. Super. at 179-80, 379 A.2d at 298; *see also* Rawlings v. D.M. Oliver Co., 97 Cal. App. 3d 890, 159 Cal. Rptr. 119 (1977) (manufacturer may be held strictly liable in tort for defects in product even though it is not a mass-produced product); *but see* Oliver v. Superior Court (Regis Builders), 211 Cal. App. 3d 86, 259 Cal. Rptr. 160 (1989) (strict doctrine limited to mass-produced homes).

Strict liability may also be applicable to an owner/developer when a building material or component is defective and causes injury to a user or consumer.[52]

§ 13.14 —Owner's Defenses

The main defense to a claim for strict liability is that the cause of action is inapplicable outside of nonresidential construction. The analogy to mass production of products simply does not hold when one constructs and sells a one-of-a-kind structure.

Recovery in a strict liability action also may be barred when the claimant discovers the defect, is aware of the danger, and nevertheless proceeds unreasonably to make use of the product in the face of such danger, because in that circumstance the claimant has assumed the risk of the defect.[53]

Strict liability may be applicable only when the product is defective for its intended use. Therefore, if the plaintiff misuses the product, recovery in strict liability may be barred.[54] Further, if the product is expected to or does reach the user or consumer with substantial change made by the owner, strict liability is inapplicable[55] against the owner.

§ 13.15 Application of Liability Rules to Hypothetical

In the hypothetical in § 13.1, the owner's implied warranties do not extend to the subcontractors or tenants because of the lack of privity. If the specifications for concrete were defective, the contractor would have a viable counterclaim or cross-claim against the owner for breach of implied warranty. However, such claim may be effectively barred by a no damages for delay or notice provision in the prime contract. None of the plaintiffs could maintain a claim against the owner under the UCC because the owner is not a seller. A negligence claim would also be inappropriate unless the owner somehow participated in the placement and curing of the concrete. However, under the *Patitucci* case (see § 13.13), the tenants could maintain a strict liability claim against the owner, but only if their units incurred some physical harm.

---

[52] *See, e.g.,* State Stove Mfg. Co. v. Hodges, 189 So. 2d 113 (Miss. 1986) (strict liability applicable to manufacturer of product and builder who constructs and sells house with the product in it).

[53] Am. Jur. 2d *Products Liability* §§ 954–59 (1984).

[54] *Id.* at § 543.

[55] *Id.* at § 551.

CLAIMS AGAINST THE GENERAL CONTRACTOR

§ 13.16 Breach of Warranties

Unlike the architect and the owner, the general contractor often provides express warranties in its contract for construction. The current AIA standard owner-contractor agreement, AIA Document A201-1987, provides the following warranty clause:

> 3.5 WARRANTY
>
> 3.5.1 The Contractor warrants to the Owner and Architect that materials and equipment furnished under the Contract will be of good quality and new unless otherwise required or permitted by the Contract Documents, that the Work will be free from defects not inherent in the quality required or permitted, and that the Work will conform with the requirements of the Contract Documents. Work not conforming to these requirements, including substitutions not properly approved and authorized, may be considered defective. The Contractor's warranty excludes remedy for damage or defect caused by abuse, modifications not executed by the Contractor, improper or insufficient maintenance, improper operation, or normal wear and tear under normal usage. If required by the Architect, the Contractor shall furnish satisfactory evidence as to the kind and quality of materials and equipment.

This clause provides significant protection to the owner. It warrants material and equipment as well as labor provided by the contractor. The material and equipment must not only be new, it must be of "good quality" unless otherwise permitted by the contract documents. Likewise, the work must not be "defective" unless such defects are inherent in the quality permitted. This language would appear to impose liability on the contractor for poor quality material and equipment, even if the poor quality could not be discovered with due diligence, and for defective work, even if the work was performed with reasonable care.[56]

The courts normally hold that the contractor also makes certain implied warranties, unless there are express provisions to the contrary. The contractor impliedly warrants that its work will be done in a skillful manner, that the material and equipment will be free from defects, and that the work will be in accordance with the plans and specifications.[57] As can

---

[56] *But see* E.D. Wesley Co. v. City of New Berlin, 62 Wis. 2d 668, 215 N.W.2d 657 (1974) (contractor does not absolutely guarantee its work).

[57] *See, e.g.,* Petersen v. Hubschman Constr. Co., 76 Ill. 2d 31, 389 N.E.2d 1154 (1979); George v. Goldman, 333 Mass. 496, 131 N.E.2d 772 (D. Colo. 1956).

be seen, these are not very different from the express warranties in the current AIA standard owner-contractor agreement. However, the implied warranty that the material and equipment will be free from defects is generally inapplicable to latent defects if the contractor had no knowledge of the defect, acted in good faith, and exercised reasonable care and skill.[58]

§ 13.17 —Contractor's Defenses

If the contractor fully complies with the plans and specifications provided by the owner and the defect or failure is a result of inadequacy in those plans and specifications, the contractor is not liable.[59] However, when the contractor does not comply with the plans and specifications provided by the owner, the contractor may be held liable, notwithstanding the fact that the plans and specifications were defective.[60]

Subsequent purchasers of buildings generally cannot recover from the contractor on the basis of breach of warranty because privity of contract does not exist[61] between these parties.

The owner's acceptance of the work ordinarily constitutes a waiver of defective work, material, or equipment when such defects are patent.[62] *Acceptance* may be express or it may be implied from the conduct of the owner.[63] The owner's failure to object to defects within a reasonable time after it knows or has reason to know of any defects may constitute an acceptance.[64] Final payment with knowledge of the defects may also constitute an acceptance.[65] However, the owner's occupancy or use of a building

---

[58] *See, e.g.,* Wisconsin Red Pressed Brick Co. v. Hood, 67 Minn. 329, 69 N.W. 1091 (1897); Whaley v. Milton Constr. & Supply Co., 241 S.W.2d 23 (Mo. App. 1951). *See generally* Annotation, *Liability of Builder or Subcontractor for Insufficiency of Building Resulting from Latent Defects in Materials Used,* 61 A.L.R.3d 792 (1975).

[59] Annotation, *Construction Contractor's Liability to Contractee for Defects or Insufficiency of Work Attributable to the Latter's Plans and Specifications,* 6 A.L.R.3d 1394 (1966).

[60] W.H. Lyman Constr. Co. v. Gurnee, 131 Ill. App. 3d 87, 475 N.E.2d 273 (1985); Burke County Schools Bd. of Educ. v. Juno Constr. Corp., 50 N.C. App. 238, 273 S.E.2d 504 (1981).

[61] Foxcraft Townhome Owners Assoc. v. Hoffman Rosner Corp., 105 Ill. App. 3d 951, 435 N.E.2d 210 (1985).

[62] Appeal of Federal Constr. Co., ASBCA No. 17,599, 73-1 B.C.A. (CCH) ¶10,003 (1973).

[63] Newport v. Hedges, 358 S.W.2d 441, 447 (Mo. Ct. App. 1962).

[64] Weinberg v. Wilensky, 26 N.J. Super. 301, 97 A.2d 707 (___ 1953).

[65] Guschl v. Schmidt, 266 Wis. 410, 63 N.W.2d 759 (1954).

or structure alone does not constitute an acceptance,[66] nor does partial payment.[67]

Claims for latent defects are not barred by acceptance, however.[68] Latent defects are those which are not discoverable by the exercise of reasonable care and diligence and of which the owner was ignorant at the time of the acceptance.[69] The owner's right to inspection does not imply a duty to inspect for defects.[70] Contracts between owners and contractors often provide a guarantee clause, such as AIA Document A201-1987 ¶ 12.2.2, in which the contractor agrees to correct any defects in materials or workmanship within one year of substantial completion of the job. Under most contracts such provisions do not constitute exclusive remedies.[71]

The contractor also is not responsible for defects arising from performing the work in the manner directed by the owner.[72] Further, when the contractor is required to perform work using materials furnished by the owner, the contractor is not liable for defects contained in them,[73] unless it knows of such defects and makes no objection before using them.[74]

A contractor is not liable for defects arising from changes made with the owner's consent.[75]

When a contractor performs work independently of other contractors, and is not responsible for the general direction of the work, it is not liable for defects caused by the other contractors. The contractor is not liable for defects caused by a subcontractor it did not hire.[76]

§ 13.18 UCC

The courts have generally held that the UCC is inapplicable to contracts for construction because they involve the sale of services rather than

---

[66] Kuhike Constr. Co. v. Mobley, Inc., 159 Ga. App. 777, 285 S.E.2d 236 (1981).

[67] McDonald v. Supple, 96 Or. 486, 190 P. 315 (1920).

[68] Kaminer Constr. Corp. v. United States, 488 F.2d 980, 203 Cl. Ct. 182 (1973).

[69] Maloney v. Oak Builders, Inc., 224 So. 2d 161 (4th Cir. 1968).

[70] Kaminer Constr. Corp. v. United States, 488 F.2d 980, 203 Cl. Ct. 182 (1973).

[71] *See, e.g.,* Bender-Miller Co. v. Thomwood Farms, Inc., 211 Va. 585, 179 S.E.2d 636 (1971).

[72] Gamm Constr. Co. v. Townsend, 32 Ill. App. 3d 848, 336 N.E.2d 592 (1975).

[73] Wood-Hopkins Contracting Co. v. Masonry Contractors, Inc., 235 So. 2d 548 (Fla. 1970); Murphy Corp. v. Petrochem Maintenance, Inc., 180 So. 2d 716 (La. 1965).

[74] City of New Orleans v. Vicon, Inc., 529 F. Supp. 1234 (E.D. La. 1982).

[75] Cueroni v. Coburnville Garage, 315 Mass. 135, 52 N.E.2d 16 (1943).

[76] Beacon Plaza Shopping Center, Inc. v. Tri-Cities Constr. & Supply Co., 2 Mich. App. 415, 140 N.W.2d 531 (1987).

goods.[77] When there is a mixed contract for the sale of goods and construction services, the UCC is inapplicable if the predominant purpose is to render construction services.[78] Purpose is determined from the intent of the parties as shown by the terms of the contract. For example, in *Nitrin, Inc. v. Bethlehem Steel Corp.*,[79] a contract for the construction of an ammonia plant was found to be a contract predominantly for services because the parties were referred to in the contract as "owner" and "contractor" rather than "buyer" and "seller," the contractor never had title to any components for the plant, and ultimate control for purchasing the components lay with the owner.

The UCC is inapplicable to most mixed contracts involving construction and services.[80] However, when the construction involves mere installation of preconstructed or manufactured goods, the UCC is usually applicable.[81]

§ 13.19 Negligence

Contractors typically perform a substantial quantity of work themselves on construction projects. They are liable to all persons who foreseeably may be injured by their work if that work was negligently performed. Thus, direct actions against contractors have been allowed by owners,[82] by subcontractors, and even by mere bystanders.[83]

As with other negligence claims, defenses of lack of foreseeability and contributory negligence are available to the contractor.

---

[77] Aranao Constr. Co. v. Success Roofing, Inc., 46 Wash. App. 314, 730 P.2d 720 (1986).

[78] *See, e.g.,* Ranger Constr. Co. v. Dixie Floor Co., 433 F. Supp. 442 (D.S.C. 1977).

[79] 35 Ill. App. 3d 577, 342 N.E.2d 65 (1976).

[80] *See, e.g.,* Lincoln Pulp & Paper Co. v. Dravo Corp., 436 F. Supp. 262 (D. Me. 1977) (UCC inapplicable to engineering and construction contract); Allied Indus. Serv. Corp. v. Kasle Iron & Metals, Inc., 62 Ohio App. 2d 144, 405 N.E.2d 307 (1977) (UCC inapplicable to contract for design and installation of pollution control system); Schenectady Steel Co. v. Bruno Trimpoli Gen. Constr. Co., 43 A.D.2d 234, 350 N.Y.S.2d 920 (1974) (UCC inapplicable to contract to furnish and erect structural steel).

[81] Meyers v. Henderson Constr. Co., 147 N.J. Super. 77, 370 A.2d 547 (1977) (UCC applicable to contract for purchase and installation of prefabricated overhead doors); Lipson v. Hawthorne Indus., Inc., 148 Ga. App. 751, 252 S.E.2d 639 (1979) (UCC applicable to contract for sale and installation of carpeting); Bonebrake v. Cox, 499 F.2d 951 (8th Cir. 1974) (UCC applicable to sale and installation of bowling alley equipment).

[82] Hanna v. Fletcher, 231 F.2d 469 (D.C. Cir. 1958).

[83] Russell v. Arthur Whitcomb, Inc., 100 N.H. 171, 121 A.2d 781 (1956) (adjoining property owner stated cause of action for damages to his property resulting from contractor's negligent failure to properly fill excavation in street).

§ 13.20 Strict Liability

Generally, a contractor cannot be strictly liable in tort for building a defective structure because the contractor ordinarily is not engaged in the business of selling the completed structure.[84] That is the role of the owner/developer. However, liability for defective components or material is a different story. When the contractor supplies the material or components for the project, the contractor may be the one who is engaged in the business of selling such items.[85] In at least one case, strict liability was applied to a contractor who installed, but did not sell, a building component.[86] Under UCC § 402A, the contractor would be liable to the ultimate user or consumer for physical harm caused by a defective product, even if the defect was caused by the manufacturer or supplier.

The same defenses to strict liability as discussed in § 13.14 are applicable for the contractor. However, as to claims for defective material or components, the defense of substantial change is particularly applicable. If there is no evidence that the product or material was defective when it left the control of the contractor, the contractor would not be liable.[87]

§ 13.21 Application of Liability Rules to Hypothetical

In the hypothetical in § 13.1, the owner could maintain a claim against the contractor for breach of express and implied warranties for using poor quality or defective concrete and for failing to perform in a skillful manner. However, if the defects were a result of faulty specifications or if the concrete deficiencies were patent and accepted by the owner, the contractor would have valid defenses. Claims against the contractor under the UCC would be inappropriate because the contract is predominantly for construction services. All plaintiffs could maintain direct claims against the contractor for negligence. Again, the tenants may have difficulty establishing foreseeability and, depending on the jurisdiction, the plaintiffs

---

[84] *See* 63 Am. Jur. 2d *Products Liability* § 569 (strict liability does not apply to one who merely constructs items, but is not engaged in the business of selling them).

[85] Niffenegger v. Lakeland Constr. Co., 95 Ill. App. 3d 420, 420 N.E.2d 262 (1981) (paving subcontractor engaged in the business of selling asphalt).

[86] *See* Brannon v. Southern Ill. Hosp. Corp., 69 Ill. App. 3d 1, 386 N.E.2d 1126 (1978) (installer of dumbwaiter). *But see* Neumann v. Davis Water & Waste, Inc., 433 So. 2d 559 (Fla. 1983) (strict liability not available against installer of sewage treatment tank).

[87] Miltz v. Boroughts-Shelving, Div. of Lear Siegler, Inc., 203 N.J. Super. 451, 497 A.2d 516 (1985) (contractor and subcontractor not strictly liable to employee or owner of building for injury from defective stairs when no evidence showed that stairs were defective when they left control of defendants).

may not recover purely economic damages. Finally, if the tenants suffered any physical harm, they might recover against the contractor in strict liability for selling a defective component (concrete).

CLAIMS AGAINST A MANUFACTURER/SUPPLIER

§ 13.22 Breach of Warranties and UCC

Manufacturers and suppliers generally sell construction material and equipment as opposed to performing construction work. Warranties as to such products are governed by the UCC, and the UCC also preempts common-law implied warranties.[88]

Article 2 of the UCC provides for express and implied warranties in the "transaction of goods." *Transactions* under the code include sales of goods, that is, the passing of title from a seller to a buyer for a price.[89] *Goods* is defined by the code as "all things (including specially manufactured goods) which are movable at the time of identification to the contract for sale."[90]

In determining which things constitute "goods" under the UCC, courts have focused on their movability. For example, the UCC is inapplicable to goods after they become fixtures to the real property. However, the UCC may be applicable to goods prior to the time they become fixtures.[91] Examples of things which qualify as goods are building materials,[92] furniture and carpeting,[93] industrial machinery,[94] heating systems,[95] and pumping stations.[96] When the contract calls for the rendition of services as well as the sale of goods, the UCC is inapplicable if the rendition of services predominates.[97]

---

[88] *See, e.g.,* R. Clinton Constr. Co. v. Bryant & Reaves, Inc., 442 F. Supp. 838 (N.D. Miss. 1977).

[89] U.C.C. § 2-106(1). It also includes the trade, barter, or exchange of goods. U.C.C. § 2-304(1). It does not include the leasing of goods. Security Life Ins. Co. v. Executive Car Leasing Co., 443 S.W.2d 915 (Tex. 1968).

[90] U.C.C. § 2-105(1) (1977).

[91] Thomas v. Bove, 687 P.2d 534 (Colo. 1984) (UCC applicable to sale of heating system for home where sale occurred prior to installation).

[92] Tracor, Inc. v. Austin Supply & Drywall Co., 484 S.W.2d 446, (Tex. 1972).

[93] Sturgis Co. v. H.D. Baker Co., 11 Wash. App. 597, 524 P.2d 413 (1974).

[94] Greco v. Bucciconi, 283 F. Supp. 978 (W.D. Pa. 1967).

[95] Thomas v. Bove, 687 P.2d 534 (Colo. 1984).

[96] USEMCO, Inc. v. Marbro Co., 60 Md. App. 351, 483 A.2d 88 (1984).

[97] See **§ 13.18**.

§ 13.22 BREACH OF WARRANTIES AND UCC

The express warranties provided by the UCC are set forth in § 2-313, entitled Express Warranties by Affirmation, Promise, Description, Sample. It requires the seller to conform its goods to the standard of prior affirmation, promise, description, or sample.

UCC § 2-314 creates an implied warranty of merchantability. It requires that the goods conform to normal commercial standards. To be applicable, the seller must be a merchant with respect to goods of that kind.[98]

Section 2-315 creates an implied warranty that the goods will be fit for a particular purpose. It is different from the implied warranty of merchantability in that: (1) it warrants fitness for a specific rather than a general purpose,[99] (2) it arises only when the seller has reason to know of the buyer's particular purpose and the buyer relies on the seller's judgment to select suitable goods,[100] and (3) the sale need not be made by a merchant.[101]

A seller is liable for its own warranties as well as those of any manufacturer which it adopts. The UCC does not require vertical privity, that is, a direct contractual relationship between injured buyer and the seller or manufacturer who provides the warranty.[102] But the UCC does require certain horizontal privity, in that it places limitations on which persons other than a buyer may sue for breach of warranty. In most jurisdictions, warranty protection is limited to buyers and to certain of their household members and guests.[103]

---

[98] U.C.C. § 2-314(1) (1977).

[99] Schenk v. Pelkey, 176 Conn. 245, 405 A.2d 665 (1978).

[100] Lathrop v. Tyrrell, 128 Ill. App. 3d 1067, 471 N.E.2d 1049 (1984).

[101] Thompson Farms, Inc. v. Corno Feed Prods., 173 Ind. App. 682, 366 N.E.2d 3 (1977).

[102] Thus, the ultimate buyer may have a direct claim against the seller or manufacturer whose warranty was breached notwithstanding the fact that there were intermediate sellers. *See, e.g.,* Henningsen v. Bloomfield Motors, Inc., 32 N.J. 358, 161 A.2d 69 (1960).

[103] The UCC has three optional forms for adoption by the states pertaining to horizontal privity. The first option provides that a warranty extends to any natural person who is in the family or household of the buyer or who is a guest in its home if it is reasonable to expect that such person may use, consume, or be affected by the goods and who is injured in its person by breach of the warranty. U.C.C. § 2-318, Alternate A. The second option provides that a warranty extends to any natural person who may reasonably be expected to use, consume, or be affected by the goods and who is injured in its person by breach of the warranty. *Id.,* Alternate B. The third option provides that a warranty extends to any person who may reasonably be expected to use, consume, or be affected by the goods and who is injured (not necessarily in its person) by the breach of warranty. *Id.,* Alternate C. The majority of states has adopted the first option.

394 CAUSES OF ACTION

The types of losses recoverable for breach of warranty include personal injury[104] and may extend to property damage[105] and economic loss when such loss is within the contemplation of the parties.[106]

§ 13.23 —Manufacturer/Supplier Defenses

Under the UCC, sellers may limit their warranties by inserting disclaimer clauses in their contracts with the buyer.[107] Such clauses are disfavored and are strictly construed.[108] To be valid, disclaimers must strictly follow the UCC requirements of conspicuousness.[109]

If the plaintiff is too remote from the purchaser, the plaintiff's claim will be barred by the lack of horizontal privity.[110]

A statute of limitations defense is possible, because under the UCC, an action for breach of warranty must be commenced within four years after the cause of action accrues.[111]

§ 13.24 Negligence

Manufacturers and suppliers owe a duty of reasonable care in the manufacture and handling of their products. The duty of the manufacturer may include inspection and testing of the product,[112] the design of the product,[113] and the giving of warnings concerning product hazards.[114] The duties are also applicable to those who assemble or process products manufactured by another.[115] The nonmanufacturing seller has a duty of ordinary care to see that the product is not in such a condition that in its

---

[104] U.C.C. § 2-316 (1977). *See, e.g.,* Roberts v. General Dynamics, Convair Corp., 425 F. Supp. 688 (S.D. Tex. 1977) (airplane passenger could maintain action against plane manufacturer for personal injuries).

[105] Continental Cooper & Steel Indus., Inc. v. E.C. Red Cornelius, Inc., 104 So. 2d 50 (Fla. 1958) (defective underground cable); Spence v. Three Rivers Builders & Masonry Supply, Inc., 353 Mich. 120, 90 N.W.2d 873 (1958) (cracked cinder blocks used to construct cottage).

[106] Santor v. A.M. Karagheusian, Inc., 44 N.J. 52, 207 A.2d 305 (1965) (defective carpet).

[107] U.C.C. § 2-316 (1977).

[108] Singer Co. v. E.I. Du Pont de Nemours & Co., 579 F.2d 433 (8th Cir. 1978).

[109] U.C.C. § 2-316(2) (1977).

[110] See § 13.22.

[111] U.C.C. § 2-725(a) (1977).

[112] 63 Am. Jur. 2d *Products Liability* §§ 305–12 (1984).

[113] *Id.* at §§ 359–74.

[114] *Id.* at §§ 324–52.

[115] *Id.* at § 305.

normal use it would become dangerous.[116] Generally, these duties are owed to anyone who might be expected to use the product or be endangered by its probable use.[117] As with other kinds of negligence actions, a successful plaintiff in a products liability action based on negligence is entitled to compensation for his actual injury.[118] However, there is a split of authority over the recoverability of economic damages. The majority of jurisdictions that have considered the issue has not permitted recovery of purely economic loss, absent personal injury or property damage.[119] A principal rationale for denying recovery of purely economic damages is that a malfunctioning product causing only a loss of income violates only a contractual expectation.[120]

The defenses normally available in negligence actions, such as unforeseeable plaintiff,[121] lack of proximate cause,[122] and misuse of the product,[123] are available to the manufacturer or supplier.

§ 13.25 Strict Liability

Manufacturers and suppliers may be strictly liable for injury resulting from defective products. In fact, products liability is the area in which strict liability has undergone most of its development. The elements of a claim for strict liability for defective products are set forth in the Restatement (Second) of Torts § 402A which was discussed in § 13.7 in connection with defective construction. When these elements have been established, strict liability of the manufacturer or supplier for a defective product will carry through to the ultimate user or consumer despite a lack of privity and defendants' reasonable care.[124] The scope of

---

[116] *Id.* at §§ 313–23.

[117] *Id.* at §§ 335–74 (1984).

[118] 22 Am. Jur. 2d *Damages* § 80 (1988).

[119] *See, e.g.,* Hart Eng'g Co. v. FMC Corp., 593 F. Supp. 1471 (D.R.I. 1984) (in an action by engineering company against supplier of machinery used in waste water treatment plant, costs for removal, shipment, and reinstallation of defective machinery were economic losses for which negligence or strict liability were available remedies); *see also* Jones & Laughlin Steel Corp. v. Johns Manville Sales Corp., 626 F.2d 280 (3d Cir. 1980) (defective fireproofing materials). *Contra,* Pisano v. American Leasing, 146 Cal. App. 194, 194 Cal. Rptr. 77 (1983) (defective sander, recovery of loss profits and business opportunities allowed).

[120] Berg v. General Motors Corp., 87 Wash. 2d 584, 555 P.2d 818 (1976).

[121] Drackett Prods. Co. v. Blue, 152 So. 2d 463 (Fla. 1963).

[122] Walker v. Wittenberg, Deloney & Davidson, Inc., 241 Ark. 525, 412 S.W.2d 626 (1967) (falling concrete slab, and misuse of the product).

[123] 63 Am. Jur. 2d *Products Liability* § 380 (1984).

[124] *See, e.g.,* Vaughn v. Edward M. Chadbourne, Inc., 462 So. 2d 512 (Fla. 1985) (manufacturer of defective asphalt strictly liable to plaintiffs injured in auto accident).

recoverable damages is similar to that of products liability claims based on negligence.

The various defenses to strict liability available to the owner and contractor discussed in §§ 13.14 and 13.20 may also be available to manufacturers and suppliers.

§ 13.26 Application of Liability Rules to Hypothetical

In the hypothetical in § 13.1, the owner could maintain a claim in negligence against the cement manufacturer and concrete supplier. To succeed, it must prove that they breached a duty of care. The tenants, contractor, and subcontractors probably could not maintain such a claim because their damages are purely economic. Moreover, the tenants are probably unforeseeable plaintiffs.

Under the UCC, the owner and contractor, as buyers, could maintain claims against the manufacturer and supplier for breach of implied (and perhaps express, depending on representations made) warranties. Unlike claims brought under negligence, no proof of breach of the duty of care is required. The tenants and subcontractors could not maintain claims under the UCC because of lack of horizontal privity.

Delay damages also could be recoverable by the owner and contractor if within the contemplation of the parties.

As a user or consumer, the owner can maintain an action against the manufacturer or supplier under strict liability principles. Again, no proof of fault or privity are required. The contractor is also a user or consumer, but its damages are purely economic, which would not be recoverable under strict liability. The tenants and subcontractors are neither users nor consumers.

CLAIMS AGAINST THE SURETY, INSURER, AND LENDER

§ 13.27 Types of Bonds

Normally on any sizeable construction project, the contractor would be required to obtain three types of bonding: a bid bond, a performance bond, and a payment bond. Each bond may be required specifically by statute, local regulation, or the contract between the parties. It is in the interest of the owner to obtain such bonds from the contractor and its surety as a means to protect against a wide variety of contractor defaults. In the realm of defect and failure claims, an owner's ability to recover against a surety is usually limited by the applicable provisions and specific exclusions of the bond issued to the contractor. However, the surety would most probably not be liable for the conduct of the contractor unless the liability arose from the express terms of any surety's bond or the existence of language in the form of an indemnification agreement in the contract for construction which, by reference, is incorporated into the surety's obligation. The bid, performance, or payment bond is, therefore, not intended to protect the owner from loss resulting from personal injury or property loss claims; they only ensure a type of performance by the contractor, for which the surety will become liable in the event of the contractor's default.

§ 13.28 All Risk Policies

The contract for construction rather than the payment or performance bond addresses the means by which protection is to be afforded to the owner against any claim for personal injury or property damage occurring as a result of a defect or failure in the construction or construction process. A performance or payment bond would be the wrong vehicle for such coverage, but instead the contract typically indicates that either the owner or the contractor is to procure what is known in the industry as an all risk insurance policy to protect itself from claims for damages because of bodily injury, personal injury, or property damage, even that flowing from its own employees.[125] Because of such insurance provisions and the all risk nature of the policy, physical loss or damage to the property is usually covered. At times, the contract for construction also requires that the contractor maintain the liability insurance in the name of the owner as an additional insured. Further, in certain municipalities and jurisdictions of the United States, and because of the self-insured nature of such municipalities and jurisdictions, applicable local, municipal, and state regulations may require that the contract specify that the owner be named as an additional insured in the all risk policy.

Unfortunately, certain risks may specifically be excluded from the all risk policy and, therefore, injury or property damage resulting from defective design or work may not be covered. The specific list of exclusions from coverage in the all risk policy should be carefully reviewed to ensure that adequate coverage exists for the type and scope of project sought to be constructed.

---

[125] AIA A201-1986, art. 11.

§ 13.29 Indemnification Clauses

A careful review of the contract for construction must also be made to determine whether it contains an indemnification clause by which the contractor is obligated to hold the owner and architect harmless from all liability except when the architect's fault, as attributable to the owner, is the primary cause of damage.[126] If there is an indemnification clause and if personal injury or property damages are excluded from the all risk insurance policy, the contractor will have to prove the architect was the professional at fault in causing the construction defect or failure, through professional negligence, in order to preclude the possibility of having to indemnify the owner for the loss.

The purpose of the indemnification provision is of course to shift the risk of personal injury or property loss from the owner to the contractor because the contractor is presumed to have control of the work. If a catastrophic failure occurs either during the course of construction or after project completion, and either an employee of the contractor or a member of the public is injured, a claim will be brought against the contractor. The owner and the architect will most probably be made parties to such litigation as well. Therefore, the indemnification provision is an attempt to negate any potential recovery from the owner or the architect.

However, the indemnification provision is not unlimited and usually only covers injury or damage arising from the negligent acts or omissions of the contractor. Also, the indemnification provision does not operate to give the owner of a structure an additional cause of action when the owner claims that the contractor has failed to construct a building in accordance with the plans, specifications, and other contract documents. Therefore, the indemnification of the owner by the contractor only applies to specified claimants and specified property and does not cover the property or structure that is the subject of the construction contract.

§ 13.30 Subcontractor's Payment or Performance Bond Construed to Extend Coverage for Defects and Failures

In the event that the contract for construction executed by the subcontractor contains an indemnification provision, the owner as well as the general contractor may be indemnified for loss arising from personal injury or property damage claims against the subcontractor when the indemnification language of the subcontract agreement is held by a given court to

---

[126] John B. Tieder, Jr. & Julian F. Hoffar, Proving Construction Contract Damages 75 (1988).

extend the coverage afforded by the subcontractor's performance and/or payment bond. Generally, however, the enforceability of an indemnity agreement depends upon the specific provisions contained in the contract for construction and the identity of the negligent party in relation to the specific construction contract and bond. As a rule, the surety is usually liable to an injured third party for personal injury or property damage only if the construction contract can reasonably be interpreted to be for the injured third party's benefit.[127]

§ 13.31 Claims Against the Lender

In the private sector, usually dealing with residential or commercial construction, a lender may become liable for claims of the owner, contractor, or subcontractors, if the lender's activities constitute negligence or if such ties between the contractor and the lender exist that the lender becomes vicariously liable for the acts of the contractor in completing the defective construction.

In *Connor v. Great Western Savings & Loan Association*,[128] the California Supreme Court held that a construction lender was liable to subdivision home purchasers for construction defects in their properties. Great Western had financed the contractor, but it had failed to utilize the normal and customary procedure when it performed the examination of foundation specifications. Thereafter, when the defectively constructed foundations cracked, the owners brought suit against several defendants, including the lender, Great Western. The court concluded that, because of its extensive involvement in the project, Great Western, the lender, had a duty to the owners to exercise reasonable care in preventing defective construction. In short, the extensive nature of other involvement by the lender as opposed to the mere making of routine construction inspections appeared to be the most compelling basis for the court to impose liability.

Generally, the making of routine inspections by a construction lender during the course of construction is assumed to be for the lender's benefit alone and does not form the basis of a claim by an owner that an inspection was carelessly performed by the lender when a defect later diminishes the value of the owner's structure.[129] In contrast, in *Rudolph v. First Southern Savings & Loan Association*,[130] the court found that a sufficient basis existed for allowing the jury to decide whether a lender had assumed

---

[127] Howard, Needles, Tammen & Bergendoff v. Steers, Perini & Pomeroy, 312 A.2d 621 (Del. 1973).

[128] 69 Cal. 2d 850, 447 P.2d 609, 73 Cal. Rptr. 369 (1969).

[129] Butts v. Atlanta Fed. Sav. & Loan Ass'n, 152 Ga. 40, 262 S.E.2d 230 (1979).

[130] 414 So. 2d 64 (Ala. 1982).

a duty to make an inspection for the owner's benefit, when the lender had its own inspector review the work for proper and adequate performance prior to funding a draw to the contractor.[131] Therefore, in the event that a construction lender performs inspections on properties under construction, the owner, in addition to having a cause of action against the contractor for defective construction, may have a cause of action against the construction lender.

Liability to the construction lender may also arise in tort actions brought against lenders involved in construction projects when construction defects contribute to personal injuries. Also, government agencies have been known to cite lenders for building code violations if such lenders go beyond the usual scope of financing, such as being active participants either as joint venturers with a contractor or as equity participants with the general contractor.

In addition, government agencies that make direct loans to home buyers or guaranty or insure loans made by private lenders have been held liable for construction defects. Normally, inspections are utilized by a particular agency to insure that, if the agency is forced to market a property in the event of foreclosure, such agency would have sufficient security in the property. Unfortunately, purchasers of properties have now also claimed that, when a government inspection is carelessly and improperly completed, a cause of action accrues against the governmental agency for damages in the amount necessary to correct the defects that the government inspection reasonably should have disclosed. The most successful approach in pursuing a government agency has been that the government negligently assumed and performed a duty. In such an event, even though the government agency may not have had a duty to inspect, if it voluntarily does so inspect, it may thereafter be liable for negligently carrying out such an inspection.[132] Although the Federal Tort Claims Act (FTCA) was once thought to bar claims on the basis of misrepresentation, the FTCA fell under scrutiny in *Block v. Neal*.[133] Fortunately, Congress has enacted legislation providing for monetary relief to home buyers who purchase new homes utilizing various federal loan programs and later discover that there are structural defects in the buildings.[134] Also, in 1980, amendments to Federal Housing Authority regulations attempted to exclude any duty of careful inspection, whether based in contract or tort. Further, various courts have construed the various inspection clauses to mean that the dwelling unit, when completed, reasonably conforms to government regulations, in terms of lending guidelines. However, approval of the structure

---

[131] *Id.* at 71.
[132] Block v. Neal, 460 U.S. 289, 103 S. Ct. 1089 (1983).
[133] *Id.*
[134] 42 U.S.C.A. § 1479(c).

§ 13.32   DAMAGES GENERALLY   401

by any such agency is not conclusive on the question of whether the structure was completed properly or in accordance with the original plans and specifications.[135]

In assessing the damages against the construction lender or a given governmental agency, the owner would submit the actual costs of correcting the defective construction to either the lender or the agency for payment. Thereafter, quite likely it will become the duty of the construction lender or the agency to pursue recovery directly from the original building contractor.

DAMAGES AVAILABLE TO OWNER AS PLAINTIFF

§ 13.32   Damages Generally

Having reviewed the legal theories of liability and the need to establish causation, the question then becomes: what are the various types of costs and damages beyond those briefly referenced in §§ 13.2 through 13.31 that the owner, the contractor, the architect, the engineer, the subcontractor, or the surety would be likely to collect or to have to pay in the event that a given plaintiff is successful at trial? The types of costs and damages to be recovered fall into two main categories: those available to the contractor and those available to the owner. The recovery obtained by any other member of the construction team flows from the position of these two contracting parties. The owner and the contractor therefore should be mindful of the restrictions on the recoverability of the cost and damage categories identified, depending on the specific contract terminology, applicable case law, and, of course, federal and state statutory provisions.

Assuming that the owner has been successful in proving the liability of the contractor on one or more legal theories, the issue then becomes what damages the owner is owed from the other breaching or tortiously performing party or parties. There are two broad categories of damages ex contractu: direct damages and consequential damages. *Direct damages,* also known as "general damages," are those which naturally arise from the breach of contract; they are damages which, in the ordinary course of human experience, can be expected to result from a breach.[136] On the other hand, *special damages,* or "consequential damages," are those which may not naturally flow from the breach, but which may reasonably be held to have been within the contemplation of the parties at the time of

---

[135] Cash v. United States, 571 F. Supp. 513 (N.D. Ga. 1983).
[136] Roanoke Hosp. Ass'n v. Doyle & Russell, Inc., 215 Va. 796, 214 S.E.2d 155, 160 (1975).

contracting.[137] Consequential damages are those which arise from the intervention of "special circumstances" not ordinarily predictable.[138] If damages are determined to be direct, they are compensable. If damages are determined to be consequential, they are compensable only if it is determined that the special circumstances were within the contemplation of both contracting parties.

Whether special circumstances were within the contemplation of the parties is a question of fact. Therefore, prior to the owner's successfully recovering any of the cost or damage categories described in **§§ 13.33** through **13.46**, a determination must be made as to whether such damages are direct, such as in the case of the cost of corrective work, or consequential, thus requiring consideration of whether such damages were within the contemplation of the contracting parties at the time of construction contract formation. Further, whether or not the damages were contemplated by the parties defines whether or not such damages were foreseeable by the parties and, therefore, recoverable.[139]

§ 13.33 Recovery of the Cost of Corrective Work

If a contractor intentionally abandons performance after the owner discovers a defect in the original performance, or if the contractor is financially incapable of correcting the defective work, or if there is a construction failure, the owner is entitled to the cost of completion of the work specified in the original construction contract.[140] When the work has been completed by the contractor and is defective, the owner can proceed to recover the cost of correcting or replacing the work under one of two alternate theories. The owner may choose to sue for the repair or replacement value of the work[141] or the owner can sue for the diminished value of the structure, defined to be the difference between the value of the defective structure and that of the structure had it been completed in accordance with the plans and specifications.[142] When the cost of correcting the defective work would greatly exceed the structure's diminished value, or result in economic waste, the diminished value theory is usually

---

[137] Hadley v. Baxendale, 9 Ex. 341, 156 Eng. Rep. 145 (1854); Morris v. Mosby, 227 Va. 517, 317 S.E.2d 493 (1984).

[138] Roanoke Hosp. Ass'n v. Doyle & Russell, Inc., 215 Va. 796, 214 S.E.2d 155, 160 (1975).

[139] Restatement (Second) of the Law of Contracts § 351 (1981).

[140] Robinhorne Constr. Corp. v. Snyder, 47 Ill. 2d 349, 265 N.E.2d 670 (1973).

[141] Bloomsburg Mills, Inc. v. Sordoni Constr. Co., 401 Pa. 358, 164 A.2d 201 (1960).

[142] George A. Shegda, Inc. v. Standard Merchandising Co., 231 Pa. Super. 194, 332 A.2d 498 (1974).

§ 13.34 LIQUIDATED DAMAGES

applied.[143] Also, when the cost of correcting the defective work is not clearly disproportionate to the loss of value, the contractor cannot insist that the owner's damages be measured in terms of the diminution of fair market value.[144]

§ 13.34 Liquidated Damages Clause

In the event that a defect or failure in the construction delays the completion of the building, the owner may take advantage of a liquidated damages clause in the construction contract, if one exists. The parties to a construction contract often stipulate to a fixed sum (the liquidated damages) to be paid for each day of delay in the completion beyond an agreed upon date. Typical language in a liquidated damages clause might read:

> If the Contractor fails to comply within the time for completion of the Work, or if applicable, within the time as prescribed herein for any particular section of the Work, the Contractor shall pay to the Owner the sum of $_____ per day as liquidated damages for such default and not as a penalty, for every day or part of a day which shall elapse between the relevant time for completion and the date stated in an unconditional Certificate of Occupancy.[145]

Owners must ensure that the daily liquidated damages rate is not so high that it acts to impose a penalty on the contractor. The liquidated damages amount must not be unreasonable in relation to the actual damages suffered by the owner, otherwise the liquidated sum would be ruled by a court to be a penalty and, therefore, unrecoverable.[146] Because of the difficulty in quantifying construction damages, however, courts usually accept the liquidated damages amount as reasonable and strictly enforce the contractual provisions for the benefit of the owner, but only if the owner is innocent of any wrongdoing or fault contributing to the defective work and resulting delay.

Typically, the owner withholds the liquidated damages amount from the contractor's final payment or retainage. If a construction contract specifies the submission of monthly progress reports and monthly payments based on them, the owner should be careful to withhold only the liquidated damages amount which represents the amount of the delay projected to occur by the contractor as a result of correcting the defective work. Otherwise, the owner might find itself in the position of withholding anticipated delay

---

[143] Groves v. Wunder Co., 205 Minn. 163, 286 N.W. 235 (1939).

[144] Douglass v. Licciardi Constr. Co., Inc., ___ Pa. Super. ___, 562 A.2d 913 (1989).

[145] Modified language from the form contract drafted by the Federation Internationale Des Ingeieurs-Counseils (FIDIC).

[146] Barr & Sons v. Cherry Hill Centre, Inc., 90 N.J. Super. 358, 217 A.2d 631 (1966).

amounts during the job progress when the contractor has already made provision for bringing the project back on schedule through alternate means or resource rescheduling, which would then render the withholding of liquidated damages by the owner wrongful. Depending on the construction contract, such action could result in a material breach of contract entitling the contractor to recovery.

The payment by the contractor or the proper deduction of liquidated damages from any payment due to the contractor does not preclude the owner from pursuing its other contractual remedies.

§ 13.35   Recovery of the Cost to Complete

In *Lurlee, Inc. v. Pernoshal-39 Co.*,[147] an owner and contractor had entered into a contract for the construction of apartments on fully developed property. Some months after construction began, the contractor abandoned further work on the project and unilaterally abrogated the contract alleging that the owner-developer had breached the contract by refusing to make further payments. Notwithstanding the contract price of $630,360, the contractor had been paid the sum of $647,692. There was concern that the degree of completeness would prohibit permanent financing, which concern apparently was justified. Because of the default in the payment of the construction loan, the construction lender ultimately foreclosed on the property. The contractor claimed an unpaid balance of over $100,000 and asserted quantum meruit as grounds for recovery. Thereafter, the owner-developer sought counterclaims against the contractor seeking to recover damages for the cost of replacing the defective work, the cost of completing the overall project, and lost rent revenues. The court concluded that the owner's costs for correcting the defective construction work were recoverable, as well as the cost of completion to the point at which the project could have obtained a degree of completeness "that would not endanger the permanent financing commitment."[148] This holding is typical of cases in which a contractor abandons work on the project or is terminated because of its defective performance.

The contractor's liability for breach of contract is expressly stated in its agreement with the owner. The contractor has a duty to carry out the construction in a workmanlike manner and to use "good materials."[149] The contractor also has been held to a standard of ordinary skill prevailing at the time and place of the construction, which would include a warranty

---

[147] 135 Ga. App. 724, 218 S.E.2d 701 (1975).
[148] *Id.*
[149] Lincoln v. Pohly, 325 S.W.2d 170 (Tex. Civ. App. 1959).

§ 13.36 INTEREST RATE DIFFERENTIAL

that the building is free of major structural defects.[150] Substantial, unapproved deviations from the plans and specifications can constitute defective performance that is a breach of contract, even if an architect has negligently approved the work.[151] Further, if a contractor is guilty of tortious conduct, such as concealing a deviation from plans and specifications or actively concealing a construction defect, the contractor may be liable in tort as well as contract.[152]

Regardless of the reasons for the project's not being completed by the original contractor, the measure of the owner's damages against the contractor is the amount of funds paid in excess of the original price, which is an item of direct, actual damage. As such, the question of foreseeability or special circumstance is not at issue.[153]

§ 13.36 Recovery of Interest Rate Differential

In *Roanoke Hospital Association v. Doyle & Russell, Inc.*,[154] the court considered the issue of interest on construction financing as a compensable damage. The court indicated that direct contract damages are those which naturally or ordinarily and predictably arise from a breach of contract. The court found that direct damages were compensable, while consequential damages would be compensable only if there were special circumstances which placed the damages within the contemplation of the parties. In the construction setting, a defect or failure of construction ordinarily necessitates delay in the completion of the project. The delayed completion date results in additional interest costs on construction financing, and interest revenue also may be lost on invested funds. In *Roanoke Hospital,* the court concluded that the owner's claim for interest paid on the construction loan in excess of what it would have paid had the construction been accomplished on time was recoverable from the contractor because it naturally arose from the contractor's breach. Further, the court concluded that the interest that would have been earned by the owner during the period of delay on the investment of its funds in the project was also recoverable from the contractor. The interest expense that the owner was able to show it was going to incur on a payment loan at a higher rate of interest than it would have incurred on the originally projected permanent loan at a lower rate of interest was recoverable. The

---

[150] Hartley v. Ballow, 286 N.C. 51, 209 S.E.2d 776 (1974).
[151] First Nat'l Bank v. Cann, 503 F. Supp. 419 (N.D. Ohio 1980).
[152] Morris Clark v. D.L. Aenchbacher, 143 Ga. App. 282, 238 S.E.2d 442 (1977).
[153] 135 Ga. App. 724, 218 S.E.2d 701 (1975).
[154] 215 Va. 796, 214 S.E.2d 155 (1975).

differential in the permanent loan interest was computed over the life of the originally projected permanent loan.

The court further held that both the incremental construction interest costs and incremental permanent interest costs were "special circumstances" and as such were consequential damages. Normally, incremental construction interest costs are held to be within the contemplation of the parties at the time of contract formation and, thus, are recoverable as consequential damages. However, incremental permanent interest costs have been widely held as not being within the contemplation of the parties at the time of original contract formation and therefore are probably not recoverable, even though at times they are clearly incurred.

Construction interest must be carefully identified and segregated into the specific types of interest incurred. Specifically, a request for incremental construction interest must be segregated into two components: the interest accruing on any outstanding construction loan, and the interest accruing on the funds that a given owner or developer may have invested in a particular project. An aggregation of the two separate interest quantities may result in the nonrecovery of the aggregate amount, if submitted as such to the trier of fact.[155]

§ 13.37 Recovery of Land Value

One of the most difficult and contested awards for special damages to the owner-developer from the contractor is for the actual cash loss suffered on an original tract of land on which the construction lender has foreclosed after the contractor's defective performance or failure to complete the project. In *Lurlee, Inc. v. Pernoshal-39 Co.*,[156] a jury was authorized to conclude that an owner-developer was entitled to recover its actual cash loss for land acquired for a given project when the land was later taken back by the construction lender through foreclosure. The allowed recovery in this situation would be the actual cash or capital expenditure for the acquisition of the land, as opposed to its fair market value appraised at the time of the project's initiation or construction loan closing, or as appraised as of the date of foreclosure. However, for the contractor to be liable to the owner for the loss of the property value, the abandonment or defective performance of the project by the contractor must be the proximate cause of the lender's initiating a legally and technically correct foreclosure sale.

---

[155] *Id.*

[156] 135 Ga. App. 724, 218 S.E.2d 701 (1975).

§ 13.38 Recovery for Loss of Permanent Financing Commitment

In the event that the private owner or developer has to place a new permanent loan commitment as a result of the defective and, therefore, delayed performance by the contractor, the contractor may be liable to the private owner-developer for the loss of the permanent lending commitment. The loss of a permanent financing commitment is generally held not to be recoverable, because it is not foreseeable at the time of contract formation. However, if there were "special circumstances" or if the loss was foreseeable by the contractor, such as through adequate warning or notice from the owner, the expenses of negotiating a substitute permanent loan are recoverable[157] and any increased costs resulting from a differential of higher interest rates on a new permanent loan are also possibly recoverable.[158]

§ 13.39 Recovery of Lost Revenues

Often the owner-developer loses rent or lease revenues from either an office complex or residential rental project as a result of defective construction performed by the contractor. In *Camper v. W.J. McDermott*,[159] an owner sued the contractor on a private project for loss of rental revenues as a result of the contractor's breach of contract.[160] The owner maintained that the work and the material used in the construction of a pool as part of an apartment complex was defective. Further, because the work on the pool was defective and, thus, was not completed on time, the owner alleged a loss of rental revenues for the entire apartment project. The court concluded that damages were recoverable, but that "in the absence of the elimination of other reasons or proof the loss of rental revenues in a specific amount cannot be attributed to damage flowing from a failure to abide by a contract that a swimming pool would be installed within a reasonable time."[161]

Therefore, although recovery of lost rent revenues may at first glance seem appropriate, a causal connection between the contractor's defective performance and the loss of rent or lease revenues must be established in order for recovery to be allowed. An owner's or developer's attempt to

---

[157] Bohemian-American Workingman's Gymnastic Ass'n (SOKOL) v. Northern Bank, 120 N.Y.S. 134 (Sup. Ct. 1909).

[158] Shurtleff v. Oxidental Bldg. & Loan Ass'n, 105 Neb. 557, 181 N.W. 374 (1921).

[159] 266 Cal. App. 2d 41, 71 Cal. Rptr. 590 (1968).

[160] *Id.*; Psaty & Fuhrman v. Housing Auth., 76 R.I. 87, 68 A.2d 32 (1949).

[161] Camper v. W.J. McDermott, 266 Cal. App. 2d 41, 71 Cal. Rptr. 590 (1968).

recover lost rent or lease revenues may also be prohibited on the grounds that the rent revenues are too remote to have been foreseeable at the time that the construction contract was originally let or, in any event, too speculative to be awarded as damages. In *Lurlee, Inc. v. Pernoshal-39 Co.*,[162] lost rental revenues were recovered from the time that the contractor abandoned the project, after defectively completing a portion of the work, until the time a foreclosure action extinguished the rights of the owner-developer in the project. Also, in an action against a surety for failure of structural components in a warehouse, the issue of the surety's liability for lost rents was submitted to the jury for determination as to whether the policy rental endorsement covered all of the owner's claims, including those for lost rents.[163]

§ 13.40 Recovery for Loss of Use

Recovery for the loss of use of a structure has been held to be a proper subject of damages when the loss is a direct and proximate result of a structural defect and such damages were within the contemplation of the parties at the time of contract formation.[164] In *Northern Petrochemical Co. v. Thorsen & Thorshov, Inc.*,[165] major structural flaws occurred in a building, consisting of large cracks in the walls and floor. The owner was awarded both the reconstruction costs for replacing the defective portions of the building plus the loss in value of the structure as a result of the defect. The owner also was entitled to recover its lost profits and excess operating costs, and the court ruled that all such costs constituting the loss of use of the building were not controlled by a stipulated damages clause contained in the construction contract.

Loss of use damages are thus defined to include loss of profits and excess operating costs which occur during an "extraordinary" delay resulting from the defective construction and/or design and necessary reconstruction of a building.[166] Therefore, the courts have carved out a distinction between ordinary delays that may be within the contemplation of the parties or may fall within the scope of a no damages for delay or stipulated damages clause, and "extraordinary delays" that require a more

---

[162] 135 Ga. App. 724, 218 S.E.2d 701 (1975).
[163] Fidelity & Guar. Ins. Underwriters, Inc. v. Allied Realty Co., 238 Va. 458, 384 S.E.2d 613 (1989).
[164] Marshall v. Marvin H. Anderson Constr. Co., 283 Minn. 320, 167 N.W.2d 724 (1969).
[165] 297 Minn. 118, 211 N.W.2d 159 (1973).
[166] *Id.*

§ 13.41 Recovery of Additional Supervision Costs

Although the general contractor is usually considered the one who experiences additional overhead costs, both at the home office as well as at the jobsite, as a result of correcting defective work, the owner also experiences additional overhead costs as a result of the contractor's defective performance.[168] As an example, there are additional personnel costs during correction of defective work, including expenditures for the owner's representatives, security personnel, and clerical staff. Additional site office costs could include the owner's site office, equipment rental charges, utilities, telephone, insurance premiums,[169] sales or leasing office charges, and equipment or construction materials storage charges.[170] These charges have all been held to be compensable for the owner.

§ 13.42 Lost Profits

Whenever a building or other facility is not delivered on time through the fault of the contractor, the profits that the owner would have derived from the use of the completed structure may be recoverable. Lost profits are typically recoverable "when such profits are proven with reasonable certainty and when they reasonably may have considered to have been within the contemplation of the parties"[171] at the time of entry into the construction contract. Further, lost profits relating to other projects of the owner may be recoverable when there is a causal relationship established between the activities of the contractor in delivering one structure late and losses experienced on other projects of the owner, such as when revenues

---

[167] Indian Head Truck Line, Inc. v. Hvidsten Transp., Inc., 268 Minn. 176, 128 N.W.2d 334 (1964).

[168] Burnett & Doty Dev. Co. v. Phillips, 84 Cal. App. 3d 384, 148 Cal. Rptr. 569, 570–71 (1978); Samuel B. Harper, Jr., and Sigmund E. Florsheim d/b/a Harper & Florsheim, ASBCA No. 4959, 59-1 B.C.A. (CCH) ¶ 2986 (1959).

[169] Roanoke Hosp. Ass'n v. Doyle & Russell, Inc., 215 Va. 796, 214 S.E.2d 155, 160 (1975).

[170] Fairfax County Redev. & Hous. Auth. v. Hurst & Assocs., Inc., 231 Va. 164, 343 S.E.2d 294 (1986).

[171] Just's Inc. v. Arrington Constr. Co., 99 Idaho 462, 583 P.2d 997 (1978); W-V Enters. Inc. v. Federal Sav. & Loan Ins. Corp., 234 Kan. 354, 673 P.2d 1112 (1983).

from the operations of the business or leasing activity in the first structure were to be utilized in the other structures and operations.

In certain cases, courts have found that contemplated profits from projected developments have been based only on "speculation and conjecture."[172] The court in *Coastland* distinguished the subject real estate development as a new business venture with no proven profit record, to be compared with an established business with an established earning capacity.[173] In other cases, courts have denied recovery from profits expected on the sale of completed buildings, noting that the profit expected by the plaintiff would only depend upon the project's successful completion and other economic factors, such as a certain percentage of occupancy, which were speculative at best.[174]

However, although not specifically dealing with a defect or failure claim in reference to a terminated condominium project, the court in *Kissel Co. v. Gresly*[175] awarded lost profits based on the owner's testimony and evidence of the profits experienced on similar projects. The presentation of the lost profits claim must clearly show the relationship with the amount claimed from specific similar examples in order to obtain recovery. In *Farm Crop Energy, Inc. v. Old National Bank*,[176] the court ruled that the expert witness for the borrower-owner lacked the requisite technical knowledge to testify as to the potential for profit from the subject business enterprise. The testimony did not provide examples of the profits made by similar businesses or operations in the same area operating under substantially the same conditions, and it failed to show sample profits made in the same area by a business with a production level on a par with the production level that the expert used in the assessment of lost profits. Therefore, in the event that the private owner-developer seeks to recover lost profits because of the defective work on the project as completed by a contractor, specific evidence of comparable projects in a given community coupled with adequate technical knowledge by the expert witness for the owner-developer is essential.

§ 13.43 Recovery of Additional Professional Services

In the event that additional professional services are necessitated by a construction defect or failure to complete construction, the owner should be

---

[172] *See, e.g.,* Coastland Corp. v. Third Nat'l Mortgage Co., 611 F.2d 969 (4th Cir. 1979).
[173] *Id.* at 977.
[174] Stanish v. Polish Roman Catholic Union, 484 F.2d 713, 722 (7th Cir. 1973).
[175] 591 F.2d 47 (9th Cir. 1979).
[176] 109 Wash. 2d 923, 750 P.2d 231 (1988).

able to recover the actual extra amount expended for fees for the engineer, the architect, the construction manager, or other consultants required to correct the defect. The express provisions of the construction contract should provide for the owner to be able to recover the actual amount of professional fees expended for correction of a defect.

§ 13.44 Recovery of Attorneys' Fees

Generally, attorneys' fees are not recoverable as an element of damages in a breach of contract or other tort action, unless statutory authority specifically provides for recovery of such fees. However, most contracts for construction provide for the recovery of attorneys' fees by the party to the construction contract who is forced to resort to litigation and is thereafter successful in such litigation. Most jurisdictions hold that such a provision for the payment of attorneys' fees by the party found to be in breach of contract is valid and enforceable and not against public policy, unless specific statutory enactments exist to the contrary.[177] In most cases, the amount of attorneys' fees incurred by the party having to enforce the specific terms of a construction contract are ordered by the court to be part of the legal consequences of the other party's original wrongful act and, therefore, they are payable as a part of the damages, if so provided in the contract. In the absence of any contractual or statutory liability, however, the general rule is that attorneys' fees and expenses incurred by a plaintiff in litigation in pursuit of a claim against a defendant are not recoverable as an item of damages.[178]

§ 13.45 Recovery of Depreciation

In the event that a defect or failure in the construction results in a significant period of delay for the owner, the value of the owner's building or other equipment may depreciate during the delay. If the structure or equipment is not used during such a period, the owner should present proof of actual depreciation to the trier of fact. Further, in *Calandro Development, Inc. v. R.M. Butler Contractors, Inc.*,[179] the court apparently would have allowed for recovery of depreciation on subdivision lots against a surety but for the limitations delineated in the bond provisions. Therefore, the owner should always present proof of the depreciation method actually

---

[177] *See, e.g.,* United States ex rel. Micro-King Co. v. Community Science Technology, 574 F.2d 1292 (5th Cir. 1978).

[178] Chesapeake & Potomac Tel. Co. v. Sisson & Ryan, 234 Va. 492, 362 S.E.2d 723 (1987).

[179] 249 So. 2d 254 (La. Ct. App. 1971).

utilized on the project allegedly damaged through the defective performance of the contractor.

§ 13.46 Recovery for Loss of Professional Reputation

At times, recovery of damages for loss of business reputation by the owner can be obtained; however, such a claim normally depends upon whether the claim contains the elements of malice, fraud, or intentional interference with contractual relations or libel and slander. Recovery for damages to reputation usually requires willful and intentional conduct and knowledge on the part of the contractor, as opposed to simple negligence resulting in a construction failure or defect. In *Downy v. United Weatherproofing*,[180] the court stated that the malicious and intentional commission of a harmful act without justification or excuse had to be proven before damages would be recoverable for defamatory remarks. However, in *Calandro Development v. R.M. Butler Contractors, Inc.*,[181] the court considered the issue of recovering damages to the owner's professional reputation and seemed to indicate that, but for the specific bond limitation provisions, recovery of such damages could be had against the surety, as well as the contractor. Therefore, in instances of construction defects resulting in the nontimely completion of a project, an owner should allege and attempt to recover damages for loss of business reputation, but such damages most likely will have to be proven with great specificity. Further, these losses or damages may be pleaded in the format of an inability to comply with other contractual obligations to third persons or lessees.[182]

DAMAGES AVAILABLE TO CONTRACTOR AS PLAINTIFF

§ 13.47 Compliance with Contractual Notice Provisions

Whenever a contractor takes the position that the owner or its representatives are the culpable parties in a defect or failure claim, the contractor

---

[180] 363 Mo. 852, 253 S.W.2d 976, 980 (1953).
[181] 249 So. 2d 254 (La. Ct. App. 1971).
[182] Mark Keshishian & Sons v. Washington Square, Inc., 414 A.2d 834 (D.C. 1980).

must be extremely diligent in order to ensure compliance with provisions requiring timely notification of the owner, subcontractor, or material of the defect that has invoked the need for additional time or costs. Notice requirements are normally mandatory, and failure to comply with them can result in a complete waiver of the claim or a restriction of the contractor's right to pursue a claim for damages. Also, as a general rule in public contracts, there are follow-up notice clauses which require documentation and specific processing of the claim, including a detailed statement of the claim.

§ 13.48 Determining Actual Damages

When the owner or its representatives are responsible for a defect or failure claim, exact certainty as to the amount of actual damages assessed by the contractor has not been required by the United States Supreme Court. In *Story Parchment Co. v. Patterson Parchment Paper Co.,*[183] the Supreme Court stated that the rule which precludes the recovery of uncertain damages does not apply to those damages which were definitely attributable to the wrong and only are uncertain in their amount. "It will be enough if the evidence shows the extent of the damages as a matter of just and reasonable inference, although, the result be only proximate."[184]

The rule regarding definitive proof of damages has carried over into construction litigation. After it has been established that a contractor is entitled to recover a claim, damages will not be refused the contractor solely because it has not proven the amount to a "mathematical certainty." Most courts are willing to accept a reasonable estimate or a logical method of calculation as long as the mathematical results can be achieved with a reasonable degree of certainty.[185] In *International Air Craft Services, Inc.,*[186] for example, the court stated that the determination of damages for breach of contract is an inexact science and the sum reached by whatever method used will never be more than an approximation. Alternate methods have been developed to determine damages for breach of contract and have been utilized by contractors to meet the exigencies of a given situation. Normally, the test is reasonableness as to the method of cost quantification and damage calculation.

---

[183] 282 U.S. 555, 562 (1931).
[184] *Id.* at 562–63.
[185] Luria Bros. & Co. v. United States, 369 F.2d 701 (1966). *See also* Rusciano Constr. Corp. v. State, 37 A.D.2d 745, 323 N.Y.S.2d 21 (1971).
[186] ASBCA No. 839, 65-1 B.C.A. (CCH) ¶ 4793 (1965).

§ 13.49 Recovery of Damages for Defective Plans and Specifications

Perhaps the most common type of a contractor's defect or failure claim against the owner revolves around the incompleteness of plans and specifications for a given project. A contractor has been held to a duty to seek clarification of patent ambiguities in specifications before submitting a bid.[187] However, the construction contract may in fact disclaim design specification defects, thus precluding any damage recovery even though liability might be established for a defective design specification.[188]

Further, if the finishing of defective plans and specifications was the cause of a delay, in addition to recovering its exact cost resulting from the defective plans, the contractor also has a cause of action against the owner for delay damages,[189] even if it was the government as owner supplying the plans.[190] It is beyond question that delay damages, more specifically discussed in **Chapter 5**, are recoverable. As a general rule, "a contractor of public work who, acting reasonably, is misled by incorrect plans and specifications issued by public authorities as the basis for bids and who, as a result, submits a bid which is lower than he would have otherwise made may recover in a contract action for extra work or expenses necessitated by the conditions being other than originally represented."[191] However, a contractor must demonstrate misrepresentation or concealment of a material fact in the plans and specifications in order to be awarded recovery.

§ 13.50 —Economic Loss Doctrine

If the contractor attempts to bring an action against either the architect or the engineer for inadequately prepared plans and specifications, using causes of action sounding in tort as opposed to breach of contract, the contractor may find its ability to recover impacted by the economic loss defense. The courts have held that tort law is not intended to deal with a party's commercial interest but instead is intended to further recovery for personal injury or property damage. In *Moorman Manufacturing Co. v.*

---

[187] Blinderman Constr. Co. v. United States, 17 Cl. Ct. 860 (1989).

[188] Hydro Dredge Corp., E&GBCA No. 5303 (Oct. 16, 1989).

[189] Jasper Constr., Inc. v. Foothill Jr. College Dist., 91 Cal. App. 3d 1, 153 Cal. Rptr. 767 (1979).

[190] Laburnum Constr. Corp. v. United States, 325 F.2d 451 (Ct. Cl. 1965).

[191] Jasper Constr., Inc. v. Foothill Jr. College Dist., 91 Cal. App. 3d 1, 8, 153 Cal. Rptr. 767, 770 (1979).

National Tank Co.,[192] the Supreme Court of Illinois used the economic loss doctrine to rule that a plaintiff cannot recover solely economic losses under tort theories of strict liability, negligence, and/or misrepresentation. However, in a more recent decision, the economic loss defense was rejected in the field of architect/engineer liability.[193] There is apparently a trend toward holding architects and engineers liable for economic damages incurred by third parties, defined as those not in contract with the architect or engineer and yet foreseeably and proximately damaged as a result of acts or omissions by the architect or engineer. Therefore, it is conceivable that a contractor may be prohibited from recovery against an architect or engineer for defective plans and specifications because the losses occasioned by the contractor are economic in nature.

§ 13.51 Recovery Despite No Damages for Delay Clause

A contractor seeking to prosecute a defect or failure claim must be ever mindful of the clauses in its contract for construction. Courts have at times held that formal compliance with contractual provisions, irrespective of the existence or proof of liability, may preclude recovery of provable costs and damages. Failure to comply with various types of contractual provisions thus can negate or void an otherwise viable claim. A contractor must ensure that it follows the contractual provisions even in the face of what would appear to be direct owner noncompliance, such as the production of defective plans or specifications.

One typical contractual provision is the no damages for delay clause, which attempts to preclude recovery from the owner for damages caused by delays of any sort, even those specifically caused by the owner. The courts have routinely upheld no damages for delay clauses, although at times, they have sought to avoid unnecessarily harsh and possibly inequitable decisions that would result from strict construction of the contract language.[194] A typical no damages for delay contract provision states that "no payment or compensation of any kind shall be made to the contractor for damages because of hindrance or delay from any cause in the progress of the work, whether such hindrance or delays be avoidable."[195] The courts have found circumstances in which this exculpatory clause would not relieve the owner from liability for its own actions which

---

[192] 91 Ill. 2d 69, 435 N.E.2d 443 (1982).
[193] Donnelly Constr. Co. v. Oberg/Hunt/Gilleland, 139 Ariz. 184, 677 P.2d 1292 (1984).
[194] Cunningham Bros., Inc. v. City of Waterloo, 254 Iowa 659, 117 N.W.2d 46 (1962).
[195] Ippolito—Lutz, Inc. v. Cohoes Hous. Auth., 22 A.D. 990, 254 N.Y.S.2d 783 (1964).

caused the delay.[196] In order to avoid the effects of a no damages for delay clause, the courts have carved out essentially four exceptions:

1. Any delay that is the result of the owner's active interference, or involves fraud or bad faith on the part of the owner who is seeking the benefit of the no damages for delay provision
2. A delay of unreasonable length, justifying the abandoning of the contract by the party delayed
3. A delay not specifically enumerated as that particular sort of delay to which the no damages for delay clause is to apply
4. A delay which was not intended or contemplated by the parties to be within the preclusionary realm of the no damages for delay provision.

Active Interference

As a general rule, a no damages for delay clause does not relieve an owner from liability for damages caused by its own active interference with a contractor's work.[197] In *Kalisch-Jarcho, Inc. v. City of New York*,[198] for example, the court considered whether a contractor could recover for approximately 28 months of delays attributed to the city's "endless revisions" of defective plans and specifications.

Bad Faith or Fraud

Even though a particular delay spawned by a defect in the construction may fall within the literal parameters of a no damages for delay clause, the provision will not be enforced if the delay resulted from the owner's active misrepresentation.[199] Therefore, if an owner solicits bids and enters into a construction contract by representing that the plans and specifications are complete and defects later arise in the construction as a result of irregularities in the plans and specifications known to the owner at the time of contracting, any delays occasioned by required revisions of the plans and specifications will be compensable by the owner.

---

[196] *See, e.g., id.*

[197] Kalisch-Jarcho, Inc. v. City of N.Y., 58 N.Y.2d 377, 448 N.E.2d 413, 461 N.Y.S.2d 746 (1983); Peter Kiewit Sons' Co. v. Iowa S. Utils. Co., 355 F. Supp. 376, 397 (S.D. Iowa 1973).

[198] 58 N.Y.2d 377, 448 N.E.2d 413, 461 N.Y.S.2d 746 (1983).

[199] American Bridge Co. v. State, 245 A.D. 535, 283 N.Y.S. 577 (1935); Warner Constr. Corp. v. City of Los Angeles, 2 Cal.3d 285, 466 P.2d 996, 85 Cal. Rptr. 444 (1970).

§ 13.52 PERFORMANCE SPECIFICATIONS 417

Unreasonable Delay

The impact and operation of a no damages for delay clause is largely determined by the particular provisions of the construction contract in question and the particular circumstances of each case. However, unreasonable delays in furnishing defective plans and specifications may render the no damages for delay clause unenforceable, thus allowing the contractor to recover. For instance, in *Brady v. Board of Education*,[200] the court stated that a delay of eight months in the commencement of performance of a contract was sufficiently unreasonable to justify the contractor's abandoning the contract, notwithstanding a no damages for delay clause.

Beyond the Contemplation of the Parties

In keeping with the foreseeability and reasonableness requirements of a claim for damages, the courts have determined that a no damages for delay clause is inapplicable when the delay was caused by an occurrence which was not within the contemplation of the parties. In *J.R. Stevenson Corp. v. County of Westchester*,[201] the court ruled that whether or not revisions to plans and specifications were within the contemplation of the parties was an issue to be determined by the trier of fact. When the owner's plans and specifications are ultimately determined to be so erroneous that the delays occasioned were not originally contemplated by the parties at the time of contract formation, the no damages for delay clause will not prohibit recovery of reasonable delay costs by the contractor.[202]

§ 13.52 Recovery for Defective Performance Specifications

When an owner provides performance specifications, only a general description of the end product is delineated and only the operational characteristics desired for the item to be constructed are specified by the owner. The owner is only interested in end performance and, therefore, leaves the general discretion and specification of detail to the contractor. The owner, however, reserves the right of final acceptance or rejection of the end product. On occasion, contractors have been allowed to recover from an owner for defective performance specifications when the specified end result or product turned out to be impossible to achieve by any

---

[200] 222 A.D. 504, 226 N.Y.S. 707, 709 (1926).
[201] 113 A.D.2d 918, 493 N.Y.S.2d 819 (1985).
[202] Buckley & Co. v. State, 140 N.J. Super. 289, 356 A.2d 56 (1975).

available method.[203] Provided that the contractor did not assume the risk of impossibility, such as by assuring the owner of special expertise in its ability to meet performance specifications, the contractor who proves impracticability and/or commercial impossibility can recover its costs.[204] It is also necessary that neither party anticipated the difficulties encountered by the contractor at the time of contract formation.[205] In such a case of defective performance specifications, the contractor would be entitled to recovery of its reasonable, actual costs and damages, as long as they were foreseeable at the time of contract formation.

§ 13.53 Recovery of Consequential Costs and Damages

Normally, in the defect and failure setting, the contractor is entitled to compensation from the owner for the owner's failure to provide either plans or specifications free from defect or defect-free materials, as promised under the construction contract. Also, the owner's failure to provide and approve shop drawings within a reasonable time, to properly, adequately, or timely prepare the construction site, to properly coordinate the work of multiple contractors, or to timely obtain building permits would justify additional compensation to the contractor.

However, when the contractor is ordered to discontinue work in order to correct work it defectively performed, the contractor cannot recover additional costs incurred from the delay for making the requisite correction. The contractor is not only precluded from recovery of the actual costs, but also from recovery of consequential costs and damages resulting from the defective performance delay.

Because of the foreseeability requirement as to consequential damages and their recovery, unless the contractor (1) establishes a causal link between the breach and the increased cost or damage and (2) adequately proves the actual amount of additional costs or damages incurred, the contractor cannot recover. Although the calculation of specific delay claims and damages appears in **Chapter 5**, several consequential cost categories are of note.

Labor Escalation

If the wages that a contractor is required to pay its labor force should rise during the period of the delay spawned by the owner's defective

---

[203] Maxwell Dynamometer Co. v. United States, 386 F.2d 855 (Ct. 1967).
[204] Tombigee Constructors v. United States, 420 F.2d 1037 (Ct. Cl. 1970).
[205] *Id.*

performance, the contractor is entitled to recover the difference between the original wage rates and the increased wage rates. The requisite data to prove the claim for increased costs caused by labor escalation include the anticipated labor schedule, the actual labor schedule, the wage rates actually paid over the life of the project, and, most importantly, the time periods during which specific wage rates were paid.

Worker loading schedules or other resource schedules should always be prepared before the start of work. However, these schedules may need to be developed after the fact, as an element of proof of the consequential labor escalation costs. The anticipated schedules must reflect only actual worker hours expended in accordance with the originally anticipated curve and should not reflect recovery of any cost differential between anticipated worker hours and actual worker hours.

Because each trade usually has its own wage rate, worker loading curves should be prepared on a trade-by-trade basis first, and then overlaid into a composite exhibit, thus rendering both the trade-by-trade and composite exhibits available as elements of proof.

In order to calculate the wage rates to be used in the preparation of the worker loading curves, on nonunion projects the actual wage rates can be obtained from the contractor's payroll records, and on union projects wage rates can be obtained from the union agreements in effect at the time that the work was performed. In the event that both union and nonunion trades performed tasks on the project, or a combination of union and nonunion laborers performed varying tasks on a given project, weighted averages can be computed by totalling all of the labor costs on the project for a specific period and dividing by the number of labor hours worked during a particular time frame. Of course, prior to developing the escalated labor cost analysis, the contractor should segregate those additional hours expended by the contractor as a result of its own inefficiency.

Materials and Supplies

Typically, the contractor may be delayed in ordering at fixed price quotes because the owner provided defective specifications. However, in order to collect for material and supply escalation costs, it is essential that the contractor establish the rationale as to why the purchase did not occur earlier, that most typically being the lack of adequate specifications.

In order to prove the pricing differential, contractors should present item-by-item purchase orders, invoices, or other of the contractor's own actual records, thereby verifying all price increases. Again, only the differential in the cost experienced by the contractor as a result of the price escalation is recoverable, and not additional costs for increased quantities because of the contractor's own original estimate and in efficiency. If the price of materials and supplies dropped during a given period, it is possible that an owner could claim an equitable reduction in the cost of the overall structure as an offset against other contractor claims for the owner's defective plans or specifications.

Loss of Bonding Capacity

In some instances, the contractor's surety may limit the contractor from obtaining bonding for new work when an owner has delayed the performance of a contract through its own defective performance. In order to recover on such a claim, the contractor must prove that (1) other projects were available during the period of defective performance by the owner, (2) the contractor was unable to obtain bonds for such other available work during the period, and (3) if such other work had been obtained, the contractor would have realized a profit.[206] As a rule, the government has denied recovery for the loss of additional profits caused by the loss of bonding capacity of the contractor on the theory that the damages were not foreseeable at the time of contracting.

Loss of Profits

Lost profits as an element of actual damages can be claimed on a defaulted contract. Further, profits on claims for additional, different, or extra work are also at times recoverable. However, a claim that profits on a completed project were less than what they would have been but for the owner's defective performance that caused the delay on the project, are rarely successful and have generally been held to be either not foreseeable or not sufficiently proven to be recovered. However, in considering a loss of profits, one court relatively recently indicated that a recovery for lost business opportunities was appropriate.[207] Also, if the contractor seeks to prove a loss of profits on other projects, it must prove that the damages are not speculative and that there is a direct causal relationship between the performance of the delayed project and profits to be derived from other projects or lost opportunities.[208]

---

[206] Cruse v. Clawson, 137 Mont. 439, 352 P.2d 989 (1960).
[207] Eastern Tunneling Corp. v. Southgate Sanitation Dist., 487 F. Supp. 109, 115–16 (D. Colo. 1980).
[208] Burnett & Doty Dev. Co. v. Phillips, 84 Cal. App. 3d 384, 148 Cal. Rptr. 569 (1978).

OTHER CONSIDERATIONS

§ 13.54 Statutes of Limitations and Repose

Because of the length of time that major construction projects take to complete, as well as the inherent potential for latent structural defects to remain unidentified for extensive periods after completion, the statute of limitations provides protection against unending liability of the contractor, the owner, and other parties to the construction project. There is relative uniformity in the states in that, normally, a statute of limitations begins to run when the construction project is "substantially completed," or when the contractual obligations of the contractor to the owner have been fully performed and payment has been made thereon.

If there is an attempt to conceal a construction defect that impairs the quality of the structure, the statute of limitations runs from the date of the discovery of the concealed defect. Also, in a contract action by a contractor brought before the expiration of the statute of limitations, the owner may offset a claim for a defect in the work, notwithstanding that the claim might otherwise be precluded by the statute of limitations' time for filing requirement.

Statutes of repose differ from statutes of limitations in that *statutes of repose* establish a time period after which no cause of action may accrue. In *Yarborough v. Hilton Hotels Corp.*,[209] the Colorado Supreme Court found that, because 10 years had elapsed after "substantial completion" of a building prior to the filing of a wrongful death lawsuit, the architects were immune from suit, as the state statute specified a period of 10 years after which no cause of action would accrue.[210] Also, a claim by an owner for alleged latent defects in the deficiency of design or construction which arose more than 10 years after "substantial completion" of the project was precluded by California state statutory provisions.[211] When the cause of action accrues more than 10 years after substantial completion, a claim will not arise as against the surety, either.

While a statute of limitations addresses actions filed between all parties to the construction project, statutes of repose have been enacted primarily as a means of protecting the construction professional, the architect, or the engineer from liability for construction defects in the design which did not become apparent for a substantial period of time after

---

[209] 655 P.2d 822 (Colo. 1983).
[210] 655 P.2d at 825.
[211] Regents of Univ. of Cal. v. Hartford Accident & Indem. Co., 59 Cal. App. 3d 675, 131 Cal. Rptr. 112 (1976).

a given structure was completed. For the foregoing reasons, an owner must aggressively pursue any potential cause of action or run the risk of having recovery and possibly the action itself precluded by the combined effect of the statute of limitations and any statute of repose in effect in the jurisdiction in which the owner seeks recovery.

§ 13.55 Claims Preparation and Settlement Considerations

A *claim* as opposed to formal litigation has been defined as "an adjustment or interpretation of contract terms, payment of money, extension of time or other relief."[212] In public contracting, the federal government has also recognized a claim as opposed to formal litigation and has further recognized that an invoice or request for payment not acted upon within a reasonable time may be converted to a claim.[213] The emphasis of a claim in the defect and failure setting is not necessarily on the actual damages incurred by the owner or the contractor, but may instead focus on the consequential damages experienced by the owner. Much of the expense of developing a claims package or pursuing formal litigation is a result of the complexity of factual issues which must be proven by the parties, especially in the consequential damages setting. It is imperative that proof of the proximate causation and consequential damages be undertaken by persons with technical expertise, including counsel, expert witnesses, and other members of the claim or litigation team. The sheer number of parties to a complex construction project requires that only qualified experts and personnel be utilized in pursuing the claim through to formal litigation if required. The records containing the requisite elements of proof may be so voluminous that conventional discovery methods, as contained in the Federal Rules of Civil Procedure, may break down.

An expert witness in construction litigation is critical. The expert witness should participate in the preparation of any claims package prepared for submission in furtherance of settlement, arbitration, or formal litigation. As a general rule, the claims package should not include citation to cases or legal authorities unless the claim is only to be reviewed by opposing counsel. Recitation of legal authorities in the claims package may prompt the opposing party to submit the entire matter to corporate counsel, thereby precluding or delaying meaningful management level negotiations. However, specific statutory authority, for example relating to interest recovery or statutory authority for the filing of liens, is probably best included.

---

[212] AIA A201-1987, General Conditions of the Contract for Construction.
[213] Contract Disputes Act of 1978, 41 U.S.C. § 61.

Formal audit reports should not be submitted in the initial claims package, although a formal audit should be performed in order to develop the requisite independent credibility of the claims package prior to its submission. Submission of the formal audit report in the claims package would tend to invite a detailed formal audit response, which may again preclude meaningful negotiations at the management level. However, if both parties have employed auditors, those auditors should participate in settlement negotiations in order to isolate categories of specific costs upon which agreement can be reached. A settlement can be built by listing line item cost categories upon which agreement can be reached.

Preparation of a claims package and accompanying audit are highly recommended because of the potential sanctions that can be levied against the contractor for submission of an improper claim in either the private or public setting. The sanctions in the public contract setting are far more onerous and can entail preclusion of the contractor firm from future public projects.[214] Also, sanctions in the form of attorneys' fees may be leveled against the attorney and the client for pursuing what is later determined to be a factually incorrect claim.[215]

If negotiations are unsuccessful after preparation of the claims package and settlement, before beginning protracted litigation, the parties should first determine whether resort to arbitration—because of the highly skilled nature of arbitration panels that are now available—is a more advantageous course of conduct. A strong claims package should greatly assist in the resolution of the claim in either the arbitration or litigation setting.

---

[214] *Id.;* The False Claims Act, 18 U.S.C. § 231; The False Statements Act, 18 U.S.C. § 1001 *et seq.;* The Truth in Negotiation Act, 10 U.S.C. § 2306(f) (1962).

[215] Fed. R. Civ. P. 11.

CHAPTER 14

CONSTRUCTION DELAY CLAIMS

John N. Chapin, Jr.
Stephen D. Hurst, Esquire*

John N. Chapin, Jr. is Coopers & Lybrand's partner-in-charge of Litigation & Claims Services for the central United States. He has been responsible for this practice since 1979 and has testified as an expert witness in several matters. Mr. Chapin holds a B.S.B.A. and an M.B.A. from Washington University. He is a certified fraud examiner and also a certified management consultant.

Stephen D. Hurst is a trial lawyer and partner in the Chicago office of Hinshaw, Culbertson, Moelmann, Hoban & Fuller. He received his B.A. in 1976 from Wofford College and his J.D. in 1981 from ITT/Chicago-Kent College of Law. Mr. Hurst has represented owners, designers, contractors, and insurers at trial, in arbitration, and on appeal. He is a member of the American Bar Association's Forum Committee on the construction industry.

*Kay Crider, at Hinshaw & Culbertson, and Mark Hosfield, at Coopers & Lybrand, contributed substantially to this chapter. The authors appreciate their assistance.

§ 14.1 Introduction

DEVELOPING THE CLAIM

§ 14.2 Preparing the Claim
§ 14.3 Identifying Potential Defendants
§ 14.4 Obtaining Construction Documentation
§ 14.5 Identifying the Causes of Delay
§ 14.6 Focusing the Claim on the Responsible Party

DEVELOPING LEGAL THEORIES OF LIABILITY

§ 14.7 Compensable and Noncompensable Delays
§ 14.8 Anticipating Potential Defenses
§ 14.9 Mitigating Damages

DEVELOPING ELEMENTS OF DAMAGES

§ 14.10 Contracts with Liquidated Damages Provisions
§ 14.11 Contracts without Liquidated Damages Provisions
§ 14.12 —Direct Damages
§ 14.13 —Consequential Damages

DEVELOPING LITIGATION STRATEGIES

§ 14.14 Selecting a Forum
§ 14.15 Presenting the Claim
§ 14.16 Conclusion
§ 14.17 Bibliography

§ 14.1 Introduction

In order to pursue a delay claim against a designer or contractor the owner must gather voluminous construction documentation, organize the documents into a manageable form, and synthesize the documented facts into a graphic summary which can be easily understood. This chapter will discuss practical methods for accomplishing these tasks.

Every construction project should be approached with the understanding that some sort of delay will probably occur. Preventing delay starts at the bargaining table; the owner should include as many discouragements for delay as possible in the construction contract itself. After ground is broken it is important for the owner to closely monitor construction progress on a systematic basis. The goal is to detect and limit the delay before it grows so large that litigation becomes inevitable. The costs and frustrations of presenting a delay damage claim on behalf of an owner are substantial and filing a lawsuit will not expedite the construction process. To the contrary, a lawsuit will more likely escalate hostilities and further aggravate the delay. As the party in control of construction funds, the owner also possesses a number of effective self-help remedies. Threatening termination of the nonperforming contractor, making demands upon the surety, or simply withholding payments for labor and materials otherwise due the delinquent contractor can be effective solutions to delay problems. Self-help remedies can sometimes result in the contractor's suing the owner for nonpayment. Even so, they should not be overlooked. The owner's primary goal, after all, is to occupy and utilize the new structure at the earliest available date. But for exceptional circumstances, litigation should not be pursued until this goal has been accomplished.

However, once the decision to pursue a delay claim has been reached, even if during the construction phase, preparation of the claim should begin immediately while sources of evidence and witnesses are available on the jobsite. Obtaining the same information from the opponent's lawyer later will be much more difficult and expensive.

DEVELOPING THE CLAIM

§ 14.2 Preparing the Claim

Generally speaking, an owner's claim for delay damages depends on the contract provisions, what was known at the bargaining table, and the owner's ability to prove damages with reasonable certainty. The success of the claim hinges on the owner's ability to prove that:

1. The project was delayed
2. The delays are compensable
3. The defendants caused the delay
4. Damages resulted.

Delay claims must be thoroughly prepared before the complaint is filed. Proper preparation requires close cooperation between the owner and the owner's attorneys and accountant. The accountant must be retained and consulted at the earliest possible date in order to streamline the investigative process and avoid duplicative work. The accountant will also be needed to determine the financial impact of the delay upon the owner by calculating both the direct and consequential damages. Factors involved in the calculation include evaluation of interest rates, inflation, additional overhead, lost profits, and loss of business value. An accountant's expertise is absolutely essential for a full understanding and development of these factors. The accountant's work can then be integrated into the legal analysis performed by the owner's attorney, so that the owner is reasonably well informed as to the strength of the potential delay claim. If there is anything worse for an owner than a delayed construction project, it is learning that a delay claim is not well-founded after substantial sums have been expended in litigation.

Working together, the owner, the architect, and the accountant should develop the claim along the lines discussed in §§ 14.3 through 14.6.

§ 14.3 Identifying Potential Defendants

In assessing any delay situation, the owner must be able to identify the specific people or entities who are participating in the construction project. That information is not always easy to obtain. If the owner has contracted separately with an architect and a series of contractors, the necessary roster of players may be set forth in the contract documents. But if the owner has contracted with a single entity to provide all design and construction services related to the project, the traditional roles played by designers and contractors may overlap and a party's title may not be particularly descriptive of the services provided on the site. It is therefore vitally important for the owner to determine the identity of the people who perform the following functions:

1. Construction. Is the owner dealing with a developer who is constructing a building itself, or with a general contractor who has delegated the construction work to several subcontractors?
2. Design. Who are the architects, engineers (structural, mechanical, and electrical), and other designers (landscape, interior, and industrial) involved on the project? Were they hired by the owner, by the developer, or by a constructor? Who are the designers' consultants?
3. Construction Management. Has an independent construction manager been retained to coordinate and schedule jobsite activities? If not, is construction management being handled by the architect, by the general contractor, or by some other entity? What role, if any, does the owner play in construction management?
4. Governmental Regulation. What government agencies have issued permits, conducted inspections, or otherwise been involved in the project? Which employees at these agencies are responsible for permits and inspections and the like?
5. Specialty and Trade Contractors. Who has been responsible for each discrete portion of the construction work, such as masonry, carpentry, roofing, drywall, piping, and insulation? Sub-subcontractors must be identified.
6. Bonds and Insurance. What companies have provided bid bonds, performance bonds, and payment bonds applicable to the project?

§ 14.4 OBTAINING DOCUMENTATION 429

Have any commercial insurance carriers issued business interruption insurance to the owner? What policies of insurance potentially provide coverage to the designers and the contractors?

7. Suppliers. Who has furnished the raw materials and equipment used in the construction process? Who retained them? Did they participate in the design process by preparing shop drawings?

It should also be recognized that in some instances a single company may be handling more than one aspect of the project. The corporate structure of the potential defendant should be fully understood before filing the complaint. Additionally, litigation can have a substantial impact on all entities possessing an equity interest in the property under construction. An owner contemplating litigation should seek the advice and consent of all business partners, co-owners, financiers, and insurers before filing suit.

§ 14.4 Obtaining Construction Documentation

As noted in § 14.1, construction delay claims are usually document-intensive. However, many construction documents can usually be obtained before filing suit without the expense of formal discovery. There is nothing improper in visiting the jobsite during the construction process to interview workers who may become witnesses, to take photographs, and to gather documents which may be used as evidence at trial.[1] Construction workers, especially if hired out of a union hall on a temporary basis, can be surprisingly forthcoming about construction progress and the cause of delays.

In addition to identifying potential witnesses, the owner should attempt to obtain as many of the following documents as possible:

1. Construction Schedules. The construction schedule is a document that plays an equally important role both in the bidding process and during the entire time when construction is progressing. The construction schedule can be as simple as a single paragraph in a construction contract setting forth the dates for commencement and completion of the project, or as complex as a computer-generated schedule based upon the Critical Path Model. Either way, the schedule, and all revisions to the schedule, must be obtained.

---

[1] These types of source materials can sometimes be obtained by visiting the site while the construction is in progress and asking the site supervisor for the documents. Although these documents can also be obtained through formal legal discovery after the lawsuit has been filed, the owner should try to obtain them in a more informal manner before deciding whether to file suit at all.

2. **Bidding Documents.** All bidding documents must be assembled, including materials submitted by unsuccessful bidders. A comparison between the successful and unsuccessful bids can often be quite enlightening. In addition, all bidding documents furnished by the owner, including plans, specifications, and standard contract documents, should be obtained and evaluated.

3. **Contractors' Logs.** Contractors typically keep detailed logs memorializing events that occur on the site. An owner will not be provided with these materials unless they are specifically requested. These logs can vary in form from a bound volume maintained in the construction trailer to a spiral note pad that the superintendent carries in his pocket. These logs are often considered to be personal property of the superintendent and may not be contained in the contractor's office records. The logs should be obtained from the construction personnel directly if possible.

4. **Progress Reports.** Contractors and architects typically provide periodic reports on the progress of the work. On larger projects, progress reports may be furnished to the owner on a daily basis. It is important to recognize that the reporting form is often executed in duplicate, and that the form that the contractor or architect retains in its records often contains more information than the form provided to the owner during the course of construction. Complete copies of these progress reports should be obtained.

5. **Delivery Tickets.** Material and equipment suppliers making deliveries to the jobsite typically execute documents verifying delivery dates and the quality and quantity of supplies delivered. Although such deliveries are typically summarized by contractors, the summaries are often unreliable and incomplete. The original delivery tickets should be obtained if possible.

6. **Union Records.** Construction unions retain a surprising amount of information concerning the work force on a given project. Because contractors make payments to each individual worker's union pension and welfare fund, records of these funds (if they can be obtained) can provide a very detailed breakdown of the labor force on site at any given point in time.

7. **Design Documents.** The owner should attempt to obtain a complete set of design documents, including all revisions. The design documents should be furnished not only by the architect but also by all of the architect's consultants. An effort should also be made to obtain shop drawings submitted by contractors, subcontractors, and equipment suppliers. Equipment suppliers often have a surprising amount of influence on the ultimate design, and that influence must be evaluated in each instance. Although architects often retain shop drawings, it may be necessary to obtain shop drawings

§ 14.5   IDENTIFYING CAUSES

from individual contractors. The architect's shop drawing log usually provides an outline of the parties who submitted shop drawings for review and should be obtained if possible.

8. Contract Documents. All contracts, including the general conditions to the contract, supplemental conditions to the contract, subcontracts, consulting contracts, insurance contracts, and contracts for materials and equipment, should be gathered.
9. Progress Photographs. Construction superintendents often keep their own photographic record of problems on the jobsite, in addition to whatever formal progress photographs may be contained in the contractor's home office. All photographs available should be obtained.
10. Change Order Proposals and Requests for Time Extensions. All correspondence from the contractor referencing requests for change orders and for extensions of time should be collected and organized.
11. Test Reports. All test reports, such as concrete slump tests, curtain wall water tests, and soils analyses, should be obtained.
12. Government Records. Efforts should be made to obtain the complete records of all governmental agencies who issued permits or conducted inspections of the jobsite. Any relevant building codes should also be obtained.

Other useful documentation may include payroll records, fabrication records from equipment and component suppliers, applicable insurance policies and certificates of insurance, almanacs and weather data, telephone memos, fax and telecopier logs, and equipment usage logs. This documentation should be obtained and evaluated by the owner's lawyer and accountant at the earliest available date because it will aid in the identification of the potential defendants. This step is crucial in order to ensure that the proper parties are sued within the applicable statute of limitations.[2] This information is also necessary to progress to the next step of analysis.

§ 14.5   Identifying the Causes of Delay

Virtually every party involved in the construction project, including the owner, can take actions which result in delay. Each significant event must be investigated and analyzed by the accountant and the lawyer to

---

[2] Many state legislatures have enacted special legislation establishing specific periods of limitation and repose for construction litigation. *See, e.g.*, Ill. Rev. Stat. ch. 110A, para. 13-214 (1977). The governing statute of limitations should be determined at the earliest possible date.

determine what effect, if any, it had on the overall completion of the construction project.

Some of the more common causes of delay on construction projects are:

1. Errors in Plans and Specifications. Construction design has a great impact on construction progress. Contractors typically rely blindly upon architectural plans and specifications, and errors or inconsistencies in those documents often result in delays.

2. Errors in Bidding Documents. Aside from those situations in which an unscrupulous contractor knowingly submits a low bid for purposes of securing work, honest errors in preparing bids are a fact of life. An owner should recognize that construction estimating is not an exact science and that contractors typically submit bids relying upon handshakes with their consultants, subcontractors, and suppliers. Clerical errors or a change in price by a supplier can have a devastating effect upon the contractor's profit and can materially effect the contractor's time for performance on the project. Although bid errors are sometimes the result of deficiencies in plans and specifications relied upon by the contractor in preparing the bid, the contractor's own errors in preparing the bids can be a touchstone for delay.

3. Delays in Construction Commencement. In the real world of construction, the commencement date specified by a contract often has no relationship to the date that construction activity begins in earnest. Neither does the date of formal ground-breaking ceremonies. Contractors often experience problems during the initial process of mobilizing forces. Weather can also be a factor in the delay.

4. Site Conditions. Actual conditions at the jobsite may cause delays at the outset, particularly with reference to subsurface conditions. Every construction contractor has encountered situations in which utility lines exist where none are marked on the plans and specifications, old foundations or archaeologically significant relics are unearthed, bedrock is found in an area in which compacted soil is expected, or the moisture content of the soil does not comport with the percentage reflected in the soil analysis.

5. Submittal Review. Architects and designers have an obligation to review shop drawing submittals and to answer requests for clarification in a timely manner. Although the time to do so reasonably varies depending on the volume and complexity of submittals, slow submittal review can delay construction.

6. Governmental Interference. Delays can be encountered in securing permits from local governmental authorities.

7. Change Orders. Changes are almost inevitable in most construction projects. Construction contracts typically provide that the change

§ 14.5 IDENTIFYING CAUSES

initiated by the owner will result in an equitable adjustment of the time allotted for completion of the project. Each change order authorized by the owner or initiated by the contractor should be closely evaluated.

8. Effects of Acceleration. In renovations to existing structures, the construction project is typically scheduled to proceed in stages so that the owner can continue to occupy the structure or some portion of it during renovation. Delays are often encountered when, for one reason or another, the planned flow of construction is disrupted, thereby requiring contractors to work in more than one stage of the project at the same time. Providing sufficient management during periods of acceleration is often a problem.

9. Effects of Construction Cycles. The effects of weather, the availability of skilled labor, and other problems occur cyclically in the construction industry. These cyclical changes can result in delays that may not be readily apparent.

10. Poor Construction Management. The function of construction management is, generally, to provide a line of communication between the owner and the tradespeople who ultimately perform the work. When the construction coordinator does not relay communications in a timely fashion or relays inaccurate or incomplete information, delays inevitably result.

11. Default by Subcontractors. The skilled tradespeople responsible for a discrete portion of the construct portion of the construction work can fail to perform. Bankruptcy can sometimes be the cause of such a default and is an increasingly common problem in the construction industry. It is important to inquire whether the surety responded in a timely fashion in these types of cases.

Obviously, every delay claim develops along its own set of facts. It should also be recognized that a delay in one discrete portion of the project, especially if the delay occurs early in the construction project, may have a substantial impact on overall job performance.[3] One day of delay during a critical time in the project can multiply itself exponentially as the job progresses.

---

[3] Tracking construction delays is a topic beyond the scope of this chapter. Excellent commentaries concerning the Critical Path Method can be found in Wickwire, Hurlburt, & Lerman, *Use of Critical Path Method Techniques in Contract Claims: Issues and Developments, 1974 to 1988,* 18 Pub. Constr. L.J. 338 (1989); Wickwire & Smith, *The Use of Critical Path Method Techniques in Contract Claims,* 7 Pub. Constr. L.J. 1 (1974).

§ 14.6 Focusing the Claim on the Responsible Party

Filing suit against both the contractor and the architect should be avoided as a general rule. Delay claims usually do not result in quick settlements, so suing both the contractor and the architect merely for the purpose of increasing the number of defendants is not likely to bring good results. To the contrary, should evidence developed through discovery indicate that the defendant was not sued in good faith after proper investigation, modern court rules may impose substantial attorneys' fees and court costs as a penalty against the erring owner.[4]

Although some learned commentators disagree,[5] from a practical standpoint there are good reasons why owners should hesitate to file suit for delay against both an architect and a contractor. A complaint filed by an owner against an architect may, in some circumstances, be construed as a judicial admission against the owner's interests in the owner's case against the contractor should the architect be found to have been acting as the owner's agent. More fundamentally, once named as defendants, both the architect and the contractor likely will refuse to cooperate should the owner require their testimony at trial. Architects and contractors are natural antagonists, and each can be a damning witness against the other so long as the owner does not alienate them.[6] Thus, unless good claims clearly exist against both the architect and the contractor, the owner should resist the temptation to bring claims against both in the same lawsuit.[7]

DEVELOPING LEGAL THEORIES OF LIABILITY

§ 14.7 Compensable and Noncompensable Delays

Although the attorney must take the lead in developing theories of legal liability for the owner's delay claim, the accountant's advice concerning damages (discussed in §§ 14.10–14.13) is essential to guide the lawyer in selecting the appropriate theory of recovery. In order to do so, the lawyer

---

[4] *See, e.g.,* Fed. R. Civ. P. 11; Ill. Rev. Stat. ch. 110A, para. 2-611 (1985).

[5] *See, e.g.,* Crewdson, R. L., *Recovering Delay Damages Against an Architect: An Owner's Approach,* 9 Constr. Law. 24 (1989).

[6] *See, e.g.,* Berman v. Rubin, 138 Ga. App. 849, 227 S.E.2d 802 (1976); Dresco Mechanical Contractors, Inc. v. Todd-CEA, Inc., 531 F.2d 1292 (5th Cir. 1976).

[7] *See* Hunt v. Blasius, 74 Ill. 2d 203, 384 N.E.2d 368 (1979), in which the court held that a contractor cannot be held liable for following plans and specifications unless those plans are so obviously incorrect that a reasonable contractor would not follow them.

and the accountant should work together in evaluating the compensable and noncompensable delays and potential defenses.

Only some of the extra costs incurred from delays can be recovered by the owner. A *noncompensable delay* is one that occurs without the fault of either contracting party.[8] Examples of noncompensable delays, sometimes referred to as "excusable" delays, are acts of God and disasters such as earthquakes, tornados, fires, floods, and strikes. As a result of a noncompensable delay the owner must grant a time extension to the contractor, but the contractor is not entitled to be reimbursed for its delay costs.[9] In a noncompensable delay situation, the owner may not assess liquidated damages. The owner may also not default the contractor and may not demand that the contractor adhere to its original schedule of completion by accelerating. The owner's time and money should not be wasted pursuing claims for these types of delays.[10]

A *compensable delay* is caused by one of the parties and allows the party harmed to recover damages as a result of that delay. Examples of such compensable delays include a failure of the contractor to properly staff a job, poor workmanship, failure to order materials and equipment in a timely fashion, failure of the general contractor to properly coordinate its subcontractors, and any contractor-caused delay which is not explained.[11] A contractor is thus generally responsible to the owner for delays caused by its own forces, its subcontractors, and its suppliers, and also for foreseeable delays caused by other parties or outside forces and circumstances except to the extent that the contract excuses such delays.[12]

§ 14.8 Anticipating Potential Defenses

Contract defenses available to potential defendants must be identified and evaluated at the earliest available date. For example, construction contracts commonly provide that the owner will purchase business risk insurance. Some courts have held that, by purchasing such insurance, the owner has waived its potential claims for business interruption and loss of

---

[8] Ryan III & Diestler, *Procedures and Preparations of Proof for Damage Claims Based on Construction Delays and Failures,* in Construction Litigation 4–17 (Illinois Institute for Continuing Legal Education 1987).

[9] B. Bramble & M. Callahan, Construction Delay Claims 83 (John Wiley & Sons 1987).

[10] *Id.* at 92.

[11] *See* Ryan III & Diestler, *Procedures and Preparations of Proof for Damage Claims Based on Construction Delays and Failures,* in Construction Litigation 4-19 (Illinois Institute for Continuing Legal Education 1987).

[12] Royal Ornamental Iron, Inc. v. Devon Bank, 32 Ill. App. 3d 101, 336 N.E.2d 105 (1975).

use of the premises or building[13] to the extent such claims are covered by insurance.

In addition to contract defenses, the owner should recognize that it is under an implied duty to cooperate in the performance of a contract.[14] Thus, if a contractor is able to show that the owner made it practically impossible for the builder to complete the construction in a timely fashion or contributed to the delays, the owner may not be able to recover damages from the contractor. Again, substantial savings of time and money may be realized by not pursuing claims which are clearly defensible. Focus on the claims that are strong and ignore those that are not. Weak claims will only cloud the issues.

§ 14.9 Mitigating Damages

The owner who suffers damages is required to do whatever is reasonable and foreseeable to mitigate or reduce those damages.[15] To the extent the owner had opportunities to mitigate damages and clearly made a decision not to do so, the damages may be reduced by the amount that could have been saved.[16]

Mitigation of damages should be quantified and documented, if possible, by listing savings achieved by reducing costs or postponing purchases. These types of actions may assist in completing the construction project more rapidly and in strengthening the owner's claim when litigated.

DEVELOPING ELEMENTS OF DAMAGES

§ 14.10 Contracts with Liquidated Damages Provisions

Developing the damages portion of the case requires close cooperation between the attorney and the accountant. Although the accountant will take the lead in quantifying the damages, the theory of liability selected by the lawyer can have a profound and sometimes limiting effect on the damages that may be recovered.

---

[13] *See, e.g.,* Portland Freight v. Canadian Imperial Bank of Commerce, 97 Or. App. 304, 776 P.2d 35 (1988); *see also* AIA B151-1987 ¶ 9.4.

[14] Gulf, Mobile & Ohio R.R. v. Illinois Cent. R.R., 128 F. Supp. 311 (D. Ala. 1954), *aff'd,* 225 F.2d 816 (5th Cir. 1955).

[15] B. Bramble & M. Callahan, Construction Delay Claims 232 (John Wiley & Sons 1987).

[16] *See, e.g.,* Gundersons, Inc. v. Tull, 678 P.2d 1061 (Colo. Ct. App. 1983), *remanded,* 709 P.2d 940 (Colo. 1985).

§ 14.10   LIQUIDATED DAMAGES PROVISIONS 437

Most modern construction contracts contain a liquidated damages provision. These provisions set out fixed amounts of money, usually assessed per day, that have been agreed upon by the parties as the amount of damages to be paid in the event of a breach of the contract. The purpose of agreeing to damage amounts in advance is to avoid litigation concerning the actual amount of damages sustained. Thus, although the actual damages may exceed or fall short of the amount stated in the liquidated damages provision, the sum agreed to in the contract usually governs the amount recoverable. Such a result is not always beneficial to the owner. When the actual damages greatly exceed the liquidated damages provided for by contract, pleading a claim based upon a liquidated damages provision probably forecloses the owner from recouping the full extent of its losses. A better strategy would probably involve a pleading demonstrating why the liquidated damages clause of the contract is not enforceable. Only in the exceptional case is an owner entitled to recover both liquidated damages and actual damages.[17]

If the owner causes the delay, the contractor is not liable for liquidated damages.[18] If both the contractor and the owner are partially at fault, the majority view is that no recovery will be allowed under the liquidated damages provision.[19] A minority view, however, allows the owner a partial recovery for the portion of the delay attributable to the contractor.[20]

Courts generally give effect to liquidated damages provisions and do not treat them as illegal penalties if the parties have expressed their agreement in clear and explicit terms.[21] If the clause has been inserted as a

---

[17] Traditionally, an owner could recover either liquidated or actual damages, but not both. Rex Trailer Co. v. United States, 350 U.S. 148 (1956). Although this is still the general rule, a few cases have allowed recovery of both liquidated and actual damages. *See, e.g.,* Hillsborough County Aviation Auth. v. Cone Bros. Constr. Co., 285 So. 2d 619 (Fla. Dist. Ct. App. 1973), in which the court held that an actual damages clause was to be given effect to the extent that it provided for recovery not covered by the liquidated damage clause. For a more in-depth discussion of liquidated damage provisions, see Clarkson, Miller, & Muris, *Liquidated Damages v. Penalties: Sense or Nonsense, 1978* Wis. L. Rev. 351 (1978), and Goetz & Scott, *Liquidated Damages, Penalties and the Just Compensation Principle: Some notes on an enforcement model and a theory of efficient breach,* 77 Colum. L. Rev. 554 (1977).

[18] United States *ex rel.* Gillioz v. John Kerns Constr. Co., 140 F.2d 792 (8th Cir. 1944).

[19] L.A. Reynolds Co. v. State Highway Comm'n, 271 N.C. 40, 155 S.E.2d 473 (1967); Kiewit & Sons' Co. v. Pasadena City Junior College Dist., 59 Cal. 2d 241, 379 P.2d 18, 28 Cal. Rptr. 714 (1963); Barnard-Curtiss Co. v. United States, 257 F.2d 565 (10th Cir. 1958); United States v. Texas Constr. Co., 224 F.2d 289 (5th Cir. 1955).

[20] Aetna Casualty & Sur. Co. v. Butte-Meade Sanitary Water Dist., 500 F. Supp. 193 (D.S.D. 1980); Southwest Eng'g Co. v. United States, 341 F.2d 998 (8th Cir.), *cert. denied,* 382 U.S. 819 (1965).

[21] Pick Fisheries, Inc. v. Berns Elecs. Sec. Servs., 35 Ill. App. 3d 467, 342 N.E.2d 105 (1976).

deterrent or penalty, however, simply to prevent a party from breaching the contract, the clause will be deemed unenforceable.[22]

§ 14.11 Contracts without Liquidated Damages Provisions

When a liquidated damage clause is not included in the contract or is not enforceable, the owner may recover *actual damages,* which encompass both direct and consequential damages.[23] In these types of situations, competent legal and accounting advice is essential to establish legally cognizable elements of damage.

Regardless of the theory behind recovery, it is important to remember that an owner is entitled to both direct and consequential damages. It is equally important to remember that proof of direct damages is generally simpler than proof of consequential damages.

§ 14.12 —Direct Damages

Direct damages are those that arise naturally and ordinarily from a breach of contract; they are damages which, in the ordinary course of human experience, can be expected to result from a breach.[24] Courts have held that such items as lost rental value, interest, the escalation and depreciation of the cost of materials, additional management fees and overhead, and damage claims for others constitute direct damages.

Lost rental value. If construction is delayed by the contractor, the owner is entitled to the reasonable rental value it would have received had the construction been completed on time.[25] The owner can recover such rental value despite the fact that more work is required to complete the project.[26] Rental values can be established using actual leases or comparable leases.

Interest. Construction delays often result in excess interest costs to owners. This is because, while costs have been expended towards construction

---

[22] *Id.*
[23] B. Bramble & M. Callahan, Construction Delay Claims 228-32 (1987).
[24] Roanoke Hosp. Assoc. v. Doyle & Russell, Inc., 215 Va. 796, 214 S.E.2d 155, 160 (1975).
[25] Marshall v. Carl F. Shultz, Inc., 438 So. 2d 533, 534 (Fla. Dist. Ct. App. 1983); Gregory v. Weber, 51 Or. App. 547, 626 P.2d 392, 396 (1981); J. Clutter Custom Digging v. English, 181 Ind. App. 603, 393 N.E.2d 230, 235 (1979); Burnett & Doty Dev. Co. v. Phillips, 84 Cal. App. 3d 384, 148 Cal. Rptr. 569 (1978); Herbert & Brooner Constr. Co. v. Golden, 499 S.W.2d 541, 549 (Mo. Ct. App. 1973).
[26] *See, e.g.,* Ryan v. Thurmond, 481 S.W.2d 199 (Tex. Ct. App. 1972) *writ ref'd n.r.e.*

§ 14.12 DIRECT DAMAGES 439

and interest is being paid on loans to finance those costs, the revenue generating ability of that property is not available. If the interest costs for the project stand alone and can be isolated readily, then the interest during the delay period is an appropriate base to start with in determining damages. Additional interest may also result from money used to finance inventory.

Added interest costs that arise from the owner's extended borrowing as a result of the delay are classified as direct damages. These interest costs as well as the interest revenue lost during such an extended term are predictable results of the delay and are, therefore, compensable direct damages.[27] Interest costs that are attributable to higher interest rates during the delay period are considered to be consequential damages, however, and are compensable only if they were within the contemplation of the parties at the time the contract was executed.[28] In addition, the typical construction loan is at a higher interest rate than the permanent loan. The difference in interest rates between the construction loan and the permanent loan may also be recoverable for the delay period.

It is important for the owner to analyze actual out-of-pocket costs, anticipated interest costs, and the forecasted start-up of the revenue-generating capability of the project. There may be times when delayed construction actually results in an interest savings to the owner.

Storage and cost of materials—escalation and depreciation. If materials, furniture, or fixtures have been purchased and stored in anticipation of the completion date and this was known or reasonably contemplated by both parties, the storage costs during the delay period may be recoverable. In addition, if the purchase of materials was delayed in anticipation of the completion of the project, the owner should be able to recover damages for any price increases that occurred during the period of delay. Actual depreciation may also be recovered by the owner if the building, furniture, fixtures, or installed equipment cannot be used until the building is completed. The key word here, however, is "actual." The owner must be able to show that the assets are worth less as a result of the period of delay.[29]

Additional management fees and overhead. Excess costs can be sought by an owner when a project is delayed as a result of the need to fulfill

---

[27] Roanoke Hosp. Assoc. v. Doyle & Russell, Inc., 215 Va. 796, 214 S.E.2d 155, 161 (1975). *See also,* Herbert & Brooner Constr. Co. v. Golden, 499 S.W.2d 541 (Mo. Ct. App. 1973).

[28] *Id.*

[29] *See, e.g.,* County Excavation, Inc. v. New York, 44 Misc. 2d 1057, 255 N.Y.S.2d 708 (Ct. Cl. 1964).

commitments to customers or clients.[30] These costs may consist of outside warehousing, subcontracting of production or service capability at a higher cost, providing replacement rental or lease space or temporary housing for customers, excess consulting and project management costs, excess maintenance and utility costs, and any other costs incurred by the owner during the delayed construction period. Such costs should be documented with actual invoices or bills and a description of the services provided or other reasons for expenditures. The owner must carefully compare these additional costs to the originally anticipated costs for these services if they were to be provided anyway, to determine whether these costs are in fact damages as opposed to cost savings.

Damage claims for others. In the case of a project with multiple prime contractors, the owner may be faced with a delay claim from several contractors caused by yet another contractor. These claims can include all of the items of damages typical of a contractor delay claim and therefore should become a part of the owner's claim against the contractor who caused the delay. Proof of these damages can be obtained through the use of the submitted claims of the contractors or in judgments against the owner through trial or arbitration.

The owner should be cautious when settling disputes with contractors for delay damages when the owner has a claim against another contractor who caused the delay. Settlement agreements should be reviewed to be sure they will not impair the owner's claim against the contractor at fault.

§ 14.13 —Consequential Damages

Consequential damages are those that arise from special circumstances that were within the contemplation of the parties at the time the contract was made.[31] In the words of one court, "if the damages suffered do not usually flow from the breach, then it must be established that the special circumstances giving rise to them should have reasonably been anticipated at the time the contract was made."[32] This rule is codified in § 351 of the *Restatement (Second) of Contracts,* which states:

(1) Damages are not recoverable for loss that the party in breach did not have reason to foresee as a probable result of the breach when the contract was made.

---

[30] *See, e.g.,* Northern Petrochem. Co. v. Thorsen & Thorshov, Inc., 297 Minn. 118, 211 N.W.2d 159, 166 (1973).

[31] Roanoke Hosp. Assoc. v. Doyle & Russell, Inc., 215 Va. 796, 214 S.E.2d 155, 160 (1975).

[32] Spang Indus., Inc. v. Aetna Casualty & Sur. Co., 512 F.2d 365, 368 (2d Cir. 1975).

(2) Loss may be foreseeable as a probable result of a breach because it follows from the breach (a) in the ordinary course of events, or (b) as a result of special circumstances, beyond the ordinary course of events that the party in breach had reason to know.

Although the determination of whether damages are direct or consequential is a question of law, the determination of whether special circumstances were within the contemplation of the parties at the time of the contract is a question of fact.[33] The elements of proof are, accordingly, more complex than those required in a case of direct damages.

The owner's ability to recover consequential damages may be the product of several factors, including the relationship between the owner and the contractor, the special circumstances of the contract, and, most importantly, the contractor's knowledge or reasonable expectation of the owner's need for an intended use of the project. Examples of recoverable consequential damages are lost profits and sales from either existing or new businesses, increased production costs, and loss of business value.

Lost profits and sales from an existing business. When the structure being built is for an existing business, and the business has been interrupted or interfered with for a longer-than-anticipated period of time, lost profits can be measured by looking at the revenue generation and profitability of the business prior to, during, and after the construction. The estimated effect of outside market factors and economic conditions must also be taken into consideration.

Measuring the damages resulting from business interruption caused by a delayed construction project can be done in a number of ways. One of the most common methods is to develop standards or trends on each of the components of the income statement prior to construction, and analyze these same components during the construction period to determine the differences. There may be both a permanent and a temporary impact on lost profits. The temporary impact is the actual change during the delayed construction period.[34] The permanent impact may be a change that occurred during this same construction period but which is not recoverable following the completion of construction. This type of impact may consist of a reduction in price, profit margin, or market share, or lost opportunities because of the inability to supply current and future customers with products or services by the contractual completion date.[35]

Lost sales can be measured in the same manner as lost profits: evaluate sales prior to the construction period and analyze the effect or reduction

---

[33] Roanoke Hosp. Assoc. v. Doyle & Russell, Inc., 215 Va. 796, 214 S.E.2d 155, 160 (1975).

[34] For an excellent treatment of lost profits, see R. Dunn, Recovery of Damages for Lost Profits (1987).

[35] *Id.*

in sales during the construction period. Specifically, this can be accomplished by analyzing trends in sales and sales on a customer-by-customer or product-by-product basis, if the information is available. It is helpful to estimate the effect of outside market conditions, economic factors, and any other conditions which were not the result of the construction delay, in order to more clearly determine the changes in sales that resulted directly from the delay.

A permanent or semipermanent decline in sales may result from being unable to bid on a supply contract to customers in a long-lead-time or a once-a-year order-type environment. Lost sales for a product can be measured by looking at the total market for that product, the company's prior market share or expected market share for the period, and the orders and prices that competitors obtained, if available. There may be other more appropriate means of estimation, depending on the information available to the owner.

Lost profits from a new business. In a new business venture when the owner experiences a delayed start-up, the owner is faced with the task of measuring what profits the business would have generated had the construction project been completed on time, because there will be no prior-to-construction business records to analyze. Lost sales from a delayed new business can be measured by the actual performance of the company subsequent to the completion of the construction, adjusted for damages resulting from the late start.[36] Lost sales for a new business that failed altogether because of the delay can be determined by estimating the sales the company would have had if its business had opened on the original contract completion date. In these types of lost profit cases, a judge or jury may rely on probable and inferential as well as direct and positive proof.[37]

The owner must be careful to consider comparable economic and financial information as well as the specific features of the business in measuring lost sales or revenues. Growth rates for similar companies in that industry may be appropriate as a means of measuring sales or revenue. The owner should be sure to adjust any sales and growth rates by factors

---

[36] Harsha v. State Sav. Bank, 346 N.W.2d 791, 798 (Iowa 1984); Cook Assocs., Inc. v. Warnick, 664 P.2d 1161, 1166 (Utah 1983); Multivision Northwest, Inc. v. Jerrold Elecs. Corp., 356 F. Supp. 207, 217 (N.D. Ga. 1972); Ferrel v. Elrod, 63 Tenn. App. 129, 469 S.W.2d 678, 686 (1971); Pace Corp. v. Jackson, 284 S.W.2d 340, 348 (Tex. 1955).

[37] American Bldgs. Co. v. DBH Attachments, Inc., 676 S.W.2d 558, 562 (Tenn. Ct. App. 1984); Delahanty v. First Pa. Bank, 318 Pa. Super. 90, 464 A.2d 1243 (1983); Fields Eng'g Equip. v. Cargill, Inc., 651 F.2d 589, 593 (8th Cir. 1981). For more cases specifically involving building construction, *see* Burnett & Doty Dev. Co. v. Phillips, 84 Cal. App. 3d 384, 148 Cal. Rptr. 569, 571 (1978); Natco, Inc. v. Williams Bros. Eng'g Co., 489 F.2d 639, 640 (5th Cir. 1974).

such as the regional market it had intended to and was capable of serving, any customer or client relationships which would have caused its experience to deviate from average statistics, and any special services or features of the product which would have allowed the company to enjoy greater or less than average sales.

Changes in production costs. In a product manufacturing environment, the company needs to measure the anticipated cost of production or changes to the cost of production during the construction period. In the case of a business interruption, cost of production trends can be measured in a manner similar to sales, except that production costs are made up of several components. These components can include the cost of materials, labor, freight, utilities, plant overhead, and general and administrative expenses. Fluctuation in labor costs may stem from labor contract agreements that are modified during the construction as well as changes that occur in the experience level of the labor force because of the delay. Changes in the type of materials available or a cost increase in the materials originally called for may result from a construction delay. Similarly, utility costs may also be affected by outside rate increases during the delay; however, the increase may simply be a result of the additional use of these utilities during the construction period. If this increased use of utilities is extended because of the delay, the owner may be able to obtain damages.

Changes in costs of factory or plant overhead will probably contain both a variable and fixed component. These costs should be broken down into detailed components, if possible, to try to develop an understanding of their behavior prior to and during the construction period. Although cost components can be analyzed independently, they can also be combined in order to estimate their effect on the increased cost of production as a result of the delay.

General and administrative expenses are made up of the ongoing expenses of running a company. They too can be broken down into fixed and variable components. Although not always necessary or possible, it is helpful to analyze these costs at a line item level to determine the behavior of each line during, prior to, and subsequent to the construction period. These types of costs can be analyzed by using trends and taking into consideration any outside factors affecting their behavior.

Loss of business value. When the completion of a construction project was delayed and a business failed or suffered a reduction in value, some or all of this failure/reduction may be attributable to the delay. The owner must measure what the business value could have been but for the delay and once again isolate the effects of outside factors and owner actions on the business value to determine the damages actually caused by the delay.

The business can be valued using a variety of methods. The owner should examine the available methods to determine which one is most appropriate for each specific case. One method is the income revenue and cost forecast approach, which uses a capital asset pricing model.[38] Another method would be to measure goodwill.[39] Any method used should take into consideration a discount factor in order to value the company at the appropriate period of time.

DEVELOPING LITIGATION STRATEGIES

§ 14.14 Selecting a Forum

The forum used to resolve construction disputes can be critically important. Assuming that a contractor has not already set the forum by initiating a delay claim against the owner, the owner should weigh the relative merits of proceeding in arbitration or, alternatively, in a state or federal court of law. Generally speaking, arbitration proceedings are less expensive and less time-consuming, but may result in a compromised verdict. Actions usually proceed more rapidly in federal courts than in state courts, especially in urban areas in which state court dockets are congested. Finally, a decision must be made as to whether the claim will be presented to a judge in a bench trial or to a jury. Again, generally, jury trials tend to be more protracted and expensive, but sometimes result in more dramatic verdicts.

§ 14.15 Presenting the Claim

The use of graphs and charts is the only effective way to present evidence in a delay claim. In addition to their visual impact, well-prepared charts and graphs provide several tactical advantages for the trial lawyer. First, working closely with the accountant to prepare these charts and graphs forces the lawyer to organize his thoughts and proofs in a concise fashion. Secondly, the chart serves as a blueprint for the lawyer's direct examination of the accountant and as a memory aid for the accountant when testifying as a witness. Third, once admitted into evidence these exhibits eliminate the need for the trial lawyer to rely on the reams of documentation used to prepare these charts and graphs.

---

[38] J.C. Van Horne, Financial Management and Policy (1986).
[39] *Id.*

§ 14.15 PRESENTING CLAIM

Of course, a chart or graph will not be allowed into evidence until the proper foundation has been presented. Laying the foundation for the charts and graphs is a tedious but necessary process. All of the documents relied on by the lawyer and the accountant in preparing the charts and graphs must be authenticated, properly identified, and admitted into evidence. To assist in this process, it is a good idea to prepare a well-organized and detailed volume summarizing and describing each single document according to a numerical or computerized index system. The volume should be neatly bound and cross-indexed to the graphs and charts which have been created from the construction documents. The bound volume should be presented to the court and opposing counsel at the start of the trial in the same manner as a trial brief. The volume will assist the lawyers and the judge (or arbitrator) in tracking the documents which have been marked for use in testimony and admitted into evidence. Once the raw documents have been admitted into evidence, the bound volume will be devastatingly effective in laying the foundation for the charts and graphs.

The presentation of delay damages should be made with a simple exhibit by using a graph showing major categories, as in **Figure 14-1**. Each category can have its own series of exhibits to further explain the calculations. The cause of delay and concurrent delays can be illustrated using a simple barchart exhibit, as in **Figure 14-2**. Simple exhibits are typically more effective than complex ones; exhibits that are too complex can become confusing. A different set of charts may be needed to explain the effect of the delay on a company's performance, as, for example, in **Figure 14-3**.

Another method for demonstrating lost revenue, changes in costs of production, and other company costs to determine lost profits is a forecasted income statement for the delay period. The income statement should be prepared in graphic form to demonstrate what the company's income would have looked like had the construction been completed on time. Factors such as reduced interest costs resulting from lower levels of

Damages	
Lost Profits	$1,000,000
Consulting Fees	65,000
Equipment Storage	25,000
Interest	370,000
Total Damages	$1,460,000

Figure 14-1.

DELAY CLAIMS

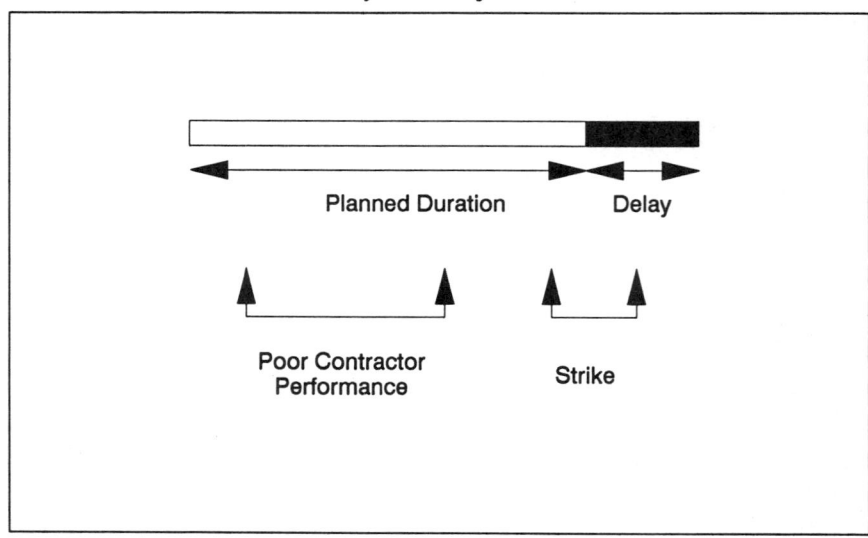

Figure 14–2.

Figure 14–3.

working capital must be considered in measuring company income. When the final income statement has been constructed, the owner can subtract the actual income from the estimated income to demonstrate damages. The owner should then attempt to isolate the changes in income caused by outside factors such as owner's actions, price changes, competitive conditions, and economic factors to determine those damages strictly attributable to the construction delay.

Still another method for measuring and presenting lost profits is to individually add up incidents of lost sales or reduced orders and to detail changes and costs. Rather than estimating the difference between what could have been and what actually occurred, this method attempts to isolate individual components of damage and then presents their total. It is important when measuring lost profits to treat each case individually and to use the information available for that case to develop the appropriate methodology. There may be other appropriate methods of measurement depending on the specific circumstances.

§ 14.16 Conclusion

Bringing a delay claim against a designer or contractor can be an expensive and time-consuming process for an owner. Litigation of the claim should be pursued only as a last resort after all self-help remedies have been exhausted. An owner should take time to select competent professionals to assist in the careful preparation and analysis of available evidence before filing suit. The key to recovering delay damages for an owner is close cooperation between the owner, the accountant, and the owner's attorney.

§ 14.17 Bibliography

Annotation, *Modern Status of Rule as to Whether Cost of Correction or Difference in Value of Structures is Proper Measure of Damages for Breach of Construction Contract,* 41 A.L.R.4th 131 (1985).

Annotation, *Recovery of Punitive Damages for Breach of Building or Construction Contract,* 40 A.L.R.4th 110 (1985).

Annotation, *Insurance: Subrogation of Insurer Compensating Owner or Contractor for Loss Under "Builder's Risk" Policy Against Allegedly Negligent Contractor or Subcontractor,* 22 A.L.R.4th 701 (1983).

Annotation, *Contractual Provision for Per Diem Payments for Delay in Performance as One for Liquidated Damages or Penalty,* 12 A.L.R.4th 891 (1982).

Annotation, *Coverage under Builder's Risk Insurance Policy,* 97 A.L.R.3d 1270 (1980).

Annotation, *Validity and Construction of "No Damage" Clause with Respect to Delay in Building or Construction Contract,* 74 A.L.R.3d 187 (1976).

Annotation, *Breach or Repudiation of Contract as Affecting Right to Enforce Arbitration Clause Therein,* 32 A.L.R.3d 377 (1970).

Annotation, *Building Contractor's Liability, Upon Bond or Other Agreement to Indemnify Owner For Injury or Death of Third Persons Resulting From Owner's Negligence,* 27 A.L.R.3d 663 (1969).

Annotation, *Builder's Risk Insurance Policies,* 94 A.L.R.2d 221 (1964).

Annotation, *Construction and Effect of a "Changed Conditions" Clause in a Public Works or Construction Contract,* 85 A.L.R.2d 211 (1962).

Annotation, *Liability of Contractor or Owner of Building Being Demolished for Injuries to Infant On Premises,* 64 A.L.R.2d 972 (1959).

Annotation, *Construction and Effect of Provision in Private Building and Construction Contract that Work Must Be Done to Satisfaction of Owner,* 44 A.L.R.2d 1114 (1955).

Annotation, *Liability of Building or Owner of Building Under Construction for Injuries Received On Premises by Infant,* 44 A.L.R.2d 1253 (1955).

Balkin, *Recovery of Damages for Delay Against Sureties on Public Works Bonds,* 20 Forum 640 (1985).

Callahan & Hohns, *Construction Schedules.* The Mitchie Co., 1983.

Closen & Weiland, *The Construction Industry Bidding Cases: Application of Traditional Contract, Promissory Estoppel, and Other Theories to the Relations Between General Contractors and Subcontractors,* 13 J. Marshall L. Rev. 565 (1980).

Cox, *Owner Damages,* Construction Briefings, Mar. 1983.

Crewdson, R. L., *Recovering Delay Damages Against an Architect: An Owner's Approach,* 9(1) Constr. Law 24 (1989).

Cushman, *Proving Liability and Damage to the Owner in Construction and Design Claims* (ABA 7th annual fall meeting, Nov. 1983).

Foster, *Construction Management and Design-Build/Fast Track Construction: A Solution which Uncovers a Problem For the Surety,* 46 Law & Contemp. Probs. 95 (1983).

Gerber, *Building Contracts-Breach-Damages-Mitigation-Appropriate Measure for Damages,* 57 Austl. L.J. 583 (1983).

§ 14.17 BIBLIOGRAPHY

Goldberg, *The Owner's Duty to Coordinate Multi-Prime Construction Contractors, a Condition of Cooperation,* 28 Emory L.J. 377 (1979).

Goldenhersh, *Pitfalls to Avoid in Construction Contracts,* 14 Colo. Law. 217 (1985).

Groton, *Should the Owner Sign an Arbitration Agreement in a Construction Contract?,* 43 Arb. J. 5 (1988).

Hagberg, *Mistake in Bid, Including New Procedures Under Contract Disputes Act of 1978,* 13 Pub. Constr. L.J. 257 (1983).

Handelman, *When Owners Should Pay for Construction Delay: Corinno Civetta Construction Corporation and No-Damage-For-Delay Clause,* 9 Cardozo L. Rev. 879 (1987).

Handley, *Liability of Design Professionals Arising Out of Shop Drawing Review,* 52 Ins. Couns. J. 311 (1985).

Harrington, Thum, & Clark, *The Owner's Warranty of the Plans and Specifications for a Construction Project,* 14 Pub. Cont. L.J. 240 (1984).

Howell, *Unusually Severe Weather in Government Construction Contracting,* 23 A.F. L. Rev. 156 (1982–83).

King & Epstein, *Owner's Counsel Reviews (and suggests changes to) the New (1987) AIA General Conditions of the Contract for Construction,* 5 Prac. Real. Est. Law. 9 (1989).

Kiwala, *Introduction to Construction Litigation,* 3 Prac. Real. Est. Law. 21 (1987).

Leslie, *When Trouble Comes—Contractor Default, Representing the Owner,* 8 Forum 493 (1973).

Lesser, *The No Damage for Delay Clause: Avoiding Delay Claims in Construction,* 10 Trial Diplomacy J. 16 (1987).

Liddle, *Authorization for Extra Work Under Building Contract,* 7 Am. Jur. P.O.F.2d 239 (1975).

Locke, *Building and Construction Contracts—Waiver of Provision Requiring Written Change Orders,* 25 Am. Jur. P.O.F.2d 561 (1981).

Locke, *Building Contractor's Justification For Walking Off Jobs,* 2 Am. Jur. P.O.F.2d 321 (1974).

Mairose, *Subsurfco, Inc. v. B.Y. which District: Applying the Diminution in Value Measure of Damages for Breach of Rural Water Construction Contracts,* 29 S.D.L. Rev. 445 (1984).

Marcus, *Building a Construction Contract that Works: The Owner's Role,* 44 Arb. J. 3 (1989).

McCarter, *Construction Litigation,* 45 J. Mo. B. 125 (1989).

McClure, *Differing Site Conditions: Evaluating the Material Difference,* 15 Pub. Cont. L.J. 138 (1984).

Murphy, *Promissory Estoppel: Subcontractors' Liability in Construction Bidding Cases,* 63 N.C.L. Rev. 387 (1985).

Musa, *Doing Business with Insolvent Developers or Contractors,* 12 Real Est. L.J. 3 (1983).

Nielsen & Galloway, *Proof Development for Construction Litigation,* 7 Am. J. Trial Advoc. 433 (1984).

Pitcher, *Owner's Unreasonable Rejection of Building Contractor's Work,* 8 Am. Jur. P.O.F.2d 717 (1976).

Pitcher, *Responsibility for Delay in Completing Building Construction,* 5 Am. Jur. P.O.F.2d 371 (1975).

Plotnick, *Understanding & Proving Construction Delay Claims,* 2 Prac. Real. Est. Law. 67 (1986).

Ryan & Diestler, *Procedures and Preparations of Proof for Damage Claims Based on Construction Delays and Failures,* Construction Litigation. Illinois Institute for Continuing Legal Education, 1987.

Schenk, *Construction Wars: The Truce Lies in the Bargain,* 19 Real Prop. Prob. & Tr. J. 766 (1984).

Sklar, *Pitfalls of the Construction Contract,* 33 Decalogue J. 14 (1987).

Treacy, *No Pay for Delay,* 21 Pub. Cont. 12 (1986).

Wickwire, Hurlburt, & Lerman, *Use of Critical Path Method Techniques in Contract Claims: Issues and Developments, 1974 to 1988,* 18 Pub. Cont. L.J. 338 (1989).

Wickwire & Smith, *The Use of Critical Path Method Techniques in Contract Claims,* 7 Pub. Cont. L.J. 1 (1974).

CHAPTER 15

CLAIMS FOR LOST PROFIT

Herman M. Braude, Esquire
George D. Sergio, CPA

Herman M. Braude received a B.S. in Civil Engineering in 1961 from Cooper Union, and continued his education at Georgetown University, where he was editor of the Law Journal from 1963 to 1964. He earned his J.D. while working fulltime as a construction engineer, specification writer, and contract specialist for various government agencies. After graduation he interned as a law clerk to Judge Samuel E. Whitaker, sitting on the United States Court of Claims and by designation on the United States Court of Appeals for the Fifth Circuit from 1964 to 1965. He is presently the senior partner in the Washington-Baltimore law firm of Braude & Margulies, P.C., which specializes in construction law. Mr. Braude has been a visiting lecturer on construction contracting at the National Law Center at George Washington University since 1985 and has written over a dozen articles on the subject.

George D. Sergio is executive vice president and chief financial officer of WWF Paper Corporation, a large independent paper distributor. His broad based experience spans 22 years including a diversified financial consulting background in the areas of construction, bankruptcies and troubled companies. Prior to joining WWF Paper Corporation, George served as partner-in-charge of the Philadelphia Real Estate, Construction, and Environmental Services Group for Coopers & Lybrand.

§ 15.1   Introduction
§ 15.2   Foreseeability
§ 15.3   Proximate Cause

§ 15.4   Reasonable Certainty
§ 15.5   Calculation of Profits
§ 15.6   —Impairment of Corporate Value
§ 15.7   Unestablished Business versus Established Business
§ 15.8   Proof Techniques
§ 15.9   Hypothetical Lost Profits Case
§ 15.10 Conclusion

§ 15.1 Introduction

Lost profits are generally considered special or consequential damages and as such they must be specifically pled and proven.[1] In order to receive damages for lost profits, it is generally accepted that the owner must prove that the lost profits (1) were foreseeable by the contracting parties at the time of the contracting (see § 15.2),[2] (2) were proximately caused by the breach of the contract (see § 15.3),[3] and (3) were calculated with reasonable certainty (see § 15.4).[4] Some jurisdictions distinguish the lost profits of an established business from the lost profits of an unestablished business (see § 15.6).[5] The dwindling number of courts that distinguish the latter instance do so because they view lost profits for an unestablished business, with no history of past profits, as always being speculative

---

[1] Cox, Tieder, & Denian, Owner Damages—Basic Principles and Guidelines (Mar. 1983); Comment, *Remedies-Lost Profits as Contract Damages for an Unestablished Business: The New Business Role Becomes Outdated,* 56 N.C.L. Rev. 695-96 (1978) [hereinafter Comment, *Remedies-Lost Profits*].

[2] Burnett & Doty Dev. Co. v. Phillips, 84 Cal. App. 3d 384, 148 Cal. Rptr. 569, 571 (1978); Fields Eng'g & Equip. v. Cargill, Inc., 651 F.2d 589, 593 (8th Cir. 1981); Kenford Co. v. County of Erie, 67 N.Y.2d 257, 493 N.E.2d 234, 235, 502 N.Y.S.2d 131 (1986); Drews Co. v. Ledwith-Wolfe Assocs., Inc., 296 S.C. 207, 371 S.E.2d 532, 535 (1980).

[3] Fields Eng'g & Equip. v. Cargill, Inc., 651 F.2d 589, 593 (8th Cir. 1981); McDermott v. Middle East Carpet Co., 811 F.2d 1422 (11th Cir. 1987); Dobbs, Handbook on the Law of Remedies 798 (1973).

[4] McDermott v. Middle East Carpet Co., 811 F.2d 1422 (11th Cir. 1987); Fields Eng'g & Equip. v. Cargill, Inc., 651 F.2d 589 (8th Cir. 1981); Drews Co. v. Ledwith-Wolfe Assocs., Inc., 296 S.C. 207, 371 S.E.2d 532 (1988); Evergreen Amusement Corp. v. Milstead, 206 Md. 610, 112 A.2d 901 (1955); S. Jon Kreedman & Co. v. Meyers Bros. Parking-West Corp., 58 Cal. App. 3d 173, 130 Cal. Rptr. 41 (1976); Knier v. Azores Constr. Co., 78 Nev. 20, 368 P.2d 673 (1972); Alpine Indus., Inc. v. Gohl, 30 Wash. App. 750, 637 P.2d 998 (1981). *See, generally,* Note, *The Requirement of Certainty in the Proof of Lost Profits,* 64 Harv. L. Rev. 317 (1950).

[5] *See* Evergreen Amusement Corp. v. Milstead, 206 Md. 610, 112 A.2d 901, 904 (1955).

and, therefore, not recoverable.[6] Evidence of lost profits may be proffered by using such proof techniques as showing prior or subsequent profits of a particular business, showing analogies to the profits of a similar business, and showing lost profits through the use of expert witnesses.[7]

This chapter will review the three elements necessary to prove lost profits and the methods for calculating them. In addition, it will review the distinction courts make between established and unestablished businesses and will offer suggestions on proof techniques to show lost profits in either instance.

§ 15.2 Foreseeability

In order to recover lost profits, an owner must show that lost profits were reasonably foreseeable at the time of contracting as a natural consequence of a breach.[8] Whether the parties actually foresaw that loss of profits would result from breach is not an issue.[9] It is enough that lost profits were reasonably foreseeable at the time of contracting. Whether lost profits were reasonably foreseeable is a question of fact.[10]

Although foreseeability must be proven in order to recover lost profits, as a practical matter foreseeability of lost profits to the owner is not a difficult element to prove in construction cases.[11] For example, in *Burnett & Doty Development Co. v. Phillips,*[12] a developer brought suit against a contractor for lost profits resulting from delay. The California Court of Appeals held that the fact that the defendant was aware that a profit was anticipated upon the completion of the project and the fact that the contract contained a specific date of completion was sufficient evidence to support a finding that lost profits were reasonably foreseeable.

---

[6] *See id.;* Comment, *Remedies-Lost Profits* at 1.

[7] *See* Dunn, Recovery of Damages for Lost Profits § 7.1 (1978) [hereinafter Dunn].

[8] Burnett & Doty Dev. Co. v. Phillips, 84 Cal. App. 3d 384, 148 Cal. Rptr. 569, 571 (1978); Fields Eng'g & Equip. v. Cargill, Inc., 651 F.2d 589, 593 (8th Cir. 1981); Kenford Co. v. County of Erie, 67 N.Y.2d 257, 493 N.E.2d 234, 235, 502 N.Y.S.2d 131 (1986); Drews Co. v. Ledwith-Wolfe Assocs., Inc., 296 S.C. 207, 371 S.E.2d 532, 535 (1980).

[9] Sears, Roebuck & Co. v. Goudie, 290 A.2d 826, 832, 833 (D.C. 1972); Cox, Tieder & Denian, *Owner Damages—Basic Principles and Guidelines,* 87-3 Construction Briefings 147 (Mar. 1983).

[10] Roanoke Hosp. Ass'n v. Doyle & Russell, Inc., 215 Va. 796, 214 S.E.2d 155 (1975); Cox, Tieder & Denian, *Owner Damages—Basic Principles and Guidelines,* 87-3 Construction Briefings 147 (Mar. 1983).

[11] *See* Comment, *Remedies-Lost Profits* at 700.

[12] 84 Cal. App. 3d 384, 148 Cal. Rptr. 569 (1978).

The Supreme Court of Pennsylvania in *Exton Drive-in, Inc. v. Home Indemnity Co.*[13] employed similar reasoning to hold that lost profits were foreseeable. In *Exton,* the owner of a drive-in movie theater brought suit against a paving contractor for loss of profits as a result of the contractor's delay. The contract contained a clause requiring completion of the contract on a specified date and stated that "time was of the essence."[14] The *Exton* court asserted that "[t]his provision as to timely performance gave [the contractor] notice that a delay in completion of the contract would delay the opening of the theater . . . . Losses caused by [the contractor's] failure to timely perform were accordingly foreseeable."[15] *Exton* illustrates how contract provisions may be used to prove foreseeability.

It is also possible to look outside the contract provisions to show that lost profits are reasonably foreseeable. In *Sears, Roebuck & Co. v. Goudie,*[16] the parties entered into a contract to install air conditioning in a building. Part of the building was a coffeehouse operated by Goudie under a corporation that she owned. The rest of the building was Goudie's private residence. When the air conditioning system proved to be inadequate, Goudie brought suit against the contractor and the subcontractor for loss of "net" profits that would have been paid to Goudie as her income from the coffeehouse. The *Goudie* court asserted that lost profits from the coffeehouse business were foreseeable because the contractor was aware that Goudie operated the coffeehouse, from discussions that took place at the time of contracting, from the contractor's inspection of the premises, and from the fact that the architectural plans of the building were made available to the contractor. The *Goudie* case illustrates how special circumstances made known to the contractor at the time of contracting may be used to prove foreseeability.

Some cases suggest that when lost profits extend for a substantial length of time they may not be foreseeable. For example, in *Fields Engineering & Equipment, Inc. v. Cargill,*[17] an owner entered into a contract for the construction of an overhead grain conveyor-belt system. When the contractor failed to complete the system, the Eighth Circuit Court of Appeals held that lost profits for the following season were not foreseeable because the owner had sufficient time to complete the remaining construction.

In *Kenford Co. v. County of Erie,*[18] a landowner brought a multimillion dollar suit for lost profits against the County of Erie for failure to construct a domed stadium under a contract in which the owner was to

---

[13] 436 Pa. 480, 261 A.2d 319, 323–24 (1969).

[14] *Id.,* 261 A.2d at 323.

[15] *Id.*

[16] 290 A.2d 826 (D.C. 1972).

[17] 651 F.2d 589 (8th Cir. 1981).

[18] 67 N.Y.2d 257, 493 N.E.2d 234, 502 N.Y.S.2d 131 (1986).

receive a 20-year management contract. The New York Court of Appeals denied recovery, in part because lost profits under the 20-year management contract were not foreseeable. The court asserted that

> the provisions in the contract do not suggest or provide such a heavy responsibility on the part of the county. In the absence of any provision for such an eventuality, the common sense rule to apply is to consider what the parties would have concluded had they considered the subject. The evidence here fails to demonstrate that liability for loss of profits over the length of the contract would have been in the contemplation of the parties at the relevant times.[19]

In *Kenford,* the substantial period of time and the contract provisions were used together to show lack of foreseeability.

§ 15.3 Proximate Cause

Proximate cause is the second element that the owner must prove in order to recover lost profits.[20] The proof of proximate cause is achieved by showing that the wrongdoer's actions causally relate to the owner's lost profits.[21] In other words, the owner must show that "profits must have been prevented or lost as a natural consequence of the breach of the contract."[22] Because the facts and circumstances of each case determine whether the wrongdoer's actions were in fact the proximate cause of the owner's lost profits, a summary of some existing cases is the best way to explain the workings of this element.

In most situations, proving that the owner's lost profits flowed directly from the contract breach is not a difficult task. In *Alpine Industries, Inc. v. Gohl,*[23] an owner alleged that the contractor's delay in completing construction of a new plant on time caused lost profits. The owner had anticipated an early February completion and had solicited new clients. After February had passed, Alpine could not meet the demand of its old and new clients and, therefore, stopped seeking new business. At trial, a certified public accountant, who was an expert witness on construction delays causing lost profits, testified regarding cash flow forecasting based on Alpine's previous share of the market as well as its current diminished

---

[19] *Id.* 493 N.E.2d at 236.
[20] Fields Eng'g & Equip., Inc. v. Cargill, Inc., 651 F.2d 589, 593 (8th Cir. 1981); McDermott v. Middle East Carpet Co., 811 F.2d 1422 (11th Cir. 1987); Dobbs, Handbook on the Law of Remedies 798 (1973).
[21] *See* Koncinsky v. Smith, 390 So. 2d 1377, 1383 (La. Ct. App. 1980).
[22] Drews Co. v. Ledwith-Wolfe Assocs., Inc., 296 S.C. 207, 371 S.E.2d 532, 535 (1988).
[23] 30 Wash. App. 750, 637 P.2d 998, 1001 (1981).

share. Furthermore, Alpine personnel and clients testified as to cancelled orders and production problems. The Court of Appeals of Washington found that this evidence was sufficient to prove the causation of lost profits. The court held that the plaintiff, Alpine, was not required to prove that the entire loss was the contractor's fault and the trier of fact could exercise informed discretion in apportioning the loss of profits to the defendant's breach.

In some situations, the intervention of other contractors and the effect of owner stop orders may be factors which effect causation of lost profits. In *Fields Engineering & Equipment, Inc. v. Cargill*,[24] the Eighth Circuit Court of Appeals found that breach of a construction contract by a millwright contractor caused the owner to lose a portion of profits in the owner's business. The contract involved the construction of a grain conveyor-belt system between two grain silos. When the completion deadline in the contract was not met and the agreed maximum contract amount was surpassed by the contractor's charges, the owner ordered the contractor to leave the jobsite and subsequently hired another contractor to finish the work. The owner sued for breach of contract seeking damages which included lost profits. The court of appeals did not award the complete amount sought for lost profits, however, because the owner did not "show with reasonable certainty that all of the loss flowed directly from the breach."[25] The court questioned why it had taken the second contractor on the job one year to complete a small percentage of the contract ($38,000 of the total $270,000 contract price), when the entire contract was to have been completed in nine months. Consequently, the Eighth Circuit Court of Appeals only allowed the owner profits claimed for one season, because the original contractor's delay only caused the loss of additional profits for that season. The decision in *Fields Engineering* indicates that a court may be flexible in determining the proximate cause of an owner's lost profits.

Several courts have been more rigid and denied the owner recovery of lost profits because the wrongdoer's acts were not shown to be the proximate cause of the lost profits. In *Exton Drive-In, Inc. v. Home Indemnity Co.*,[26] the owner claimed, inter alia, that the contractor's delay in grading and paving a drive-in had caused the owner to lose anticipated profits. The Supreme Court of Pennsylvania determined, however, that the grading and paving contractor's delay in finishing his work did not delay the theater's opening because the theater screen was not installed until just before the scheduled opening date, thereby making it impossible to open

---

[24] 651 F.2d 589, 593 (8th Cir. 1981).
[25] *Id.* at 590.
[26] 436 Pa. 480, 261 A.2d 319 (1969), *cert. denied,* 400 U.S. 819 (1970).

§ 15.3 PROXIMATE CAUSE

sooner. Consequently, the owner could not recover lost profits which he claimed resulted from a delay in opening.

Similarly, in *Koncinsky v. Smith*,[27] the Court of Appeals of Louisiana found that the owner's failure to establish proximate cause necessitated a denial of claimed lost profits. The owner-developer in *Koncinsky* claimed that the defective installation of a subdivision's water system caused him to lose profits. The owner alleged that a decrease in lot sales over a three-year period and a decrease in his taxable income over the same period indicated that the lots had lost their desirability because of the defective water system. The court pointed out that many factors could have contributed to this decrease in lot sales; therefore, the owner had failed to show a causal relationship between the decreased sales and the contractor's conduct.

Exton and *Koncinsky* indicate that the owner must not take the proximate cause element for granted. *Exton* shows how a delay is not always the proximate cause of lost profits. *Koncinsky* shows that, even when a causal relationship may exist between the lost profits and the wrongdoer's act, the direct causal relationship must be made clear to a court or lost profits will be denied. Consequently, in order to prove the proximate cause element in a lost profits claim, the owner must make sure the wrongdoer did, in fact, cause lost profits as well as make sure that the lost profits cannot be blamed on other factors.

Various other techniques could have been employed in *Koncinsky* to establish a direct causal relationship. A market analysis of comparable lot sales in the area is one such technique. That analysis would consider such things as lot sizes and prices in competing subdivisions, market demographics, highway construction, zoning, and other factors that would impact property value. It would be prudent to utilize appropriate experts such as independent real estate appraisers or valuation consulting firms to testify in support of the analysis. Further, if market comparisons indicate that significant sales occurred at competing subdivisions during the three-year period in question, market comparisons showing lot sales market percentages both before and after that three-year period would demonstrate whether Koncinsky experienced adverse trends that could be related to the wrongdoer's actions.

It also would be advisable to analyze Koncinsky's books and records to more specifically demonstrate that other external influences, like new or ongoing significant development projects, did not adversely impact operating results. An analysis of such projects should be performed to establish that they were progressing as planned and that sales and profits were

---

[27] 390 So. 2d 1337 (La. Ct. App. 1980).

consistent with internally developed estimates. These techniques should assist a plaintiff in establishing a claim for lost profits.

§ 15.4 Reasonable Certainty

The final element for recovery of lost profits is that lost profits must be proven with reasonable certainty.[28] It is not necessary that an owner provide an exact measurement of lost profits.[29] It is generally considered unfair to allow a wrongdoer to escape liability simply because damages cannot be measured exactly.[30] Courts will allow recovery if the owner can provide a reasonable method of calculating lost profits.[31] The *Restatement (Second) of Contracts* § 331 states that:

> Damages are recoverable for losses caused or for profits and other gains prevented by the breach only to the extent that the evidence affords a sufficient basis for estimating the amount in money with reasonable certainty. . . . Where the evidence does not afford a sufficient basis for a direct estimation of profits, but the breach is one that prevents the use and operation of the property from which profits would have been made, damages may be measured by the rental value of the property or by interest on the value of the property.[32]

The certainty requirement in purpose and in practice operates to preclude excessive damage awards as a convenient means of keeping within bounds of reasonable expectation the risk which litigation imposes upon commercial ventures.[33]

---

[28] McDermott v. Middle East Carpet Co., 811 F.2d 1422 (11th Cir. 1987); Fields Eng'g & Equip. v. Cargill, Inc., 651 F.2d 589 (8th Cir. 1981); Drews Co. v. Ledwith-Wolfe Assocs., Inc., 296 S.C. 207, 371 S.E.2d 532 (1988); Evergreen Amusement Corp. v. Milstead, 206 Md. 610, 112 A.2d 901 (1955); S. Jon Kreedman & Co. v. Meyers Bros. Parking-West Corp., 58 Cal. App. 3d 173, 130 Cal. Rptr. 41 (1976); Knier v. Azores Constr. Co., 78 Nev. 20, 368 P.2d 673 (1972); Alpine Indus. v. Gohl, 30 Wash. App. 750, 637 P.2d 998 (1981); *see, generally,* Note, *The Requirement of Certainty in the Proof of Lost Profits,* 64 Harv. L. Rev. 317 (1950).

[29] Alpine Indus. v. Gohl, 30 Wash. App. 750, 637 P.2d 998, 1001 (1981); Noble v. Tweedy, 90 Cal. App. 2d 738, 203 P.2d 778, 782 (1949); Hoag v. Jenan, 86 Cal. App. 2d 556, 195 P.2d 451, 455 (1948); Landmark, Inc. v. Stockman's Bank & Trust Co., 680 P.2d 471, 476 (Wyo. 1984).

[30] Hoag v. Jenan, 86 Cal. App. 2d 556, 195 P.2d 451, 455 (1948); Noble v. Tweedy, 90 Cal. App. 2d 738, 203 P.2d 778, 782 (1949).

[31] Landmark, Inc. v. Stockman's Bank & Trust Co., 680 P.2d 471, 476 (Wyo. 1984); Drews Co. v. Ledwith-Wolfe Assocs., Inc., 296 S.C. 207, 371 S.E.2d 532, 536 (1988); Hoag v. Jenan, 86 Cal. App. 2d 556, 195 P.2d 451, 455 (1948).

[32] Restatement of Contracts § 331 (1932).

[33] C. McCormick, Handbook on the Law of Damages 105 (1935).

§ 15.5 Calculation of Profits

The *Restatement (Second) of Contracts* defines *profits* as "the net pecuniary gain from a transaction."[34] This pecuniary gain is calculated by determining the difference between the owner's gross profits and the owner's business expenses.[35] Consequently, when an owner seeks damages for lost profits, it must plead and prove lost net profits.[36] Furthermore, factors such as depreciation and overhead may effect lost profits.

In *Drews Co. v. Ledwith-Wolfe Associates, Inc.*,[37] the South Carolina Supreme Court denied recovery of lost profits to a restaurant owner in his action regarding the breach of a construction contract because the owner had not provided an adequate basis for determining net profits. The owner had provided evidence of gross profits as well as his own testimony that he expected to receive net profits of one-third his gross profits. The court asserted that the "[o]wner's expectations, unsupported by any particular standard or fixed method for establishing net profits, were wholly insufficient to provide the jury with a basis for calculating profits lost with reasonable certainty."[38]

The amount of net profits that is recoverable may be affected by depreciation. In *Oliver B. Cannon & Sons, Inc. v. Dorr-Oliver, Inc.*,[39] the trial court found that a painting subcontractor was liable to the owner of a plant for faulty tank lining applied by the subcontractor. The Delaware Supreme Court concluded that the calculation of lost profits would properly encompass the actual losses including a "decline in value or appreciation of the plant while it was out of production."[40] However, the court concluded that the straight-line method of accounting used by the owner to calculate lost profits was not acceptable. Although the court recognized the applicability of straight-line depreciation in accounting procedures, it viewed this "arbitrary accounting" method as improper for determining actual depreciation losses in an action for lost profits. Hence, the issue was remanded to the court that made the original decision in order to determine actual depreciation, which would encompass factors such as increased life expectancy resulting from the wrongdoer's repairs of his faulty work. *Dorr-Oliver* indicates that the owner seeking lost profits for work done on an existing business should include a provable calculation of depreciation to

---

[34] Restatement (Second) of Contracts § 331 comment b.

[35] *See* Comment, *Remedies-Lost Profits*.

[36] Sears, Roebuck & Co. v. Goudie, 290 A.2d 826, 833 (D.C. 1972); Comment, *Remedies-Lost Profits*.

[37] 296 S.C. 207, 371 S.E.2d 532 (1980).

[38] *Id.*, 371 S.E.2d at 536.

[39] 394 A.2d 1160, 1161 (Del. Super. Ct. 1978).

[40] *Id.* at 1163.

its loss of profit damages when the wrongdoer's actions render its equipment or its plant inoperative for a period of time.

To better understand the merits of the court's position in *Dorr-Oliver,* it is necessary to address the "arbitrary accounting" comment in that case. Under generally accepted accounting principles, costs incurred in constructing or acquiring a fixed asset are capitalized and depreciated over the asset's estimated useful life. Accountants do not normally perform ongoing analyses of "net realizable value" of such assets to determine whether the asset's book value should be written down. The only time "recorded asset cost" becomes an issue, particularly for "long-term" or "fixed" assets, is when realization of historic costs amounts is brought into question because the company's continuation as a "going concern" is doubtful. In addition, although accountants normally look to cost versus sales price of an inventory item to ensure inventory is stated at its net realizable value (selling price less selling costs should equal or exceed recorded cost), such an analysis is not performed for fixed assets. Accordingly, it is understandable that the *Dorr-Oliver* court concluded that recorded values of long-term assets and related depreciation are, by themselves, an inappropriate measurement tool for determining diminution in value. A more appropriate measure would be to analyze production quantity and quality of a fixed asset versus contracted or product specified performance during the period of the defect.

In *Northern Petrochemical Co. v. Thorsen & Thorshov, Inc.,*[41] in addition to an award for lost profits, a jury awarded damages of $140,000 for a building's diminution in value caused by defective construction and design. The diminution in value was based on unchallenged testimony that the value of the building, even after reconstruction, was $140,000 less than it would have been had the building been properly constructed. Reconstruction efforts to restore the building as contracted would have entailed unreasonable economic waste, and the costs incurred were necessary to make the building safe and usable.

The owner also should be aware that damages for overhead will most likely not be awarded if lost gross profits are awarded. As the California Court of Appeals pointed out in *Burnett & Doty Development Co. v. C.S. Phillips,*[42] the owner could not recover fixed or overhead costs for the five-month discontinuance of business caused by a site work contractor. According to the court, an award of lost gross profits and overhead for this period of time would constitute double recovery and, therefore, overhead was properly excluded from the damages.

In any calculation of lost profits, lost profits should be reduced by an amount representing all costs saved because of the wrongdoer's action.

---

[41] 297 Minn. 118, 211 N.W.2d 159 (1973).
[42] 84 Cal. App. 3d 384, 148 Cal. Rptr. 569 (1978).

§ 15.6   IMPAIRMENT OF VALUE   461

Accordingly, "lost profits damages are usually defined as lost net profits; all costs must be deducted."[43] There are situations in which gross and net profits are essentially the same. The following hypothetical example and court case should better illustrate the point.

Hypothetical: In the operation of A&B Mall, an owner/operator of several area retail malls seeks recovery of gross profits in its claim of lost profits because of construction delays. That claim would be appropriate if the owner/operator were able to demonstrate that no additional corporate costs would have been incurred or that fixed/overhead expenses did not vary. In this example, the owner can demonstrate that lost profits are recoverable and that corporate administrative costs would not vary because existing staff was adequate to perform the necessary monitoring, recordkeeping, and property management functions of all owned malls, including A&B Mall. Consequently, cost reductions to lost profits should be limited to those costs directly related to on-site management of A&B Mall.

In *Schatz v. Abbott Labs, Inc.,*[44] the plaintiff theatre owners sought to recover lost profits to their business because the defendant's factory emitted noxious odors which adversely affected attendance at the plaintiffs' movie house, which was located near the factory. An award for lost profits was based on loss of gross receipts which was upheld on appeal because fixed expenses did not vary with increased attendance. Additionally, although film rental costs to the plaintiffs were based on a percentage of gross receipts, and increased as attendance increased, it also was demonstrated that refreshment sales and profits increased proportionately and offset the added film rental expense.

A&B Mall and *Schatz v. Abbott* are good examples of factors to consider in utilizing gross profits as a basis for calculating lost profits.

§ 15.6   —Impairment of Corporate Value

Another approach in calculating lost profits is one that measures impairment of corporate value. The corporate value calculation contemplates the impact of a wrongdoer's actions on the past, present, and prospective operations of an enterprise by comparing earnings and cash flows that would have occurred without the wrongdoer's actions to actual earnings and cash flows.

The following set of facts illustrates this point. A competitor's production facilities are acquired by a toy manufacturer who plans to consolidate products presently manufactured at several of the acquired facilities

---
[43] Dunn at 155.
[44] 51 Ill. 2d 143, 281 N.E.2d 323 (1972).

(which will be closed and sold) into one facility to both improve production efficiency and to reduce debt. After several years, the consolidation is completed and the toy manufacturer puts the now-idle factories on the market. A buyer is found and, prior to consummating the sale, the plaintiff must meet newly enacted state guidelines and have the land and buildings of the idle facilities tested to ensure the properties are free of environmental contamination.

Testing discloses that one of the sites is contaminated, and the manufacturer must clean up the contaminated site prior to the sale. The contamination is traced back to the original plant owner who had then sold the facilities to the previous owner, the competitor. The sale of the one facility is not consummated and, further, the significant amounts of cash required to maintain that facility have resulted in a serious ongoing cash drain to the existing owner's other operations.

The current owner has the following potential claims:

1. Recovery of actual costs incurred in maintaining the idle facility during the period subsequent to the discovery of the contamination
2. Loss of use of the proceeds to be realized from sale of the idle facility in the ongoing business
3. If the sale would not have been consummated for reasons unrelated to the contamination, recovery would be sought for loss of rental and other income that would be derived from the idle facility absent the contamination
4. Impairment of corporate value resulting from costs incurred in maintaining that facility.

The first area is a matter of factual record. The current owner did or did not incur certain costs directly related to maintaining the facility. Whatever the amount of those costs, they should be recovered. The areas most pertinent to lost profits relate to loss of use of sale proceeds to be realized from the sale or, if the building would not have been sold for reasons other than the contamination, loss of rental and other income, and impairment of corporate value resulting from costs diverted from ongoing operations to maintain the contaminated facility.

Loss of sale proceeds also is a matter of fact and assumes that the sale would have been consummated were it not for the contamination. Loss of rental income, although calculable, would require using appropriate real estate experts to address arguments related to rental value, occupancy, rental terms, and the like. Impairment of corporate value would require use of qualified valuation experts who could analyze and present an overall assessment of adverse business enterprise impact caused by the contamination. In performing the whole analysis, historic profit results of the business as well as projected results would be compared to determine the impairment to corporate value.

The valuation experts would calculate loss of corporate value using industry-established approaches to estimate the value of a business enterprise as a going concern. Two conventional and accepted approaches to estimate the value of a business enterprise as a going concern are the market approach and the income approach.

The *market approach* establishes fair market value through an analysis of comparable company financial statistics, including multiples of market price to various earnings parameters, and application of the derived multiples to the appropriate earnings measurements of the subject company.

The *income,* or discounted cash flow, *approach,* estimates the value of a business by evaluating the company's future cash flows that are available for distribution and discounting these cash flows to their present worth at an appropriately derived rate of return.

Loss of rental and other income would include a calculation of interest on related debt and costs associated with renting the building such as brokerage commissions, management fees, vacancy, and contingencies. However, with respect to loss of corporate value, no additional reductions other than those directly related to the idle facility (which will be recovered separately) should be necessary. In this example, the toy manufacturer is incurring costs which would not have been incurred were it not for the contamination caused by the original plant owner; the fixed expenses related to the manufacturing process are not relevant to the claim and are appropriately excluded in arriving at the lost profits claimed.

§ 15.7 Unestablished Business versus Established Business

The traditional rule still in effect in some courts is that lost profits are not recoverable for an unestablished business.[45] Lost profits for an unestablished business are considered too speculative.[46] For example, in *Evergreen Amusement Corp. v. Milstead,*[47] an owner sought lost profits damages for delay in the construction of a drive-in theater. The court denied recovery

---

[45] Evergreen Amusement Corp. v. Milstead, 206 Md. 610, 112 A.2d 901, 904 (1955); *see generally* Comment, *Remedies-Lost Profits;* Note, *The New Business Rule and the Denial of Lost Profits,* 48 Ohio St. L.J. 855 (1987).

[46] Evergreen Amusement Corp. v. Milstead, 206 Md. 610, 112 A.2d 901, 904 (1955); Exton Drive-In Inc. v. Home Indemnity Co., 436 Pa. 480, 261 A.2d 319 (1970); Knier v. Azores Constr. Co., 78 Nev. 20, 361 P.2d 673, 675 (1972); *In re* Phoenix Restoration Specialists, 14 Bankr. 115, 123, 124 (D. Mass. 1981).

[47] 206 Md. 610, 112 A.2d 901 (1955).

because the theater was an unestablished business and because lost profits were too speculative. The court asserted that

> [u]nder the great weight of authority, the general rule clearly is that loss of profit is a definite element of damages in an action for breach of contract or in an action for harming of an established business which has been operating for a sufficient length of time to afford the basis of estimation with some degree of certainty as to probable loss of profits, but that on the other hand, loss of profits from a business which has not gone into operation may not be recovered because they are merely speculative and incapable of being ascertained with the requisite degree of certainty.[48]

In *Exton Drive-In, Inc. v. Home Indemnity Co.*,[49] also involving the construction of a drive-in theater, the owner sought lost profits not from delay (the theater opened on time) but on the ground that the theater was operated in an unfinished condition. The court asserted that "the anticipated profits of a new and untried business which were attributable solely to the unfinished condition of the business premises were too speculative to provide a basis for an award of damages."[50]

In jurisdictions that deny recovery for lost profits to unestablished businesses, the pivotal question is often whether the business was established.[51] In *In re Phoenix Restoration Specialists, Inc.*,[52] the owner of a chain of retail stores sought recovery of lost profits from a contractor who had abandoned work on construction of a new store. The issue in *Phoenix Restoration* was whether a new store in an existing chain was an unestablished business. The United States Bankruptcy Court in Massachusetts held that the store was a new business and denied recovery of lost profits, asserting that "[t]he fact that the defendant had similar stores in other locations does not help the defendant."[53] However, some courts have held that the new business rule does not apply when the new business is part of an existing chain.[54]

Courts have also split on the issue of how long a business must be in operation before it is considered established.[55] For example, in *Berlin Development Corp. v. Vermont Structural Steel Corp.*,[56] the lessee-operators of businesses sought recovery of lost profits from a construction contractor

---

[48] *Id.*, 112 A.2d at 904.

[49] 436 Pa. 480, 261 A.2d 319 (1969).

[50] *Id.*, 261 A.2d at 324.

[51] Dunn at 122–130; Comment, *Remedies-Lost Profits* at 702–704.

[52] 14 Bankr. 115 (D. Mass. 1981).

[53] *Id.* at 124.

[54] *See* Dunn at 125.

[55] *Id.* at 122–24.

[56] 127 Vt. 367, 250 A.2d 189 (1968).

§ 15.7 UNESTABLISHED BUSINESS

for damage caused by a roof that leaked. The businesses began operation in October and the leaking occurred over the course of the following two winters. The court held that the businesses were new businesses and could not recover lost profits. However, other jurisdictions have held that one month or less may be sufficient experience to treat a business as established under the new business rule.[57]

The modern trend is for cases to view the new business rule as a rule of evidentiary sufficiency rather than an automatic bar to recovery of lost profits of a new business.[58] The great majority of jurisdictions does not now deny recovery of lost profits for new businesses when profits can be established with reasonable certainty. For example, in *S. Jon Kreedman & Co. v. Meyers Bros. Parking-West Corp.*,[59] a lessee sought recovery of lost profits for breach of a contract to construct a parking garage. The California Court of Appeals asserted that the issue was not whether the business was established but "rather, whether the damages can be calculated with reasonable certainty."[60] The *Kreedman* court held that, because the parking business, itself, was an established business, the owner was experienced, and "the operation of a parking garage is a relatively simple operation with sufficiently few decision points . . . the prediction of profits [was] reasonably possible."[61] *Kreedman* suggests that the nature of the business may be used to establish reasonable certainty of lost profits.

In *McDermott v. Middle East Carpet Co.*,[62] the owner sought recovery for lost profits against consultants who had contracted to construct a carpeting manufacturing plant. The Eleventh Circuit Court of Appeals held that lost profits were recoverable despite the fact that it was an unestablished business. The court asserted that "[w]here . . . the claimant can show a 'track record' of profitability, evidence can be considered and lost profits can be awarded."[63] In this case, the subsequent profitability of the plant was sufficient to establish a track record of profitability. *McDermott* illustrates how the owner's subsequent experience can be used to meet the reasonable certainty requirements.

---

[57] *See* Dunn at 123.

[58] Dunn at § 4.2; Comment, *Remedies-Lost Profits* at 695; Note, *The New Business Rule*, 48 Ohio St. L.J. 855, 859 (1987); McDermott v. Middle East Carpet Co., 811 F.2d 1422, 1426, 1427 (11th Cir. 1987); S. Jon Kreedman & Co. v. Meyers Bros. Parking-West Corp., 58 Cal. App. 3d 173, 130 Cal. Rptr. 41, 49 (1976); Landmark, Inc. v. Stockman's Bank & Trust Co., 680 P.2d 471, 476 (Wyo. 1984); Drews Co. v. Ledwith-Wolfe Assocs., Inc., 296 S.C. 207, 371 S.E.2d 532, 534 (1988).

[59] 58 Cal. App. 3d 173, 130 Cal. Rptr. 41 (1976).

[60] 130 Cal. Rptr. at 49.

[61] *Id.*

[62] 811 F.2d 1422 (11th Cir. 1987).

[63] *Id.* at 1428.

§ 15.8 Proof Techniques

There are numerous techniques that the owner may use to establish the fact that actual lost profits were experienced. The *Restatement (Second) of Contracts* states that lost profits "may be established with reasonable certainty with the aid of expert testimony, economic and financial data, market surveys and analyses, business records of similar enterprises, and the like."[64] Because lost profits are often denied as a result of their speculative nature, the owner should try to utilize as many of the following techniques as possible in an effort to convince the trier of fact that the profits were, in fact, lost. Proof must pass the realm of conjecture, speculation, or opinion not founded on facts and, therefore, proof must consist of actual facts from which a reasonably accurate conclusion regarding the cause and the amount of the loss can be logically and rationally drawn. Because the proof of lost profits is dependent upon the different facts pertinent to each case, review of individual cases will not be focused upon in this section. Instead, this section will offer different techniques accepted by courts to prove lost profits so the reader has an idea of the methods that can be used, depending upon the facts of a particular case.

In today's business environment, business forecasts and projections have become more common. In January, 1986, the American Institute of Certified Public Accountants issued a *Guide for Prospective Financial Statements* (AICPA Guide). The AICPA Guide provides an appropriate framework for the preparation and presentation of financial forecasts and projections and provides guidance to accountants in performing services and determining the reasonableness of underlying assumptions to prospective financial data. Although subsequent performance must be based on a comparable period, experts can provide objective, independent testimony as to the reasonableness of assumptions being used and whether factors other than the wrongdoer's actions caused the loss. The AICPA Guide should be a reference source for supporting or attacking an expert's analysis and association with projections and their underlying assumptions.

The best proof of lost profits is a showing of the owner's profits that were achieved prior to the wrongdoer's acts that caused the loss.[65] A problem in using this proof technique, however, is determining the length of time that the owner had to have achieved profits prior to the incident. If profits have not been achieved for a long period of time, a court may find

---

[64] Restatement (Second) of Contracts § 352 at 146 (1981).

[65] Burnett & Doty Dev. Co. v. Phillips, 84 Cal. App. 3d 384, 148 Cal. Rptr. 569 (1978) (holding developer supported proof of lost profits); Sears, Roebuck & Co. v. Goudie, 290 A.2d 826, 833 (D.C. 1972) (allowing average of dollar amount estimated lost); Hoag v. Jenan, 86 Cal. App. 2d 556, 195 P.2d 451, 456 (1948) (stating provable data of actual experience provide the basis for estimation of profits with a satisfactory degree of definiteness); Dunn at § 5.5 (1978).

them too speculative to carry over into the period during which the loss is claimed. See § **15.7**. Another obvious problem is that a new business has no prior profits.

If an established business did not achieve profits for an interim period prior to the incident, it may be appropriate to use an expert to identify the prior loss as an isolated, nonrecurring event. For instance, a company may have had a history of profits but attempted introduction of a new product which was not accepted, or the company could have experienced a strike at a major facility or incurred one-time costs in the shutdown of an older facility.

New businesses, as well as established businesses, may attempt to prove lost profits by showing the existence of profits subsequent to the incident that caused the lost profits.[66] It should be noted that, if the jurisdiction adheres to the view that lost profit recovery is not allowable for an unestablished business,[67] the court may not admit evidence of subsequent profits. See § **15.7**.

When the owner attempts to prove lost profits by showing prior profits or by showing subsequent profits, it must be sure that there is a uniformity in the comparison. In other words, the time during which profits were made must be directly comparable to the times during which the profits were lost.[68] For example, economists have proven that a manufacturing plant can be expected to lose profits during the first few years of production before hitting a break-even point and subsequently becoming profitable.[69] Hence, the first few years of operation would theoretically not be comparable to the latter years. The profits of a business may also be affected by seasonal changes. For example, if an owner claims an air conditioning contractor's breach has caused a decline in her summer business and therefore a decline in her profits, she should show what the profits were in the prior or subsequent summers.[70] Pointing out these types of common sense comparisons will contribute greater credibility to the owner's evidence of lost profits.

The owner may also proffer evidence of profits made by businesses similar to the owner's. This is sometimes referred to as the "yardstick" method.[71] The comparable business may be a similar business operated by

---

[66] McDermott v. Middle East Carpet Co., 811 F.2d 1422, 1429 (11th Cir. 1987); Drews Co. v. Ledwith-Wolfe Assocs., Inc., 296 S.C. 207, 371 S.E.2d 532, 535–36 (1980).

[67] *See* Evergreen Amusement Corp. v. Milstead, 206 Md. 610, 112 A.2d 901, 904 (1955); *see generally* Comment, *Remedies-Lost Profits;* Note, *The New Business Rule and the Denial of Lost Profits,* 48 Ohio St. L.J. 855 (1987).

[68] Guide for Prospective Financial Statements (American Institute of Certified Public Accountants 1986); Dunn at §§ 5.5–5.6.

[69] McDermott v. Middle East Carpet Co., 811 F.2d 1422, 1427 (11th Cir. 1987).

[70] *See* Sears, Roebuck & Co. v. Goudie, 290 A.2d 826 (D.C. 1972).

[71] Comment, *Remedies-Lost Profits* at 713.

the owner in a different location[72] or it may be a franchise branch in a different location. The comparable business could also be the owner's competition in the same location.[73] Once again, the owner must bring out a direct comparison between the businesses, which could focus on such factors as location and market surveys.[74] Even though the profits of similar businesses are extrinsic to the profits of the business in question, this evidence carries probative value if sufficient parallels can be established and, therefore, it should be proffered to the trier of fact.[75]

For new businesses, in addition to using the yardstick method in direct comparisons of businesses, subsequent comparisons of actual versus projected results, as well as the reasonableness of underlying assumptions given the guidelines established in the AICPA Guide, should serve to reinforce or raise doubts about a plaintiff's claim for lost profits. For example, assume a start-up household goods manufacturer were to seek recovery of lost profits caused by construction delays in converting and upgrading a facility to conform to the manufacturer's production process. Recovery in this instance would depend on documentation related to projected results, the backlog of customer orders, the presence of other outside causes, the extent to which costs could have been mitigated, and, after production begins, the comparison of subsequent performance to projections.

In a review of projected versus actual results, the likelihood of recovery is enhanced if assumptions are appropriate in the circumstances and are adequately supported by underlying documentation, if the calculations provided are accurate, and if actual performance either favorably compares to projected results, or differences, if any, are identified and adequately explained.

As the cases illustrate, courts generally have held that an injured party should receive lost profits, calculated as revenues less costs. Although there are exceptions to every general rule, fixed or overhead expenses which increase as a result of a wrongdoer's nonperformance should be included in arriving at lost profits.

§ 15.9 Hypothetical Lost Profits Case

Consider the following illustrative case, in which additional costs are being incurred by an owner's organization because of construction delays. An owner of a chain of 60 regional shoe stores has engaged in a

---

[72] *In re* Phoenix Restoration Specialists, Inc., 14 Bankr. 115, 124 (D. Mass. 1981); S. Jon Kreedman & Co. v. Meyers Bros. Parking-West Corp., 58 Cal. App. 3d 173, 130 Cal. Rptr. 41, 49 (1976); Dunn at § 5.7.

[73] *See* Evergreen Amusement Corp. v. Milstead, 206 Md. 610, 112 A.2d 901 (1955).

[74] *Id.*

[75] Comment, *Remedies-Lost Profits* at 713; Dunn at 149.

§ 15.9 HYPOTHETICAL CASE 469

modernization and product line expansion of his stores. Store modernization and product line expansion are based on appropriate market studies and separate store projections. To date, a total of 40 stores have been modernized and expanded. Sales results of those stores reflect a 30 percent increase in sales and a 20 percent increase in operating profit margin, which are fairly consistent with the owner's market survey and projections. In contemplation of the modernization and product line expansion program, the owner incurred additional debt, upgraded his information processing systems both at the stores and the corporate offices, brought in additional purchasing agents, increased the number of sales personnel at the stores, increased warehouse storage capacity, and brought in additional corporate administrative personnel. Construction delays occurred because of contractor mismanagement, improper scheduling of material delivery, and high turnover of contractor personnel. The remaining 20 stores were not completed as scheduled; the average delay was 12 months. Of the 20 stores not completed, five stores had significant construction activity which adversely impacted normal operations for the 12 months, as well as delayed introduction of the new shoe lines.

In **Table 15–1**, unabsorbed fixed costs of 15 unrenovated stores are identified by comparing the financial results of renovated stores to that of unrenovated stores. Further, assume that the five stores experiencing the average 12-month delay had substantially reduced sales as shown in **Table 15–2**. The owner's claim for lost profits should include, at a minimum:

1. The lost profits for unrenovated versus existing stores where the changeover lagged for 12 months (see **Figure 15–1**) is calculated as follows:

Actual loss before taxes	$ (3,778)
Additional fixed expenses	
Not absorbed	11,468
Number of stores	× 15
	$228,690

2. The lost profits for the five stores whose operations were disrupted for 12 months, with resultant reduced sales (see **Figure 15–2**) is calculated as follows:

Actual loss before taxes	$ (16,280)
Profit obtained for	
renovated stores	12,500
Total claim	28,780
Number of stores	× 5
	$143,900

Table 15–1

Unabsorbed Fixed Costs for Unrenovated Stores

	Unrenovated	Renovated
Revenues	$384,500	$500,000
Direct Expenses	355,278	454,500
Operating Profit	29,222	45,500
Operating Profit Percentage	7.6%	9.1%
Fixed Expenses		
Pre Expansion	21,532	
Post Expansion	11,468	33,000
Earnings (Loss) Before Income Taxes	$ (3,778)	$ 12,500
Earnings (Loss) as a Percentage of Sales	(.9%)	2.5%

Table 15–2

Lost Profits for Five Unrenovated Stores

Revenues for the Year	$220,000
Direct Expenses	203,280
	16,720
Fixed Expenses	33,000
Loss Before Income Taxes	$(16,280)
(Loss) As a Percentage of Sales	(7.4%)

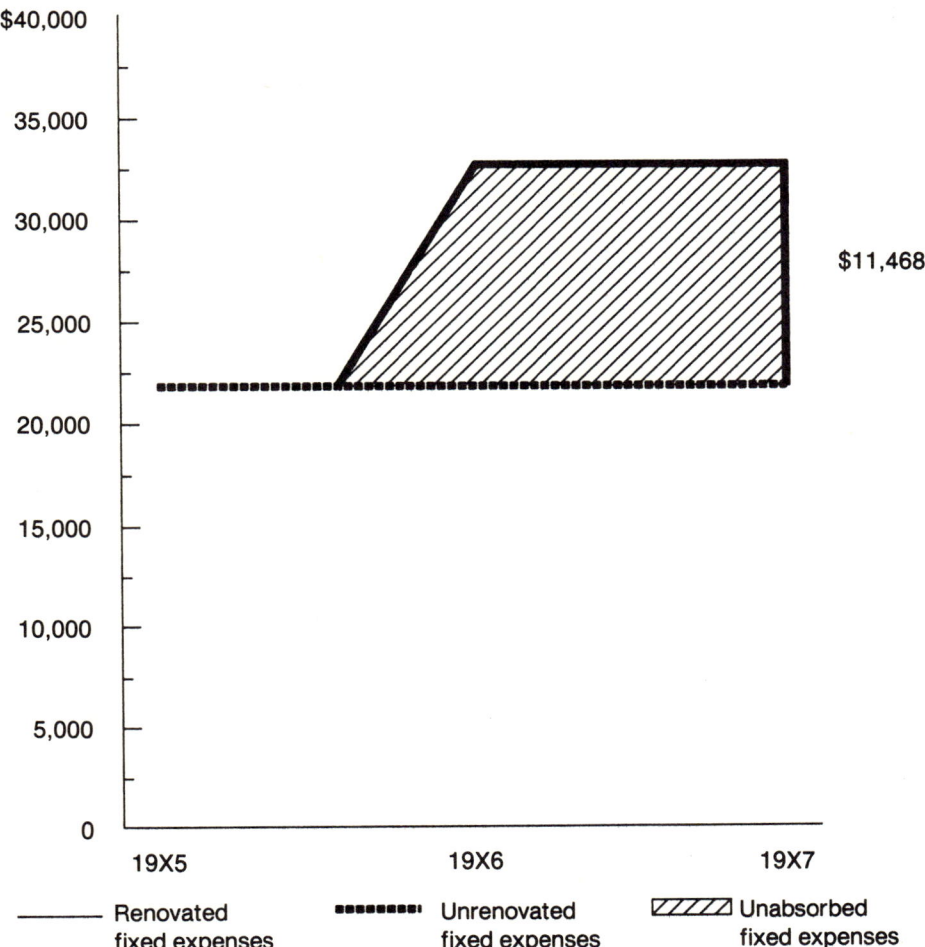

Figure 15–1. XYZ shoe sales unabsorbed fixed costs for the period 19X5 through 19X7.

An argument can be made that the loss of profits calculation should simply be loss of profits of the 20 stores based on profits obtained at the 40 stores which were renovated. The claim using that approach would increase $15,480 to $388,070 versus the $372,590, as determined above.

The above example does not reduce the lost profits claimed for related income taxes. The reasoning here is the company should be in the same position it would have been in but for the actions of the wrongdoer. Because the owner will pay taxes on lost profits recovered from the defendant, that recovery should not be diminished by any tax calculations in its claim for lost profits.

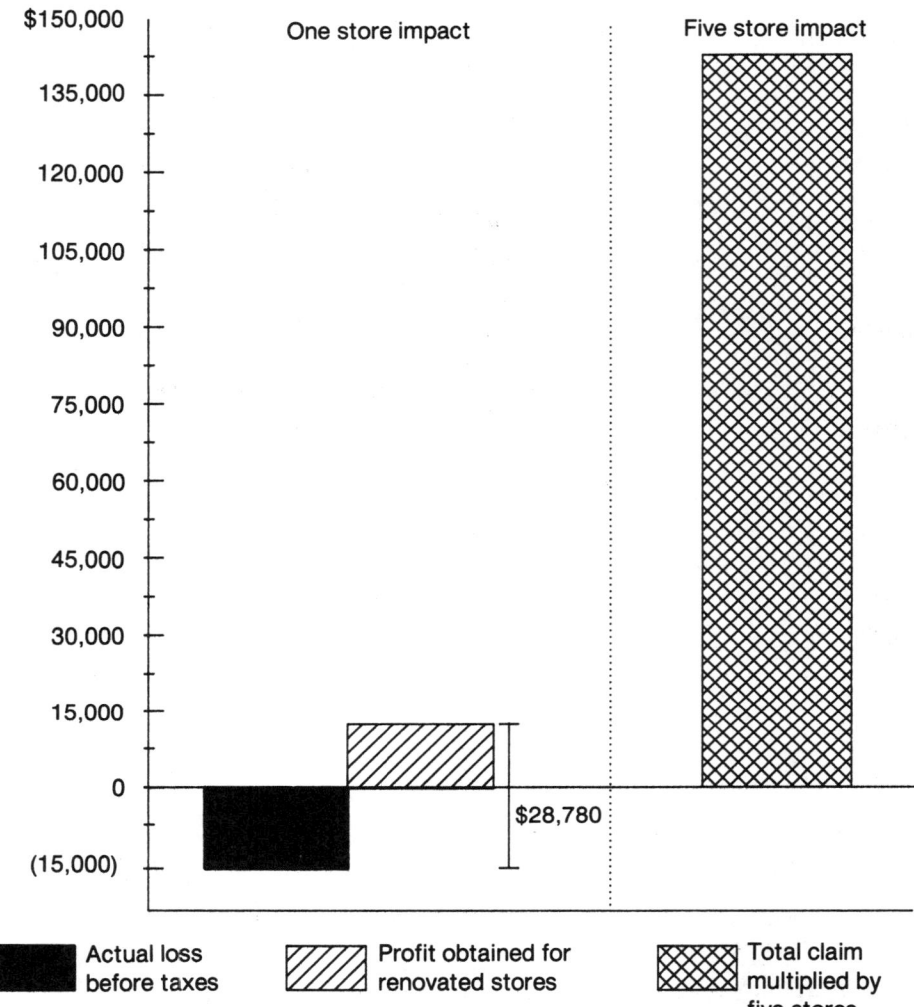

Figure 15–2. XYZ shoe sales impact for one store versus five stores calculated for the 12 months ending 19X7.

Finally, in addressing lost profits for an unestablished business, the same example can be modified. Assume the facts above remain the same but the owner decides to increase the number of area shoe stores to 80, and of the 40 stores that were completed, 10 were new stores. Because they are unestablished businesses, the question arises as to whether the owner would have a problem recovering anticipated profits on the 10 new stores which experienced construction delays. The fact pattern as established provides the necessary data to establish the amount of loss with reasonable accuracy and should enable the owner to recover lost profits for those stores.

§ 15.10 Conclusion

An owner's lost profits must be specifically pled and must be proven by showing foreseeability, proximate cause, and reasonable certainty. Although some courts distinguish lost profits of established businesses from those of unestablished businesses, the trend today is to allow recovery of lost profits in either instance when the owner can prove said losses with reasonable certainty. The owner will be most successful in obtaining lost profits when its net lost profits are shown by its past and subsequent profits, testimony from experts and lay witnesses, and development of direct relationships between the expected profits of its business and the realized profits of similar businesses.

TABLE OF CASES

Case	Book §
AAAA Enters., Inc., ASBCA No. 28172, 86-1 B.C.A. (CCH) ¶ 18,628 (1986)	§ 8.15
Accent General, Inc., ASBCA No. 28813, 87-2 B.C.A. (CCH) ¶ 19,689 (1987)	§ 8.10
A.C.E.S., Inc., ASBCA No. 21417, 79-1 B.C.A. (CCH) ¶ 13,809 (1979)	§ 6.5
Acme Process Equip. Co. v. United States, 347 F.2d 509, 535 (Ct. Cl. 1965)	§ 5.7
Aerojet Gen. Corp., ASBCA No. 17171, 74-2 B.C.A. (CCH) ¶ 10,863 (1974)	§§ 11.11, 11.16
Aetna Casualty & Sur. Co. v. Board of Trustees, 223 Cal. App. 2d 337, 35 Cal. Rptr. 765 (1963)	§ 10.22
Aetna Casualty & Sur. Co. v. Butte-Meade Sanitary Water Dist., 500 F. Supp. 193 (D.S.D. 1980)	§ 14.10
Afgo Eng'g Corp., VACAB No. 1236, 79-2 B.C.A. (CCH) ¶ 13,900 (1979)	§ 8.13
Air Cool, Inc., 88-1 B.C.A. (CCH) ¶ 20,399 (ASBCA 1988)	§ 10.3
Alaska's Children's Servs. v. Smart, 677 P.2d 899 (Alaska 1984)	§ 10.31
Albert Elia Bldg. v. New York State Urban Dev., 54 A.D.2d 337, 388 N.Y.S.2d 462 (1976)	§§ 9.2, 9.4
Alder v. Drudis, 30 Cal. 2d 372, 182 P.2d 195 (1947)	§ 10.1
Algernon Blair, Inc., GSBCA No. 4072, 76-2 B.C.A. (CCH) ¶ 12,073 (1976)	§ 6.15
Al Johnson Constr. Co. v. Missouri Pac. R.R., 426 F. Supp. 639 (E.D. Ark. 1976), aff'd, 553 F.2d 103 (8th Cir. 1977)	§§ 8.18, 9.3, 9.8
Al Johnson Constr. Co. v. United States, 854 F.2d 467 (Fed. Cir. 1988)	§§ 8.17, 9.3
Allen v. Gardner, 126 Cal. App. 2d 335, 272 P.2d 99 (1954)	§ 11.11
Allen Constr. Co. v. United States, 646 F.2d 487 (Ct. Cl. 1981)	§ 9.13
Allen-Howe Specialties v. U.S. Constr., Inc., 611 P.2d 705 (Utah 1980)	§ 5.4
Allied Indus. Serv. Corp. v. Kasle Iron & Metals, Inc., 62 Ohio App. 2d 144, 405 N.E.2d 307 (1977)	§ 13.18
Alpine Constr. Co. v. Water Works Bd., 377 So. 2d 954 (Ala. 1979)	§ 10.22

Case	Book §
Alpine Indus., Inc. v. Gohl, 30 Wash. App. 750, 637 P.2d 998 (1981)	§§ 15.1, 15.3, 15.4
American Bldgs. Co. v. DBH Attachments, Inc., 676 S.W.2d 558 (Tenn. Ct. App. 1984)	§ 14.13
American Bridge Co. v. State, 245 A.D. 535, 283 N.Y.S. 577 (1935)	§§ 5.4, 5.11
American Casualty Co., *In re,* 851 F.2d 794 (6th Cir. 1988)	§ 11.3
American Dredging Co. v. United States, 107 Ct. Cl. 1010 (1975)	§ 8.10
American Elec., 76-2 B.C.A. (CCH) ¶ 12,151 (ASBCA 1976)	§ 10.2
American-Hawaiian Eng'g & Constr. Co. v. Butler, 165 Cal. 497, 133 P. 280 (1913)	§§ 10.13, 10.40
American Sanitary Sales Co. v. State, 178 N.J. Super. 429, 429 A.2d 403 (1981)	§ 7.6
Anderson Dev. Corp. v. Coastal States, 543 S.W.2d 402 (Tex. Civ. App. 1976)	§ 7.4
Aranao Constr. Co. v. Success Roofing, Inc., 46 Wash. App. 314, 730 P.2d 720 (1986)	§ 13.18
Arc Elec. Constr. Co. v. George A. Fuller Co., 24 N.Y.2d 99, 247 N.E.2d 111, 299 N.Y.S.2d 129 (1969)	§§ 10.1, 10.2, 10.3
Archie & Allen Spiers Inc. v. United States, 296 F.2d 757, 155 Ct. Cl. 614 (1961)	§ 8.20
Arctic Contractors, Inc. v. State, 564 P.2d 30 (Alaska 1977)	§ 10.24
Armor & Co. v. Nard, 463 F.2d 8 (8th Cir. 1972)	§ 10.18
A.R. Moyer, Inc. v. Graham, 285 So. 2d 397 (Fla. 1973)	§ 13.5
Arntz Contracting Co., EBCA No. 187-12-81, 84-3 B.C.A. (CCH) ¶ 17,604 (1984)	§ 5.6
Arthur Painting Co., ASBCA No. 20267, 76-1 B.C.A. (CCH) ¶ 11,894 (1976)	§ 6.11
Atlantic Elec. Co., 83-2 B.C.A. (CCH) ¶ 16,671 (GSBCA 1983)	§ 9.3
Austin-Griffith, Inc. v. Goldberg, 224 S.C. 372, 79 S.E.2d 447 (1953)	§§ 10.24, 10.25
Automatic Extruding & Packaging, GSBCA No. 4036, 74-2 B.C.A. (CCH) ¶ 10,949 (1974)	§ 6.4
Avey v. Leather Prods. Co., 73 Ohio App. 245, 55 N.E.2d 813 (1942)	§ 10.24
Avner v. Longridge Estates, 272 Cal. App. 2d 607, 77 Cal. Rptr. 633 (1969)	§ 13.13
Aydin Corp., 89-3 B.C.A. (CCH) ¶ 22,044 (EBCA 1989)	§ 10.3
Bagwell Coatings, Inc. v. Middle S. Energy, Inc., 797 F.2d 1298 (5th Cir. 1986)	§ 9.5
Baldwin-Lima-Hamilton Corp. v. United States, 434 F.2d 1371 (Ct. Cl. 1970)	§ 1.6
Baltimore Contractors v. United States, 12 Cl. Ct. 328 (1987)	§§ 8.12, 8.19

CASES

Case	Book §
Barnard-Curtiss Co. v. United States, 257 F.2d 565 (10th Cir. 1958)	§ 14.10
Barr & Sons v. Cherry Hill Centre, Inc., 90 N.J. Super. 358, 217 A.2d 631 (1966)	§ 13.34
Bartlow-Hope Elec. Corp. v. Herzog, 692 S.W.2d 404 (Mo. Ct. App. 1985)	§ 5.14
Barton & Son Co., 65-2 B.C.A. (CCH) ¶ 4,874 (1965)	§ 5.9
Basle v. Commissioner, CA-3, 58-2 U.S. Tax Cas. (CCH) ¶ 9748, 256 F.2d 581, *aff'g* 16 T.C.M. 745 (1957)	§ 2.2
Bates & Rogers Constr. Corp. v. North Shore Sanitary Dist., 128 Ill. App. 3d 962, 471 N.E.2d 915 (1984)	§ 13.5
Bause v. Anthony Pools, Inc., 205 Cal. App. 2d 606, 23 Cal. Rptr. 265 (1962)	§ 10.18
B.C. Richter Contracting Co. v. Continental Casualty Co., 230 Cal. App. 2d 491, 41 Cal. Rptr. 98 (1964)	§§ 10.12, 10.36, 10.39
Beacon Plaza Shopping Center, Inc. v. Tri-Cities Constr. & Supply Co., 2 Mich. App. 415, 140 N.W.2d 531 (1987)	§ 13.17
BECO Corp., ASBCA No. 27090, 82-2 B.C.A. (CCH) ¶ 16,124 (1982)	§ 7.6
Beco Corp. v. Roberts & Sons Constr. Co., 114 Idaho 704, 760 P.2d 1120 (1988)	§ 8.10
Bender-Miller Co. v. Thomwood Farms, Inc., 211 Va. 585, 179 S.E.2d 636 (1971)	§ 13.17
Benham v. World Airways, 296 F. Supp. 813 (D. Haw. 1969)	§ 9.9
Bennett v. United States, 371 F.2d 859 (Ct. Cl. 1967)	§§ 5.15, 9.7, 9.9
Berg v. General Motors Corp., 87 Wash. 2d 584, 555 P.2d 818 (1976)	§ 13.24
Berley Indus., Inc. v. City of N.Y., 45 N.Y.2d 683, 385 N.E.2d 281, 412 N.Y.S.2d 589 (1978)	§ 5.14
Berlin Dev. Corp. v. Vermont Structural Steel Corp., 127 Vt. 367, 250 A.2d 189 (1968)	§ 15.7
Berman v. Rubin, 138 Ga. App. 849, 227 S.E.2d 802 (1976)	§ 14.6
Bernard McMenamy Contractor, Inc., ENGBCA No. 3413, 77-1 B.C.A. (CCH) ¶ 12,335 (1977)	§ 8.10
Betsy Ross Flag Co., ASBCA No. 12124, 67-2 B.C.A. (CCH) ¶ 6688 (1967)	§ 6.4
Big Boy Drilling Corp. v. Etherridge, 44 Cal. App. 2d 114, 11 P.2d 953 (1941)	§ 11.3
Big 4 Constr. Co., DOTCAB No. 75-18, 76-2 B.C.A. (CCH) ¶ 12,029 (1976)	§ 6.7
Biggin & Co. v. Permanite, Ltd., [1951] 2 K.B. 314 (C.A.)	§ 4.18
Bill's Janitor Serv., ASBCA No. 10345, 65-2 B.C.A. (CCH) ¶ 4916 (1965)	§ 6.4
Black Lake Pipe Line Co. v. Union Constr. Co., 538 S.W.2d 80 (Tex. 1976)	§§ 9.4, 9.11

Case	Book §
Blake Constr. Co., ASBCA No. 20747, 83-1 B.C.A. (CCH) ¶ 16,410 (1983)	§ 8.11
Blake Constr. Co. v. Coakley Co., 431 A.2d 569 (D.C. 1981)	§ 5.4
Blankenship Constr. Co. v. North Carolina State Highway Comm'n, 28 N.C. App. 593, 222 S.E.2d 452 (1976)	§ 8.14
Blinderman Constr. Co. v. United States, 695 F.2d 552 (Fed. Cir. 1982)	§ 5.5
Blinderman Constr. Co. v. United States, 17 Cl. Ct. 860 (1989)	§ 13.49
B.L. Metcalf Gen. Contractor, Inc. v. Earl Erne, Inc., 212 Cal. App. 2d 689, 28 Cal. Rptr. 382 (1963)	§§ 10.9, 10.18
Block v. Neal, 460 U.S. 289 (1983)	§ 13.31
Bloomsburg Mills, Inc. v. Sordoni Constr. Co., 401 Pa. 358, 164 A.2d 201 (1960)	§ 13.33
B&M Roofing & Painting Co., ASBCA No. 26998, 86-1 B.C.A. (CCH) ¶ 18,833 (1986)	§ 8.11
Bohemian-American Workingman's Gymnastic Ass'n (SOKOL) v. Northern Bank, 120 N.Y.S. 134 (Sup. Ct. 1909)	§ 13.38
Bo McAllister & Lloyd Thompson, IBCA No. 2144, 88-1 B.C.A. (CCH) ¶ 20,329 (1988)	§ 8.15
Bonebrake v. Cox, 499 F.2d 951 (8th Cir. 1974)	§ 13.18
Boston, City of v. New England Sales & Mfg., Inc., 386 Mass. 820, 438 N.E.2d 68 (1982)	§ 10.25
Brady v. Board of Educ., 222 A.D. 504, 226 N.Y.S. 707 (1926)	§ 13.51
Brannon v. Southern Ill. Hosp. Corp., 69 Ill. App. 3d 1, 386 N.E.2d 1126 (1978)	§ 13.20
Bregman Constr. Corp., ASBCA No. 9000, 64 B.C.A. (CCH) ¶ 4426 (1964)	§ 8.9
Bresler v. Commissioner, 65 T.C. 182 (1975)	§ 2.1
Brick's, Inc., DOTBCA No. 1906, 89-1 B.C.A. (CCH) ¶ 21,381 (1989)	§ 6.13
Brinderson Corp. v. Hampton Rds. Sanitation Dist., 825 F.2d 41 (4th Cir. 1987)	§ 8.22
Brookhaven Landscape & Grading Co. v. J.F. Barton Contracting Co., 677 F.2d 516 (11th Cir. 1982)	§ 9.5
Bruce Constr. Corp. v. United States, 324 F.2d 516, 163 Ct. Cl. 97 (1963)	§§ 5.13, 9.13
Bruno Preski v. Warchol Constr. Co., 111 Ill. App. 3d 641, 444 N.E.2d 1105 (1985)	§ 9.9
Bryant & Bryant, ASBCA No. 27910, 88-3 B.C.A. (CCH) ¶ 20,923 (1988)	§ 6.5
Bryant Elec. Co. v. City of Fredericksburg, 762 F.2d 1192 (4th Cir. 1985)	§ 13.5
Buckley & Co. v. State, 140 N.J. Super. 289, 356 A.2d 56 (1975)	§§ 5.7, 6.2, 6.16, 9.6, 13.51

Case	Book §
Building Maintenance Co., ENGBCA No. 4115, 83-2 B.C.A. (CCH) ¶ 16,629 (1983)	§ 11.3
Bulk Oil (U.S.A.) v. Sun Oil Trading Co., 697 F.2d 481 (2d Cir. 1983)	§ 3.3
Burke County Pub. Schools Bd. of Educ. v. Juno Constr. Corp., 50 N.C. App. 238, 273 S.E.2d 504 (1981)	§ 13.17
Burnes Estate v. Fidelity & Deposit Co., 96 Mo. App. 467, 70 S.W. 518 (1902)	§ 9.19
Burnett & Doty Dev. Co. v. Phillips, 84 Cal. App. 3d 384, 148 Cal. Rptr. 569 (1978)	§§ 9.7, 13.41, 13.53, 14.12, 15.1, 15.2, 15.5, 15.8
Burns v. Hanover Ins. Co., 454 A.2d 325 (D.C. App. 1982)	§§ 7.4, 10.22, 10.24
Busch v. Wilcox, 82 Mich. 315, 46 N.W. 940 (1915)	§ 8.18
Butts v. Atlanta Fed. Sav. & Loan Ass'n, 152 Ga. 40, 262 S.E.2d 230 (1979)	§ 13.31
Calandro Dev. v. R.M. Butler Contractors, Inc., 249 So. 2d 254 (La. Ct. App. 1971)	§ 13.46
California Farm & Fruit Co. v. Schiappa-Pietra, 151 Cal. 732, 91 P. 593 (1907)	§ 10.13
California Lettuce Growers, Inc. v. Union Sugar Co., 45 Cal. 2d 474, 289 P.2d 785 (1955)	§ 10.29
Call v. Alcan Pac. Co., 251 Cal. App. 2d 442, 59 Cal. Rptr. 763 (1967)	§ 10.12
Camper v. W.J. McDermott, 266 Cal. App. 2d 41, 71 Cal. Rptr. 590 (1968)	§ 13.39
Canon Constr. Corp., ASBCA No. 16142, 72-1 B.C.A. (CCH) ¶ 9404 (1972)	§ 5.11
Cantrell v. Woodhill Enters., Inc., 273 N.C. 490, 160 S.E.2d 476 (1968)	§ 10.12
Capital Elec. Co. v. United States, 729 F.2d 743 (Fed. Cir. 1984)	§§ 5.12, 5.13
Capitol Coal Sales Corp., ASBCA 16551, 73-1 B.C.A. (CCH) ¶ 9779 (1973)	§ 6.4
Carl M. Halvorson, Inc. v. United States, 461 F. 2d 1337, 198 Ct. Cl. 882 (1972)	§ 8.17
Carolina Metal Prods. Corp. v. Larson, 389 F.2d 490 (5th Cir. 1968)	§ 9.2
Carrig v. Gilbert-Varker Corp., 314 Mass. 351, 50 N.E.2d 59 (1943)	§ 10.16
Carrols Equities Corp. v. Villnave, 57 A.D. 2d 1044, 395 N.Y.S.2d 800 (1977)	§ 12.15
Carter v. Sherburne Corp., 132 Vt. 88, 315 A.2d 870 (1974)	§ 5.2
Cash v. United States, 571 F. Supp. 513 (N.D. Ga. 1983)	§ 13.31
Castagna & Son Inc. v. City of N.Y., Index No. 3803/81 (N.Y. Sup. Ct. 1989)	§ 5.4
Certified Corp. v. Hawaii Teamsters & Allied Workers, Local 996, 597 F.2d 1269 (10th Cir. 1979)	§ 1.7

Case	*Book §*
C.E. Wylie Constr. Co., ASBCA No. 26545, 85-1 B.C.A. (CCH) ¶ 17,933 (1985)	§ 8.1
CFI Constr. Co., DOT No. 1782, 1801, 87-1 B.C.A. (CCH) ¶ 19,547 (1987)	§ 8.15
Champagne-Webber, Inc. v. City of Fort Lauderdale, 519 So. 2d 696 (Fla. Dist. Ct. App. 1988)	§ 8.17
Chaney Bldg. Co., Inc. v. Sunnywide School Dist. No. 12, 147 Ariz. 270, 709 P.2d 904 (1985)	§ 9.5
Chaney & James Constr. Co. v. United States, 421 F.2d 728 (Ct. Cl. 1970)	§§ 1.6, 5.4
Chaplin v. Hicks, [1911] 2 K.B. 786 (C.A.)	§ 4.18
Charles D. Weaver v. United States, 538 F.2d 346 (Ct. Cl. 1976)	§ 9.13
Charles T. Parker Constr. Co. v. United States, 433 F.2d 77, 193 Ct. Cl. 320 (1970)	§ 8.22
Chesapeake Indus., Inc. v. Togova Enters., Inc., 149 Cal. App. 3d 901, 197 Cal. Rptr. 348 (1983)	§ 11.11
Chesapeake & Potomac Tel. Co. v. Sisson & Ryan, 234 Va. 492, 362 S.E.2d 723 (1987)	§ 13.44
Christie v. United States, 237 U.S. 234 (1915)	§ 8.17
Chrysler Corp., 75-1 B.C.A. (CCH) ¶ 11,236 (ASBCA 1975)	§ 10.22
C.J. Langenfelder & Son, Inc., MDOT 1000, 1-Md. Inst. for Cont'g Prof. of Ed. Law 97 (1980), *aff'd sub nom.* Maryland Port Admin. v. C.J. Langenfelder, 50 Md. App. 525, 438 A.2d 1374 (1982)	§ 9.3
Clack v. United States, 395 F.2d 773, 184 Ct. Cl. 40 (1968)	§§ 8.1, 8.22
Clemente Constr. Corp. v. P.T. Cox Contracting Co., 172 Misc. 904, 16 N.Y.S.2d 483 (1939)	§§ 10.24, 10.25
C. Norman Peterson Co. v. Container Corp., 172 Cal. App. 3d 628, 218 Cal. Rptr. 592 (1985)	§ 10.37
COAC, Inc. v. Kennedy Eng'rs, 67 Cal. App. 3d 916, 136 Cal. Rptr. 890 (1977)	§ 10.41
Coastal, Inc., PSBCA No. 1728, 88-1 B.C.A. (CCH) ¶ 20,272 (1988)	§ 8.6
Coastland Corp. v. Third Nat'l Mortgage Co., 611 F.2d 969 (4th Cir. 1979)	§ 13.42
Coatesville Contractors & Eng'rs v. Borough of Ridley Park, 509 Pa. 553, 506 A.2d 862 (1986)	§§ 1.6, 6.5
Coath & Goss, Inc., ASBCA No. 20949, 76-1 B.C.A. (CCH) ¶ 11,887 (1976)	§ 8.10
Coath & Gross, Inc. v. United States, 101 Ct. Cl. 702 (1944)	§ 5.5
Coleman Eng'g Co., Inc. v. North Am. Aviation, Inc., 65 Cal. 2d 396, 420 P.2d 713, 55 Cal. Rptr. 1 (1966)	§ 10.12
Coley Properties Corp. v. United States, 593 F.2d 380 (Ct. Cl. 1979)	§§ 5.9, 9.6
College Point Boat Corp. v. United States, 267 U.S. 12 (1925)	§ 10.11

CASES

Case	Book §
Columbus, City of v. Clark-Dietz & Assocs. Eng'rs, 550 F. Supp. 610 (N.D. Miss. 1982)	§§ 13.5, 13.6, 13.9
Commerce Int'l Co. v. United States, 338 F.2d 81 (Ct. Cl. 1964)	§ 5.6
Commercial Contractors, Inc. v. United States Fidelity & Guar. Co., 524 F.2d 944 (5th Cir. 1975)	§ 10.12
Commissioner v. Tellier, 66-1 U.S. Tax Cas. (CCH) ¶ 9319, 383 U.S. 687 (1966)	§ 2.5
Commonwealth v. Acchioni & Canuso, Inc., 14 Pa. Commw. 596, 324 A.2d 828 (1974)	§ 8.19
Commonwealth Highway & Bridge Auth. v. General Asphalt Paving Co., 46 Pa. Commw. 114, 405 A.2d 1138 (1979)	§ 5.4
Community Power Suction Furnace Cleaning Co., ASBCA No. 13803, 69-2 B.C.A. (CCH) ¶ 7963 (1969)	§ 8.11
Competex, S.A. v. Labow, 783 F.2d 333 (2d Cir. 1986)	§ 3.5
Conner v. Southern Nev. Paving, Inc., 741 P.2d 800 (Nev. 1987)	§§ 10.17, 10.20, 10.22
Connor v. Great W. Sav. & Loan Ass'n, 69 Cal. 2d 850, 447 P.2d 609, 73 Cal. Rptr. 369 (1969)	§ 13.30
Consolidated Diesel Elec. Corp., ASBCA No. 10486, 67-2 B.C.A. (CCH) ¶ 6669 (1967)	§ 8.17
Consolidated Pac. Eng'g v. Greater Anchorage Area Borough, 563 P.2d 251 (Alaska 1979)	§ 1.13
Continental Consol. Corp., ASBCA No. 10662, 67-1 B.C.A. (CCH) ¶ 6127 (1967)	§§ 6.5, 6.6
Continental Copper & Steel Indus., Inc. v. E.C. Red Cornelius, Inc., 104 So. 2d 50 (Fla. 1958)	§ 13.22
Continental Heller Corp., 84-2 B.C.A. (CCHO) 717,276 (GSBCA 1984)	§ 9.1
Contract Master Servs., Inc. v. United States, 225 Ct. Cl. 735 (1980)	§ 8.22
Contracting & Material Co. v. City of Chicago, 20 Ill. App. 3d 684, 314 N.E.2d 598 (1974), *rev'd on other grounds,* 64 Ill. 2d 21, 349 N.E.2d 389 (1976)	§ 7.1
Cook v. Jacklitch & Sons, Inc., 315 N.W.2d 660 (N.D. 1982)	§ 10.26
Cook v. Oklahoma Bd. of Pub. Affairs, 736 P.2d 140 (Okla. 1987)	§ 8.15
Cook Assocs., Inc. v. Warnick, 664 P.2d 1161 (Utah 1983)	§ 14.13
Cornell Wrecking v. United States, 184 Ct. Cl. 289 (1968)	§ 5.12
Corrino Civetta Constr. Corp. v. City of New York, 67 N.Y.2d 297, 493 N.E.2d 905, 502 N.Y.S.2d 681 (1986)	§ 5.4
Cosmic Constr. Co., ASBCA Nos. 24041, 24036, 88-2 B.C.A. (CCH) ¶ 20,623 (1988)	§ 6.13
Cosmo Constr. Co. v. United States, 439 F.2d 160, 194 Ct. Cl. 559 (1971)	§ 6.16

Case	Book §
Costanza Constr. Corp. v. City of Rochester, 147 A.D.2d 929, 537 N.Y.S.2d 394 (1989)	§ 8.15
County Excavation, Inc. v. New York, 44 Misc. 2d 1057, 255 N.Y.S.2d 708 (Ct. Cl. 1964)	§ 14.11
Crawford Painting & Drywall Co. v. J.W. Bateson Co., 857 F.2d 981 (5th Cir. 1988)	§ 6.2
CRF v. United States, 624 F.2d 1054 (Ct. Cl. 1980)	§ 6.5
Cross Constr. Co., ENGBCA No. 3676, 79-1 B.C.A. (CCH) ¶ 13,707 (1979)	§ 8.1
Cruse v. Clawson, 137 Mont. 439, 352 P.2d 989 (1960)	§ 13.53
Cueroni v. Coburnville Garage, 315 Mass. 135, 52 N.E.2d 16 (1943)	§ 13.17
Cumberland Perry Vocational Technical School Auth. v. Bogar & Bink, 261 Pa. Super. 350, 396 A.2d 433 (1978)	§ 1.13
Cunningham Bros., Inc. v. City of Waterloo, 254 Iowa 659, 117 N.W.2d 46 (1962)	§ 13.51
CWC, Inc., ASBCA No. 26432, 82-1 B.C.A. (CCH) ¶ 15,907 (1982)	§ 5.9
Dale's Sand & Gravel Co. v. Westwood Constr. Co., 62 Or. App. 570, 661 P.2d 1378 (1983)	§ 9.3
Dallas, City of v. Hubbell, 325 S.W.2d 880 (Tex. Civ. App. 1959)	§ 6.5
Daniel Hamm Drayage Co. v. Waldinger Corp., 508 F. Supp. 390 (E.D. Mo.), *aff'd in part and modified in part,* 666 F.2d 1213 (8th Cir. 1981)	§§ 8.17, 8.18
David J. Tiernay, GSBCA Nos. 7107, 6198, 88-2 B.C.A. (CCH) ¶ 20,806 (1988)	§ 6.15
Davidson v. McKown, 157 Kan. 217, 139 P.2d 421 (1943)	§ 8.18
Davidson & Jones, Inc. v. County of New Hanover, 41 N.C. App. 661, 255 S.E.2d 580 (1979)	§ 13.5
Dawco Constr., Inc. v. United States, No. 450-86C, 1989 U.S. Cl. Ct. (LEXIS 238) (Nov. 17, 1989)	§ 6.15
Degenarrs Co. v. United States, 2 Cl. Ct. 490 (1983)	§ 5.12
Dehnert v. Arrow Sprinklers, Inc., 705 P.2d 846 (Wyo. 1985)	§ 9.10
Delahanty v. First Pa. Bank, 318 Pa. Super. 90, 464 A.2d 1243 (1983)	§ 14.13
De La Hoya v. Slim's Gun Shop, 80 Cal. App. 3d Supp. 6, 146 Cal. Rptr. 68 (1978)	§ 11.3
Delco Elecs. Corp. v. United States, 17 Cl. Ct. 302 (1989)	§ 6.15
Deutsche Bank Filiale Nürnberg v. Humphrey, 272 U.S. 517 (1926)	§§ 3.3, 3.4
D. Federico Co. v. New Bedford Redev. Auth., 723 F.2d 122 (1st Cir. 1983)	§ 8.10
D.H. Dave, Inc. v. Gerben Contracting Co., ASBCA No. 13995, 73-2 B.C.A. (CCH) ¶ 10,191 (1973)	§ 5.11
Dick Corp. v. State Pub. School Bldg. Auth., 27 Pa. 498, 365 A.2d 663 (1976)	§ 9.5

Case	Book §
District Concrete Co. v. Bernstein Concrete Corp., 418 F.2d 1030 (D.C. Cir. 1980)	§ 9.4
D. Joseph DeVito v. United States, 413 F.2d 1147, 188 Ct. Cl. 979 (1969)	§ 10.12
D.K. Meyer Corp. v. Brevco, 206 Neb. 318, 292 N.W.2d 773 (1980)	§ 9.5
Donnelly Constr. Co. v. Oberg/Hunt/Gilleland, 139 Ariz. 184, 677 P.2d 1292 (1984)	§§ 13.2, 13.5, 13.50
D'Onofrio Bros. Constr. Corp. v. New York City Bd. of Educ., 72 A.D.2d 760, 421 N.Y.S.2d 377 (1979)	§§ 9.5, 9.10
Doral Country Club, Inc. v. Curcie Bros. Inc., 174 So. 2d 749 (Fla. Dist. Ct. App. 1965)	§ 9.5
Douglass v. Licciardi Constr. Co., Inc., ___ Pa. Super. ___, 562 A.2d 913 (1989)	§ 13.33
Downy v. United Weatherproofing, 363 Mo. 852, 253 S.W.2d 976 (1953)	§ 13.46
Drackett Prods. Co. v. Blue, 152 So. 2d 463 (Fla. 1963)	§ 13.24
Dravo Corp., ENGBCA No. 3901, 80-2 B.C.A. (CCH) ¶ 14,757 (1980)	§§ 8.8, 8.21, 8.23
Dravo Corp. v. Commonwealth, 564 S.W.2d 16 (Ky. Ct. App. 1977)	§ 8.19
Dravo Corp. v. United States, 594 F.2d 842, 219 Ct. Cl. 416 (1979)	§ 11.11
Dresco Mechanical Contractors, Inc. v. Todd-CEA, Inc., 531 F.2d 1292 (5th Cir. 1976)	§ 14.6
Drews Co. v. Ledwith-Wolfe Assocs., Inc., 296 S.C. 207, 371 S.E.2d 532 (1988)	§§ 15.1, 15.2, 15.3, 15.4, 15.5, 15.7, 15.8
Durkee R.J. v. Commissioner, CA-6, 47-1 U.S. Tax Cas. (CCH) ¶ 9279, 162 F.2d 184, *aff'g* 14 T.C.M. 96 (1955)	§ 2.2
Eastern Tunneling Corp. v. Southgate Sanitation Dist., 487 F. Supp. 109 (D. Colo. 1980)	§§ 13.10, 13.53
Eastway Constr. Corp. v. City of N.Y., 762 F.2d 243 (2d Cir. 1985)	§ 9.7
E.B. Jones Constr. Co. v. County of Denver, 717 P.2d 1009 (Colo. Ct. App. 1986)	§ 9.20
E.C. Ernst Co. v. Koppers Co., 476 F. Supp. 729 (W.D. Pa. 1979)	§ 9.20
E.C. Ernst Co. v. Koppers Co., 626 F.2d 324 (3d Cir. 1980), *on remand, aff'd in part and rev'd in part*, 520 F. Supp. 830 (W.D. Pa. 1981)	§§ 5.14, 7.10, 9.13, 11.3, 11.11
E.C. Ernst, Inc. v. Manhattan Constr. Co., 387 F. Supp. 1001 (S.D. Ala. 1974), *aff'd in part, rev'd in part*, 551 F.2d 1026 (5th Cir. 1977)	§§ 5.4, 6.5, 13.2, 13.3, 13.5, 13.10
Ecko Enters., Inc. v. Remi Fortin Constr., Inc., 118 N.H. 37, 382 A.2d 368 (1978)	§ 9.5

Case	Book §
E.C. Nolan v. State, 58 Mich. App. 294, 227 N.W.2d 323 (1975)	§ 5.4
Economy Swimming Pool Co. v. Freeling, 236 Ark. 888, 370 S.W.2d 438 (1963)	§§ 10.11, 10.12, 10.26
Ed Goetz Painting Co., DOTCAB No. 1168, 83-1 B.C.A. (CCH) ¶ 16,134 (1983)	§ 6.11
Ed Hackstaff Concrete, Inc. v. Powder Condominium "A" Owners' Ass'n, 679 P.2d 1112 (Colo. Ct. App. 1984)	§ 10.26
E.D. Wesley Co. v. City of New Berlin, 62 Wis. 2d 668, 215 N.W.2d 657 (1974)	§ 13.16
Eggers & Higgins v. United States, 403 F.2d 225, 185 Ct. Cl. 765 (1968)	§§ 6.16, 9.5
E.H. Morrell Co. v. State, 65 Cal. 2d 787, 423 P.2d 551, 56 Cal. Rptr. 479 (1976)	§ 8.20
Eichleay Corp., 60-2 B.C.A. (CCH) ¶ 2,688 (1960)	§ 5.12
Eickhoff Constr. Co., ASBCA No. 20049, 77-1 B.C.A. (CCH) ¶ 12,398 (1977)	§ 5.9
Electronic & Missile Facilities v. United States, 306 F.2d 554 (5th Cir. 1962), *rev'd on other grounds,* 374 U.S. 167 (1963)	§ 1.12
Ell-Dorer Contracting Co. v. State, 197 N.J. Super. 175, 484 A.2d 356 (1984)	§ 8.19
Eller Bros. v. Home Fed. Sav. & Loan Ass'n, 623 S.W.2d 624 (Tenn. Ct. App. 1981)	§ 10.22
Elte, Inc. v. S. S. Mullen, Inc., 469 F.2d 1127 (9th Cir. 1972)	§ 7.4
Ely v. Bottini, 179 Cal. App. 2d 287, 3 Cal. Rptr. 756 (1960)	§ 10.22
Emerald Forest Util. Dist. v. Simonsen Constr. Co., 679 S.W.2d 51 (Tex. Ct. App. 1984)	§§ 8.14, 13.9
Employers Ins. v. Construction Management Eng'rs, 297 S.C. 354, 377 S.E.2d 119 (Ct. App. 1989)	§ 9.19
Episcopal Hous. Corp. v. Federal Ins. Co., 273 S.C. 181, 255 S.E.2d 451 (1979)	§ 1.13
Essential Constr. Co., ASBCA No. 18,706, 89-___ B.C.A. (CCH) ¶ ___ (1989)	§ 7.10
Essential Constr. Co., ASBCA No. 18,491, 78-2 B.C.A. (CCH) ¶ 13,314 (1978)	§ 6.4
Evergreen Amusement Corp. v. Milstead, 206 Md. 610, 112 A.2d 901 (1955)	§§ 15.1, 15.4, 15.7, 15.8
Exber, Inc. v. Sletten Constr. Co., 92 Nev. 721, 558 P.2d 517 (1976)	§ 1.13
Excavation Constr. Inc. v. WMATA, 624 F. Supp. 582 (D.D.C. 1984)	§ 5.12
Exton Drive-In, Inc. v. Home Indem. Co., 436 Pa. 480, 261 A.2d 319 (1969), *cert. denied,* 400 U.S. 819 (1970)	§§ 10.21, 15.2, 15.3

Case	Book §
Fairfax County Redev. & Hous. Auth. v. Hurst & Assocs., Inc., 231 Va. 164, 343 S.E.2d 294 (1986)	§ 13.41
Fairlane Estates, Inc. v. Carrico Constr. Co., 228 Cal. App. 2d 65, 39 Cal. Rptr. 35 (1964)	§ 10.16
Fanning & Doorley Constr. Co. v. Geigy Chem. Corp., 305 F. Supp. 650 (D.R.I. 1969)	§ 8.19
Farm Crop Energy, Inc. v. Old Nat'l Bank, 109 Wash. 2d 923, 750 P.2d 231 (1988)	§ 13.42
Fattore Co. v. Metropolitan Sewerage Co., 505 F.2d 1 (7th Cir. 1974)	§ 5.14
Feather v. United Mine Workers, 711 F.2d 530 (3d Cir. 1983)	§ 11.11
Federal Constr. Co., Appeal of, ASBCA No. 17,599, 73-1 B.C.A. (CCH) ¶ 10,003 (1973)	§ 13.17
Fehlhaber Corp. v. United States, 151 F. Supp. 817, 138 Ct. Cl. 571 (1957)	§ 8.19
Fergerson Contracting Co. v. Charles E. Story Constr. Co., 417 S.W.2d 228 (Ky. 1967)	§ 9.19
Ferguson Contracting Co. v. State, 237 N.Y. 186, 142 N.E. 580 (1923)	§ 9.4
Ferrel v. Elrod, 63 Tenn. App. 129, 469 S.W.2d 678 (1971)	§ 14.13
Feuerland-Werkstatten GmbH, ASBCA No. 32,970, 87-___ B.C.A. (CCH) ¶ ___ (1987)	§ 7.10
Fidelity & Deposit Co. v. Harris, 360 F.2d 402 (9th Cir. 1966)	§ 10.41
Fidelity & Deposit Co. v. Stool, 607 S.W.2d 17 (Tex. Civ. App. 1980)	§ 10.18
Fidelity & Guar. Ins. Underwriters, Inc. v. Allied Realty Co., 238 Va. 458, 384 S.E.2d 613 (1989)	§ 13.39
Fields Eng'g Equip. v. Cargill, Inc., 651 F.2d 589 (8th Cir. 1981)	§§ 14.13, 15.1, 15.2, 15.3, 15.4
First Nat'l Bank v. Cann, 503 F. Supp. 419, (N.D. Ohio 1980)	§ 13.35
First Nat'l Bank v. Indian Indus., Inc., 600 F.2d 702 (8th Cir. 1979)	§§ 10.1, 10.12, 10.26
Fishbach & Moore Corp., ASBCA No. 18146, 77-1 B.C.A. (CCH) ¶ 12,300 (1977)	§§ 5.6, 6.8
Flex-Y-Plan Indus., GSBCA No. 4117, 76-1 B.C.A. (CCH) ¶ 11,713 (1976)	§ 6.10
Flippen Materials Co. v. United States, 312 F.2d 408, 160 Ct. Cl. 357 (1963)	§§ 8.17, 8.19
Folk Constr. Co. v. United States, 2 Cl. Ct. 681 (1983)	§ 6.7
Forest Constr. Inc. v. Farrell Creek Steel Co., 484 So. 2d 40 (Fla. Dist. Ct. App. 1986)	§ 9.2
Fortec Constructors v. United States, 8 Cl. T. 490 (1985)	§ 5.8
Fort Sill Assocs. v. United States, 183 Ct. Cl. 301 (1968)	§ 8.11

Case	*Book §*
Foster Constr. C.A. & Williams Bros. Co., A Joint Venture v. United States, 435 F.2d 873, 193 Ct. Cl. 587 (1970)	§§ 8.1, 8.6
Fountain v. Semi-Tropic Land & Water Co., 99 Cal. 677, 34 P. 497 (1893)	§ 10.11
Fox v. United States, 7 Cl. Ct. 60 (1984)	§ 8.6
Foxcraft Townhome Owners Assoc. v. Hoffman Rosner Corp., 105 Ill. App. 3d 951, 435 N.E.2d 210 (1985)	§ 13.17
Fox Valley Eng'g, Inc. v. United States, 151 Ct. Cl. 228 (1960)	§ 9.3
Frank R. Ragonese v. United States, 120 F. Supp. 768 (Ct. Cl. 1954)	§ 8.21
Frank W. Miller Constr. Co., ASBCA No. 22347, 78-1 B.C.A. (CCH) ¶ 13,039 (1978)	§ 8.1
Frank W. Whitcomb Constr. Corp. v. Cedar Constr. Co., 142 Vt. 541, 459 A.2d 985 (1983)	§ 10.41
Franklin Contracting v. State, 134 N.J. Super. 198, 338 A.2d 875 (Law Div. 1975)	§ 5.4
Fred A. Arnold, Inc., ASBCA No. 18915, 75-2 B.C.A. (CCH) ¶ 11,496 (1975)	§§ 6.5, 6.6
Free-Flow Packaging Corp., GSBCA No. 3992, 75-1 B.C.A. (CCH) ¶ 11,105 (1975)	§ 6.4
Freund v. United States, 260 U.S. 60 (1922)	§ 9.4
Frommeyer v. L&R Constr. Co., 261 F.2d 879 (3d Cir. 1958)	§ 9.2
Gable v. Silver, 258 So. 2d 11 (Fla. Dist. Ct. App. 1972)	§ 13.11
Gallegos v. Graff, 32 Colo. App. 213, 508 P.2d 798 (1973)	§ 13.9
Gamm Constr. Co. v. Townsend, 32 Ill. App. 3d 848, 336 N.E.2d 592 (1975)	§ 13.17
G.C.S., Inc. v. Foster Wheeler Corp., 437 F. Supp. 757 (W.D. Pa. 1975)	§ 5.4
General Builders Supply Co. v. MacArthur, 228 Md. 320, 179 A.2d 868 (1962)	§ 12.15
General Builders Supply Co. v. United States, 187 Ct. Cl. 477, 409 F.2d 246 (1969)	§ 6.7
General Dynamics Corp., DOTCAB 1232, 83-1 B.C.A. (CCH) ¶ 16,386 (1983)	§ 11.7
General Ins. Co. v. Hercules Constr. Co., 385 F.2d 13 (8th Cir. 1967)	§ 5.14
General Ry. Signal Co. v. Washington Metro. Area Transp. Auth., 598 F. Supp. 595 (1984)	§ 9.13
George v. Goldman, 333 Mass. 496, 131 N.E.2d 772 (D. Colo. 1956)	§ 13.16
George A. Fuller Co. v. United States, 69 F. Supp. 409 (Ct. Cl. 1947)	§§ 5.4, 6.5
George A. Shegda, Inc. v. Standard Merchandising Co., 231 Pa. Super. 194, 332 A.2d 498 (1974)	§ 13.33

CASES

Case	Book §
George Hyman Constr. Co. v. Washington Metro. Area Transit Auth., 816 F.2d 753 (D.C. Cir. 1987)	§ 5.12
Gherdi v. Board of Educ., 53 N.J. Super. 349, 147 A.2d 535 (App. Div. 1979)	§ 5.4
Gilbert v. Powell, 165 Ga. App. 504, 301 S.E.2d 683 (1983)	§ 9.13
Giuliani Contracting Co., ASBCA No. 33341, 87-2 B.C.A. (CCH) ¶ 19,743 (1987)	§ 8.7
Glasgow v. Pennsylvania Dep't of Transp., 529 A.2d 576 (Pa. Commw. Ct. 1987)	§ 6.14
Glassman Constr. Co. v. Maryland City Plaza, Inc., 371 F. Supp. 1154 (D. Md. 1974)	§§ 5.5, 5.7
G.M. Shupe, Inc. v. United States, 5 Cl. Ct. 662 (1984)	§ 5.6
Gogo v. Los Angeles County Flood Control Dist., 45 Cal. App. 2d 334, 114 P.2d 65 (1941)	§ 5.7
Golf Landscaping Inc. v. Century Constr. Co., 39 Wash. App. 895, 696 P.2d 590 (1985)	§ 5.12
Gordon H. Ball, Inc., ENGBCA No. 3563, 78-1 B.C.A. (CCH) ¶ 13,055 (1978)	§ 8.5
Gore v. Daniel O'Connell's Sons, Inc., 17 Mass. App. 645, 461 N.E.2d 256 (1984)	§ 13.6
Great Lakes Dredge & Dock Co. v. United States, 96 F. Supp. 923 (Ct. Cl. 1951)	§ 5.9
Greco v. Bucciconi, 283 F. Supp. 978 (W.D. Pa. 1967)	§ 13.22
Greeman v. Yuba Power Prods., Inc., 59 Cal. 2d 57, 377 P.2d 897 (1962)	§ 13.6
Green v. City of N.Y., 283 A.D.2d 485, 128 N.Y.S.2d 715 (1954)	§ 9.1
Green Builders, Inc., ASBCA No. 35518, 88-2 B.C.A. (CCH) ¶ 20,734 (1988)	§ 5.9
Gregory v. Weber, 51 Or. App. 547, 626 P.2d 392 (1981)	§ 14.11
Grossman Holdings, Ltd. v. Hourihan, 414 So. 2d 1037 (Fla. 1982)	§ 10.18
Grover-Diamond Assocs. v. American Arbitration Ass'n, 297 Minn. 324, 211 N.W.2d 787 (1973)	§ 1.13
Groves v. Wunder Co., 205 Minn. 163, 286 N.W. 235 (1939)	§ 13.33
Gulf, Mobile & Ohio R.R. v. Illinois Cent. R.R., 128 F. Supp. 311 (D. Ala. 1954), *aff'd,* 225 F.2d 816 (5th Cir. 1955)	§ 14.8
Gundersons, Inc. v. Tull, 678 P.2d 1061 (Colo. Ct. App. 1983), *remanded,* 709 P.2d 940 (Colo. 1985)	§ 14.9
Guschl v. Schmidt, 266 Wis. 410, 63 N.W.2d 759 (1954)	§ 13.17
Gust K. Newberg, Inc. v. Illinois State Toll Highway Auth., 153 Ill. App. 3d 918, 506 N.E.2d 658 (1987)	§ 5.4
Guy F. Atkinson Co., ENGBCA No. 4693, 87-3 B.C.A. (CCH) ¶ 19,971 (1987)	§ 8.6
Hadley v. Baxendale, 156 Eng. Rep. 145 (Ex. 1854)	§§ 10.34, 11.11, 13.2, 13.32

Case	Book §
Haener v. Ada County Highway Dist., 108 Idaho 170, 697 P.2d 1184 (1985)	§ 8.10
Hallman Bros. v. United States, 68 F. Supp. 204, 107 Ct. Cl. 555 (1946)	§ 8.1
Haney v. United States, 676 F.2d 584, 230 Ct. Cl. 148 (1982)	§ 5.8
Hanna v. Fletcher, 231 F.2d 469 (D.C. Cir. 1958)	§ 13.19
Hansel Phelps Constr. Co., ENGBCA No. 3368, 74-2 B.C.A. (CCH) ¶ 10,728 (1974)	§ 6.5
Harlan Fuel Co. v. Wiggerton, 203 Ky. 546, 262 S.W. 957 (1924)	§ 13.9
Harsha v. State Sav. Bank, 346 N.W.2d 791 (Iowa 1984)	§ 14.13
Hart Eng'g Co. v. FMC Corp., 593 F. Supp. 1471 (D.R.I. 1984)	§ 13.24
Hartley v. Ballow, 286 N.C. 51, 209 S.E.2d 776 (1974)	§ 13.35
Hartline-Thomas, Inc. v. H.W. Ivey Constr. Co., 161 Ga. App. 91, 289 S.E.2d 196 (1982)	§ 9.5
Havens Steel Co. v. Randolph Eng'g Co., 613 F. Supp. 514 (W.D. Mo. 1985), aff'd, 813 F.2d 186 (8th Cir. 1987)	§§ 6.5, 11.11
Hawley v. Orange County Flood Control Dist., 211 Cal. App. 2d 708, 27 Cal. Rptr. 478 (1963)	§ 5.4
Head Painting Contractor, ASBCA No. 26249, 82-1 B.C.A. (CCH) ¶ 15,886 (1982)	§ 6.4
Heinkel v. City of Corvallis, 13 Or. App. 375, 510 P.2d 579 (1973)	§ 10.22
Helene Curtis Indus., Inc. v. United States, 312 F.2d 774, 160 Ct. Cl. 437 (1963)	§ 8.17
Henningsen v. Bloomfield Motors, Inc., 32 N.J. 358, 161 A.2d 69 (1960)	§ 13.22
Henry Shenk Co. v. Erie County, 319 Pa. 100, 178 A. 662 (1935)	§ 5.4
Herbert & Brooner Constr. Co. v. Golden, 499 S.W.2d 541 (Mo. Ct. App. 1973)	§ 14.11
Herbert W. Jaeger & Assoc. v. Slovak Am. Charitable Assoc., 156 Ill. App. 3d 106, 507 N.E.2d 863 (1987)	§ 9.13
Hercules Constr. Co., VACAB No. 2508, 88-2 B.C.A. (CCH) ¶ 20,527 (1988)	§ 8.12
Hicks v. Guinness, 269 U.S. 71 (1925)	§ 3.4
Highland Constr. Co. v. Union Pac. R.R., 683 P.2d 1042 (Utah 1984)	§ 10.37
Hillsborough County Aviation Auth. v. Cone Bros. Constr. Co., 285 So. 2d 619 (Fla. Dist. Ct. App. 1973)	§ 14.10
H.N. Bailey & Assocs. v. United States, 449 F.2d 376, 196 Ct. Cl. 166 (1971)	§ 8.18
Hoag v. Jenan, 86 Cal. App. 2d 556, 195 P.2d 451 (1948)	§§ 15.4, 15.8
Hoffman v. Pfingsten, 260 Wis. 160, 50 N.W.2d 369 (1951)	§ 9.9
Hollerbach v. United States, 233 U.S. 165 (1914)	§§ 8.17, 8.19

CASES

Case	Book §
Honolulu Roofing Co. v. Felix, 49 Haw. 578, 426 P.2d 298 (1967)	§ 9.19
Housing Auth. v. E.W. Johnson Constr. Co., 264 Ark. 523, 573 S.W.2d 316 (1978)	§§ 5.9, 7.4
Housing Auth. v. Hubbell, 325 S.W.2d 880 (Tex. Civ. App. 1959)	§ 5.4
Howard, Needles, Tammen & Bergendoff v. Steers, Perini & Pomeroy, 312 A.2d 621 (Del. 1973)	§ 13.30
Hoyer Constr. Co., ASBCA No. 21616, 84-2 B.C.A. (CCH) ¶ 17,249 (1984)	§ 8.11
H.T.C. Corp. v. Olds, 486 P.2d 463 (Colo. App. 1971)	§§ 9.4, 9.5
Huang Int'l Inc. v. Foose Constr. Co., 734 P.2d 975 (Wyo. 1987)	§ 9.5
Huber, Hunt & Nichols v. Moore, 67 Cal. App. 3d 278, 136 Cal. Rptr. 603 (1977)	§ 9.13
Human Advancement, Inc., HUDBCA 77-215, 81-2 B.C.A. (CCH) ¶ 25,317 (1981)	§ 6.16
Hunt v. Blasius, 74 Ill. 2d 203, 384 N.E.2d 368 (1979)	§ 14.6
Hurlen Constr. Co., ASBCA No. 31069, 86-1 B.C.A. (CCH) ¶ 18,690 (1986)	§ 8.12
Hyde Constr. Co., ASBCA No. 8393, 1963 B.C.A. (CCH) ¶ 3,911 at 19,391 (1963)	§ 7.8
Hydro Dredge Corp., E&GBCA No. 5303, 89-___ B.C.A. (CCH) ¶ ___ (Oct. 16, 1989)	§ 13.49
Ida Grove Roofing v. City of Storm Lake, 378 N.W.2d 313 (Iowa Ct. App. 1985)	§ 9.5
Ideker, Inc. v. Missouri State Highway Comm'n, 654 S.W.2d 617 (Mo. Ct. App. 1983)	§ 8.6
Illinois State Toll Highway Auth. v. Gust K. Newberg, Inc., 177 Ill. App. 3d 6, 531 N.E.2d 982 (1989)	§ 10.22
Illinois Structural Steel Corp. v. Pathman Constr. Co., 23 Ill. App. 3d 1, 318 N.E.2d 232 (1974)	§ 1.20
Indian Head Truck Line, Inc. v. Hvidsten Transp., Inc., 268 Minn. 176, 128 N.W.2d 334 (1964)	§ 13.40
Industrial Indem. Co. v. Wick Constr. Co., 680 P.2d 1100 (Alaska 1984)	§§ 10.22, 10.24
Inman v. Binghamton Hous. Auth., 3 N.Y.2d 137, 143 N.E.2d 895, 164 N.Y.S.2d 699 (1957)	§ 13.5
Insulation Contracting & Supply v. Kravco, Inc., 209 N.J. Super. 367, 507 A.2d 754 (App. Div. 1986)	§ 10.41
Integrated, Inc. v. Alec Fergusson Elec. Contractor, 250 Cal. App. 2d 287, 58 Cal. Rptr. 503 (1967)	§§ 10.11, 10.12, 10.13, 10.40, 11.3
International Air Craft Servs., ASBCA No. 839, 65-1 B.C.A. (CCH) ¶ 4793 (1965)	§ 13.48
Ippolito—Lutz, Inc. v. Cohoes Hous. Auth., 22 A.D.2d 990, 254 N.Y.S.2d 783 (1964)	§ 13.51
Jack Morehouse dba Morehouse Painting, ICBA No. 2087, 86-3 B.C.A. (CCH) ¶ 19,014 (1986)	§ 8.1

Case	Book §
Jackson-Swindell-Dressler, A Joint Venture, ENGBCA No. 3614, 76-2 B.C.A. (CCH) ¶ 12,222 (1976)	§ 8.10
Jacobs & Young, Inc. v. Kent, 230 N.Y. 239, 129 N.E. 889 (1921)	§ 10.18
J.A. Jones Constr. Co. v. United States, 395 F.2d 783, 184 Ct. Cl. 1 (1968)	§ 8.17
James Julian, Inc. v. Town of Elkton, 341 F.2d 205 (4th Cir. 1965)	§§ 8.10, 8.20
James McHugh Constr. Co., ENGBCA No. 4600, 82-1 B.C.A. (CCH) ¶ 15,682 (1982)	§ 8.15
James Stuart Polsek & Assocs. v. Bergen County Iron Works, 142 N.J. Super. 516, 362 A.2d 63 (1976)	§ 1.13
James Walford Constr. Co., GSBCA 6498, 83-1 B.C.A. (CCH) ¶ 16,277, 25 G.C. ¶ 196 (1983)	§ 7.4
Jasken v. Sheehy Constr. Co., 642 P.2d 58 (Colo. Ct. App. 1982)	§ 10.30
Jasper Constr., Inc. v. Foothill Junior College Dist., 91 Cal. App. 3d 1, 153 Cal. Rptr. 767 (1979)	§§ 10.22, 13.49
J.A. Tobin Constr. Co. v. Kemp, 239 Kan. 240, 718 P.2d 302 (1986)	§ 9.5
J.B. Williams Co. v. United States, 450 F.2d 1379 (Ct. Cl. 1979)	§ 6.5
J. Clutter Custom Digging v. English, 181 Ind. App. 603, 393 N.E.2d 230 (1979)	§ 14.11
J.D. Hedin Constr. Co. v. United States, 347 F.2d 235 (Ct. Cl. 1965)	§§ 5.11, 5.15, 13.9
J.D. Hedin Constr. Co. v. United States, 408 F.2d 424, 171 Ct. Cl. 70 (1969)	§ 5.2
J.D. Shotwell Co., 65-2 B.C.A. (CCH) ¶ 5243 (ASBCA 1965)	§ 10.3
J.E. Hathaway & Co. v. United States, 249 U.S. 460, 39 S. Ct. 346 (1919)	§ 10.24
J.F. Shea Co. v. United States, 4 Cl. Ct. 46 (1983)	§ 6.5
J.F. Whalen, ENGBCA No. 2859, 69-1 B.C.A. (CCH) ¶ 7519 (1969)	§ 8.6
J.L. Simmons Co. v. United States, 412 F.2d 1360, 188 Ct. Cl. 684 (1969)	§ 13.9
J.M. Hollis Constr. Co., 641 S.W.2d 354 (Tex. Ct. App. 1982)	§ 5.14
J&M Lumber, Inc., ASBCA No. 25951, 82-1 B.C.A. (CCH) ¶ 15,500 (1982)	§ 6.4
John E. Green Plumbing & Heating v. Turner Constr., 742 F.2d 965 (6th Cir. 1984), cert. denied, 105 S. Ct. 2328 (1985)	§ 6.16
John F. Burke Eng'g & Constr., ASBCA No. 8182, 1963 B.C.A. (CCH) ¶ 3713 (1963)	§ 5.9
John F. Harkins Co. v. School Dist., 313 Pa. Super. 425, 460 A.2d 260 (1983)	§ 5.10
John L. Gregory & Sons v. A. Guenther & Sons Co., 147 Wis. 2d 298, 432 N.W.2d 584 (1986)	§ 6.2

CASES

Case	Book §
Johnson Controls, Inc. v. National Valve & Mfg. Co., 569 F. Supp. 758 (E.D. Okla. 1983)	§§ 7.3, 7.10
Johnson & Son Erectors, ASBCA No. 24564, 81-1 B.C.A. (CCH) ¶ 15,082 (1981)	§ 5.9
Jones & Laughlin Steel Corp. v. Johns Manville Sales Corp., 626 F.2d 280 (3d Cir. 1980)	§ 13.24
Joseph Bell v. United States, 404 F.2d 975, 186 Ct. Cl. 189 (1968)	§ 11.11
Joseph F. Trionfo & Sons v. Board of Educ., 41 Md. App. 103, 395 A.2d 1207 (1979)	§ 8.18
Joseph Morton Co. v. United States, 3 Cl. Ct. 120, *aff'd*, 757 F.2d 1273 (Fed. Cir. 1985)	§ 10.11
J.R. Stevenson Corp. v. County of Westchester, 113 A.D.2d 918, 493 N.Y.S.2d 819 (1985)	§§ 5.4, 13.51
J.S.&H. Constr. Co. v. Richmond County Hosp. Auth., 473 F.2d 212 (4th Cir. 1973)	§ 1.12
J&T Constr. Co., DOTCAB Nos. 73-4, 75-2 B.C.A. (CCH) ¶ 11,398 (1975)	§ 8.10
Juengel Constr. Co. v. Mt. Etna, Inc., 622 S.W.2d 510 (Mo. Ct. App. 1981)	§§ 10.29, 10.34
Just's Inc. v. Arrington Constr. Co., 99 Idaho 462, 583 P.2d 997 (1978)	§ 13.42
Kahaluu Constr. Co., ASBCA No. 31187, 89-1 B.C.A. (CCH) ¶ 21,308 (1989)	§§ 8.11, 8.12
Kaiser Indus. Co. v. United States, 340 F.2d 322, 169 Ct. Cl. 310 (1965)	§ 8.18
Kalisch-Jarcho, Inc. v. City of N.Y., 58 N.Y.2d 377, 448 N.E.2d 413, 461 N.Y.S.2d 746 (1983)	§§ 6.16, 13.51
Kaminer Constr. Corp. v. United States, 488 F.2d 980, 203 Ct. Cl. 182 (1973)	§ 13.17
Keco Indus., Inc., ASBCA Nos. 15184, 15547, 72-2 B.C.A. (CCH) ¶ 9576 (1972)	§ 11.11
Keel v. Titan Constr. Corp., 639 P.2d 1228 (Okla. 1981)	§ 13.5
Kenford Co. v. County of Erie, 67 N.Y.2d 257, 493 N.E.2d 234, 502 N.Y.S.2d 131 (1986)	§§ 15.1, 15.2
Kenneth Reed Constr. Corp., ENGBCA Nos. 2748, 2749-50, 2861, 72-1 B.C.A. (CCH) ¶ 9,407 (1972)	§ 7.5
Kenworthy v. State, 236 Cal. App. 2d 378, 46 Cal. Rptr. 396 (1965)	§ 10.22
Kiewit & Sons' Co. v. Pasadena City Junior College Dist., 59 Cal. 2d 241, 379 P.2d 18, 28 Cal. Rptr. 714 (1963)	§ 14.10
Kinetic Builders, Inc., ASBCA No. 32627, 88-2 B.C.A. (CCH) ¶ 20,657 (1988)	§§ 8.6, 8.12
Kingery Constr. Co. v. Scherbarth Welding, Inc., 186 Neb. 653, 185 N.W.2d 857 (1971)	§ 5.2
Kirkpatrick v. Temme, 98 Nev. 523, 654 P.2d 1011 (1982)	§ 10.17
Kissel Co. v. Gresly, 591 F.2d 47 (9th Cir. 1979)	§ 13.42

Case	Book §
Klingensmith, Inc. v. United States, 731 F.2d 805 (Fed. Cir. 1984)	§ 5.5
Knier v. Azores Constr. Co., 78 Nev. 20, 368 P.2d 673 (1972)	§§ 15.1, 15.4, 15.7
Koncinsky v. Smith, 390 So. 2d 1377 (La. Ct. App. 1980)	§ 15.3
Korshoj Constr. Co., IBCA No. 321, 63 B.C.A. 3848 (1963)	§ 9.5
Kos Kam, Inc., ASBCA No. 34684, 88-1 B.C.A. (CCH) ¶ 20,246 (1988)	§ 8.12
Kriegler v. Eichler Homes, Inc., 269 Cal. App. 2d 224, 74 Cal. Rptr. 749 (1969)	§ 13.13
Kuhike Constr. Co. v. Mobley, Inc., 159 Ga. App. 777, 285 S.E.2d 236 (1981)	§ 13.17
Laburnum Constr. Corp. v. United States, 325 F.2d 451 (Ct. Cl. 1965)	§ 5.15
La Crosse Garment Mfg. Co. v. United States, 432 F.2d 1377 (Ct. Cl. 1970)	§ 1.6
LaFarge Conseils et Etudes, S.A. v. Kaiser Cement & Gypsum Corp., 791 F.2d 1334 (9th Cir. 1986)	§ 11.3
Lambert Constr. Co., DOTBCA No. 77-9, 78-1 B.C.A. (CCH) ¶ 13,221 (1978)	§ 6.4
Landmark, Inc. v. Stockman's Bank & Trust Co., 680 P.2d 471 (Wyo. 1984)	§§ 15.4, 15.7
Lane Constr. Corp., 84-2 B.C.A. (CCH) ¶ 17,490 (ENGBCA 1988)	§ 9.9
L.A. Reynolds Co. v. State Highway Comm'n, 271 N.C. 40, 155 S.E.2d 473 (1967)	§ 14.10
Lass v. Montana State Highway Comm'n, 157 Mont. 121, 483 P.2d 699 (1971)	§§ 5.15, 11.8
Lathrop v. Tyrrell, 128 Ill. App. 3d 1067, 471 N.E.2d 1049 (1984)	§ 13.22
Lawson v. Durant, 213 Kan. 772, 518 P.2d 549 (1974)	§§ 10.22, 10.24
Lea County Constr. Co., ASBCA No. 13964, 72-1 B.C.A. (CCH) ¶ 9,298 (1972)	§ 11.21
Leavell & Co., C.H., ASBCA No. 18625, 74-2 B.C.A. (CCH) ¶ 10,885 (1974)	§ 8.7
Leavell & Co., Appeal of, ASCBCA No. 16099, 72-2 B.C.A. (CCH) ¶ 9,694 (1972)	§ 9.5
Lee Hoffman v. United States, 340 F.2d 645, 166 Ct. Cl. 39 (1964)	§ 8.1
Lemire Contracting, 89-2 B.C.A. (CCH) ¶ 21,763 (IBCA 1989)	§ 10.3
Levan v. Richter, 152 Ill. App. 3d 1082, 504 N.E.2d 1373 (1987)	§ 9.13
Lewis-Nicholson, Inc. v. United States, 550 F.2d 26 (Ct. Cl. 1977)	§ 6.5
Lichter v. Mellon-Stuart Co., 305 F.2d 216 (3d Cir. 1962)	§ 5.6
Lincoln v. Pohly, 325 S.W.2d 170 (Tex. Civ. App. 1959)	§ 13.35

CASES

Case	Book §
Lincoln Pulp & Paper Co. v. Dravo Corp., 436 F. Supp. 262 (D. Me. 1977)	§ 13.18
Linneman Constr., Inc. v. Montana-Dakota Util. Co., 504 F.2d 1365 (8th Cir. 1974)	§ 9.5
Lipson v. Hawthorne Indus., Inc., 148 Ga. App. 751, 252 S.E.2d 639 (1979)	§ 13.18
Lisbon Contractors, Inc. v. United States, 828 F.2d 759 (Fed. Cir. 1987)	§ 10.3
Little Rock School Dist. v. Matson, Inc., 576 S.W.2d 709 (Ark. 1979)	§ 13.7
L.L. Hall Constr. Co. v. United States, 379 F.2d 559 (Ct. Cl. 1966)	§ 5.12
Longboat Key, Town of v. Carl E. Widell & Son, 362 So. 2d 719 (Fla. Dist. Ct. App. 1978)	§ 8.10
Looker v. Gulf Coast Fair, 203 Ala. 42, 81 So. 832 (1919)	§ 13.2
Loomis & Loomis, Inc. v. Stecker & Colavecchio Architects, Inc., 6 Conn. App. 88, 503 A.2d 181 (1986)	§ 7.3
Loranger Constr., ASBCA No. 9643, 65-2 B.C.A. (CCH) ¶ 5071 (1965)	§ 8.1
Louis Lyster Gen. Contractor, Inc. v. City of Las Vegas, 83 N.M. 138, 489 P.2d 646 (1971)	§§ 10.24, 10.25
Louis M. McMasters, Inc., ASBCA No. 80-159-4, 86-3 B.C.A. (CCH) ¶ 19,067 (1986)	§ 8.10
Luria Bros. & Co. v. United States, 369 F.2d 701, 177 Ct. Cl. 676 (1966)	§§ 5.11, 6.7, 6.13, 8.23, 13.48, 13.53
Lurlee, Inc. v. Pernoshal—39 Co., 135 Ga. App. 724, 218 S.E.2d 701 (1975)	§§ 13.35, 13.37, 13.39
MacPherson v. Buick Motor Co., 217 N.Y. 382, 111 N.E. 1050 (1916)	§ 13.5
Macri v. United States, 353 F.2d 804 (9th Cir. 1965)	§ 9.5
Main Cornice Works, Inc. v. National Union Fire Ins. Co., 258 F. Supp. 377 (S.D. Cal. 1966)	§§ 10.38, 10.39
Major v. Commissioner, 76 T.C. 239 (1981)	§ 2.2
Malone v. Santora, 135 Conn. 286, 64 A. 51 (1949)	§ 9.19
Maloney v. Oak Builders, Inc., 224 So. 2d 161 (4th Cir. 1968)	§ 13.17
Malor Constr. Corp., IBCA No. 1688-6-83, 84-1 B.C.A. (CCH) ¶ 17,023 (1984)	§ 7.4
Manshul Constr. v. Dormitory Auth., 111 Misc. 2d 209, 444 N.Y.S.2d 792 (1981)	§ 11.19
Marathon Oil Co. v. Hollis, 167 Ga. App. 48, 305 S.E.2d 864 (1983)	§§ 10.12, 10.32, 10.34, 10.40
Mark Keshishian & Sons v. Washington Square, Inc., 414 A.2d 834 (D.C. 1980)	§ 13.46
Marlin Assocs., Inc., GSBCA No. 5663, 82-1 B.C.A. (CCH) ¶ 15,739 (1982)	§ 6.6

Case	Book §
Marshall v. Carl F. Shultz, Inc., 438 So. 2d 533 (Fla. Dist. Ct. App. 1983)	§ 14.11
Marshall v. Marvin H. Anderson Constr. Co., 283 Minn. 320, 167 N.W.2d 724 (1969)	§ 13.40
Martin v. Phillips, 122 N.H. 34, 440 A.2d 1124 (1982)	§§ 10.16, 10.17, 10.26
Martin v. Rollins, Inc., 238 Ga. 119, 231 S.E.2d 751 (1977)	§§ 10.13, 10.27, 10.40
Massachusetts Bonding & Ins. Co. v. John R. Thompson Co., 88 F.2d 825 (8th Cir. 1937)	§ 9.19
Maverick Diversified, Inc., ASBCA No. 19838, 76-2 B.C.A. (CCH) ¶ 12,104 (1976)	§ 8.10
Max Drill, Inc. v. United States, 427 F.2d 1233 (Ct. Cl. 1970)	§ 9.9
Maxwell Dynamometer Co. v. United States, 386 F.2d 855 (Ct. Cl. 1967)	§ 13.52
McClung Constr. Co. v. Muncy, 65 S.W.2d 786 (Tex. Civ. App. 1933)	§ 8.18
McCormick Constr. v. United States, 13 Cl. Ct. 496 (1987)	§ 8.23
McDermott v. Middle East Carpet Co., 811 F.2d 1422 (11th Cir. 1987)	§§ 15.1, 15.3, 15.4, 15.7, 15.8
McDevitt & Street Co. v. Marriott Corp., 713 F. Supp. 906 (E.D. Va. 1989)	§§ 7.4, 7.10
McDonald v. Supple, 96 Or. 486, 190 P. 315 (1920)	§ 13.17
McLaurin v. Holley, 484 So. 2d 807 (La. Ct. App. 1986)	§ 9.5
McNutt Constr. Co., ENGBCA No. 4724, 85-3 B.C.A. (CCH) ¶ 18,397 (1985)	§ 7.3
M.D. Activities, IBCA No. 2113, 88-1 B.C.A. (CCH) ¶ 20,328 (1988)	§ 8.14
Meathe v. State Univ. Constr. Fund, 65 A.D.2d 49, 410 N.Y.S.2d 702 (1978)	§ 9.20
Merritt-Chapman & Scott Corp. v. State, 54 A.D.2d 37, 386 N.Y.S.2d 894 (1976), aff'd, 43 N.Y.2d 690, 371 N.E.2d 790, 401 N.Y.S.2d 28 (1977)	§§ 7.4, 7.11
Merritt-Chapman & Scott Corp. v. United States, 528 F.2d 1392 (Cl. Ct. 1976)	§ 5.6
Metro Indus. Painting Corp. v. Terminal Constr. Co., 287 F.2d 382 (2d Cir. 1961)	§ 1.12
Metropolitan Paving, 9 CCF ¶ 72,357, 325 F.2d 241 (Ct. Cl. 1963)	§ 5.9
Metropolitan Sewerage Comm'n v. R.W. Constr., Inc., 72 Wis. 2d 365, 241 N.W.2d 371 (1976)	§§ 8.6, 8.10, 8.15
Metropolitan Sewerage Comm'n v. R.W. Constr., Inc., 78 Wis. 2d 451, 255 N.W.2d 293 (1977)	§ 9.13
Meyer Schwartz Plumbing Co. v. Shelby Constr. Co., 338 S.W.2d 781 (Mo. 1960)	§ 9.4
Meyers v. Henderson Constr. Co., 147 N.J. Super. 77, 370 A.2d 547 (1977)	§ 13.18
Miliangoes v. George Frank (Textiles) Ltd., [1975] 3 All E.R. 801 (H.L.)	§ 3.5

CASES

Case	Book §
Miltz v. Boroughts-Shelving, Div. of Lear Siegler, Inc., 203 N.J. Super. 451, 497 A.2d 516 (1985)	§ 13.20
Minmar Builders, Inc., ASBCA No. 3430, 72-2 B.C.A. (CCH) ¶ 9,599 (1972)	§ 7.6
M.J. Womack, Inc. v. House of Representatives, 509 So. 2d 62 (La. Ct. App. 1987)	§ 13.5
Mobile Chem. Co. v. Blount Bros. Corp., 809 F.2d 1175 (5th Cir. 1987)	§§ 6.7, 7.1, 7.3, 7.4, 7.8
Mo Co., ASBCA No. 21403, 78-2 B.C.A. (CCH) ¶ 13,313 (1978)	§ 8.12
Monmouth Pub. Schools, Dist. 38 v. D.H. Rouse Co., 153 Ill. App. 3d 901, 506 N.E.2d 315 (1987)	§ 10.11
Montgomery Ross Fisher & H.A. Lewis, 85-2 B.C.A. (CCH) ¶ 18,108 (GSBCA 1985)	§ 9.10
Montgomery-Ross-Fisher, Inc., PSBCA Nos. 1033, 1096, 84-2 B.C.A. (CCH) ¶ 17,492 (1984)	§ 5.9
Moorehead Constr. Co. v. City of Grand Forks, 508 F.2d 1008 (8th Cir. 1975)	§§ 5.10, 9.13
Moorman Mfg. Co. v. National Tank Co., 91 Ill. 2d 69, 435 N.E.2d 443 (1982)	§ 13.50
Morgan v. Crowley, 91 Ga. App. 58, 85 S.E.2d 40 (1954)	§ 9.5
Morris v. Mosby, 227 Va. 517, 317 S.E.2d 493 (1984)	§ 13.32
Morris Clark v. D.L. Aenchbacher, 143 Ga. App. 282, 238 S.E.2d 442 (1977)	§ 13.35
Morris Mechanical Enters., Inc. v. United States, 554 F. Supp. 433, 1 Cl. Ct. 50 (1982)	§§ 5.3, 5.8
Morrison-Knudsen Co. v. United States, 345 F.2d 535, 170 Ct. Cl. 712 (1965)	§ 8.21
Moses H. Cone Memorial Hosp. v. Mercury Constr. Corp., 460 U.S. 1 (1983)	§ 1.12
Mound Bayon, City of v. Ray Collins Constr. Co., 499 So. 2d 1354 (Miss. 1986)	§ 9.5
Moundsview Indep. School Dist. v. Buetow & Assoc., Inc., 253 N.W.2d 836 (Minn. 1977)	§ 13.3
M.S.I. Corp., GSBCA No. 2429, 69-1 B.C.A. (CCH) ¶ 7,750 & B.C.A. (CCH) ¶ 7,377 (1968)	§§ 7.7, 7.8
Multivision Northwest, Inc. v. Jerrold Elecs. Corp., 356 F. Supp. 207 (N.D. Ga. 1972)	§ 14.13
Murphy Corp. v. Petrochem Maintenance, Inc., 180 So. 2d 716 (La. 1965)	§ 13.17
M.W. Goodell Constr. v. Monadnock Skating Club, 121 N.H. 320, 429 A.2d 329 (1981)	§ 10.18
M&W Masonry Constr., Inc. v. Head, 562 P.2d 957, (Okla. Ct. App. 1976)	§§ 10.9, 10.12, 10.37
Nat Harrison Assocs. v. Gulf States Utils. Co., 491 F.2d 578 (5th Cir. 1974)	§ 7.1
Natco, Inc. v. Williams Bros. Eng'g Co., 489 F.2d 639 (5th Cir. 1974)	§ 14.13

Case	Book §
National Bonding & Accident Ins. Co., 83-3 B.C.A. (CCH) ¶ 16,863 (ENGBCA 1983)	§ 9.10
National Farm Workers Serv. Center v. Caratan, Inc., 146 Cal. App. 3d 796, 194 Cal. Rptr. 617 (1983)	§§ 11.3, 11.11
National Shawmut Bank v. Amsterdam Casualty Co., 411 F.2d 843 (1st Cir. 1969)	§ 9.19
Natkin & Co. v. George A. Fuller Co., 347 F. Supp. 17 (W.D. Mo. 1972)	§§ 7.2, 7.3, 7.6, 7.8
Nebraska Pub. Power Dist. v. Austin Power, Inc., 773 F.2d 960 (8th Cir. 1985)	§§ 5.9, 5.12, 5.14, 11.11
Nello L. Teer Co. v. Washington Metro. Area Transit Auth., 695 F. Supp. 583 (D.D.C. 1988)	§§ 7.5, 7.8
Nelse Mortensen & Co. v. Group Health Coop., 17 Wash. App. 703, 566 P.2d 560 (1977)	§ 5.4
Nelson v. Commonwealth, 235 Va. 228, 368 S.E.2d 239 (1988)	§ 6.11
Nestos Painting Co., 86-2 B.C.A. (CCH) ¶ 18,993 (GSBCA 1986)	§ 10.40
Nether Providence Township School Auth. v. Thomas M. Durkin & Sons, 505 Pa. 42, 476 A.2d 904 (1984)	§ 9.10
Neumann v. Davis Water & Waste, Inc., 433 So. 2d 559 (Fla. 1983)	§ 13.20
New Haven v. National Steam Economizer Co., 79 Conn. 482, 65 A. 959 (1907)	§ 9.19
New Orleans, City of v. Vicon, Inc., 529 F. Supp. 1234 (E.D. La. 1982)	§§ 13.9, 13.17
New Pueblo Constructors, Inc. v. State, 144 Ariz. 95, 696 P.2d 185 (1985)	§§ 9.5, 9.13, 10.31
New York Shipbuilding Co., ASBCA No. 16164, 76-2 B.C.A. (CCH) ¶ 12,300 (1976)	§ 6.7
New York Shipbldg. Co., a Div. of Merritt-Chapman & Scott Corp., 73-1 B.C.A. (CCH) ¶ 9852 (ASBCA 1973)	§ 10.3
Newport v. Hedges, 358 S.W.2d 441 (Mo. Ct. App. 1962)	§ 13.17
Newsom v. United States, 676 F.2d 647 (Ct. Cl. 1982)	§ 9.9
Niffenegger v. Lakeland Constr. Co., 95 Ill. App. 3d 420, 420 N.E.2d 262 (1981)	§ 13.20
Nitrin, Inc. v. Bethlehem Steel Corp., 35 Ill. App. 3d 577, 342 N.E.2d 65 (1976)	§ 13.18
Noble v. Tweedy, 90 Cal. App. 2d 738, 203 P.2d 778 (1949)	§ 15.4
Nogler Tree Farm, AGBCA No. 81-104-1, 81-2 B.C.A. (CCH) ¶ 15,315 (1981)	§ 6.4
Nomellini Constr. Co. v. State *ex rel.* Dep't of Water Resources, 19 Cal. App. 3d 240, 96 Cal. Rptr. 682 (1971)	§§ 5.7, 10.22
Norair Eng'g Corp., ENGBCA No. 3568, 77-1 B.C.A. (CCH) ¶ 12,225 (1977)	§ 8.10

CASES

Case	Book §
Norair Eng'g Corp. v. United States, 666 F.2d 546 (Ct. Cl. 1981)	§§ 7.3, 7.4, 7.7, 7.8, 7.10, 9.3
North Hempstead, Town of v. Sea Crest Constr. Corp., 119 A.D.2d 744, 501 N.Y.S.2d 156 (1986)	§ 10.24
North Slope Technical Ltd. v. United States, 14 Cl. Ct. 242 (1988)	§§ 6.5, 8.15
Northeast Clackamas County Elec. Coop. v. Continental Casualty Co., 221 F.2d 329 (9th Cir. 1955)	§§ 5.4, 6.5
Northern Helix Co. v. United States, 455 F.2d 546, 199 Ct. Cl. 998 (1972)	§ 11.7
Northern Petrochem. Co. v. Thorsen & Thorshov, Inc., 297 Minn. 118, 211 N.W.2d 159 (1973)	§§ 13.40, 14.11, 15.5
Northwest Ironwork, Inc. v. Rippling River Corp., 71 Or. App. 144, 691 P.2d 111 (1984)	§ 9.20
Nyhus v. Travel Management Corp., 466 F.2d 440 (D.C. Cir. 1972)	§ 9.5
Oliver v. Superior Court (Regis Builders), 211 Cal. App. 3d 86, 259 Cal. Rptr. 160 (1989)	§ 13.13
Oliver B. Cannon & Sons, Inc. v. Dorr-Oliver, Inc., 394 A.2d 1160 (Del. Super. Ct. 1978)	§ 15.5
Oliver-Finnie Co. v. United States, 279 F.2d 498 (Ct. Cl. 1960)	§ 5.15
Ordinance Research, Inc. v. United States, 609 F.2d 462 (Ct. Cl. 1979)	§ 1.6
Oregon State Highway Comm'n v. DeLong Corp., 9 Or. App. 550, 495 P.2d 1215 (1972), *cert. denied,* 411 U.S. 965, *reh'g denied,* 412 U.S. 944 (1973)	§§ 10.12, 10.24, 10.25
Owen v. Dodd, 431 F. Supp. 1239 (N.D. Miss. 1977)	§ 13.5
Owen Constr. Co. v. Iowa State Dep't of Transp., 274 N.W.2d 304 (Iowa 1979)	§ 5.4
Owen L. Schwam Constr. Co., ASBCA No. 22407, 79-2 B.C.A. (CCH) ¶ 13,919 (1979)	§ 5.9
Ozark Dam Constructors v. United States, 127 F. Supp. 187 (Ct. Cl. 1955)	§ 5.4
Paccon, Inc., ASBCA No. 7890, 1963 B.C.A. (CCH) ¶ 3,659 (1963)	§ 6.6
Paccon, Inc., ASBCA No. 7890, 65-2 B.C.A. (CCH) ¶ 4996 (1965)	§§ 5.11, 5.15, 6.5, 6.6
Pace Corp. v. Jackson, 284 S.W.2d 340 (Tex. 1955)	§ 14.13
Pacific Alaska Contractors, Inc. v. United States, 436 F.2d 461, 193 Ct. Cl. 850 (1971)	§ 8.6
Pacific Employers Ins. Co. v. City of Berkeley, 158 Cal. App. 3d 145, 204 Cal. Rptr. 387 (1984)	§§ 10.22, 10.25
Pan-Artic Corp., ASBCA No. 20133, 77-1 B.C.A. (CCH) ¶ 12,514 (1977)	§ 6.5
Parem Contracting Corp. v. Welch Constr. Co., 128 N.H. 259, 512 A.2d 1104 (1986)	§ 10.17

Case	Book §
Pathman Constr. Co., ASBCA No. 23392, 85-2 B.C.A. (CCH) ¶ 18,096 (1985)	§ 5.8
Pathman Constr. Co., ASBCA No. 22003, 82-1 B.C.A. (CCH) ¶ 15,790 (1982)	§ 6.5
Pathman Constr. Co. v. Hi-Way Elec. Co., 65 Ill. App. 480, 382 N.E.2d 453 (1978)	§ 5.6
Patitucci v. Drelich, 153 N.J. Super. 177, 379 A.2d 297 (1977)	§ 13.13
Paul E. McCollum, Sr., 81-2 B.C.A. (CCH) ¶ 15,310 (ASBCA 1981)	§ 10.3
Paul Hardeman, Inc. v. Arkansas Power & Light Co., 380 F. Supp. 298 (E.D. Ark. 1974)	§§ 10.11, 10.12, 10.28, 10.30, 10.37, 10.39
Paul N. Howard v. Puerto Rico Aqueduct Sewer, 744 F.2d 880 (1st Cir. 1984)	§ 8.23
Pebble Building Co. v. G.J. Hopkins, Inc., 223 Va. 188, 288 S.E.2d 437 (1982)	§ 6.11
Pennsylvania Eng'g Corp. v. Islip Resource Recovery Agency, 710 F. Supp. 456 (E.D.N.Y. 1986)	§ 1.12
Perry v. Thomas, 482 U.S. 483 (1987)	§ 1.12
Peter Kiewit Sons' Co. v. Iowa S. Utils. Co., 355 F. Supp. 376 (S.D. Iowa 1973)	§§ 5.4, 13.51
Peter Kiewit Sons' Co. v. Summit Constr. Co., 422 F.2d 242 (8th Cir. 1969)	§§ 5.15, 6.5, 6.9, 6.12, 9.4
Petersen v. Hubschman Constr. Co., 76 Ill. 2d 31, 389 N.E.2d 1154 (1979)	§ 13.16
Phoenix Coal Co. v. Commissioner, CA-2, 56-1 U.S. Tax Cas. (CCH) ¶ 9366, 231 F.2d 420 (9th Cir. 1956)	§ 2.2
Phoenix Contracting Corp. v. New York City Health, 118 A.D.2d 477, 499 N.Y.S.2d 953 (1986)	§ 5.4
Phoenix Restoration Specialists, Inc., *In re,* 14 Bankr. 115 (D. Mass. 1981)	§§ 15.7, 15.8
Pick Fisheries, Inc. v. Berns Elecs. Sec. Servs., 35 Ill. App. 3d 467, 342 N.E.2d 105 (1976)	§ 14.10
Pickard's Sons Co. v. United States, 532 F.2d 739, 209 Ct. Cl. 643 (1976)	§ 9.13
Pike Indus., Inc. v. Middlebury Assocs., 140 Vt. 67, 436 A.2d 725 (1981), *cert. denied,* 455 U.S. 947 (1982)	§ 10.41
Pine Bluff Hotel Co. v. Monk & Ritchie, 122 Ark. 308, 183 S.W.2d 761 (1916)	§ 13.9
Pinkerton & Laws Co. v. Roadway Express, Inc., 650 F. Supp. 1138 (N.D. Ga. 1986)	§ 8.18
P.J. Maffei Bldg. & Wrecking v. United States, 732 F.2d 913 (Fed. Cir. 1984)	§§ 8.5, 8.8
Plumbers & Fitters Local 761 v. Matt J. Zaich Constr. Co., 418 F.2d 1054 (9th Cir. 1969)	§ 9.13
Pollard v. Saxe & Yolles Dev. Co., 12 Cal. 3d 374, 525 P.2d 88, 1115 Cal. Rptr. 648 (1974)	§ 13.11

CASES

Case	Book §
Porter v. Arrowhead Reservoir Co., 100 Cal. 500, 35 P. 146 (1893)	§ 10.12
Portland Freight v. Canadian Imperial Bank of Commerce, 97 Or. App. 304, 776 P.2d 35 (1988)	§ 14.8
Post Constr. Co., 84-1 B.C.A. (CCH) ¶ 16,959 (HUDBCA 1983)	§ 9.5
Pots Unltd., Ltd. v. United States, 600 F.2d 790, 220 Ct. Cl. 405 (1979)	§ 10.11
Powell's Gen. Contracting Co., DOTCAB No. 1088, 80-2 B.C.A. (CCH) ¶ 14,680 (1980)	§ 8.18
Prather v. Latshaw, 188 Ind. 204, 122 N.E. 721 (1919)	§ 6.4
Prestex, Inc. v. United States, 3 Cl. Ct. 373 (1983)	§ 6.4
Prichard Bros. v. Grady Co., 436 N.W.2d 460 (Minn. Ct. App. 1989)	§ 6.14
Psaty & Fuhrman v. Housing Auth., 76 R.I. 87, 68 A.2d 32 (1949)	§ 13.39
P.T. Perusahaan Pelayaran Samudera Trikora Lloyd v. Salzachtal, 373 F. Supp. 267 (E.D.N.Y. 1974)	§ 3.3
PT&L Constr. Co. v. New Jersey Dep't of Transp., 108 N.J. 539, 531 A.2d 1330 (1987)	§§ 8.15, 8.18, 8.19
Public Constructors, Inc. v. State, 55 A.D.2d 368, 390 N.Y.S.2d 481 (1977)	§ 8.20
Quality Seeding, Inc., 88-3 B.C.A. (CCH) ¶ 21,020 (IBCA 1988)	§ 10.3
Queensbury Union Free School Dist. v. Jim Walter Corp., 91 Misc. 2d 804, 398 N.Y.S.2d 832 (1977)	§ 13.2
Rachlin & Co. v. Tra-Mor, Inc., 33 A.D.2d 370, 308 N.Y.S.2d 153 (1970)	§ 3.3
Ragnar Benson, Inc. v. Bechtel Power Corp., 651 F. Supp. 962 (D. Pa. 1986)	§ 9.13
Ranger Constr. Co. v. Dixie Floor Co., 433 F. Supp. 442 (D.S.C. 1977)	§ 13.18
Ratcliffe v. Evans, [1892] 2 Q.B. 524 (C.A.)	§ 4.18
Rawlings v. D.M. Oliver Co., 97 Cal. App. 3d 890, 159 Cal. Rptr. 119 (1977)	§ 13.13
Ray v. William G. Eurice & Bros., 201 Md. 115, 93 A.2d 272 (1952)	§ 10.16
Raymey Constr. Co. v. Apache Tribe, 673 F.2d 315 (10th Cir. 1982)	§ 13.6
Raymond Int'l, Inc. v. Baltimore County, 45 Md. App. 247, 412 A.2d 1296 (1980)	§ 8.15
Raytheon Prod. Corp. v. Commission, CA-1, 44-2 U.S. Tax Cas. (CCH) ¶ 9424, 144 F.2d 110 (1944), *cert. denied,* 323 U.S. 779	§ 2.1
R. Clinton Constr. Co. v. Bryant & Reaves, Inc., 442 F. Supp. 838 (N.D. Miss. 1977)	§ 13.22
Reed Paving, Inc. v. Glen Ave. Builders, 148 A.D.2d 934, 539 N.Y.S.2d 173 (1989)	§ 6.5

Case	Book §
Regan/Nager Constr. Co., 85-1 B.C.A. (CCH) ¶ 17,778 (PSBCA 1984)	§ 9.6
Regents of Univ. of Cal. v. Hartford Accident & Indem. Co., 59 Cal. App. 3d 675, 131 Cal. Rptr. 112 (1976)	§ 13.54
Reif v. Smith, 319 N.W.2d 815 (S.D. 1981)	§ 9.5
Reliance Enters., ASBCA No. 20808, 76-1 B.C.A. (CCH) ¶ 11,831 (1976)	§ 6.5
Reliance Ins. Co. v. Colbert, 365 F.2d 530 (D.C. Cir. 1966)	§ 9.19
Renel Constr. Inc. v. Brooklyn Coop. Meat Distrib. Center, 59 A.D.2d 391, 399 N.Y.S.2d 429 (1977)	§ 13.3
Rex Trailer Co. v. United States, 350 U.S. 148 (1956)	§ 14.10
R.J. Daigle & Sons Contractors, Inc. v. Sampey Bros. Gen. Contractors, Inc., 424 So. 2d 270 (La. Ct. App. 1983)	§ 9.3
R.M. Hollingshead v. United States, 111 F. Supp. 285 (Ct. Cl. 1953)	§ 1.6
Roanoke Hosp. Ass'n v. Doyle & Russell, Inc., 215 Va. 796, 214 S.E.2d 155 (1975)	§§ 11.18, 13.32, 13.36, 13.41, 14.11, 14.13, 15.2
Robert McMullan & Sons, Inc., ASBCA No. 19129, 76-2 B.C.A. (CCH) ¶ 12,072 (1976)	§ 5.9
Roberts v. General Dynamics, Convair Corp., 425 F. Supp. 688 (S.D. Tex. 1977)	§ 13.22
Roberts v. Security Trust Sav. Bank, 196 Cal. 557, 238 P. 673 (1925)	§§ 9.5, 9.19
Robinhorne Constr. Corp. v. Snyder, 47 Ill. 2d 349, 265 N.E.2d 670 (1973)	§ 13.33
Robinson v. United States, 261 U.S. 486 (1923)	§§ 5.7, 10.22
Roger J. Au & Son, Inc. v. Northeastern Ohio Regional Sewer Dist., 29 Ohio App. 3d 284, 504 N.E.2d 1209 (1986)	§ 8.14
Rogers Excavating, AGBCA 79-180-4, 7 C.C. ¶ 391 (1984)	§ 7.10
Roy Storm Excavating & Grading Co. v. Miller Davis Co., 149 Ill. App. 3d 1093, 509 N.E.2d 105 (1986)	§ 8.15
Royal Ornamental Iron, Inc. v. Devon Bank, 32 Ill. App. 3d 101, 336 N.E.2d 105 (1975)	§ 14.7
R&R Constr. Co., VABCA No. 1101, 74-2 B.C.A. (CCH) ¶ 10,857 (1974)	§ 6.4
R&R Constr. Co. v. Junior College Dist. No. 529, 55 Ill. App. 3d 115, 370 N.E.2d 599 (1977)	§§ 9.1, 9.8
R.S. Noonan, Inc. v. Morrison-Knudsen Co., 522 F. Supp. 1186 (E.D. La. 1981)	§ 6.5
Rudolph v. First S. Sav. & Loan Ass'n, 414 So. 2d 64 (Ala. 1982)	§ 13.31
Rusciano Constr. Corp. v. State, 37 A.D.2d 745, 323 N.Y.S.2d 21 (1971)	§ 13.48
Russell v. Arthur Whitcomb, Inc., 100 N.H. 171, 121 A.2d 781 (1956)	§ 13.19
Ryan v. Thurmond, 481 S.W.2d 199 (Tex. Civ. App. 1972)	§§ 10.20, 14.11

Case	Book §
Saddler v. United States, 287 F.2d 411 (Ct. Cl. 1961)	§ 9.4
St. Charles Floor Co. v. Hoelzer, 545 S.W.2d 844 (Mo. Ct. App. 1978)	§ 9.8
Salt Lake City Contractors, Ltd., VABCA No. 1362, 80-2 B.C.A. (CCH) ¶ 14,713 (1980)	§ 5.13
Samuel B. Harper, Jr. and Sigmund E. Florsheim d/b/a Harper & Florsheim, ASBCA No. 4959, 59-1 B.C.A. (CCH) ¶ 2986 (1959)	§ 13.41
Samuel N. Zarpas, Inc., ASBCA No. 4722, 59-1 B.C.A. (CCH) ¶ 2170 (1961)	§ 5.11
Sandwich Islands Constr., ASBCA No. 35244, 88-3 B.C.A. (CCH) ¶ 21,143 (1988)	§ 8.7
Sandy Hites Co. v. State Highway Comm'n, 347 Mo. 954, 149 S.W.2d 828 (1941)	§§ 8.17, 13.9
Santa Fe Eng'rs, Inc., PSBCA No. 902, 84-2 B.C.A. (CCH) ¶ 17,377 (1984)	§ 6.4
Santor v. A.M. Karagheusian, Inc., 44 N.J 52, 207 A.2d 305 (1965)	§ 13.22
Sasso Contracting Co. v. State, 173 N.J. Super. 486, 414 A.2d 603 (1980)	§ 8.19
Savin Bros., Inc. v. State, 62 A.D.2d 511, 405 N.Y.S.2d 516 (1978), *aff'd*, 47 N.Y.2d 934, 393 N.E.2d 1641, 419 N.Y.S.2d 969 (1979)	§ 9.8
Schatz v. Abbott Labs, Inc., 51 Ill. 2d 143, 281 N.E.2d 323 (1972)	§ 15.5
Schenectady Steel Co. v. Bruno Trimpoli Gen. Constr. Co., 43 A.D.2d 234, 350 N.Y.S.2d 920 (1974)	§ 13.18
Schenk v. Pelkey, 176 Conn. 245, 405 A.2d 665 (1978)	§ 13.22
Schiapper v. Levitt & Sons, 44 N.J. 70, 207 A.2d 314 (1965)	§ 13.13
Sea Ledge Properties, Inc. v. Dodge, 283 So. 2d 55 (Fla. Dist. Ct. App. 1973)	§ 10.17
Sears, Roebuck & Co. v. Enco Assocs., Inc., 43 N.Y.2d 389, 372 N.E.2d 555, 401 N.Y.S.2d 767 (1977)	§ 13.2
Sears, Roebuck & Co. v. Goudie, 290 A.2d 826 (D.C. 1972)	§§ 15.2, 15.5, 15.8
Security Life Ins. Co. v. Executive Car Leasing Co., 443 S.W.2d 915 (Tex. 1968)	§ 13.22
Seifford v. Housing Auth., 192 Neb. 643, 223 N.W.2d 816 (1974)	§§ 7.4, 9.3
Service Steel Erectors Co. v. SCE, Inc., 573 F. Supp. 177 (W.D. Va. 1983)	§ 9.5
Shamrock Constr. Co., *In re*, No. B-621-73 slip op. (Bankr. D.N.J. June 30, 1980)	§ 9.20
Shank-Artukovich v. United States, 13 Cl. Ct. 346 (1987), *aff'd without opinion*, 848 F.2d 1245 (1988)	§§ 8.11, 8.23
Sheehan v. Pittsburgh, 213 Pa. 133, 62 A.2d 642 (1905)	§ 5.4
Shoffer Indus. v. W.B. Lloyd Constr. Co., 42 N.C. App. 259, 275 S.E.2d 50 (1979)	§ 13.5

Case	*Book §*
T.F. Scholes, Inc., ASBCA No. 5009, 59-2 B.C.A. (CCH) ¶ 2,375 (1959)	§ 7.4
Thackrah v. Hass, 119 U.S. 499 (1886)	§ 10.13
Thomas v. Bove, 687 P.2d 534 (Colo. 1984)	§ 13.22
Thomas H. Welch v. Helvering, 3 U.S. Tax Cas. (CCH) ¶ 1164, 290 U.S. 111 (1933)	§ 2.5
Thompson Farms, Inc. v. Corno Feed Prods., Div. of Nat. Oats Co., 173 Ind. App. 682, 366 N.E.2d 3 (1977)	§ 13.22
Thompson Ramo Woodridge, Inc. v. United States, 361 F.2d 222, 175 Ct. Cl. 527 (1966)	§ 13.9
Thorn Constr. Co. v. Utah Dep't of Transp., 598 P.2d 365 (Utah 1979)	§ 9.13
Time Contractors, J.V., DOTBCA Nos. 1669, 1691, 87-1 B.C.A. (CCH) ¶ 19,582 (1987)	§ 6.6
Titan Atl. Constr. Corp., ASBCA No. 23588, 82-2 B.C.A. (CCH) ¶ 15,808 (1982)	§ 8.10
Titan Midwest Constr. Corp., ASBCA No. 23594, 81-1 B.C.A. (CCH) ¶ 15,067 (1981)	§ 8.8
Tombigbee Constr. Co. v. United States, 420 F.2d 1037 (Ct. Cl. 1970)	§§ 9.3, 13.52
Tracor, Inc. v. Austin Supply & Drywall Co., 484 S.W.2d 446 (Tex. 1972)	§ 13.22
Traylor v. Henkels & McCoy, Inc., 99 Idaho 560, 585 P.2d 970 (1978)	§ 10.34
Tribble & Stephens Co. v. Consolidated Serv., 744 S.W.2d 945 (Tex. Ct. App. 1987)	§ 6.5
Tri-Cor, Inc. v. United States, 458 F.2d 112 (Ct. Cl. 1972)	§§ 7.4, 7.5, 7.10
Tucker v. Bitler Bros., 197 N.Y.S.2d 899 (1960)	§ 7.4
Tull v. Gundersons, Inc., 709 P.2d 940 (Colo. 1985)	§§ 10.29, 10.30, 10.31, 10.34, 10.35
Turnkey Enters., Inc. v. United States, 597 F.2d 750, 220 Ct. Cl. 179 (1979)	§§ 8.1, 8.5
Tutor Saliba-Perini, PSBCA No. 1201, 87-2 B.C.A. (CCH) ¶ 19,755 (1987)	§ 6.15
Tuttle-Whit Constr., Inc. v. State, 371 So. 2d 1096 (Fla. 1979)	§ 6.16
Twin River Constr. Co. v. Public Water Dist. No. 6, 653 S.W.2d 682 (Mo. Ct. App. 1983)	§§ 10.24, 10.25
Umbaugh Builders, Inc. v. Parr Co., 86 Misc. 2d 1036, 385 N.Y.S.2d 698 (1976)	§ 12.1
Unicorn Management Corp. v. City of Chicago, 404 F.2d 627 (7th Cir. 1968)	§ 5.4
Unitec, Inc., ASBCA No. 22025, 79-2 B.C.A. (CCH) ¶ 13,923 (1979)	§ 8.12
United Contractors, Inc. v. United States, 368 F.2d 585, 177 Ct. Cl. 151 (1966)	§§ 8.5, 8.15
United States v. American Sur. Co., 322 U.S. 96 (1944)	§ 10.24

Case	Book §
United States v. Atlantic Dredging Co., 253 U.S. 1 (1920)	§ 8.17
United States v. Behan, 110 U.S. 338 (1883)	§§ 10.30, 10.31, 10.32
United States v. Blair, 321 U.S. 730 (1943)	§ 5.9
United States v. Callahan Walker Constr. Co., 317 U.S. 56 (1942)	§ 6.7
United States v. Citizens & S. Nat'l Bank, 367 F.2d 473 (4th Cir. 1966)	§ 6.5
United States v. Freel, 186 U.S. 309 (1902)	§ 9.19
United States v. Heckinger, 397 U.S. 203 (1970)	§ 9.9
United States v. Henke Constr. Co., 157 F.2d 13 (8th Cir. 1946)	§ 9.2
United States v. Spearin, 248 U.S. 132 (1918)	§§ 1.6, 5.4, 8.17, 13.7
United States v. Systron-Donner Corp., 486 F.2d 249 (9th Cir. 1973)	§ 10.26
United States v. Texas Constr. Co., 224 F.2d 289 (5th Cir. 1955)	§ 14.10
United States v. United Eng'g & Constr. Co., 234 U.S. 236 (1913)	§§ 5.7, 10.22
United States v. William F. Klingsmith, Inc., 670 F.2d 1227 (D.C. Cir. 1982)	§ 10.27
United States *ex rel.* Bldg. Rentals Corp. v. Western Casualty & Sur. Co., 498 F.2d 335 (9th Cir. 1974)	§§ 10.9, 10.11, 10.12, 10.36, 10.39
United States *ex rel.* Gillioz v. John Kerns Constr. Co., 140 F.2d 792 (8th Cir. 1944)	§ 14.10
United States *ex rel.* Micro-King Co. v. Community Science Technology, 574 F.2d 1292 (5th Cir. 1978)	§ 13.44
United States *ex rel.* N. Maltese & Sons, Inc. v. Juno Constr. Corp., 759 F.2d 253 (2d Cir. 1985)	§§ 10.30, 11.3
United States *ex rel.* R.W. Vaught Co. v. F.D. Rich Co., 439 F.2d 895 (8th Cir. 1971)	§ 1.20
United States *ex rel.* Susi Contracting Co. v. Zara Contracting Co., 146 F.2d 606 (2d Cir. 1944)	§§ 10.11, 10.12, 10.37, 10.39
United States Fidelity & Guar. v. Orlando Utils. Comm'n, 564 F. Supp. 962 (M.D. Fla. 1983)	§ 7.5
United States Steel Corp. v. Missouri Pac. R.R., 668 F.2d 435 (8th Cir. 1982)	§ 5.11
United Telecommunications, Inc. v. American Television & Communications Corp., 536 F.2d 1310 (10th Cir. 1976)	§ 11.18
Universal Builders, Inc. v. Moon Motor Lodge, Inc., 430 Pa. 550, 244 A.2d 10 (1968)	§§ 9.5, 9.10
Urania, Town of v. M.P. Dumesnil Constr. Co., 492 So. 2d 888 (La. 1986)	§ 13.9
U.S. Indus., Inc. v. Blake Constr. Co., 671 F.2d 539 (D.C. Cir. 1982)	§§ 6.5, 6.10
USA Petroleum Corp. v. United States, 821 F.2d 622 (Fed. Cir. 1987)	§ 1.6

Case	Book §
USEMCO, Inc. v. Marbro Co., 60 Md. App. 351, 483 A.2d 88 (1984)	§ 13.22
Varo, Inc., ASBCA No. 15000, 72-2 B.C.A. (CCH) ¶ 9,717 (1972)	§ 7.8
Vaughn v. Edward M. Chadbourne, Inc., 462 So. 2d 512 (Fla. 1985)	§ 13.25
V.L. Nicholson Co. v. Transcom Inv. & Fin., Ltd., 595 S.W.2d 474 (Tenn. 1980)	§§ 5.5, 9.10
Volt Information Sciences v. Stanford Univ., ___ U.S. ___, 109 S. Ct. 1248 (1989)	§ 1.12
Walker v. Wittenberg, Deloney & Davidson, Inc., 241 Ark. 525, 412 S.W.2d 626 (1967)	§ 13.24
Walter Kidde Constructors, Inc. v. Conn., 37 Conn. Supp. 50, 434 A.2d 962 (1981)	§ 9.4
Warner Constr. Corp. v. City of Los Angeles, 2 Cal. 3d 285, 466 P.2d 996, 85 Cal. Rptr. 444 (1970)	§§ 8.18, 13.51
Watson Lumber Co. v. Gunnewig, 79 Ill. App. 2d 377, 226 N.E.2d 270 (1976)	§ 9.10
W.C. James, Inc. v. Phillips Petroleum Co., 485 F.2d 22 (10th Cir. 1973)	§ 5.4
Weeks Dredging & Contracting v. United States, 861 F.2d 728, 13 Cl. Ct. 193 (1987)	§§ 8.4, 8.6, 8.7, 8.20, 8.21, 8.22, 8.23
W.E. Garrison Grading Co. v. Piracci Constr. Co., 27 N.C. App. 725, 221 S.E.2d 512 (1975)	§ 1.7
Weichman Eng'rs v. State, 31 Cal. App. 3d 741, 107 Cal. Rptr. 529 (1973)	§ 8.17
Weinberg v. Wilensky, 26 N.J. Super. 301, 97 A.2d 707 (1953)	§ 13.17
Westcott v. State, 264 A.D. 463, 36 N.Y.S.2d 23 (1942)	§ 9.4
Western Contracting Corp. v. State Bd. of Equalization, 39 Cal. App. 3d 341, 114 Cal. Rptr. 227 (1974)	§ 8.1
Western States Painting Co., 69-1 B.C.A. (CCH) ¶ 7616 (ASBCA 1969)	§ 10.3
Westinghouse Elec. Supply Co. v. Fidelity & Deposit Co., 560 F.2d 1109 (3d Cir. 1977)	§ 9.12
Westminster Elec. Corp. v. Salem Eng'g & Constr., 712 F.2d 720 (1st Cir. 1983)	§ 10.30
Whaley v. Milton Constr. & Supply Co., 241 S.W.2d 23 (Mo. App. 1951)	§ 13.16
White Constr. & Equip. Rental v. Rinner & Garrett, Inc., 340 So. 2d 283 (La. 1976)	§ 9.2
W.H. Lyman Constr. Co. v. Gurnee, 131 Ill. App. 3d 87, 475 N.E.2d 273 (1985)	§ 13.17
Williams Eng'g, Inc. v. Goodyear, 496 So. 2d 1012 (La. 1986)	§ 9.20
Wineland v. Commissioner, 10 T.C.M. 919 (1951)	§ 2.7

CASES

Case	Book §
Wisconsin Red Pressed Brick Co. v. Hood, 67 Minn. 329, 69 N.W. 1091 (1897)	§ 13.16
Witty v. C. Casey Homes, Inc., 102 Ill. App. 3d 619, 430 N.E.2d 191 (1981)	§ 10.18
Womack & Vorhies v. United States, 389 F.2d 793, 182 Ct. Cl. 399 (1968)	§ 8.6
Wood-Hopkins Contracting Co. v. Masonry Contractors, Inc., 235 So. 2d 548 (Fla. 1970)	§ 13.17
WRB Corp. v. United States, 183 Ct. Cl. 409 (1968)	§§ 9.13, 10.37
Wrecking Corp. v. Memorial Hosp., 63 A.D.2d 615, 405 N.Y.S.2d 83 (1978)	§ 9.11
Wright v. Creative Corp., 30 Colo. App. 575, 498 P.2d 1179 (1972)	§ 13.13
Wunderlich Contracting Co. v. United States, 351 F.2d 956 (Ct. Cl. 1965)	§§ 9.4, 9.5
W-V Enters. Inc. v. Federal Sav. & Loan Ins. Corp., 234 Kan. 354, 673 P.2d 1112 (1983)	§ 13.42
X.L.O. Concrete Corp. v. John T. Brady & Co., 104 A.D.2d 181, 482 N.Y.S.2d 476 (1984), *aff'd,* 66 N.Y.2d 970, 489 N.E.2d 768, 498 N.Y.S.2d 799 (1985)	§ 10.22
Yamas Constr. Co., ASBCA No. 27366, 86-3 B.C.A. (CCH) ¶ 19,090 (1986)	§ 8.12
Yarborough v. Hilton Hotels Corp., 655 P.2d 822 (Colo. 1983)	§ 13.54
Yukon Constr. Corp., FAACAP No. 66-4, 65-2 B.C.A. (CCH) ¶ 5005 (1965)	§ 8.9
Zenith Constr., ASBCA No. 33576, 89-3 B.C.A. (CCH) ¶ 21,894 (1989)	§ 8.15
Zinger Constr. Co., GSBCA No. 6568, 84-3 B.C.A. (CCH) ¶ 17,537 (1984)	§ 6.13
Zirin Laboratories Int'l, Inc. v. Mead-Johnson & Co., 208 F. Supp. 633 (E.D. Mich. 1962)	§ 11.11
Zobel & Dahl Constr. v. Crotty, 356 N.W.2d 42 (Minn. 1984)	§ 11.3
Zontelli & Sons, Inc. v. City of Nashwauk, 373 N.W.2d 744 (Minn. 1985)	§ 8.15

INDEX

ABANDONMENT
   Causes of action and recovery
      §§ 13.33, 13.35, 13.51
   Changes in scope claims §§ 9.5, 9.7
   Termination claims §§ 10.1, 10.12,
      10.23, 10,25, 10.37
ACCELERATION
   Delay claims against contractors
      § 14.5
   Differing site conditions claims
      § 8.24
   Disruption claims § 6.8
   Termination claims § 10.34
ACCELERATION CLAIMS
   Generally § 7.2
   Acceleration generally § 7.1
   Acceleration in fact § 7.3
   Constructive acceleration § 7.4
   Elements of claim
   –Generally § 7.7
   –Acceleration orders § 7.8
   –Costs § 7.9
   –Time extensions § 7.10
   Proving acceleration
   –Acceleration provisions § 7.11
   –Documentation § 7.12
   –Identifying acceleration § 7.5
   –Proof § 7.6
   Proving damages
   –Generally § 7.13
   –Actual costs analysis § 7.16
   –Evidence § 7.20
   –Information available, using § 7.15
   –Measurement of damages § 7.14
   –Mitigation and cost savings § 7.19
   –Overhead and profit § 7.18
   –Reasonableness of damage figures
      § 7.17
ACCELERATION ORDERS
   Acceleration claims § 7.8
ACCEPTANCE
   Causes of action and recovery
      §§ 13.17, 13.52

ACCESS TO JOBSITE
   Acceleration claims § 7.4
   Disruption claims § 6.5
ACCOUNTANTS
   Delay claims against contractors
      § 14.2
   International claims. See
      INTERNATIONAL CLAIMS
ACCRUAL METHOD
   Tax considerations §§ 2.3, 2.6
ACTS OF GOD
   Acceleration claims § 7.4
   Delay claims against contractors
      § 14.7
   Disruption claims § 6.4
ACTUAL COST OF COMPLETION
   Termination claims § 10.17
ACTUAL COSTS
   Acceleration claims §§ 7.9, 7.16
   Cost claims § 12.1
   Disruption claims § 6.9
ACTUAL COSTS METHOD
   Changes in scope claims § 9.13
ACTUAL DAMAGES
   Contractor as plaintiff, damages
      available § 13.48
   Delay claims against contractors
      § 14.11
   Termination claims §§ 10.23.
      10.24
ACTUAL KNOWLEDGE
   Differing site conditions claims
      §§ 8.14, 8.22
ACTUAL NOTICE
   Changes in scope claims § 9.5
ADDITIONAL COSTS
   International claims § 4.11
ADDITIONAL WORK
   Changes in scope claims. See
      CHANGES IN SCOPE CLAIMS
   Extra work. See EXTRA WORK
ADJUDICATION
   Tax considerations § 2.2

INDEX

ADVERSE WEATHER
  Acceleration claims § 7.4
  Disruption claims §§ 6.3, 6.4, 6.13
AGC DOCUMENT 600
  Payment delay claims §§ 11.2, 11.4, 11.6
AGREEMENT TO ARBITRATE
  Pricing and proving claims § 1.12
AIA DOCUMENT A201
  Causes of action and recovery §§ 13.16, 13.17
  Changes in scope claims § 9.10
  Differing site conditions claims § 8.3
  Payment delay claims §§ 11.2, 11.4, 11.5
  Pricing and proving claims §§ 1.2, 1.5, 1.8, 1.13
  Termination claims § 10.5
AICPA GUIDE
  Lost profit claims § 15.8
ALLEGHENY METHOD
  Delay claims against owners § 5.19
ALL RISK POLICIES
  Surety, insurer, claims against § 13.28
ALTERNATIVE DISPUTE RESOLUTION
  International claim resolution inside U.S. § 3.1
AMBIGUITY
  Changes in scope claims § 9.9
AMIABLE COMPOSITEUR
  International claims § 4.18
ANOMALOUS GEOLOGICAL CONDITIONS
  Differing site conditions claims § 8.12
ANTICIPATORY PROFIT
  Termination claims §§ 10.3, 10.30, 10.31
APPEALS
  Arbitration § 1.11
  Pricing and proving claims, arbitration § 1.11
APPORTIONMENT OF DAMAGES
  Termination claims § 10.22
APPORTIONMENT OF DELAY
  Delay claims against owners §§ 5.6, 5.7
ARBITRATION
  Delay claims against contractors § 14.14

ARBITRATION *(Continued)*
  International claim resolution inside U.S. §§ 3.6, 3.8, 3.9
  International claim resolution outside U.S. §§ 4.6, 4.11
  Pricing and proving claims. See PRICING AND PROVING CLAIMS
ARBITRATOR SELECTION
  Pricing and proving claims § 1.10
ARCHITECTS
  Claims against. See CAUSES OF ACTION AND RECOVERIES
  Delay claims against contractors § 14.6
ARCHITECT'S AUTHORITY
  Changes in scope claims § 9.10
ASSUMPTION OF RISK
  Causes of action and recovery § 13.14
ATTORNEYS
  Pricing and proving claims § 1.23
  Tax considerations § 2.7
ATTORNEYS' FEES
  Causes of action and recovery § 13.55
  International claim resolution inside U.S. § 3.9
  Owner as plaintiff, damages available § 13.44
  Payment delay claims § 11.3
AUDITS
  Causes of action and recovery § 13.55
  Changes in scope claims § 9.7
  Pricing and proving claims § 1.29
AUTHORIZATION
  Changes in scope claims § 9.10
AWARD PRESERVATION
  Pricing and proving claims § 1.1

BAD FAITH
  Causes of action and recovery § 13.51
  Changes in scope claims § 9.5
  Differing site conditions claims § 8.19
BASELINE CPM
  Acceleration claims § 7.5
BASIS IN ASSET
  Tax considerations § 2.1
BENCHMARK OF PRODUCTIVITY
  Differing site conditions claims §§ 8.26–8.30

INDEX

BENCH TRIALS
Delay claims against contractors § 14.15
BETTERMENT
International claims § 4.23
BID BONDS
Causes of action and recovery § 13.27
BIDDING DOCUMENTS
Delay claims against contractors §§ 14.4, 14.5
BID PREPARATION STAGE
Differing site conditions claims, proving § 8.21
BIDS
Pricing and proving claims § 1.15
BONDS
Causes of action and recovery § 13.53
Changes in scope claims §§ 9.18, 9.19
Delay claims against contractors § 14.3
Payment delay claims § 11.18
Pricing and proving claims § 1.2
Surety, insurer, claims against §§ 13.27, 13.30
Termination claims §§ 10.19, 10.39
BOOKS AND RECORDS
Cost claims § 12.3
BREACH OF CONTRACT
International claims §§ 4.8, 4.13
BREACH OF IMPLIED WARRANTY
Differing site conditions claims § 8.17
BREACH OF STATUTORY DUTY
International claims § 4.8
BREACH OF WARRANTY
Architect, claims against §§ 13.2, 13.3
Contractor, claims against §§ 13.16, 13.17
Differing site conditions claims § 8.17
Manufacturer/supplier, claims against §§ 13.22, 13.23
Owner, claims against §§ 13.9. 13.10
BREACH-DAY RULE
International claim resolution inside U.S. §§ 3.4, 3.5
BRETTON-WOODS AGREEMENT
International claim resolution inside U.S. § 3.2

BUDGETS
Acceleration claims § 7.12
BUILDING CODE VIOLATIONS
Causes of action and recovery § 13.31
BURDEN OF PROOF
Delay claims against owners § 5.6
Termination claims § 10.3
BYSTANDERS
Causes of action and recovery § 13.19
CAPITAL GAIN
Tax considerations § 2.1
CAPITAL ITEMS
Tax considerations § 2.5
CAPITAL RETURN
Tax considerations § 2.1
CARDINAL CHANGES
Changes in scope claims §§ 9.3, 9.19
CARTERET METHOD
Delay claims against owners § 5.19
CASH RECEIPTS AND DISBURSEMENT METHOD
Cost claims § 12.3
Tax considerations §§ 2.3, 2.6
CAUSATION
Delay claims against contractors § 14.5
Payment delay claims § 11.13
CAUSE AND EFFECT
Differing site conditions claims § 8.28
Disruption claims § 6.8
CAUSES OF ACTION AND RECOVERIES
Generally § 13.1
Architect, claims against
–Breach of warranties §§ 13.2, 13.3
–Liability rules applied to hypothetical § 13.8
–Strict liability § 13.13
Claims preparation § 13.55
Contractor as plaintiff, damages available
–Actual damages § 13.48
–Consequential costs and damages § 13.53
–Economic loss § 13.50
–No damages for delay clause, recovery with § 13.51
–Notice compliance § 13.47

CAUSES OF ACTION AND RECOVERIES *(Continued)*
  –Performance specifications, recovery for defective § 13.52
  –Plans and specifications, damages for defects §§ 13.49, 13.50
  Contractor, claims against
  –Breach of warranties §§ 13.16, 13.17
  –Liability rules applied to hypothetical § 13.21
  –Negligence § 13.19
  –Strict liability § 13.20
  –UCC § 13.18
  Manufacturer/supplier, claims against
  –Breach of warranty and UCC §§ 13.22, 13.23
  –Liability rules applied to hypothetical § 13.26
  –Negligence § 13.24
  –Strict liability § 13.25
  Owner as plaintiff, damages available
  –Generally § 13.32
  –Attorneys' fees recovery § 13.44
  –Completion, cost recovery § 13.35
  –Corrective work, cost recovery § 13.33
  –Depreciation recovery § 13.45
  –Financing, recovery for loss of permanent commitment § 13.38
  –Interest rate differential recovery § 13.36
  –Land value recovery § 13.37
  –Liquidated damages § 13.34
  –Lost profits § 13.42
  –Lost revenue recovery § 13.39
  –Professional reputation, recovery for loss of § 13.46
  –Professional services, recovery for additional § 13.43
  –Supervision cost recovery § 13.41
  –Use, recovery of loss of § 13.40
  Owner, claims against
  –Breach of warranties §§ 13.9, 13.10
  –Liability rules applied to hypothetical § 13.15
  –Negligence § 13.12
  –Strict liability §§ 13.13, 13.14
  –UCC § 13.11
  Settlement considerations § 13.55

CAUSES OF ACTION AND RECOVERIES *(Continued)*
  Statutes of limitation and repose § 13.54
  Surety, insurer, lender, claims against
  –All risk policies § 13.28
  –Bonds, types § 13.27
  –Indemnification clauses § 13.29
  –Lender, claims against § 13.31
  –Subcontractor's bond as defect coverage § 13.30
  UCC
  –Generally §§ 13.4, 13.11, 13.18
  –Negligence §§ 13.5, 13.6
CERTAINTY
  Lost profit claims § 15.4
  Payment delay claims §§ 11.11, 11.20, 11.21
CERTIFICATION
  Payment delay claims § 11.15
CHANGED SITE CONDITIONS
  See DIFFERING SITE CONDITIONS
CHANGE ORDERS
  Acceleration claims §§ 7.4, 7.10, 7.12
  Changes in scope claims. See CHANGES IN SCOPE CLAIMS
  Delay claims against contractors §§ 14.4, 14.5
  Pricing and proving claims § 1.7
CHANGES
  Causes of action and recovery § 13.17
  Cost claims § 12.1
  Delay claims against owners § 5.3
  Payment delay claims §§ 11.14, 11.16
CHANGES CLAUSE
  Delay claims against owners §§ 5.2, 5.9
  Disruption claims § 6.5
CHANGES IN SCOPE CLAIMS
  Generally § 9.1
  Cardinal changes § 9.4
  Claim preparation § 9.7
  Constructive change orders § 9.3
  Cost calculations § 9.13
  –Generally § 9.8
  –Authorization § 9.10
  –Extra and additional work § 9.9
  –Reason for extra work § 9.11
  –Value of extra work § 9.12
  Fast-track contract § 9.20

INDEX

CHANGES IN SCOPE CLAIMS *(Continued)*
  Formal change orders § 9.2
  Notice and writing requirements § 9.5
  Performance bonds § 9.19
  Pricing claim § 9.6
  Recoverable costs
  –Generally § 9.14
  –Bonds, insurance and operational costs § 9.18
  –Equipment costs § 9.17
  –Labor costs § 9.15
  –Materials costs § 9.16
CHARTS AND GRAPHS
  Acceleration claims § 7.20
  Delay claims against contractors § 14.15
  Pricing and proving claims §§ 1.24, 1.25
CHOICE OF LAW
  International claims. See INTERNATIONAL CLAIMS
CLAIM DEVELOPMENT
  Pricing and proving claims. See PRICING AND PROVING CLAIMS
CLAIM DOCUMENT
  Pricing and proving claims §§ 1.25, 1.30
CLAIM PREPARATION
  Contractor as plaintiff, damages available § 13.55
COLLECTION COSTS
  Payment delay claims § 11.3
COLLECTION FEES
  Payment delay claims § 11.20
COMMERCIAL CONTRACTS
  Payment delay claims § 11.14
COMMON LAW
  International claim resolution inside U.S. §§ 3.2, 3.5
  International claim resolution outside U.S. §§ 4.8, 4.17
COMMON LAW RIGHTS IN TERMINATION
  Termination claims § 10.1
COMMUNICATIONS
  Pricing and proving claims § 1.17
COMPENSABLE DELAY
  Acceleration claims § 7.4
  Delay claims against contractors § 14.7

COMPENSABLE DELAY *(Continued)*
  Delay claims against owners § 5.4
COMPENSABLE DISRUPTION
  Disruption claims § 6.5
COMPENSATORY DAMAGES
  International claim resolution outside U.S. § 4.13
  Tax considerations § 2.5
COMPLETED CONTRACT METHOD
  Tax considerations § 2.3
COMPLETION COSTS
  Causes of action and recoveries § 13.1
  Owner as plaintiff, damages available § 13.35
  Termination claims §§ 10.18, 10.30
COMPOSITE INTEREST RATE
  Payment delay claims § 11.16
COMPUTING DELAY
  Delay claims against owners § 5.8
CONCEALED CONDITIONS
  Differing site conditions claims. See DIFFERING SITE CONDITION CLAIMS
CONCEALMENT
  Causes of action and recovery §§ 13.49, 13.54
  Differing site conditions claims § 8.18
CONCRETE
  Differing site conditions claims § 8.10
CONCURRENT DELAY
  Delay claims against owners §§ 5.5, 5.6
CONDITIONS NORMALLY INHERENT
  Differing site conditions claims § 8.11
CONDITIONS PRECEDENT
  Termination claims § 10.11
CONFIDENTIALITY
  Cost claims § 12.5
CONFLICTING INFORMATION
  Differing site conditions claims § 8.7
CONSEQUENTIAL DAMAGES
  Causes of action and recovery §§ 13.5, 13.32, 13.36, 13.53
  Contractor as plaintiff, damages available § 13.53

CONSEQUENTIAL DAMAGES
(Continued)
  Delay claims against contractors § 14.13
  International claim resolution inside U.S. § 3.9
  Lost profit. See LOST PROFIT; LOST PROFIT CLAIMS
  Pricing and proving claims § 1.27
  Termination claims §§ 10.3, 10.11–10.13, 10.22, 10.23

CONSOLIDATION
  Multi-party arbitration § 1.13

CONSTRUCTION OF LANGUAGE
  International claim resolution outside U.S. § 4.5

CONSTRUCTIVE ACCELERATION
  Acceleration claims §§ 7.1, 7.4, 7.5, 7.7, 7.10
  Changes in scope claims § 9.3

CONSTRUCTIVE CHANGE ORDERS
  Changes in scope claims § 9.3

CONSUMER PROTECTION STATUTES
  Termination claims § 10.26

CONTINUED PERFORMANCE
  Termination claims § 10.8

CONTRACT DATA
  Differing site conditions claims § 8.21

CONTRACT DISPUTES ACT
  Payment delay claims § 11.15

CONTRACT INTERPRETATION
  Pricing and proving claims § 1.6

CONTRACTOR CLAIMS AGAINST OWNERS
  Acceleration claims. See ACCELERATION CLAIMS
  Changes in scope claims. See CHANGES IN SCOPE CLAIMS
  Delay claims against owners. See DELAY CLAIMS AGAINST OWNERS
  Differing site condition claims. See DIFFERING SITE CONDITION CLAIMS
  Disruption claims. See DISRUPTION CLAIMS
  Payment delay claims. See PAYMENT DELAY CLAIMS
  Termination claims. See TERMINATION CLAIMS

CONTRACTORS
  Claims by. See CONTRACTOR CLAIMS AGAINST OWNERS
  Claims against. See CAUSES OF ACTION AND RECOVERIES
  Delay claims against contractors § 14.3

CONTRACTS
  Delay claims against contractors § 14.4
  Differing site conditions. See DIFFERING SITE CONDITIONS
  Disruption claims § 6.4
  International claim resolution inside U.S. § 3.7
  International claim resolution outside U.S. §§ 4.3, 4.19
  Payment delay claims. See PAYMENT DELAY CLAIMS
  Pricing and proving claims. See PRICING AND PROVING CLAIMS

CONTRACTUAL RIGHTS IN TERMINATION
  Termination claims § 10.1

CONTRIBUTION
  International claim resolution outside U.S. § 4.16

CONTRIBUTORY NEGLIGENCE
  Causes of action and recovery § 13.6

CONVENIENCE
  Termination. See TERMINATION CLAIMS

COOPERATION, DUTY OF
  Pricing and proving claims § 1.6

CORPORATE VALUE IMPAIRMENT
  Lost profit claims § 15.6

CORRECTION COSTS
  Termination claims § 10.18

CORRECTIVE WORK
  Causes of action and recovery § 13.32
  Owner as plaintiff, damages available § 13.33

CORRESPONDENCE
  Acceleration claims § 7.12

COST ACCOUNTANTS
  Cost claims § 12.5

COST ACCOUNTING RECORDS
  Pricing and proving claims § 1.19

COST CLAIMS
  Generally § 12.1

COST CLAIMS *(Continued)*
Books and records, contractors § 12.3
Cost plus v. fixed fee term § 12.7
Cost-plus contracting, owner's involvement § 12.14
Efficient performance, contractor's § 12.8
Investigating contractor-claimed costs § 12.2
Labor costs, reasonableness § 12.10
Material costs, reasonableness § 12.9
Office and administrative costs § 12.11
Overstated contractor costs § 12.6
Premature payments § 12.13
Profit margin § 12.12
Record collection § 12.5
Subcontractors and suppliers, contractor's relationship with § 12.4
Surety's obligations § 12.15

COST CODES
Acceleration claims § 7.16

COST PLUS CONTRACTING
Cost claims §§ 12.1, 12.7, 12.14

COST REVIEWS
Pricing and proving claims § 1.29

COSTS
Acceleration claims §§ 7.7, 7.9
Arbitration § 1.9
Differing site conditions claims § 8.4
International claim resolution inside U.S. § 3.8
International claim resolution outside U.S. § 4.6
Tax considerations § 2.1
Termination claims § 10.3

COST SAVINGS
Acceleration claims § 7.19

COSTS CALCULATION
Changes in scope claims. See CHANGES IN SCOPE CLAIMS

CREDIBILITY
Differing site conditions claims §§ 8.22, 8.23
Disruption claims § 6.13

CREDIT CHECKS
Payment delay claims § 11.20
Pricing and proving claims § 1.2

CRITICAL DELAY
Acceleration claims § 7.5

CRITICAL PATH METHOD
Acceleration claims §§ 7.5, 7.6
Changes in scope claims § 9.6
Delay claims against owners § 5.8
Disruption claims § 6.8

CURRENCY EXCHANGE RATES
International claims. See INTERNATIONAL CLAIMS

DAILY LOGS
Acceleration claims § 7.12
Changes in scope claims § 9.17
Cost claims § 12.4
Delay claims against contractors § 14.4

DAMAGES
Acceleration claims. See ACCELERATION CLAIMS
Delay claims against contractors. See DELAY CLAIMS AGAINST CONTRACTORS
Differing site conditions. See DIFFERING SITE CONDITIONS
Disruption claims § 6.7
International claims §§ 4.13, 4.14, 4.17, 4.19, 4.24
Payment delay claims. See PAYMENT DELAY CLAIMS
Pricing and proving claims. See PRICING AND PROVING CLAIMS
Tax considerations. See TAX CONSIDERATIONS
Termination. See TERMINATION CLAIMS

DAMAGES COMPUTATION
Delay claims against owners § 5.10

DEFAULT
Delay claims against contractors § 14.5
Termination. See TERMINATION CLAIMS

DEFECTIVE PERFORMANCE
Causes of action and recovery § 13.35

DEFECTIVE WORK
See POOR WORKMANSHIP

DEFENDANTS
Delay claims against contractors § 14.3

DEFENDANTS' TAXES
Tax considerations. See TAX CONSIDERATIONS

INDEX

DEFENSES
  Delay claims against contractors
    § 14.8
  Disruption claims § 6.16

DELAY
  Differing site conditions claims
    § 8.24
  Disruption claims §§ 6.2, 6.8
  Payment delay claims. See
    PAYMENT DELAY CLAIMS

DELAY CLAIMS AGAINST CONTRACTORS
  Generally § 14.1
  Bibliography § 14.17
  Claim development
  –Causes of delay, identifying § 14.5
  –Defendant identification § 14.3
  –Documentation, construction
    § 14.4
  –Preparing claim § 14.2
  –Responsible party, focusing claim
    on § 14.6
  Claim representation § 14.15
  Conclusion § 14.16
  Damages, elements
  –Consequential damages § 14.13
  –Contracts with liquidated damages
    provisions § 14.10
  –Contracts without liquidated
    damages provisions § 14.11
  –Direct damages § 14.12
  Forum selection § 14.14
  Liability, legal theories
  –Compensable and noncompensable
    delays § 14.7
  –Defenses, anticipating § 14.8
  –Mitigating damages § 14.9

DELAY CLAIMS AGAINST OWNERS
  Generally § 5.1
  Accountant's perspective
  –Cost escalation § 5.16
  –Equipment costs § 5.18
  –General conditions costs § 5.17
  –Home office overhead, unabsorbed
    § 5.19
  –Interest and cost of capital § 5.20
  Attorney's perspective
  –Burden of proof and apportionment
    § 5.6
  –Cause of delay § 5.2
  –Compensable and noncompensable
    delay § 5.4

DELAY CLAIMS AGAINST OWNERS *(Continued)*
  –Computing delay § 5.8
  –Concurrent delay § 5.5
  –Early finish, right of § 5.9
  –Excusable and nonexcusable delay
    § 5.3
  –Home office overhead § 5.13
  –Idle equipment costs § 5.12
  –Increased material and labor costs
    § 5.11
  –Liquidated damages, apportionment
    § 5.7
  –Lost profits § 5.15
  –Prejudgment interest and financing
    costs § 5.14
  –Pricing claim § 5.10
  Bibliography § 5.22
  Summary § 5.21

DELAY DAMAGES
  Causes of action and recovery
    § 13.49
  Termination claims §§ 10.23,
    10.24, 10.27, 10.34

DELAY IN USE
  Termination claims §§ 10.20,
    10.24

DELAYS
  Changes in scope claims §§ 9.6,
    9.7
  Termination claims § 10.25

DELIVERY
  Acceleration claims § 7.4
  Changes in scope claims § 9.16
  Cost claims § 12.9

DELIVERY COSTS
  Acceleration claims §§ 7.2, 7.9

DELIVERY DELAYS
  Delay claims against owners § 5.3

DELIVERY TICKETS
  Delay claims against contractors
    § 14.4

DEMAND LETTER
  Payment delay claims § 11.14

DEMOBILIZATION COSTS
  Disruption claims §§ 6.5, 6.6, 6.9
  Termination claims § 10.34

DEMONSTRATIVE EVIDENCE
  Acceleration claims § 7.20
  Delay claims against contractors
    § 14.15
  Pricing and proving claims § 1.23

INDEX

DEPRECIATION
  Delay claims against contractors § 14.12
  Lost profit claims § 15.5
  Owner as plaintiff, damages available § 13.45

DESIGN CHANGES
  Acceleration claims § 7.12
  Disruption claims § 6.2

DESIGN DEFECTS
  Causes of action and recovery § 13.28
  Changes in scope claims § 9.1

DESIGN DETAILS
  Differing site conditions claims § 8.6

DESIGN DOCUMENTS
  Delay claims against contractors § 14.4

DESIGN PROFESSIONALS
  Delay claims against contractors § 14.3
  Owner claims against. See OWNER CLAIMS

DETRIMENTAL RELIANCE
  Differing site conditions claims § 8.18

DIFFERING SITE CONDITIONS
  Generally § 8.1
  Contractual differing conditions
  –Notice § 8.14
  –Reverse differing conditions § 8.13
  –Site investigation and disclaimers § 8.15
  –Type 1, see Type 1 conditions, this heading
  –Type 2 conditions §§ 8.11, 8.12
  –Typical clauses § 8.3
  –Contract provisions § 8.2
  Delay claims against owners § 5.2
  Differing conditions without clause
  –Breach of warranty § 8.17
  –Disclaimer clauses § 8.19
  –Misrepresentation and fraud § 8.18
  –Remedies § 8.16
  Disruption claims §§ 6.3, 6.5
  Factual proof of claim
  –Bid preparation stage § 8.21
  –Encountered conditions phase § 8.21
  –Expert testimony § 8.23
  –Prebid stage § 8.20
  Proof of damages

DIFFERING SITE CONDITIONS *(Continued)*
  –Differing conditions, impact § 8.24
  –Other approaches § 8.31
  –Other costs § 8.32
  –Productivity, impacted §§ 8.25–8.30
  Type 1 conditions
  –Generally § 8.4
  –Contract indications, nature of § 8.6
  –Contract v. informational data § 8.5
  –Examples § 8.10
  –Lack of indications § 8.7
  –Material difference § 8.9
  –Reliance § 8.8

DIMINUTION OF VALUE
  Causes of action and recovery § 13.32
  Delay claims against contractors § 14.13
  Lost profit claims § 15.5
  Termination claims § 10.18

DIRECT COSTS
  Changes in scope claims §§ 9.6, 9.13
  Disruption claims § 6.7
  Pricing and proving claims § 1.19
  Termination claims § 10.39

DIRECT DAMAGES
  Causes of action and recovery §§ 13.32, 13.36
  Delay claims against contractors § 14.12
  Pricing and proving claims § 1.27

DIRECTED ACCELERATION
  Acceleration claims §§ 7.1, 7.3

DISCLAIMER CLAUSES
  Causes of action and recovery § 13.23
  Differing site conditions claims §§ 8.15, 8.19

DISCOUNTED NET PRESENT VALUE
  International claim resolution outside U.S. § 4.27

DISCOVERY
  Arbitration § 1.9
  Cost claims § 12.5

DISCRETE COST METHOD
  Acceleration claims §§ 7.14, 7.16

INDEX

DISRUPTION
  Changes in scope claims § 9.6
  Differing site conditions claims § 8.24
  Termination claims § 10.34
DISRUPTION CLAIMS
  Generally § 6.1
  Cause and effect, proving § 6.8
  Damages, nature of § 6.7
  Defenses § 6.16
  Delay v. disruption § 6.2
  Effects of disruption § 6.6
  Pricing claims
  –Generally § 6.9
  –Expert opinion § 6.13
  –Industry standards and handbooks § 6.11
  –Jury verdict method § 6.15
  –Should cost estimates § 6.10
  –Time and motion studies § 6.12
  –Total and modified total cost methods § 6.14
  Summary § 6.17
  Types
  –Generally § 6.3
  –Compensable § 6.5
  –Noncompensable § 6.4
DISRUPTION COSTS
  Delay claims against owners § 5.1
DIVERSION OF LABOR AND MATERIALS
  Cost claims § 12.3
DOCUMENTATION
  Acceleration claims §§ 7.6, 7.12
  Changes in scope claims § 9.7
  Delay claims against contractors § 14.4
  Payment delay claims § 11.3
  Pricing and proving claims. See PRICING AND PROVING CLAIMS
DOCUMENT TAX
  International claim resolution outside U.S. § 4.3
DUTY OF COOPERATION
  Pricing and proving claims § 1.6

EARLY FINISH
  Delay claims against owners § 5.9
ECONOMIC LOSS
  Causes of action and recovery §§ 13.1, 13.5, 13.22

ECONOMIC LOSS *(Continued)*
  Contractor as plaintiff, damages available § 13.50
  International claim resolution outside U.S. § 4.26
EFFICIENCY
  Cost claims § 12.8
  Lost efficiency. See LOST EFFICIENCY
EICHLEAY FORMULA
  Delay claims against owners §§ 5.13, 5.19
EJCDC DOCUMENTS
  Differing site conditions claims § 8.3
  Pricing and proving claims § 1.5
EMPLOYMENT LAW
  International claim resolution outside U.S. § 4.16
ENCOUNTERED CONDITIONS STAGE
  Differing site conditions claims, proving § 8.22
ENFORCEMENT
  Acceleration claims § 7.11
  International claim resolution outside U.S. § 4.7
ENTITLEMENT
  International claims. See INTERNATIONAL CLAIMS
ENVIRONMENTAL LAW
  International claim resolution outside U.S. § 4.8
EQUIPMENT
  Acceleration claims § 7.12
  Causes of action and recovery §§ 13.16, 13.9
  Pricing and proving claims § 1.19
EQUIPMENT COSTS
  Acceleration claims §§ 7.1, 7.2, 7.9, 7.13
  Changes in scope claims §§ 9.7, 9.17
  Cost claims § 12.11
  Delay claims against owners § 5.16
  Differing site conditions claims §§ 8.25, 8.26, 8.31
  Disruption claims §§ 6.2, 6.7
  Termination claims § 10.39
EQUITABLE ADJUSTMENT
  Changes in scope claims §§ 9.3, 9.6, 9.13

INDEX

EQUITABLE APPROACH
International claim resolution inside U.S. § 3.5
EQUITABLE RELIEF
Termination claims § 10.1
EQUITY
International claim resolution outside U.S. § 4.19
EQUITY CAPITAL
Payment delay claims §§ 11.10, 11.16
ERRORS
Design errors. See DESIGN DEFECTS
Pricing and proving claims § 1.6
Termination claims § 10.11
ESCALATION
Causes of action and recovery § 13.53
Delay claims against owners §§ 5.11, 5.16
Delay claims against contractors § 14.12
ESTABLISHED BUSINESSES
Lost profit claims § 15.7
ESTIMATES
Acceleration claims § 7.12
Payment delay claims § 11.1
Pricing and proving claims § 1.15
Termination claims § 10.31
ESTOPPEL
Changes in scope claims § 9.5
EVIDENCE
Acceleration claims § 7.20
Arbitration § 1.11
Causes of action and recovery § 13.42
Cost claims § 12.1
Differing site conditions claims § 8.20
Disruption claims § 6.13
International claims §§ 4.20, 4.22
Lost profit claims § 15.8
Payment delay claims §§ 11.13, 11.14, 11.18
Termination claims § 10.31
EX AEQUO ET BONO
International claim resolution outside U.S. § 4.18
EXCESSIVE COSTS
Cost claims. See COST CLAIMS
EXCHANGE RATES
International claim resolution outside U.S. § 4.25

EXCULPATORY CLAUSES
Causes of action and recovery §§ 13.3, 13.5, 13.10
Pricing and proving claims § 1.5
EXCUSABLE DELAY
Acceleration claims §§ 7.4, 7.5, 7.10
Delay claims against owners § 5.3
EXCUSABLE NONPERFORMANCE
Termination claims § 10.11
EXEMPLARY DAMAGES
International claim resolution outside U.S. § 4.14
EXPENSES
Termination claims § 10.3
EXPERIENCE
Differing site conditions claims §§ 8.11, 8.22
EXPERT OPINION
Disruption claims § 6.13
EXPERT WITNESSES
Acceleration claims § 7.20
Causes of action and recovery § 13.55
Differing site conditions claims § 8.23
EXPERTS
Arbitration §§ 1.10, 1.11
Causes of action and recovery §§ 13.5, 13.42
Cost claims §§ 12.7, 12.10
International claim resolution outside U.S. §§ 4.1, 4.20
Pricing and proving claims § 1.23
EXPRESS CONTRACTS
International claim resolution outside U.S. § 4.8
EXPRESS OBLIGATIONS
Disruption claims § 6.5
EXPRESS TERMS
Pricing and proving claims § 1.6
EXPRESS WARRANTIES
Causes of action and recovery §§ 13.2, 13.9, 13.22
EXTENDED DURATION COSTS
Termination claims § 10.19
EXTRACONTRACTUAL CLAIMS
International claim resolution outside U.S. § 4.8
EXTRA WORK
Causes of action and recoveries § 13.1
Changes in scope claims. See CHANGES IN SCOPE CLAIMS

EXTRA WORK *(Continued)*
   Cost claims § 12.7
   Payment delay claims. See
     PAYMENT DELAY CLAIMS
   Termination claims § 10.3

EXTRA WORK ACCOUNT SEGREGATION
   Differing site conditions claims § 8.31

FACTUAL NARRATIVE
   Pricing and proving claims § 1.25

FAILURE CLAIMS
   See DEFECT AND FAILURE CLAIMS

FAILURE TO DISCLOSE
   Differing site conditions claims §§ 8.15, 8.18

FAILURE TO PROVE CAUSATION
   Disruption claims § 6.16

FAIR-MARKET STANDARD
   Cost claims § 12.1

FAIR VALUE
   Termination claims §§ 10.26, 10.39

FAST-TRACK CONTRACTS
   Changes in scope claims § 9.20

FEDERAL ACQUISITION REGULATIONS
   Delay claims against owners § 5.13
   Differing site conditions claims § 8.3
   Payment delay claims §§ 11.2, 11.4, 11.7

FEDERAL ARBITRATION ACT
   Arbitration § 1.12

FEDERAL COMPENSATION
   Termination claims § 10.3

FEDERAL CONTRACTS
   Payment delay claims § 11.15
   Termination claims §§ 10.2, 10.4, 10.5

FEDERAL PROCUREMENT CONTRACTS
   Termination claims § 10.1

FINANCIAL ACCOUNTING STANDARDS
   International claim resolution outside U.S. § 4.22

FINANCING COSTS
   Acceleration claims § 7.4
   Changes in scope claims § 9.18
   Delay claims against owners §§ 5.14, 5.20
   Disruption claims § 6.2

FINANCING COSTS *(Continued)*
   Owner as plaintiff, damages available § 13.38
   Payment delay claims. See
     PAYMENT DELAY CLAIMS
   Termination claims § 10.20

FINES
   International claim resolution outside U.S. § 4.15

FIRE
   Acceleration claims § 7.4
   Delay claims against contractors § 14.7
   Delay claims against owners § 5.3

FIXTURES
   Causes of action and recovery § 13.22

FLOAT
   Acceleration claims § 7.5

FOLLOW-UP NOTICES
   Causes of action and recovery § 13.47

FOREIGN LANGUAGES
   International claim resolution outside U.S. §§ 4.5, 4.21

FOREIGN OBSTRUCTIONS
   Differing site conditions claims § 8.10

FORESEEABILITY
   Causes of action and recovery §§ 13.5, 13.6, 13.8, 13.19, 13.24, 13.32, 13.35, 13.38, 13.39, 13.51, 13.52
   Delay claims against contractors § 14.13
   Delay claims against owners § 5.3
   Differing site conditions claims §§ 8.22, 8.23
   Lost profit claims § 15.2
   Payment delay claims §§ 11.11, 11.17, 11.18, 11.20
   Termination claims §§ 10.21, 10.34

FORUM SELECTION
   Delay claims against contractors § 14.14
   International claim resolution inside U.S. § 3.8
   International claim resolution outside U.S. § 4.6

FRAUD
   Causes of action and recovery §§ 13.46, 13.51

INDEX

FRAUD *(Continued)*
  Cost claims § 12.1
  Differing site conditions claims
    §§ 8.17–8.19
  Termination claims §§ 10.11, 10.13
FRINGE BENEFITS
  Changes in scope claims § 9.15
  Delay claims against owners § 5.16
FRONT END LOADING
  Cost claims § 12.13
FUTURE LOSS
  International claim resolution outside
    U.S. § 4.27

GANTT CHARTS
  Acceleration claims § 7.5
GENERAL CONDITIONS COSTS
  Delay claims against owners § 5.17
GOOD FAITH
  Changes in scope claims § 9.5
  Termination claims §§ 10.11, 10.18, 10.27
GOODWILL
  Tax considerations § 2.2
GOVERNMENT AGENCIES
  Causes of action and recovery § 13.31
GOVERNMENT REGULATIONS
  Delay claims against contractors
    § 14.3
GROSS CONTRACT PRICE
  Tax considerations § 2.3
GUARANTEED MAXIMUM PRICE
  Changes in scope claims § 9.20

HANDBOOKS
  Disruption claims § 6.11
HINDRANCE
  Disruption claims § 6.1
HOME OFFICE OVERHEAD
  Delay claims against owners §§ 5.13, 5.19
HORIZONTAL PRIVITY
  Causes of action and recovery
    §§ 13.22, 13.23

IDLE EQUIPMENT COSTS
  Delay claims against owners § 5.12
ILLEGAL CONTRACT
  Termination claims § 10.11
IMPLIED OBLIGATIONS
  Disruption claims § 6.5
  Pricing and proving claims § 1.6

IMPLIED TERMS
  Pricing and proving claims § 1.6
IMPLIED WARRANTIES
  Causes of action and recovery
    §§ 13.2, 13.9, 13.16, 13.22
  Differing site conditions claims
    §§ 8.17, 8.19
  Pricing and proving claims § 1.6
INCOME APPROACH
  Lost profit claims § 15.6
INCOME STATEMENT FORECAST
  Delay claims against contractors
    § 14.15
INCREASED COSTS
  Delay claims against owners §§ 5.11, 5.16
INCREASED WAGES
  Delay claims against owners § 5.11
INDEMNIFICATION CLAUSE
  Causes of action and recovery
    § 13.30
  Surety, insurer, claims against
    § 13.29
  Pricing and proving claims § 1.5
INDIRECT COSTS
  Changes in scope claims §§ 9.6, 9.7, 9.13
  Disruption claims § 6.7
  Termination claims § 10.39
INDIRECT DAMAGES
  Causes of action and recovery
    § 13.32
INDUSTRY PRODUCTIVITY
  RATES
  Disruption claims § 6.11
INDUSTRY STANDARDS
  Changes in scope claims § 9.17
  Disruption claims § 6.11
INFLATION
  International claim resolution inside
    U.S. § 3.10
INFORMATION
  Differing site conditions claims § 8.5
INJURY TO CAPITAL
  Tax considerations § 2.1
INSPECTIONS
  Causes of action and recovery
    § 13.31
INSURANCE
  Causes of action and recovery
    § 13.41
  Changes in scope claims §§ 9.15, 9.18

INDEX

INSURANCE *(Continued)*
   Delay claims against contractors
     § 14.3
   Termination claims § 10.39
INSURER
   Claims against. See CAUSES OF
     ACTION AND RECOVERIES
INTEREST
   Changes in scope claims § 9.18
   Delay claims against contractors
     § 14.12
   Delay claims against owners §§ 5.14,
     5.20
   International claim resolution outside
     U.S. §§ 4.26, 4.29
   Payment delay claims. See
     PAYMENT DELAY CLAIMS
   Termination claims §§ 10.20, 10.34
INTEREST RATE DIFFERENTIAL
   Owner as plaintiff, damages available
     § 13.36
INTEREST RATES
   International claim resolution inside
     U.S. § 3.11
INTERFERENCE
   Acceleration claims § 7.4
   Causes of action and recovery
     §§ 13.46, 13.51
   Disruption claims § 6.5
INTERNATIONAL CLAIMS
   Accountant's matters
    –Generally § 4.19
    –Claim verification § 4.23
    –Damages, global approach § 4.24
    –Evidence style § 4.20
    –Financial evidence § 4.22
    –Foreign exchange rates § 4.25
    –Future loss, proving § 4.27
    –Interest § 4.29
    –Language barrier § 4.21
    –Purely economic loss § 4.26
    –Unjust enrichment § 4.28
   Entitlement
    –Generally § 4.9
    –Compensatory damages § 4.13
    –Contribution § 4.16
    –Fines § 4.15
    –Liquidated damages and penalties
     § 4.17
    –Loss/expense or additional costs
     § 4.11
    –Other approaches § 4.18

INTERNATIONAL CLAIMS
   (Continued)
    –Price § 4.10
    –Punitive or exemplary damages
     § 4.14
    –Quantum meruit § 4.12
   Resolution inside U.S.
    –Generally § 3.1
    –Arbitration clause, effect on
     currency fluctuations § 3.6
    –Arbitration instructions § 3.9
    –Arbitration, choice of law § 3.8
    –Conclusion § 3.12
    –Contract language, currency
     fluctuation risks § 3.7
    –Damages in foreign denominations
     § 3.4
    –Damages stated in U.S. dollars § 3.3
    –Emerging rules, currency
     fluctuations § 3.5
    –Exchange rates, damage awards
     § 3.2
    –Inflation § 3.10
    –Interest rates § 3.11
   Resolution outside U.S.
    –Generally § 4.1
    –Choice of law generally § 4.2
    –Claim types § 4.8
    –Construction § 4.5
    –Enforcement § 4.7
    –Forum § 4.6
    –Performance § 4.4
    –Validity and form § 4.3
INTERNATIONAL LAW
   International claim resolution outside
     U.S. § 4.8
INVESTIGATION
   Cost claims § 12.2
   Pricing and proving claims §§ 1.2,
     1.22
INVOICES
   Changes in scope claims § 9.16
   Cost claims § 12.3
   Payment delay claims § 11.3

JOB COST SYSTEM
   Acceleration claims § 7.12
JOB MEETINGS
   Pricing and proving claims § 1.17
JOBSITE CONDITION
   Delay claims against contractors
     § 14.5

INDEX

JOBSITE INVESTIGATIONS
   Differing site conditions claims § 8.20
JOBSITE LOGS
   Pricing and proving claims § 1.18
JOBSITE MANAGEMENT
   Acceleration claims § 7.4
   Delay claims against contractors § 14.3
JOBSITE SUPERVISION
   Delay claims against contractors § 14.5
JUDGMENT INTEREST
   Changes in scope claims § 9.7
JURY TRIALS
   Delay claims against contractors § 14.14
JURY VERDICT METHOD
   Changes in scope claims § 9.13
   Disruption claims § 6.15

LABOR AND MATERIAL PAYMENT BOND
   Pricing and proving claims § 1.2
LABOR COSTS
   Acceleration claims §§ 7.2, 7.9, 7.12
   Causes of action and recovery § 13.53
   Changes in scope claims §§ 9.7, 9.15, 9.17
   Cost claims § 12.10
   Differing site conditions claims §§ 8.25, 8.26
   Disruption claims §§ 6.2, 6.9
   Payment delay claims § 11.1
   Termination claims § 10.39
LABOR DISPUTES
   Acceleration claims § 7.4
   Delay claims against contractors § 14.7
   Delay claims against owners § 5.3
   Disruption claims §§ 6.3, 6.4
LABOR FORCE
   Termination claims, failure as grounds for §§ 10.11, 10.12
LAND VALUE
   Owner as plaintiff, damages available § 13.37
LATENT CONDITIONS
   Changes in scope claims § 9.1
LATENT DEFECTS
   Causes of action and recovery §§ 13.16, 13.17, 13.54

LAWYERS
   See ATTORNEYS
LEGAL FEES
   Tax deductability § 2.7
LENDER
   Claims against. See CAUSES OF ACTION AND RECOVERIES
LENGTH OF BID PERIOD
   Differing site conditions claims § 8.20
LIABILITY
   Delay claims against contractors. See DELAY CLAIMS AGAINST CONTRACTORS
   Tax considerations § 2.6
LICENSING
   Pricing and proving claims § 1.2
LIQUIDATED DAMAGES
   International claim resolution outside U.S. § 4.17
   Delay claims against contractors § 14.10
   Delay claims against owners §§ 5.3, 5.5, 5.7
   Owner as plaintiff, damages available § 13.34
   Termination. See TERMINATION CLAIMS
LIQUIDATED DAMAGES CLAUSE
   Delay claims against owners § 5.2
   Pricing and proving claims § 1.5
LONG-TERM CONTRACT METHOD
   Tax considerations §§ 2.3, 2.5
LOSS
   International claims §§ 4.11, 4.26, 4.27
   Tax considerations § 2.5
LOSS AMOUNT DEDUCTABILITY
   Tax considerations § 2.5
LOSS CONTRACTS
   Termination claims §§ 10.32, 10.37
LOSS OF OTHER WORK
   Payment delay claims § 11.19
LOST EFFICIENCY
   Acceleration claims § 7.9
   Disruption claims § 6.13
   Termination claims § 10.34
LOST PRODUCTIVITY
   Acceleration claims § 7.13
   Differing site conditions claims §§ 8.25–8.30

INDEX

LOST PRODUCTIVITY *(Continued)*
    Disruption claims §§ 6.6, 6.9
    Termination claims § 10.4
LOST PROFIT CLAIMS
    Generally § 15.1
    Conclusion § 15.10
    Corporate value impairment § 15.6
    Foreseeability § 15.2
    Hypothetical case § 15.9
    Profit calculation § 15.5
    Proof § 15.8
    Proximate cause § 15.3
    Reasonable certainty § 15.4
    Unestablished v. established business § 15.7
LOST PROFITS
    Acceleration claims § 7.18
    Causes of action and recovery §§ 13.1, 13.53
    Delay claims against contractors § 14.13
    Delay claims against owners § 4.21
    Differing site conditions claims § 8.18
    Disruption claims § 6.7
    International claim resolution outside U.S. § 4.26
    Owner as plaintiff, damages available § 13.42
    Payment delay claims § 11.19
    Tax considerations §§ 2.1, 2.2
    Termination claims §§ 10.21, 10.24
LOST RENT REVENUES
    Delay claims against contractors § 14.12
LOST REVENUES
    Owner as plaintiff, damages available § 13.39
LUMP SUM METHOD
    Disruption claims § 6.15

MAGNITUDE OF CONDITIONS
    Differing site conditions claims § 8.12
MALICE
    Causes of action and recovery § 13.46
MANUFACTURER
    Claims against. See CAUSES OF ACTION AND RECOVERIES
MAPS
    Differing site conditions claims § 8.5
MARKET APPROACH
    Lost profit claims § 15.6
MASS-PRODUCED HOUSING
    Causes of action and recovery § 13.13
MATERIAL
    Acceleration claims § 7.4
    Causes of action and recovery §§ 13.9, 13.13, 13.16, 13.53
MATERIAL ALTERATION
    Changes in scope claims § 9.19
    Disruption claims § 6.1
MATERIAL BREACH OF CONTRACT
    Termination claims. See TERMINATION CLAIMS
MATERIAL COSTS
    Acceleration claims § 7.13
    Causes of action and recovery § 13.20
    Changes in scope claims § 9.16
    Cost claims § 12.9
    Delay claims against owners §§ 5.11, 5.16
    Differing site conditions claims §§ 8.24, 8.32
    Disruption claims § 6.4
    Payment delay claims § 11.1
    Termination claims § 10.39
MATERIAL DIFFERENCE
    Differing site conditions claims §§ 8.1, 8.4, 8.9
MATERIAL FACT
    Differing site conditions claims § 8.17
MEASUREMENT METHODOLOGY
    Disruption claims § 6.13
MECHANIC'S LIENS
    Termination claims § 10.41
MEETINGS
    Pricing and proving claims § 1.17
METHODS OF WORK
    Changes in scope claims §§ 9.1, 9.4
MISREPRESENTATION
    Causes of action and recovery §§ 13.31, 13.49
    Differing site conditions claims §§ 8.18, 8.21
MISTAKES
    See ERRORS
MISUSE OF PRODUCT
    Causes of action and recovery §§ 13.14, 13.24

INDEX

MITIGATION OF DAMAGES
    Acceleration claims § 7.19
    Delay claims against contractors
       § 14.9
    Termination claims §§ 10.17, 10.35
MOBILIZATION COSTS
    Termination claims § 10.4
MODIFICATION OF CONTRACT
    Changes in scope claims § 9.2
    Pricing and proving claims § 1.7
MODIFIED TOTAL COST METHOD
    Acceleration claims §§ 7.14, 7.16
    Changes in scope claims § 9.13
    Disruption claims §§ 6.8, 6.14
MOTIONS TO COMPEL
    Cost claims § 12.5
MOVABLE GOODS
    Causes of action and recovery § 13.22
MULTIPLE BREACHES
    International claim resolution inside U.S. § 3.7
MULTIPLE CLAIMS
    Delay claims against contractors
       § 14.12
MULTIPLE PARTIES
    Pricing and proving claims § 1.13
MULTIPLE PRIMES
    Acceleration claims § 7.4
    Pricing and proving claims § 1.4

NEGLIGENCE
    Architect, claims against §§ 13.5, 13.6
    Contractor, claims against § 13.19
    Manufacturer/supplier, claims against § 13.24
    Owner, claims against § 13.12
NEGOTIATION
    Pricing and proving claims § 1.30
NO DAMAGES FOR DELAY CLAUSE
    Acceleration claims § 7.4
    Causes of action and recovery
       § 13.10
    Contractor as plaintiff, damages available § 13.51
    Delay claims against contractors
       § 14.1
    Delay claims against owners §§ 5.2, 5.4, 5.7
    Disruption claims §§ 6.2, 6.16
    Pricing and proving claims § 1.5
    Termination claims § 10.10

NONCOMPENSABLE DELAYS
    Delay claims against contractors
       § 14.7
    Delay claims against owners § 5.4
NONCOMPENSABLE DISRUPTION
    Disruption claims § 6.4
NONEXCUSABLE DELAY
    Acceleration claims § 7.4
    Delay claims against owners § 5.3
NOTICE
    Acceleration claims § 7.10
    Changes in scope claims §§ 9.5, 9.19
    Contractor as plaintiff, damages available § 13.47
    Differing site conditions claims
       §§ 8.14, 8.22
    Disruption claims § 6.16
    Pricing and proving claims § 1.22
    Termination claims § 10.13
NOTICE OF MODIFICATION
    Pricing and proving claims § 1.7

OCCUPANCY
    Causes of action and recovery
       § 13.17
OFFSETS
    Causes of action and recovery
       § 13.53
    Termination claims §§ 10.3, 10.27, 10.35, 10.40
OPERATIONAL COSTS
    Changes in scope claims § 9.18
ORDINARY INCOME
    Tax considerations § 2.1
OVERCROWDING
    Differing site conditions claims
       § 8.24
    Disruption claims § 6.6
OVERHEAD
    Acceleration claims §§ 7.9, 7.13, 7.18
    Causes of action and recovery
       § 13.41
    Changes in scope claims § 9.18
    Cost claims § 12.11
    Delay claims against contractors
       § 14.12
    Delay claims against owners §§ 5.13, 5.17, 5.19
    Differing site conditions claims
       § 8.18
    Disruption claims § 6.2
    Lost profit claims § 15.5

OVERHEAD *(Continued)*
    Pricing and proving claims § 1.19
    Termination claims §§ 10.19, 10.34, 10.39
OVERRUNS
    Cost claims. See COST CLAIMS
    Differing site conditions claims § 8.32
OWNER CLAIMS
    Causes of action. See CAUSES OF ACTION AND RECOVERIES
    Cost claims. See COST CLAIMS
    Defect and failure claims. See DEFECT AND FAILURE CLAIMS
    Delay claims against owners. See DELAY CLAIMS
    Lost profits. See LOST PROFIT CLAIMS
OWNER INTERPRETATIONS
    Differing site conditions claims § 8.6
OWNERS
    Claims against. See CAUSES OF ACTION AND RECOVERIES
    Pricing and proving claims § 1.2
    Termination. See TERMINATION CLAIMS

PARTIAL PAYMENT
    Causes of action and recovery § 13.17
    Payment delay claims § 11.3
PARTIES
    Pricing and proving claims. See PRICING AND PROVING CLAIMS
PAST PERFORMANCE
    Pricing and proving claims § 1.2
PATENT AMBIGUITIES
    Causes of action and recovery § 13.49
PAYMENT BOND
    Causes of action and recovery § 13.27
    Pricing and proving claims § 1.2
    Surety, insurer, claims against § 13.30
PAYMENT DELAY CLAIMS
    Generally § 11.1
    Checklist § 11.22
    Contract payment terms
    –AIA payments clauses § 11.5

PAYMENT DELAY CLAIMS *(Continued)*
    –AGC payments clauses § 11.6
    –Common terms § 11.4
    –FAR payments provisions § 11.7
    –Payment clauses § 11.3
    –Payment provisions § 11.2
    –Suggested payments clauses § 11.8
    Damages
    –Bonding, increased cost § 11.18
    –Collection fees and credit checks § 11.20
    –Foreseeability of payment delay damages § 11.17
    –Loss of other work § 11.19
    –Subcontractors' suits, defending against § 11.21
    Interest as cost
    –Generally § 11.9
    –Collectible damages § 11.11
    –Theory of § 11.10
    Interest damages
    –Commercial contracts, calculating interest § 11.14
    –Federal contracts, calculating damages for § 11.15
    –Proper interest rate determination § 11.16
    –Proving §§ 11.12, 11.13
PAYMENTS
    Cost claims § 12.4
    Delay claims against contractors § 14.1
    Termination claims §§ 10.11, 10.12, 10.35
PENALTIES
    International claim resolution outside U.S. § 4.17
PERCENTAGE MARKUP METHOD
    Delay claims against owners § 5.19
PERCENTAGE OF COMPLETION METHOD
    Tax considerations § 2.3
PERFORMANCE
    Contractor as plaintiff, damages available § 13.52
    Cost claims § 12.8
    Delay claims against owners § 5.8
    Disruption claims § 6.1
    International claim resolution outside U.S. § 4.4

INDEX

PERFORMANCE BONDS
  Causes of action and recovery
    § 13.27
  Cost claims § 12.15
  Pricing and proving claims § 1.2
  Surety, insurer, claims against
    § 13.31

PERSONAL INJURIES
  Causes of action and recovery
    §§ 13.22, 13.50

PERT CHARTS
  Acceleration claims § 7.5

PHOTOGRAPHS
  Delay claims against contractors
    § 14.4
  Pricing and proving claims § 1.18

PLAINTIFFS
  Damages available. See CAUSES OF ACTION AND RECOVERIES

PLANNING
  Pricing and proving claims. See PRICING AND PROVING CLAIMS

PLANS AND SPECIFICATIONS
  Causes of action and recovery
    §§ 13.2, 13.5, 13.9, 13.17, 13.50
  Changes in scope claims §§ 9.1, 9.3, 9.20
  Contractor as plaintiff, damages available §§ 13.49, 13.50
  Delay claims against contractors
    § 14.5
  Delay claims against owners §§ 5.4, 5.9
  Differing site conditions claims
    § 8.17
  Disruption claims §§ 6.3, 6.6
  Pricing and proving claims § 1.6

POOR WORKMANSHIP
  Acceleration claims § 7.4
  Causes of action and recoveries
    §§ 13.1, 13.16, 13.33, 13.28
  Changes in scope claims § 9.11
  Delay claims against contractors
    § 14.7
  Termination claims §§ 10.3, 10.12, 10.27

PREBID STAGE
  Differing site conditions claims, proving § 8.21

PRECEDENCE DIAGRAMMING
  Acceleration claims § 7.5

PREJUDGMENT INTEREST
  Delay claims against owners § 5.14
  Payment delay claims §§ 11.11, 11.16

PREMATURE PAYMENTS
  Cost claims § 12.13

PREPARATORY EXPENSES
  Termination claims § 10.29

PRICE
  International claim resolution outside U.S. § 4.10

PRICE EARNINGS RATIO
  International claim resolution outside U.S. § 4.27

PRICING AND PROVING CLAIMS
  Arbitration
  –Generally § 1.8
  –Arbitrator selection § 1.10
  –Enforceability of agreements to arbitrate § 1.12
  –Informality and appeals § 1.11
  –Multiple parties § 1.13
  –Time and costs § 1.9
  Avoiding risks and preserving claims
    § 1.1
  Changes in scope claims § 9.6
  Claim development
  –Generally § 1.21
  –Components of claim document
    § 1.25
  –Demonstrative evidence § 1.24
  –Early claim recognition and preparation § 1.22
  –Early expert and attorney involvement § 1.23
  Conclusion § 1.31
  Damages, calculating and proving
  –Generally § 1.26
  –Basic principles § 1.27
  –Pricing claims, methods § 1.28
  –Project cost reviews and audits
    § 1.29
  Delay claims against owners § 5.10
  Disruption claims. See DISRUPTION CLAIMS
  Negotiation and settlement § 1.30
  Participants' reliability § 1.2
  Parties and contracts
  –Generally § 1.3
  –Contract framework § 1.4
  –Express and implied terms, interpreting and applying § 1.6
  –Modification of contract § 1.7

PRICING AND PROVING CLAIMS
(Continued)
  —Standard forms and contract provisions § 1.5
  Planning
  —Communication § 1.17
  —Cost accounting records § 1.19
  —Estimating § 1.15
  —Management and documentation § 1.14
  —Project documentation § 1.18
  —Scheduling, work monitoring through § 1.20
  —Standard operating procedures § 1.16
PRIME CONTRACTOR
  Pricing and proving claims § 1.2
PRIVATE CONTRACTS
  Termination claims §§ 10.2, 10.5
PRIVITY
  Causes of action and recovery §§ 13.3, 13.5, 13.10, 13.15, 13.22
  Termination claims § 10.41
PROCESSING TIME
  Payment delay claims § 11.3
PRODUCTION COST CHANGES
  Delay claims against contractors § 14.13
PRODUCTIVITY
  Differing site conditions claims §§ 8.25–8.30
  Lost productivity. See LOST PRODUCTIVITY
PRODUCTS LIABILITY
  Causes of action and recovery §§ 13.7, 13.24, 13.25
PROFESSIONAL SERVICES
  Causes of action and recovery § 13.4
  Owner as plaintiff, damages available § 13.43
PROFIT
  Lost. See LOST PROFIT
  Pricing and proving claims § 1.29
  Termination claims § 10.3
PROFIT MARGIN
  Cost claims § 12.12
PROGRESS PAYMENTS
  Causes of action and recovery §§ 13.34, 13.35
  Payment delay claims. See PAYMENT DELAY CLAIMS

PROGRESS REPORTS
  Delay claims against contractors § 14.4
PROJECT PARTICIPANTS
  Pricing and proving claims § 1.2
PROJECT SCHEDULES
  Delay claims against owners § 5.8
PROOF
  Acceleration claims. See ACCELERATION CLAIMS
  Causes of action and recovery §§ 13.2, 13.5, 13.7
  Changes in scope claims. See CHANGES IN SCOPE CLAIMS
  Differing site conditions. See DIFFERING SITE CONDITIONS
  Lost profit claims § 15.8
  Payment delay claims §§ 11.12, 11.13
PROXIMATE CAUSE
  Causes of action and recovery § 13.24
  Lost profit claims § 15.3
PUNITIVE DAMAGES
  International claim resolution outside U.S. § 4.14
  Tax considerations § 2.5
PURE TOTAL COST METHOD
  Delay claims against owners § 5.10

QUALITY OF MATERIALS
  Cost claims § 12.9
QUANTUM MERUIT
  Acceleration claims § 7.14
  Changes in scope claims §§ 9.4, 9.13
  Cost claims § 12.1
  International claim resolution outside U.S. §§ 4.10, 4.12, 4.28
  Termination claims §§ 10.7, 10.30, 10.36, 10.39
QUASI CONTRACT
  Termination claims § 10.41

REAL ESTATE IMPROVEMENTS
  Causes of action and recovery §§ 13.7, 13.13
REASONABLE CARE STANDARD
  Causes of action and recovery §§ 13.5, 13.12, 13.24
REASONABLE CERTAINTY
  Lost profit claims § 15.4
REASONABLE COST OF COMPLETION
  Termination claims § 10.17

INDEX

REASONABLENESS
  Causes of action and recovery § 13.48
  Changes in scope claims § 9.12
  Cost claims. See COST CLAIMS
  Differing site conditions claims
    § 8.29
  Termination claims § 10.22
REASONABLE RELIANCE
  Differing site conditions claims
    § 8.18
RECORDKEEPING
  Acceleration claims § 7.6
  Cost claims § 12.3
  Disruption claims § 6.5
  Termination claims § 10.31
RECOVERY
  Causes of action available. See
    CAUSES OF ACTION AND
    RECOVERIES
REIMBURSEMENT OF LOST PROFITS
  Tax considerations § 2.1
RELIANCE
  Differing site conditions claims
    §§ 8.8, 8.17
REMOBILIZATION COSTS
  Disruption claims §§ 6.5, 6.6, 6.9
REMOTE PARTIES
  Causes of action and recovery § 13.4
RENT REVENUES
  Causes of action and recovery
    § 13.39
REPLACEMENT COSTS
  Termination claims § 10.18
REPUDIATION OF CONTRACT
  Termination claims § 10.11
REPUTATION, LOSS OF
  Owner as plaintiff, damages available
    § 13.46
RESCISSION OF CONTRACT
  Differing site conditions claims
    § 8.18
  Termination. See TERMINATION
    CLAIMS
RESTITUTION
  International claim resolution outside
    U.S. § 4.13
  Termination. See TERMINATION
    CLAIMS
REVERSE DIFFERING SITE CONDITIONS
  Differing site conditions claims § 8.13

RISK
  Pricing and proving claims § 1.1
RISK ALLOCATION
  Disruption claims § 6.4
  Pricing and proving claims § 1.3
ROCK
  Differing site conditions claims
    §§ 8.6, 8.10
SALE OR EXCHANGE
  Tax considerations § 2.1
SAVINGS CLAUSE
  Cost claims § 12.1
SCHEDULING
  Acceleration claims §§ 7.5, 7.12
  Delay claims against contractors
    § 14.4
  Disruption claims §§ 6.3-6.5
  Pricing and proving claims
    § 1.20
  Termination claims, failure as
    grounds for § 10.11
SEGREGATED COST METHOD OF PRICING CLAIMS
  Pricing and proving claims
    §§ 1.28, 1.29
SERVICES
  Causes of action and recovery
    § 13.18
SETTLEMENT
  Contractor as plaintiff, damages
    available § 13.55
  Pricing and proving claims § 1.30
  Tax considerations § 2.2
SHOP DRAWINGS
  Changes in scope claims § 9.10
  Disruption claims § 6.3
SHOULD COST ESTIMATES
  Disruption claims §§ 6.10, 6.11
SITE INVESTIGATION
  Differing site conditions claims
    §§ 8.1, 8.15, 8.23
SOIL
  Differing site conditions claims
    §§ 8.10, 8.20
SPEARIN DOCTRINE
  Pricing and proving claims § 1.6
  Causes of action and recovery
    § 13.35
SPECIAL CONDITIONS
  Differing site conditions claims
    § 8.20

SPECIALTY CONTRACTORS
  Delay claims against contractors § 14.3
SPECIFICATIONS
  Differing site conditions claims § 8.17
STANDARD CONTRACT FORMS
  Pricing and proving claims § 1.5
STANDARD OPERATING PROCEDURES
  Pricing and proving claims § 1.16
STATUTES OF LIMITATION AND REPOSE
  Causes of action and recovery §§ 13.6, 13.23
  Contractor as plaintiff, damages available § 13.54
STATUTORY INTEREST
  International claim resolution inside U.S. § 3.3
STATUTORY LAW
  International claim resolution outside U.S. § 4.8
STATUTORY RIGHTS IN TERMINATION
  Termination claims § 10.1
STOP NOTICE
  Termination claims § 10.41
STOP ORDERS
  Lost profit claims § 15.3
STORAGE
  Differing site conditions claims § 8.32
STRICT LIABILITY
  Architect, claims against § 13.7
  Contractor, claims against § 13.20
  Manufacturer/supplier, claims against § 13.25
  Owner, claims against §§ 13.13, 13.14
SUBCONTRACTORS
  Acceleration claims §§ 7.4, 7.12, 7.13
  Differing site conditions claims § 8.32
  Causes of action and recovery § 13.17
  Cost claims §§ 12.3, 12.4
  Delay claims against contractors § 14.5
  Payment delay claims § 11.21
  Pricing and proving claims §§ 1.2, 1.20

SUBCONTRACTORS *(Continued)*
  Termination claims §§ 10.3, 10.12, 10.41
SUBPOENAS
  Cost claims § 12.5
SUBSEQUENT BUYERS
  Causes of action and recovery § 13.17
SUBSTANTIAL CHANGE
  Causes of action and recovery § 13.14
SUBSTANTIAL COMPLETION
  Causes of action and recovery § 13.54
  Termination claims § 10.18
SUBSTANTIAL PERFORMANCE
  Termination claims § 10.40
SUBSTANTIATION
  Tax considerations § 2.1
SUBSURFACE CONDITIONS
  Changes in scope claims § 9.1
  Differing site conditions claims §§ 8.6, 8.15, 8.20
SUPERVISION COSTS
  Owner as plaintiff, damages available § 13.41
SUPPLIERS
  Claims against. See CAUSES OF ACTION AND RECOVERIES
  Cost claims § 12.5
  Delay claims against contractors § 14.3
SUPPLIES
  Causes of action and recovery § 13.53
  Changes in scope claims § 9.16
  Differing site conditions claims § 8.32
SURETY
  Changes in scope claims § 9.19
  Claims against. See CAUSES OF ACTION AND RECOVERIES
  Cost claims § 12.15
SUSPENSION OF WORK CLAUSE
  Delay claims against owners § 5.2
  Disruption claims §§ 6.7, 6.17

TAX CONSIDERATIONS
  Damages recovery characterization
  –Generally § 2.1
  –Substantiation § 2.2
  –Timing § 2.3

INDEX

TAX CONSIDERATIONS *(Continued)*
  Deductability of legal fees § 2.7
  Defendants' taxes
  –Generally § 2.4
  –Loss amount, treatment of § 2.5
  –Timing § 2.6
  Differing site conditions claims
    § 8.32
TAXES
  Changes in scope claims §§ 9.15, 9.16
TAX YEAR
  Tax considerations §§ 2.3, 2.6, 2.7
TERMINATION CLAIMS
  Generally § 10.1
TERMINATION CLAUSE
  Delay claims against owners § 5.2
TERMINATION OF CONTRACT
  International claim resolution outside
    U.S. §§ 4.8, 4.11
TERMINATIONS CLAIMS
  Contractor's viewpoint
  –Damage remedy, see Damage
    remedy, contractor, this heading
  –Default, recovery in § 10.40
  –Offsets against contractor § 10.35
  –Other damages available § 10.34
  –Privity, recovery by parties not in
    privity with owner § 10.41
  –Rescission, see Rescission and
    restitution, contractor, this heading
  Damage remedy, contractor
  –Generally § 10.28
  –Before work begins § 10.29
  –Completion of contract, remedy
    following § 10.33
  –During project § 10.30
  –Recordkeeping § 10.31
  –Termination of contract during
    project § 10.32
  Damage remedy, owner
  –Generally § 10.15
  –After substantial completion
    § 10.18
  –Before work begins § 10.16
  –During project § 10.17
  Liquidated damages
  –Generally § 10.22
  –Accrual of damages after
    termination § 10.25
  –Actual and liquidated damages,
    recovery § 10.24
  –Special recovery issues § 10.23

TERMINATIONS CLAIMS
(Continued)
  Owner's viewpoint
  –Damage remedy, see Damage
    remedy, owner, this heading
  –Default, owner's recovery § 10.27
  –Delay in use § 10.20
  –Extended duration and overhead
    § 10.19
  –Liquidated damages, see Liquidated
    damages, this heading
  –Lost profits § 10.21
  –Rescission or restitution favorable to
    owner § 10.26
  Rescission and restitution, contractor
  –Generally § 10.36
  –Measuring rescission § 10.39
  –Rescission as favorable § 10.38
  –Rescission as unfavorable § 10.38
  Terminations for convenience
  –Generally § 10.2
  –Conclusion § 10.6
  –Federal compensation scheme
    § 10.3
  –Private contracts § 10.5
  –Unit price settlement dangers
    § 10.4
  Terminations for default
  –Allowing or denying § 10.12
  –Favorable termination and
    rescission § 10.10
  –Nonbreaching party's options
    § 10.7
  –Option selection criteria §§ 10.8,
    10.14
  –Prerequisites for termination or
    rescission § 10.11
  –Procedures for rescission § 10.13
  –Rescission distinguished § 10.9
TERRORISM
  International claim resolution outside
    U.S. § 4.16
THIRD-PARTY BENEFICIARIES
  Causes of action and recovery
    § 13.3
  Termination claims § 10.41
THIRD-PARTY DELAY
  Acceleration claims § 7.4
TIME
  Arbitration § 1.9
TIME AND MOTION STUDIES
  Disruption claims § 6.12

TIME EXTENSIONS
  Acceleration claims §§ 7.7, 7.10
  Delay claims against contractors § 14.7
  Disruption claims § 6.4
TIME IS OF THE ESSENCE CLAUSE
  Delay claims against owners § 5.2
TORTS
  International claim resolution outside U.S. §§ 4.8, 4.13, 4.19
TOTAL COST METHOD
  Acceleration claims §§ 7.9, 7.14, 7.17
  Changes in scope claims § 9.13
  Delay claims against owners § 5.10
  Disruption claims §§ 6.8, 6.10, 6.11, 6.14
  Pricing and proving claims § 1.28
TRADE SECRETS
  Cost claims § 12.5
TYPE 1 CONDITIONS
  Differing site conditions. See DIFFERING SITE CONDITIONS
TYPE 2 CONDITIONS
  Differing site conditions claims §§ 8.11, 8.12

UNIFORM COMMERCIAL CODE
  Architect, claims against § 13.4
  Contractor, claims against § 13.18
  Manufacturer/supplier, claims against §§ 13.22, 13.23
  Owner, claims against § 13.11
  Pricing and proving claims § 1.5
UNITED NATIONS CONVENTION ON ARBITRATION
  International claim resolution inside U.S. § 3.8
UNIT PRICE
  Termination claims §§ 10.4, 10.30
UNJUST ENRICHMENT
  International claim resolution outside U.S. § 4.28
  Termination claims §§ 10.18, 10.36, 10.41
UNREASONABLE DELAY
  Causes of action and recovery § 13.51
USE, LOSS OF
  Owner as plaintiff, damages available § 13.40

UTILITIES
  Disruption claims § 6.6

VALIDITY OF CLAIM
  Changes in scope claims § 9.7
VALUATION
  International claim resolution outside U.S. § 4.27
VALUE OF EXTRA WORK
  Changes in scope claims § 9.12
VERIFICATION OF CLAIMS
  International claim resolution outside U.S. § 4.23
VERTICAL PRIVITY
  Causes of action and recovery § 13.22
VIDEOTAPES
  Pricing and proving claims § 1.18
VOLUNTARY ACCELERATION
  Acceleration claims §§ 7.1, 7.3

WAIVER
  Acceleration claims § 7.10
  Changes in scope claims § 9.5
WARRANTY
  Causes of action and recoveries § 13.2
WASTE
  Differing site conditions claims § 8.32
WATER
  Differing site conditions claims §§ 8.6, 8.10, 8.18
WEATHER
  Acceleration claims § 7.4
  Adverse weather. See ADVERSE WEATHER
  Delay claims against contractors § 14.5
  Differing site conditions claims § 8.1
WITNESSES
  Delay claims against contractors § 14.4
  Differing site conditions claims § 8.22
  International claim resolution outside U.S. § 4.20
WORK STOPPAGE
  Payment delay claims §§ 11.5–11.7

WORK SUSPENSIONS
 Disruption claims § 6.5
WORKSHEETS
 Pricing and proving claims § 1.15
WRITING REQUIREMENTS
 Changes in scope claims § 9.5

WRITTEN CONTRACTS
 Pricing and proving claims § 1.3
WRONGFUL TERMINATIONS
 Termination claims §§ 10.2, 10.14, 10.26, 10.34